23 bothersome?

merchants?

137 colonealism / imperialism
at core of modernity?

147 def of governmentality

154 order beauty harmony proportion
-156 Shaftesbury   Newton nat theol

A HISTORY OF THE
MODERN FACT

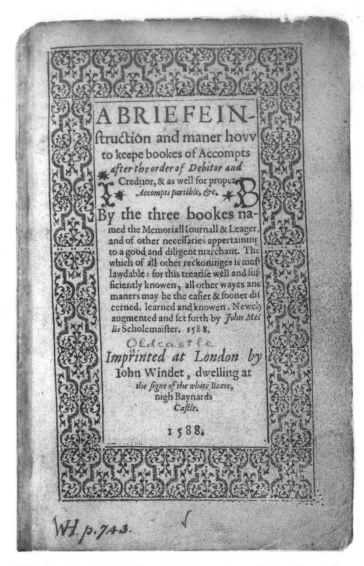

Title page from John Mellis, *A Briefe Instruction and Maner How to Keepe Bookes of Accompts after the Order of Debitor and Creditor* (London: John Windet, 1588). Courtesy, The Bancroft Library.

A HISTORY OF THE

Problems of Knowledge in the Sciences

MODERN FACT

of Wealth and Society

MARY POOVEY

The University of Chicago Press
Chicago and London

MARY POOVEY is professor of English and director of the Institute for the History of the Production of Knowledge at New York University. She is the author of *Uneven Developments: The Ideological Work of Gender in Mid-Victorian England* and *Making a Social Body: British Cultural Formation, 1830–1864*, both published by the University of Chicago Press.

The University of Chicago Press, Chicago 60637
The University of Chicago Press, Ltd., London
© 1998 by The University of Chicago
All rights reserved. Published 1998
Printed in the United States of America
07 06 05 04 03 02 01 00 99 98   1 2 3 4 5

ISBN: 0-226-67525-4 (cloth)
ISBN: 0-226-67526-2 (paper)

Library of Congress Catatloging-in-Publication Data

Poovey, Mary.
    A history of the modern fact : problems of knowledge in the sciences of wealth and society / Mary Poovey.
        p.   cm.
    Includes bibliographical references and index.
    ISBN 0-226-67525-4 (cloth : alk. paper).—ISBN 0-226-67526-2
(pbk. : alk. paper)
        1. Social sciences—Great Britain—Statistical methods—History.   2. Social sciences—Great Britain—Statistics—History.   3. Social sciences—Statistical methods—History.   4. Social sciences—Statistics —History.   I. Title.
    HA29.P6739   1998
    300'.7'2041—dc21   98-5155
                                                                                                                    CIP

The University of Chicago Press gratefully acknowledges the generous contribution of the John Simon Guggenheim Memorial Foundation toward the publication of this book.

⊚ The paper used in this publication meets the minimum requirements of the American National Standard for Information Sciences—Permanence of Paper for Printed Library Materials, ANSI Z39.48-1992.

In Memoriam

Sufi
1980–1997

CONTENTS

ACKNOWLEDGMENTS   ix

INTRODUCTION   xi – *xxv*

1. THE MODERN FACT, THE PROBLEM OF INDUCTION, AND QUESTIONS
    OF METHOD   1
    Ancient Facts, Modern Facts   7
    Methodological Considerations   16
    Thematic Overview   26

2. ACCOMMODATING MERCHANTS: DOUBLE-ENTRY BOOKKEEPING,
    MERCANTILE EXPERTISE, AND THE EFFECT OF ACCURACY   29
    "This Exquisite Deep-Diving Science"   33
    From Rhetoric to Reason of State   66

3. THE POLITICAL ANATOMY OF THE ECONOMY: ENGLISH SCIENCE
    AND IRISH LAND   92
    The Crisis in Knowledge and the Question of Method   97
    William Petty, Ireland, and Economic Matters of Fact   120
    The Authority of Mathematical Instruments   138

4. EXPERIMENTAL MORAL PHILOSOPHY AND THE PROBLEMS OF
    LIBERAL GOVERNMENTALITY   144
    Government by Taste in the Work of Defoe and Hume   157
    Experimental Moral Philosophy   175
    David Hume: From Experimental Moral Philosophy to the Essay   197

5. FROM CONJECTURAL HISTORY TO POLITICAL ECONOMY   214
    Scottish Conjectural History   218
    Description and System: The Constitution of Political Economy   236
    The Detour through Scotland: Johnson's Journey to the Western Islands   249

6. RECONFIGURING FACTS AND THEORY: VESTIGES OF
   PROVIDENTIALISM IN THE NEW SCIENCE OF WEALTH   264

   *Institutionalizing Political Economy: Dugald Stewart and the Repudiation of
   Particulars   269*
   Thomas Malthus and the Revaluation of Numerical Representation   278
   *Popularizing Political Economy: J. R. McCulloch and the Taxonomy of Modern
   Knowledge   295*

7. FIGURES OF ARITHMETIC, FIGURES OF SPEECH: THE PROBLEM OF
   INDUCTION IN THE 1830s   307

   Statistics in the 1830s   308
   *John Herschel and John Stuart Mill: Induction, Deduction, and the Limits of
   Scientific Method   317*
   Poems and Systems: The Emergence of the Postmodern Fact   325

   NOTES   329

   BIBLIOGRAPHY   387

   INDEX   409

Research for *A History of the Modern Fact* was funded in part by the John Simon Guggenheim Foundation, the National Endowment for the Humanities, the School of Social Science at the Institute for Advanced Study in Princeton, and Johns Hopkins University. I am grateful for support from all these organizations. I also thank the librarians who staff Special Collections (the Hutzler Collection of Economics Classics) at the Milton S. Eisenhower Library, Johns Hopkins University; the librarians at the Institute for Advanced Study; and the staff at Firestone Library, Princeton University.

As often happens when a project takes years to complete, I have benefited from more comments and questions about *A History of the Modern Fact* than I can possibly acknowledge. The friends and acquaintances I cite here constitute only a fraction of the silent contributors to this book, all of whom I wholeheartedly thank: John Bender, Jim Chandler, Stefan Collini, Lorraine Daston, Margie Ferguson, Peter Galison, Jonathan Goldberg, Ian Hunter, Jeffrey Minson, Joan Scott, David Simpson, and Elizabeth Weed. I am especially grateful to Tita Chico of New York University, who provided expert help with the bibliography, and to my graduate students at Johns Hopkins University, who have been a constant source of challenge and support. I also acknowledge my colleagues in the Hopkins English Department, especially Jerry Christensen, Allen Grossman, John Guillory, Jonathan Goldberg, and Ronald Paulson—whose scholarly examples set the bar a little higher. I could not have finished *A History of the Modern Fact* but for a provocative and reassuring Christmas conversation with Emily Martin. I could not have imagined this book without all the hours I spent talking with John Guillory. Without his help, I would never have learned enough about the seventeenth century to write the first parts, and without his encouragement I would have abandoned the project years ago. I also thank Alan Thomas, my editor at the University of Chicago Press, whose intelligence, patience, and commitment to this book have been invaluable.

An earlier version of chapter 7 appeared in *Critical Inquiry* 19, no. 2 (1993):

256–76, © 1993 by The University of Chicago; portions of chapter 2 were previously published in *Differences: A Journal of Feminist Cultural Studies* 8, no. 5 (1996): 1–20.

Finally, readers who do not know me may wonder why I have dedicated such a challenging book to a dog. My answer is simple: even though she did not live to see its publication, Sufi presided over all the stages of this book's research, writing, and revision. Without her imperious attendance, my work would have had less focus; without her unfailing love, my life would have had less joy.

INTRODUCTION

In the lectures, workshops, and seminars where I have discussed this project during the past few years, engaged but puzzled audiences have repeatedly asked two questions: What is "the modern fact"? and How did you decide to write such a book? It has taken me a while to understand that the answers are related, and thus I have not always had good responses to offer. This introduction is my attempt to provide more adequate answers for these important questions. I want to explain how trying to understand a feature I noticed in some early nineteenth-century texts led me to conceptualize the modern fact in such a way that I could begin to write its history.

The feature that initially caught my eye seemed to pose a historical question about conventions of representation, and thus it seemed fair game for a historically minded literary critic like me. I noticed that early nineteenth-century surveys of the newly crowded cities in Britain tended to combine interpretive accounts of neighborhoods with numerical tables that purported to describe more or less the same circumstances, apparently without analytic commentary.[1] Instead of focusing on the interpretive accounts, however, as most literary critics would have done, I became intrigued with the numbers. Instead of concentrating on what the numbers showed about urban squalor, as most historians would have done, I was curious to see what their use revealed about how Britons thought of numbers in the 1830s. What kind of representation were these protostatistical tables, I wondered, so that numbers seemed both essential and insufficient? Why did these numbers appeal to government officials in particular? What were the historical antecedents to such uses of numbers? To what semantic fields did such numerical representations belong, and how did Britons evaluate numbers in relation to the analytic descriptions that accompanied them?

For a late twentieth-century literary critic, numbers constitute something like the last frontier of representation, so I set out with the enthusiasm of a pioneer in search of the origins of that early nineteenth-century convention. What

I very gradually discovered, in the course of years of research, was that the combination of numbers and analysis in those early nineteenth-century texts spoke to a philosophical conundrum that did not necessarily involve numbers at all. In other words, I discovered I had been asking the wrong questions. Instead of exploring the historical meanings or uses of numerical representation, I needed to pose an epistemological question. I needed to ask how knowledge was understood so that it seemed to consist of both apparently noninterpretive (numerical) descriptions of particulars *and* systematic claims that were somehow derived from those particularized descriptions.

Eventually I discovered that in most of the knowledge projects produced during the long period of modernity, the two functions that appear to be separate in those early nineteenth-century texts actually coincide. Indeed, I discovered that even in those texts the two functions only *seem* to be distinct, for what look like two functions—describing and interpreting—seem to be different only because one mode of representation (the numbers) has been graphically separated from another (the narrative commentary). I discovered that in those nineteenth-century texts, as in most texts that purport to describe the material world, even the numbers are interpretive, for they embody theoretical assumptions about what should be counted, how one should understand material reality, and how quantification contributes to systematic knowledge about the world. Such figures, which simultaneously describe discrete particulars *and* contribute to systematic knowledge, constitute examples of what I have called the modern fact.

As I explain in the course of this book, numbers have come to epitomize the modern fact, because they have come to seem preinterpretive or even somehow noninterpretive at the same time that they have become the bedrock of systematic knowledge. Historically, however, there was no necessary connection between numbers and this peculiarly modern epistemological assumption, nor have numbers always seemed free of an interpretive dimension. Even though I began by thinking I was writing about the history and semantics of nineteenth-century statistics, then, I discovered that that history belonged to a larger narrative: the story of how description came to seem separate from interpretation or theoretical analysis; the story of how one kind of representation—numbers—came to seem immune from theory or interpretation. For want of a better term, I have called this narrative *A History of the Modern Fact.*

I am all too aware that "the modern fact" is an inadequate phrase for what I am trying to historicize: "the" is no doubt too definite for a concept this capacious; "modern" obviously raises too many questions about periodization; and "fact" is both too commonplace and too labile to have much meaning for most readers. Even though "the modern fact" is something of a placeholder, how-ever—designating a concept that thus far lacks both definition and name—I use the phrase because I want to put the epistemological unit it stands for into scholarly circulation; I want to encourage others to take it up, to elaborate and refine it. Unless someone applies some name to the epistemological unit that is the subject of this book, no one will be able to see how deeply embedded it has been in the ways that Westerners have come to know the world. Unless some-one calls it something, we will continue to lack a vocabulary to describe—or recognize—its passing.

As the epistemological unit that organizes most of the knowledge projects of the past four centuries, the modern fact is obviously a subject too extensive for a single book. I call the history I offer here *a* history of the modern fact be-cause many such histories could be written, and though I stand behind the par-ticular choices I have made in constructing this narrative, I realize these are not the only texts or authors I could have included, even in a study narrowed to the considerably smaller compass of the appearance of the modern fact in the British sciences of wealth and society. In the next section of this introduction I summarize the narrative I develop in this book, but though I stress its logic here and in the body of the text, I also want to signal at the outset its constructed and unconventional nature. I am convinced that the peculiar combination of num-bers and analysis that I noticed so many years ago is clarified, if not fully ex-plained, by my account of the modern fact, but I am also aware that some readers will think that the materials I have chosen do not lead directly enough to the nineteenth-century use of numerical representation. I will do my best to anticipate the objection that my choices are odd, if not arbitrary. If readers are to appreciate why I can defend my narrative as both constructed and clarifying, however, I must first explain the theoretical stakes implicit in the mode of his-torical narrative I offer here.

As I point out in chapter 1, then again at the beginning of chapter 6, the narrative I develop in *A History of the Modern Fact* departs from conventional lin-ear histories in several ways. Whereas most histories chronicle the development (or vicissitudes) of a single well-defined idea or obviously connected series of events, *A History of the Modern Fact* explicitly takes a circuitous route, rejects both the single well-defined idea and the obviously connected series of events, and pursues instead the very gradual consolidation and recurrent interrogation of an epistemological unit whose existence is almost impossible to document. I have chosen this course for three reasons. First, I want to avoid conventional histories' tendency to consolidate their objects of analysis artificially and retro-spectively. To write a history—or even a "prehistory"—of statistics, for exam-

ple, requires one to assimilate practices that predate the codification and naming of statistics *to* that nineteenth-century use of numerical representation. Although doing so enables us to see in some sense where statistics "came from," it
prevents us from understanding what eventually came to be called statistics as a
constructed (and possibly unstable) amalgam of practices, some of them developed with very different (even antithetical) agendas. This difficulty in turn mil-
itates against our recognizing the persistence of alien agendas within the
practice codified as statistics, and thus we may find it hard to fathom the contradictions that mark the early version of statistics and that led relatively quickly
to its being transformed into a mathematical enterprise.

    This situation leads to my second reason for preferring indirection. I have
chosen this circuitous route because I want both to capture something of the
absolute otherness of the past and to illuminate, at least in passing, some of the
paths not taken that were opened by the events or ideas I examine here. As a
consequence, mine is a messy history rather than a neat narrative. At times I
linger over the agendas that were *not* assimilated into what can only retrospectively be identified as the genealogical descendant of some earlier practice; and
at times I interrogate the very ideas of "development," "genealogy," "origin,"
and "outcome." I want to make it clear, however, that even though I aspire to
convey some sense of the alterity of the past, I do not believe it is possible to
confront this otherness fully; nor do I believe we are completely trapped in a
hermeneutic circle that makes everything we see a dim reflection of ourselves.
Instead of taking either of these extreme positions, I want to use my awareness
that interpretation governs all historical analysis to expand the available understandings of the materials I examine here. By juxtaposing the interpretive frame
used by the historical subjects themselves to the frame that now guides historical work, I have tried to convey both something of the way the world once
looked and something of the way it looks now, so that readers can see both how
different the past was and how deeply indebted our most commonplace ideas
are to notions that seem to belong to another era.

    Finally, I have chosen this circuitous course because the object of my analy-
sis is so abstract. For most conventional historians, I suspect, the very abstract-
ness of "the modern fact" would disqualify this epistemological unit from
historical study, for objects this abstract leave none of the traces that historians
recognize as documentable evidence. If the effect—or even the existence—of
the modern fact challenges documentation by conventional means, however,
taking such an epistemological unit as one's analytic object offers several opportunities along with the challenge. In this book, for example, focusing on this
epistemological unit has enabled me to expose the connections between
knowledge projects as different as rhetoric, natural philosophy, moral philoso-

phy, and early versions of the modern social sciences. By revealing that what connects these projects is a problematic (but symptomatic) assumption about epistemology—the assumption that systematic knowledge must draw on but also be superior to noninterpretive data collected about observed particulars— I have been able to show how a range of practices that were undeniably developed to serve different agendas also helped elaborate this assumption about knowledge. Thus we can see that even though seventeenth-century political arithmetic was not presented as a theoretical or interpretive science, as eighteenth-century conjectural history explicitly was, both of these practices actually drew on a priori assumptions to organize data so that "facts" would be available for government use. From this perspective, we can also see that the early nineteenth-century combination of numbers and analysis that initially aroused my interest was one solution to the practical problems posed by this assumption about epistemology. Separating numbers from analytic commentary opened the way for a *professional* answer to the questions raised by the supposition that systematic knowledge should be derived from noninterpretive descriptions; indeed, the professional component of this solution was implied by the apparent difference in representational kinds. Separating numbers from interpretive narrative, that is, reinforced the assumption that numbers were different in kind from the analytic accounts that accompanied them. Being different in kind, these numbers seemed to belong to a different stage of the knowledge-producing project as well, a stage that could be managed, in theory, by a different kind of expert (professional) knowledge producer.

Even if studying epistemological units such as the modern fact requires us to rethink the nature of evidence, taking up subjects like this shows us something that no conventional history can illuminate: a history of the modern fact begins to reveal the organizational principles inherent in the kinds of knowledge by which subjects of the modern world manage our relationships with each other and with society. If one assumes, as I do, that modes of representation inform what we can know, and if one also assumes, as I do, that modes of representation embody or articulate available ways of organizing and making sense of the world, then to understand the potential and limitations of what we know, we need tools to investigate the conditions that make knowledge possible. Naming the modern fact gives us such a tool. Writing even a preliminary history like this one should launch this investigation.

What follows constitutes a summary of the narrative argument of *A History of the Modern Fact*. Readers more interested in a metacommentary on the method and topics of this book might well proceed immediately to chapter 1, and readers impatient for the argument itself should begin with chapter 2. I provide this summary partly because the route I take is so circuitous and partly because my analyses are often so detailed that readers may more than occasionally want to be reminded of the shape of the overall argument.

ch 1

In the first chapter I place this book in the context of existing scholarship on the history of the social sciences, British intellectual history, and that branch of the history of science now known as science studies. In chapter 1 I also offer further reflections on the nature of the modern fact by contrasting this mode of the factual with its ancient (Aristotelian) counterpart, and I explain why I begin with double-entry bookkeeping, a practice in which the modern fact existed only in embryonic form (as opposed to being fully realized or theorized). In the course of the chapter, I develop a more sustained account of the methodology I use in the book, and I conclude with a summary discussion of some of the general themes that recur: the vicissitudes of "interest" and "disinterestedness"; the complex connections among "credit," "credibility," and "credulity"; the link between early accounts of liberal government and theories about subjectivity or desire; and some of the characteristic forms of the modern fact, including abstractions, universals, and generalizations.

Ch 2

My historical treatment of the modern fact begins in chapter 2. Here I discuss two institutions of early mercantile capitalism: double-entry bookkeeping and mercantile accommodation. As a system of writing, the former, as I have just proposed, produced a prototype of the modern fact; the latter, which constituted the informal system of agreements that underwrote merchants' willingness to accept each other's bills of exchange, implied that the rule-governed system that organized mercantile transactions provided *the* model for effective government. In emphasizing *system* over observed particulars, both double-entry bookkeeping and mercantile accommodation elaborated what one might call the theoretical dimension of the modern fact. In implying that the internally coherent systems of writing and exchange constituted signs of honesty and virtue, these two mercantile instruments demonstrated that the idea of system could carry moral connotations whose effects exceeded the referential function of mercantile writing, because one of these effects was the establishment of creditworthiness itself.

Like earlier, more informal, and more idiosyncratic methods of accounting, double-entry bookkeeping served merchants' professional agenda: the double entries of offsetting credits and debts helped them keep track of commercial transactions as well as providing records of money due and owed. When double-entry bookkeeping was codified as a printed set of rules, however, this kind of accounting took on an additional social function; positioned within the field of rhetoric by its earliest champion, Luca Pacioli, double-entry bookkeeping became a display of mercantile virtue. Proclaiming mercantile virtue was necessary in 1494, both because in the late fifteenth century merchants were generally held in low esteem and because status mattered as much as wealth, if not more, and salvation ultimately mattered more than worldly goods. The system of double-entry books displayed virtue graphically, by the balances prominently featured at the end of each set of facing ledger pages and by the order evident in the books as a system. Thus the entries in the double-entry system seemed simply to refer to the particulars of a merchant's trade, while the system as a whole, which subordinated those particulars to the all-important balance, produced meanings that exceeded even the veracity of the individual entries. Most immediately, the balances produced by this system of writing proclaimed the creditworthiness of the individual merchant; more generally, the system's formal coherence displayed the credibility of merchants as a group. In this embryonic variant of the modern fact, then, we see particulars harnessed to a general claim, whose acceptance required believing that the precision of the formal system signaled virtue itself.

Implicitly, at least, double-entry bookkeeping was both a system of writing and a mode of government, for if merchants were to benefit from the aura of credibility cast by the rectitude of the formal system, they had to obey the system's rules. In the second section of chapter 2 I take up the subject of governmentality more explicitly by turning to early merchant apologists' efforts to capitalize on the new theory of government known as reason of state. Reason-of-state arguments, which Francis Bacon actively promoted in England in the early seventeenth century, theoretically enhanced the importance of commerce, because proponents of reason of state advised the prince to strengthen the nation's resources instead of trying to govern by abstract principles. Merchant apologists like Edward Misselden and Thomas Mun took the opportunity offered by this political theory to create a position of mercantile expertise that assigned to the merchant, not the prince or his advisers, the privilege of interpreting the system of trade, which, not incidentally, only the expert merchant could see. In the writings of these merchant apologists, systems of interpretation made visible a system of trade, so that belief in the former theoretically guaranteed the coherence of the latter; and the merchant emerged as the model citizen in a society where the king's sovereignty was just beginning to be challenged in the domain of commerce.

In the second half of the seventeenth century, another group of writers began to elaborate the second dimension of the modern fact by arguing that the knowledge derived from observed particulars had nothing to do with the theoretical arguments used to support government policy, religious sectarianism, and mercantile expertise. By drawing on Francis Bacon's celebration of deracinated particulars and the method of induction, Robert Boyle and the other members of the Royal Society were able to insist that one could gather data that were completely free of any theoretical component; all one had to do was contrive experiments and assemble credible witnesses. Although Boyle and his colleagues never explained exactly how one moved from the observation of these (supposedly) theory-free particulars to general knowledge about them, their claim that one could do so opened a gap between what looked like untheorized data and systematic knowledge. At the same time, Thomas Hobbes was implicitly reinforcing this distinction between data that seemed untheoretical and systematic knowledge by arguing that political matters of fact could be generated only through *deduction,* and that the king had to authorize philosophical knowledge production in order to make political knowledge true.

The contest between Baconian induction and Hobbesian deduction constitutes the first subject of chapter 3. In this chapter I show how Bacon made a case for the untheoretical nature of the observed particular and how Boyle and the other members of the Society propped their own social authority on the claim that facts were theory-free and value-free. The subject of "interest" is of special concern to me here, for Boyle and his friends also insisted that knowledge production could (and should) be "uninterested," while Hobbes sought to mobilize the positive connotations of "interest" introduced by advocates of reason of state. One unintended consequence of the debate about interest, I submit, was a cultural revaluation of the contribution that merchants could make to systematic knowledge. Although the members of the Royal Society did not explicitly call attention to double-entry bookkeeping or mercantile writing, they did suggest that merchants constituted valuable information-gathering instruments—precisely because they had no interest in the knowledge they conveyed to natural philosophers.

The second part of chapter 3 focuses on the complex theoretical amalgam that William Petty forged by mixing Baconian induction with Hobbesian deduction. The variant of the modern fact that resulted, I propose—the economic matter of fact—begins to show why this epistemological unit eventually proved attractive to governments seeking to strengthen their national and international positions. Indeed, Petty promoted his method, which he called political arithmetic, as an explicit antidote both to the excesses of rhetoric and to the theoretical disputes that had provoked the English Civil War. Arguments based on "number, weight, and measure," he proclaimed, would compel assent as surely as mathematics did—especially if the king was willing to back the knowledge that the supposedly disinterested numbers expert produced. As I point out in this chapter, it was critical for Petty to install disinterestedness at the heart of knowledge, because nearly all the policies he recommended (and many of the numbers he conjured) would have directly supported his own interest in the estates he owned in Ireland. By implying that numbers were impartial because they could erase interest and politics, Petty helped enhance the status of numbers as a mode of (disinterested) representation. He was able to do so partly because, in the applied sciences known as the mathematicals and with the help of mathematical instruments, numbers had already begun to seem indispensable to gentlemen and government alike.

If Petty's political arithmetic, the mathematicals, and mathematical instruments helped make numerical representation seem like an effective (because impartial) instrument of rule, then why did the British government fail to sponsor the large-scale collections of numerical information that Petty promoted? What happened to the initiative he launched under the name of political arithmetic? And what connections, if any, tie the explicitly theoretical science of political economy, which Adam Smith codified in 1776, to political arithmetic, whose authority was based on its antitheoretical nature? In chapters 4 and 5 I offer my answers to these questions, which have long baffled historians of "statistical thinking" and political economy. In my account the missing disciplinary link that both connects political economy to political arithmetic and distinguishes the two is Scottish moral philosophy, particularly those subsets called experimental philosophy and conjectural history. In moral philosophy, we can see how the claim that description (particularly numerical description) is impervious to theory was tested and exposed, only finally to be reconstituted on slightly different grounds. By recovering moral philosophy as the missing link between political arithmetic and political economy, I believe, we can better understand why all systematic knowledge systems require something like a leap of faith and why all attempts to gather theory- and value-free data are marked by the very theoretical assumptions they seem to leave behind.

Chapter 4 begins by considering the effects of the Glorious Revolution on *ch.4* the relation between government and theories about subjectivity in the early eighteenth century. To understand what happened to political arithmetic in Britain, I argue, we have to appreciate the changes in governmentality wrought by the demise of the absolute monarchy, the emergence of civil society, and what has been called the financial revolution. Taking Defoe's *Essays upon Several Projects* as my primary example of an abortive scheme to implement political arithmetic in the emergent market society, I show how one legacy of the Glorious Revolution—the appearance of theories about liberal (self-) government and the mode of subjectivity that enabled individuals to govern themselves— constituted a solution to a problem that Petty introduced but the revolution made more pressing. Posed baldly, the difficulty was how to make subjects comply with a monarch whose absolute authority had been imperiled first by religious and political controversy, then by the Revolution Settlement. Paradoxically, the early eighteenth-century solution to the problem of rule involved enhancing the status of self-interest (and even party interest) and proposing that, if a disinterested account could be provided, then some segments of the state—most notably the market—could be trusted to govern themselves. As a form of self-government rather than rule by coercion, liberal governmentality elicited voluntary compliance through the mechanisms of fashion and taste; and these mechanisms did not depend on numbers in the same way that sovereignty could be said to do.

Although modern theorists of liberal government have long noted the roles of self-government and fashion, they have generally overlooked another component of the eighteenth-century form of liberal government: accounts of subjective motivation that explained why individuals emulated each other were produced against a backdrop of assumptions about universal human nature. Only because they assumed that everyone was essentially the same, in other words, and that everyone wanted the same thing(s), could philosophers explain why the dynamics of emulation and differentiation intrinsic to the market system did not lead to anarchic expressions of idiosyncratic desire. In the first half of the eighteenth century, the tasks of elaborating both the dynamics of subjectivity and the characteristics of universal "man" were taken up by Scottish moral philosophers. In the second section of chapter 4 I examine works by Francis Hutcheson and George Turnbull to show how the moral philosophers who described their work as "experimental" exposed the role that theory plays in every attempt to produce systematic knowledge from observed particulars. Indeed, in the process of "proving" their "conjectures" about "human nature" and "the human mind," these philosophers defended the presence of theory in knowledge production, for they explicitly celebrated the theoretical assumption that enabled them to explain things that were otherwise incomprehensible: the assumption that nature, like knowledge itself, embodies God's design.

In the final section of chapter 4 I turn to David Hume. Hume features prominently in *A History of the Modern Fact,* because he was the first philosopher to argue overtly that the roles theory, belief, and conjecture play in all systematic knowledge projects constitute a theoretical problematic. In so doing, Hume both formulated what philosophers now call the problem of induction *as a philosophical issue* and dismissed its practical implications. Hume was not paralyzed by the problem of induction because he converted it into a rationale for an altogether different form of knowledge production, which promoted sociality instead of generating facts. By following Hume's turn from experimental moral philosophy to the essay, I illuminate the proximity between the eighteenth-century variants of the modern fact and wisdom literature, but I also point out that government by information, the goal pursued by Bacon and Petty alike, had to compete throughout the middle part of the eighteenth century with another model of government that depended on conviviality as much as discrimination and on sociality more than on data collection.

Chapter 5 opens with a discussion of the mid-eighteenth-century branch of moral philosophy that most starkly confronted the peculiarity written into the modern fact and that lies, by extension, at the heart of the problem of induction. Retrospectively named conjectural histories, these attempts to discover the origins of commercial society had to deal with the paucity of eyewitness accounts and to theorize the "conjectures" philosophers offered in their stead. As in the work of Hutcheson and Turnbull, assumptions about human nature and Providence filled the gaps where evidence was unavailable. In works by Adam Ferguson, William Robertson, and Henry Home, Lord Kames, we see not only that the modern fact has always been susceptible to theory, but also how noncontroversial the coexistence of theoretical assumptions and (supposedly) noninterpretive descriptions was once thought to be. By turning to Hume once more, however, we also see why this coexistence soon came to seem more problematic than it appeared to the conjectural historians. In the brief conjectural history that appears in his *Treatise of Human Nature,* Hume specifically calls the theory on which assumptions about systematic knowledge rest a fiction. If theory is not (Christian) belief but fiction, then how can the philosopher claim that the philosophical system he creates is not a fiction too? Although Hume's claims about the fictive component of knowledge did not immediately undermine philosophers' confidence in the truths they produced, the observation that theory might articulate fantasy, desire, or interest eventually mandated a reconceptualizing of the grounds on which accounts of observed particulars could be said to lead to systematic claims about the world.

In the second section of chapter 5 I turn to Adam Smith's political economy, which most historians consider the prototypical modern science of wealth. I argue that political economy both drew on and superseded those reason-of-state theories of government developed in the late sixteenth and early seventeenth centuries and that it provided what looked like *both* theoretical and descriptive justifications for liberal governmentality by focusing on particulars that could be observed (and quantified) yet subordinating those particulars to abstractions that could not be seen. These abstractions—"society," "*homo economicus,*" and "the market"—constituted the basis for the new science of wealth as well as its characteristic objects of analysis. Smith's use of numbers was critical to these abstractions because he used the abstractions to show why the numbers that were available were *inadequate* to the theoretical task at hand while implying that *if* numerical information was collected in the light of philosophical theories, then numbers might be used to make these theories seem impartial—that is, merely descriptive. Thus Smith helped revalue numerical representation (which had seemed irrelevant to theorists like Hutcheson, Turnbull, and the conjectural historians), even though he was explicitly skeptical about political arithmetic, which he saw as an inferior rival to political economy.

I conclude chapter 5 with a discussion of Samuel Johnson's *Journey to the Western Islands of Scotland.* I argue that the work should be read as a conjectural history, for Johnson tried to discover what the experimental philosopher could see if he actually went to look at a prototype of commercial society. Of course by 1773 the Scottish Highlands no longer constituted the primitive society Johnson longed to find, for the forcible imposition of English laws, dress, and language after the Battle of Culloden had already begun to obliterate the indigenous Highlands culture. Even if he did not find the cradle of modern civilization in the Highlands, however, the aging Johnson did confront what he experienced as genuine cultural difference. So different were the Highlanders, in fact—both from what Johnson expected to find and from what he had always insisted were the universal traits of human nature—that he began to question the epistemological assumptions that underwrote the wisdom literature he had written all his life. Although Johnson did not theorize the cultural relativism that emerges in *Journey to the Western Islands,* his gradual acceptance of difference, along with his readiness to question the universalist assumptions he had embarked with, marks the first sign of what I take to be one of the most valuable legacies of the Enlightenment: the ability to recognize, tolerate, and even value difference. Like Hume's essays, Johnson's last work provides an example of the forms of knowledge production that rivaled numerical and philosophical knowledge in the eighteenth century. *Journey to the Western Islands* not only shows us a contemporary alternative to numbers and philosophy, however; it also explicitly calls into question the assumptions with which proponents were increasingly successfully defending political economic and numerical facts.

Chapter 6 turns to the new sciences of wealth and society to see how three early nineteenth-century writers tried to solve the epistemological problem that Hume had associated with the modern fact. I begin with the Edinburgh University professor Dugald Stewart, because Stewart both worked within the Scottish tradition of university-trained moral philosophers and personally instructed most of the young men who helped promote political economy in England. Stewart's work thus constitutes the last site at which eighteenth-century moral philosophy overlapped with the new science of wealth, which rapidly began to acquire disciplinary autonomy after 1802. In Stewart's work we see that the providentialism so prominent in eighteenth-century moral philosophy persisted well into the nineteenth century, and in such a way as to dwarf the Baconian project of constructing knowledge from observed particulars. So committed was Stewart to the notion that the philosopher should describe God's plan that he was willing to wholly discount whatever particulars departed from the order he believed God superintended. For Stewart neither the peculiarity of the modern fact nor the problem of induction constituted a difficulty, because he thought that observations should always be subordinated to belief and because particulars were irrelevant to the vision the devout philosopher could see.

Thomas Malthus, the subject of the second section of chapter 6, also tried to set aside the problem of induction; he did so not simply by subordinating induction to a philosophical practice that insisted on providential design but by claiming that people would actually become moral, and therefore close the gap between what the philosopher believed and what he saw, if they had enough information about how God had ordered the world. Malthus's attempts to instruct his readers about God's plan provoked complaints that he was immoral, however, for to romantics like Southey and Coleridge the variant of political economy that Malthus promoted seemed dangerously free of all ethical concerns. Like Petty and Smith, Malthus relied heavily on numerical representation to make his theoretical points; inevitably, Malthus's use of numbers also helped make numbers seem to be value- and theory-free, even though he did not intend to liberate either numerical representation or knowledge production more generally from ethical concerns. The way Malthus's contemporaries read his revised *Essay on the Principle of Population,* then, began to transform political economy from a moral science (which Malthus continued to consider it) to the very incarnation of the "dismal," because amoral, modern sciences that now dominate the Western world. In my reading of Malthus's *Essay,* its revisions, and its reception, I show that liberating numbers from theory and value has not always seemed like a good thing; for many of Malthus's readers, knowledge that was divorced from assumptions about Providence was amoral, whether or not it was conveyed in numerical form.

So reviled was political economy in the wake of the 1806 *Essay* that when the members of the Political Economy Club wanted to sponsor a public lecture series on the new science in 1824, they were afraid even to advertise for subscriptions. To rectify this situation, J. R. McCulloch, the young Scotsman chosen to deliver the first Ricardo Lectures, embarked on a lifelong campaign to improve the public image of political economy. By rewriting the history of the discipline, creating a canon for political economy, making reliable texts of the *Wealth of Nations* available for the first time, and placing political economy at the center of countless educational schemes, McCulloch sought to popularize the science that Malthus had rendered so disagreeable. McCulloch was not completely successful in resuscitating political economy, of course; opponents of the manufacturing system, like Dickens and Carlyle, simply turned their venom from Malthus to McCulloch and continued to lament the end of moral knowledge. After the 1830s, however, and despite the influence of Evangelicalism in Britain, the argument that systematic knowledge was by nature moral or modeled on God's design increasingly required defense. This was true in part because of another argument, which was beginning to be made successfully in the physical sciences and which McCulloch tried to appropriate: the claim that systematic knowledge was best made by experts and that the problem of induction could be solved by turning knowledge production over to professionals.

As early as 1825, McCulloch took this approach to the problem of induction by creating a taxonomy of knowledge in which two groups of professionals—political economists and politicians (or political scientists)—would be charged with producing systematic knowledge while another, less prestigious group of professionals—statisticians—would gather raw data. Separating the collection of data from the production of general knowledge would quell the animosity currently directed toward political economy, McCulloch argued, because the division would show that theories about commerce were constructed not out of self-interest or leaps of faith but out of simple observations about particulars, which were immune from interest and theoretical conjectures of any kind.

As I said at the beginning of this introduction, texts that implicitly embodied McCulloch's approach to the modern fact engaged my interest years ago, although it has taken me some time to recognize that a representational puzzle could be seen as a professional solution to an epistemological problem. In the final chapter of *A History of the Modern Fact,* I take up some of the early nineteenth-century texts where the puzzle, the solution, and the problem seem clearest. Defenses and criticisms of statistics provide particularly compelling evidence of this intersection of concerns, for statistics was both one part of McCulloch's professional solution and a site where the epistemological problem of the modern fact became visible. Indeed, at least one critic of statistics, the young subeditor for the radical *London and Westminster Review,* specifically targeted the "fact" as the heart of what could by then be seen as an epistemological problem. "There is an ambiguity in the word facts which enables the council [of the Statistical Society of London] to pass off a most mischievous fallacy," G. Robertson complained in 1838; "it either means evidences or it means anything which exists. The fact, the thing as it is without any relation to anything else, is a matter of no importance or concern whatever: its relation to what it evinces, the fact viewed as evidence, is alone important."[2] In his scathing criticism of the Society's claim to collect theory-free data, Robertson makes it clear that it was possible in the 1830s to see facts as an inherently ambiguous—or, to use my term, an epistemologically peculiar—category. In brief treatments of John Herschel's *Preliminary Discourse on the Study of Natural Philosophy* and John Stuart Mill's *Logic,* I submit both that this peculiarity continued to solicit the attention of the architects of systematic knowledge in the first half of the nineteenth century and that attempts to solve the related problem of induction were increasingly relegated to the ranks of professional knowledge producers.

The history of the modern fact does not end with the separation of statistics from political economy, of course, or with the development of more theoretically sophisticated accounts of natural and philosophical knowledge projects. Indeed, even after statistics was transformed into a mathematical practice and mathematical modeling began to surpass induction as the method of choice in the social and physical sciences, vestiges of the modern fact remained—not least in the desire that some unit of value- and theory-free representation be available for producing systematic knowledge about the social and natural worlds. That so many of us still imagine observation can be separated from systematic accounts of the world speaks to the success of the long campaign to sever the connection between description and interpretation; that numbers seem to guarantee value-free description speaks to the triumph of some of the accounts of numerical representation I chronicle here. At the same time, of course, that so many of us believe description, whether numerical or not, never was—and never can be—freed from the theoretical assumptions that seem implicit in all systematic knowledge projects implies that the campaign to free description (and numbers) from interpretation has not been a complete success. By offering a history of the modern fact, I have tried to show that this debate simply voices the peculiarity written into the epistemological unit that has dominated modernity. If I am right, then the debate will not end until the era of the modern fact has completely passed away.

The Modern Fact, the Problem of Induction, and Questions of Method

What are facts? Are they incontrovertible data that simply demonstrate what is true? Or are they bits of evidence marshaled to persuade others of the theory one sets out with? Do facts somehow exist in the world like pebbles, waiting to be picked up? Or are they manufactured and thus informed by all the social and personal factors that go into every act of human creation? Are facts beyond interpretation? Or are they the very stuff of interpretation, its symptomatic incarnation instead of the place where it begins?

In this book I do not so much answer these questions as show that they can be raised—and answered in either of the two ways that my paired questions imply—because of the peculiar role facts have been assigned in the epistemology we associate with modernity. On the one hand, because Western philosophy since the seventeenth century has insisted that the things we observe constitute legitimate objects of philosophical and practical knowledge, many people think of facts as particulars, isolated from their contexts and immune from the assumptions (or biases) implied by words like "theory," "hypothesis," and "conjecture." This is the sense of "facts" implied by Joe Friday's terse demand, "Just the facts, Ma'am." On the other hand, because philosophers have always sought to produce systematic knowledge, which has general, if not universal, application, some people think of facts as evidence that has been gathered in the light of—and thus in some sense *for*—a theory or hypothesis. According to this understanding, facts can never be isolated from contexts, nor can they be immune from the assumptions that inform theories. At the very least, what counts as a fact can never escape the idea that the knowledge that matters is systematic, not simply a catalog of observed but unrelated particulars.

Because most modern sciences in the West (including philosophy) position the category of the factual between the phenomenal world and systematic knowledge, the epistemological unit of the fact has registered the tension between the richness and variety embodied in concrete phenomena and the uniform, rule-governed order of humanly contrived systems. Because facts register this tension, in turn, they have been susceptible to the two interpretations I have described. And because they have been so susceptible, facts have become a battleground in many late twentieth-century attempts to produce systematic knowledge. Disputes over the relation between facts and values, arguments over how data are gathered and packaged, and quarrels about the very possibility of objectivity can all be seen to derive, at least in part, from the peculiarity written into the epistemological unit of the modern fact.

In this book I assume that this epistemological peculiarity constitutes a historical phase of the category of the factual and, as a consequence, that the often virulent debates I have just alluded to are best understood as historically specific symptoms of this peculiarity. Understanding these debates as historical developments or as symptoms will not resolve them, of course, nor will it diminish the stakes in what is often a consequential battle for scant resources (money, prestige). By providing a historical account of the modern fact, I seek merely to describe some of the dynamics by which what counts as a fact has become so embroiled in theoretical disputes, as well as to illuminate some of the effects of this epistemological phenomenon. If such an account can lower the temperature of current debates, this will be an unanticipated benefit; but it may well be sufficient just to shed a little light where so much heat now burns.

In the next section of this chapter and again in chapters 2 and 3, I contrast what I am calling the modern fact with its ancient (Aristotelian) counterpart; and in a moment I will also argue that the modern fact has begun to be replaced, in many domains of knowledge production and in most of the world, by a postmodern variant of the fact. (Indeed, the challenge to the modern fact that is now appearing in so many disciplines may help explain the eruption of those conflicts around the fact to which I have alluded.) My primary concern in this book, however, is with Britain and with the period from the late sixteenth century to the first half of the nineteenth. Partly I focus on Britain simply because it is the country whose intellectual, cultural, and political histories I know best. Partly I do so because the liberal form of government that emerged in England at the end of the seventeenth century encouraged private citizens and voluntary societies to initiate all kinds of knowledge-making projects, at the same time that various types of knowledge were being presented as an aid to—or even a mode of—effective state rule. Although different relationships between central governments and the production of knowledge (especially natural philosophical and demographical knowledge) existed on the Continent in this period, and even though some British writers were undeniably influenced by Continental models, I focus on Britain because the relation between knowledge production and government that was forged in eighteenth- and early nineteenth-century Britain has been particularly important for Western ideas about how governments and private organizations divide the task and reap the rewards of making systematic knowledge.

I focus on the years from the late sixteenth century to the early nineteenth because this long (and somewhat unconventional) period unites an early, as yet untheorized, incarnation of the modern fact in Britain with a variety of plans that acknowledged—and tried to professionalize—the peculiarity with which I have associated the modern fact. I place the first English manual on double-entry bookkeeping at the beginning of this study because the particularizing yet systematically arranged entries in the double-entry books constitute an early example of the tension within the fact, which was more explicitly theorized in the early seventeenth century. Because the date of the first English double-entry manual is uncertain (the surviving text dates from 1588, but this is almost certainly an adaptation of a text published in 1543), and because all sixteenth-century English manuals were rough translations of an Italian text published in 1494, I do not want to claim that the beginning of this book coincides with anything like a single or clear origin for the modern fact. Instead, I try to capture the inauguration of what can only retrospectively be called a period by showing how what looks like a variant of the modern fact appeared in a writing practice that did *not* participate in the epistemology we associate with modernity.

By the same token, I conclude with J. R. McCulloch's 1825 taxonomy of knowledge, John Herschel's 1830 attempt to deal with the problem of the fact in natural science, and John Stuart Mill's elevation of deduction over induction in the emergent social sciences not because these metatheoretical gestures constituted the end of the modern fact but because, in recognizing the tension between observed particulars and theoretical or systematic knowledge *as a problem that required a professional (or disciplinary) solution,* these three men implicitly turned the task of knowledge production in the rapidly professionalizing sciences over to so-called experts; these experts eventually introduced the reformulation that would begin to displace the modern fact after about 1870. This reformulation, which occurred at different moments in different disciplines, gradually elevated rule-governed, autonomous models over observed particulars. After the late nineteenth century, at least in the natural and social sciences, expert knowledge producers sought not to generate knowledge that was simultaneously true to nature and systematic but to *model* the *range of the normal* or sometimes simply to create the most sophisticated models from available data, often using mathematical formulas. As the units of such models, postmodern facts are not necessarily observed particulars; instead, as digital "bits" of information, the "phenomenological laws" of physics, or poststructuralist signifiers with no referent, they are themselves already modeled and thus exist at one remove from what the eye can see, although they are no less the units by which we make what counts as knowledge about our world.[2]

As the examples I have just offered suggest, one might take up the history of the fact in any or in all of those modern knowledge arrangements we call disciplines. Because I have had to limit the scope of this book, and even though I discuss rhetoric in chapter 2 and natural philosophy in chapter 3, the systems of knowledge production I am primarily concerned with in *A History of the Modern Fact* are the sciences of wealth and society—the disciplinary ancestors of what we call economics and social science. I have chosen these sciences because they originated and achieved something approaching their disciplinary form in the period I examine here, and also because, for reasons I will explain shortly, they were intimately linked to two issues with which the modern fact has historically been involved: the relation between "interest" and knowledge, and that between knowledge and government.

I have also chosen to focus on the sciences of wealth and society because, apart from mathematics, astronomy, and magic, these knowledge practices were among the first to rely on numerical representation as a critical component of knowledge production. Numerical representation is particularly central to this book because, once they were purged of the last vestiges of supernaturalism, numbers came to epitomize the peculiarity written into the modern fact. On the one hand, as signs of (what looks like or passes as) counting, numbers seem to be simple descriptors of phenomenal particulars, and because the mathematical manipulation of numbers is governed by a set of invariable rules, numbers seem to resist the biases that many people associate with conjecture or theory. On the other hand, however, because numbers also constitute the units of a system of knowledge production that is biased toward deduction—that is, mathematics—numbers inevitably carry within them the traces of a certain kind of systematic knowledge: to assign numbers to observed particulars is to make them amenable to the kind of knowledge system that privileges quantity over quality and equivalence over difference. Beyond epitomizing the peculiarity written into the modern fact, numerical representation is also important here because numbers played a crucial role in the transformations of government that occurred in the period I examine. As Michael Clanchy has argued, the collection of numerical information helped consolidate government power in England in the thirteenth and fourteenth centuries; and as I argue below, this process resumed at the end of the eighteenth century, although the British government had not collected numerical information in a theoretically informed way for most of that century.[3]

Even though numerical representation epitomizes the peculiarity I have associated with the modern fact, however, I want to stress at the outset that, historically, there was no *necessary* connection between the epistemological unit I am calling the modern fact and numbers as a specific form of representation. The disjunction between the epistemological unit and the representational form is clearest in the sixteenth century. On the one hand, this period saw numbers deployed in new ways that did *not* generate modern facts, in the sense in which I am using this term. A prime example of such an innovation is the introduction of numbers to demarcate chapters and verses of the Bible; numbers made it easier to find particular passages, but they did not contribute to a system of general knowledge. On the other hand, this period also contains examples of modern facts that did not rely on numbers. These include herbals, those catalogs of plants whose units were both observed particulars and parts of a general knowledge system but that were not numbered or quantified.[3] We need to keep in mind this disjunction between the modern fact and numerical representation, because chapters 4 and 5 detail episodes in the history of this epistemological unit in which what counted as a fact retained only a metaphorical relation to numbers.

I also want to stress, then, that the focus of *A History of the Modern Fact* is the epistemological unit, not numbers per se. I am interested in how numbers acquired the connotations of transparency and impartiality that have made them seem so perfectly suited to the epistemological work performed by the modern fact, but this book is not a history of numerical information, the uses to which numbers were put, numeracy, or mathematics. I have not tried to tell any of these stories, both because other scholarly studies of numbers, along with their collection and use, have already been written, and also because the history of mathematics is simply too huge a subject for me to take on. Moreover, the questions to be addressed in a history of numeracy (such as Who counted? What did people count? For what social and institutional purposes did people count? And with what financial backing were large-scale counting projects inaugurated?) are now and probably will always be unanswerable.[5] This book addresses numerical representation for the reasons I have already advanced—because numbers epitomized the peculiarity of the modern fact and because numerical information (eventually) became the British government's preferred mode of knowledge. Given the vicissitudes of history and the gaps in our own knowledge about the past, we may never know exactly why numbers came to play these roles, but at least we can begin to understand how the epistemological unit that now seems inseparable from numbers was forged and what conceptual problems lurk in the current conjunction of facts and numbers.

Although I concentrate primarily on early versions of the sciences of wealth and society, I am also concerned with the demise of the old status hierarchy that characterized ancient and medieval societies and the emergence of the functionally differentiated domains that we associate with modern, disciplinary knowledge.[9] Thus, implicitly, *A History of the Modern Fact* is also at least a rudimentary history of the disciplines, focusing on the stages by which the ensemble of knowledge practices that dominated the ancient world was re-ordered in such a way as to separate numerical representation from figurative language and, gradually, to elevate practices associated with numbers over those associated with metaphorical language. In the chapters that follow, I argue that the emergence of the modern fact coincided with this reordering—indeed, was instrumental to it—and that effacing this epistemological unit's character-istic peculiarity was central to creating, then sustaining, the illusion that num-bers are somehow *epistemologically* different from figurative language, that the former are somehow value-free whereas the excesses of the latter disqualify it from all but the most recreational or idealist knowledge-producing projects.

As even these introductory comments should demonstrate, *A History of the Modern Fact* participates in numerous historical projects currently under way in a variety of academic disciplines. By way of general orientation, let me identify three broad areas of research with which this project overlaps, although in the chapters that follow it will become clear that these do not exhaust the fields I have drawn on and hope to contribute to. The first project the book is obviously related to consists of various attempts to trace the prehistory or genealogy of modern economics, the social sciences, or both. These studies, inaugurated in the 1950s and 1960s by Joseph A. Schumpeter, Philip Abrams, Ronald Meek, and William Letwin, among others, were supplemented in the 1970s and 1980s with a series of specialized studies that have been indispensable to me; these in-clude treatments of individual disciplines (political arithmetic, political econ-omy), modes of representation (statistics, numbers), and disciplinary societies (the statistical societies, the Political Economy Club). More recently, Richard Olson, Eileen Janes Yeo, and Donald Levine have offered enlightening treat-ments of "social science" more generally, which have begun the crucial work of investigating the roles of gender, class, and national difference in the making of these modern disciplines.

The second general area of scholarship *A History of the Modern Fact* overlaps with can roughly be designated intellectual history. More specifically, the fields relevant to this book include the history of British political thought, Fou-cauldian attempts to devise a history of governmentality, and intellectual histo-ries of the Scottish Enlightenment. Among recent histories of political thought, I have found the work of Richard Tuck, Maurizio Viroli, Quentin Skinner, and Peter N. Miller particularly helpful; and the sensitive treatments of British intellectual history by Stefan Collini, Donald Winch, and John Burrow have been instrumental to my argument about late eighteenth- and early nine-teenth-century political theorists. Although I part company with orthodox Foucauldians in ways I describe below, I have found Foucault's theory of gov-ernmentality especially helpful for understanding eighteenth-century British liberalism; in this regard, the essays collected in *The Foucault Effect* deserve special attention, as does Mitchell Dean's *The Constitution of Poverty: Toward a Genealogy of Liberal Governance.*[9] My efforts to understand the Scottish Enlightenment have benefited particularly from the work of Nicholas Phillipson and Richard B. Sher, and I have found Andrew Skinner, Duncan Forbes, Knud Haakonssen, and Istvan Hont especially helpful in explaining the places Hume and Smith occupied in the Scottish Enlightenment.[10]

The third general scholarly project *A History of the Modern Fact* overlaps with is the subset of the history of science that Lorraine Daston has called historical epistemology. Indeed, as "a history of the categories of facticity, evidence, objectivity, and so forth," historical epistemology is the methodological label I consider most appropriate for this book, although at this point it seems more useful to expand than to limit its scholarly affiliations.[11] Insofar as historical epistemology assumes that the categories by which knowledge is organized—not only epistemological units like facts, but also institutionalized units like disciplines or professional societies—inform *what* can be known at any given time, as well as *how* this knowledge can be used, historical epistemology is a study of determinations and effects. Insofar as historical epistemology assumes that the categories by which knowledge is organized change over time, it is less a study of the inexorable march of "science" toward a fully adequate description of nature than an investigation of those developments that have increasingly made Westerners believe this march is under way.

Simply naming the scholarly projects this book overlaps with does not begin to do justice to the theoretical and methodological differences that divide the authors I have named, of course. In the chapters that follow I give considerable attention to these scholarly debates about method; and in a later section of this chapter I describe my own position within the most prominent debates. It has seemed useful at the outset to name some of the scholarly work that has contributed to *A History of the Modern Fact,* both because I want to make it clear that this is largely a synthetic work of interdisciplinary (or transdisciplinary) history, which could not have been written had more properly disciplinary histories not already existed, and because I want to set an example for readers of this book. Because it draws on and contributes to so many scholarly projects, it is necessarily a prolegomenon for future research, and I invite my readers to make of this book what I have made of others—a provocation to think disciplinary knowledge anew.

ANCIENT FACTS, MODERN FACTS

To clarify what I have identified as the peculiarity of the modern fact, it is helpful to contrast this epistemological unit with its ancient counterpart. As I point out in chapter 3, this contrast has been drawn most succinctly by Lorraine Daston and Peter Dear. According to both Daston and Dear, the two features that characterized the epistemological unit from which ancient systematic knowledge was made were universality and commonality. As Aristotle described it, in other words, genuine knowledge was not made up of discrete or observed particulars, especially if they were experienced by a single individual. Thus in the *Posterior Analytics* Aristotle declared that "sense perception must be concerned with particulars, whereas knowledge depends upon the recognition of the universal"; and in the *Metaphysics* he explained that the only experience that matters is common experience—"that which is always or that which is for the most part."[12]

To Daston, Aristotle's definition of knowledge makes "'facts,' in the sense of nuggets of experience detached from theory," an "invention" of the seventeenth century.[13] Only after Bacon complained that universals did not coalesce spontaneously out of the common experience of particulars, that they had to be constructed somehow out of the philosopher's patient observation of natural phenomena, was it possible to conceptualize a "nugget of experience detached from theory" as a valid unit of knowledge production. Because he wanted to keep philosophers from rushing to premature generalizations, moreover, and because he wanted to release the stranglehold imposed on new knowledge by Aristotle's respect for commonplaces, Bacon identified the most telling experiential "nuggets" as rarities and anomalies—the "errors, vagaries, and prodigies of nature" that the natural philosopher had to pursue or contrive in order to assemble the materials from which systematic knowledge was formed.[14]

As crucial as Daston's account is for any historical treatment of the fact, her desire to highlight the contrast between Aristotelian knowledge and the kind of knowledge associated with the scientific revolution has tended to obscure one sense in which what Bacon defined as the modern fact resembled the units ancient knowledge was made from: like their predecessors, modern facts are units of *systematic* knowledge; as such, they are inherently susceptible to the very theoretical presuppositions from which Bacon seemed to want to insulate them. This is the point Peter Dear has recently made in his discussion of the revaluation of "experience" and "experiment" that also took place in the seventeenth century. Like Daston, Dear argues that before then Western natural philosophers tended to value universal knowledge that was based on commonplace experience; in keeping with these preferences, they used reports of singular experiences simply to illustrate general knowledge claims or to compile natural histories. Dear explains that this practice began to change in the seventeenth century, and that as natural philosophers began to attach evidential weight to particular events and to use instruments to contrive unique experiences, it became increasingly difficult to muster the shared assent that had previously underwritten commonplaces. As the kind of experience the natural philosophers consulted changed, so did the epistemological significance of the singular event: "The singular experience could not be *evident*," Dear explains, "but it could provide *evidence*."[15]

This distinction constitutes the telling point of Dear's analysis: the singular experiences or observed particulars that natural philosophers began to value in the seventeenth century were not *evident*, because they were neither signifiers of anything nor self-evidently valuable; only when such particulars were interpreted *as evidence* did they seem valuable enough to collect, because only then did they acquire meaning or even, I contend, identity as facts. Although Dear does not make this point explicitly, his distinction between an evident particular and a particular that constitutes evidence helps us understand what I am calling the peculiarity of the modern fact. On the one hand, facts seem (and can be interpreted as being) simply the kind of deracinated particulars that Bacon claimed to value; on the other hand, facts seem (and can be said) to exist as identifiable units only when they constitute evidence for some theory—only, that is, when there is a theoretical reason to notice these particulars and name them as facts.

Before I continue, I need to make clear the historical specificity of my claim about the modern fact. In the seventeenth and early eighteenth centuries, natural philosophy was only one in an ensemble of knowledge-producing practices, and what I am calling the modern fact was only one of the available epistemological units called "a fact." Most important, natural philosophers worked in conjunction with—but also distinguished themselves from—natural *historians*, who *did* collect deracinated particulars and for whom *factum* retained its old connotations of "event or occurrence," "a particular truth known by actual observation" (*Oxford English Dictionary*). During the period I examine in this book, the natural historical meaning of the fact continued to be available, but it was gradually demoted to relative marginality, as the natural philosophical ideal of systematic knowledge gained prestige even among natural historians. To designate the epistemological unit whose history I am discussing "the modern fact" is not to deny the historical existence of this natural historical variant of the fact or its importance in the seventeenth and early eighteenth centuries. Instead, by placing *the* and *modern* before *fact,* I want to register a tendency, which eventually transformed the practice of natural history as well, toward privileging "facts" that were *both* observed particulars *and* evidence of some theory.

Beyond designating the modern fact as an epistemological unit of particular interest, I depart from the groundbreaking analyses of Daston and Dear in two more ways. First, I argue that Bacon's distinction between ancient commonplaces and the deracinated particulars from which modern knowledge was to be constructed was *rhetorical*—that is, Bacon insisted that the "errors, vagaries, and prodigies" of nature were immune from the theoretical assumptions that obviously informed the ancient commonplaces *because he wanted to clear a space for new knowledge and to silence academic dissent.* He sought to clear this space, in turn, by distinguishing as sharply as he could between his method and that of the Aristotelians, *even though,* as we will see in chapter 3, the distinction was most obviously a matter of *style,* not substance. Second, I differ from Daston and Dear in locating the peculiarity I associate with the modern fact—the peculiarity of seeming both true to nature (in some sense) and amenable to generalization—in a writing practice that preceded the scientific revolution of the seventeenth century. One site at which we can identify a prototype of this peculiar epistemological unit, I propose in chapter 2, was double-entry bookkeeping.[16] Although I cannot claim that published treatises on double-entry bookkeeping were the *only* places where a rudimentary form of the modern fact can be found before the seventeenth century, I do argue that this kind of mercantile writing played a crucial role in undermining the very status hierarchy from which its champions initially sought to derive their own authority. By the same token, then, and no doubt completely unintentionally, early proponents of double-entry bookkeeping helped late sixteenth-century academics elevate the observed particular to a status that rivaled those assertions about universals that had been the cornerstone of ancient knowledge.

In suggesting that Bacon's repudiation of Aristotelian commonplaces was an attempt to persuade his readers that his method was new, I do not mean to downplay the significance of this revision. Indeed, I agree with nearly all historians of science that Bacon's elevation of the observed particular looks *in retrospect* like the beginning of an epistemological revolution and that, because subsequent philosophers took Bacon at his word, the elevation of the observed particular has had consequential—and material—effects. It has, in other words, created the *effect* of the revolution that Bacon said it was. By emphasizing Bacon's *claims* about the novelty of his method rather than the innovative features of the method itself, however, in insisting that he presented a difference in style as (if it was) a substantial difference, and in maintaining that the revolutionary appearance of Bacon's innovation is at least in part a function of the way he represented his contribution (as well as how it has been read and deployed), I want to examine the dynamics by which the effect of radical difference (or even revolution) was produced. Historians like Daston and Charles Webster, who have accepted Bacon's revolutionary claims, have explored the causes of this revolu-

tion; but I want to describe how Bacon made plausible his claim to have introduced innovations.[17] Given the prestige of Aristotelian philosophy in the late
sixteenth and early seventeenth centuries, how was Bacon able to make his vision of a new philosophy compelling? Given what one can identify as the similarities between the method Bacon alleged was new and the method it
supposedly rejected, how did he emphasize novelty over resemblance, innovation over continuity?

By locating a prototype of the modern fact in double-entry bookkeeping,
I want to advance two ideas: first, that the changes in epistemology emphasized
by seventeenth-century natural philosophers like Bacon and Boyle had roots in
low as well as high cultural practices; and second, that the availability of a prototype of the modern fact in a familiar (but socially devalued) cultural practice
like commerce enabled natural philosophers to explain what kind of knowledge they wanted to produce. This is especially clear in Thomas Sprat's *History
of the Royal Society,* which cites the information merchants imported as one
model for the kind of "uninterested" knowledge that natural philosophers
wanted to offer about the natural world. When we examine the way knowledge
is produced in the double-entry bookkeeping system, however, we do not see
nature speaking for itself, as the natural philosophers claimed the laboratory enabled it to do, we see fictions being installed as props to systematic meaning and
coherence. In the double-entry bookkeeping system, in other words, certain
critical fictions—personifications like "money," numbers that have no referent,
and the equation of "book value" with price—enabled the accountant to
create the all-important balance, which produced the system's most salient
meaning: that the merchant who kept the books obeyed the order of God's harmonious world, that the merchant was creditworthy because he was honest.

The relation between systematic knowledge and these constitutive fictions
is more visible in early double-entry bookkeeping manuals than in seventeenth-century accounts of natural philosophical method. Double-entry
bookkeeping also provides a clearer example of the way systematic knowledge
could create effects beyond its explicit agenda. In addition to the obvious purpose of recording commercial transactions, double-entry bookkeeping also
displayed the merchant's moral rectitude, which was signified by the balance
and harmony so prominent in the double-entry ledger; it generalized rule-
governed behavior by encouraging merchants and their agents to reproduce in
action the orderly logic of the books; and as an effect of this generalization, it
enhanced the social status of merchants as a group.

In chapter 2, I argue that double-entry bookkeeping could produce these
effects because its first apologists positioned this system of writing within the
system of rhetoric, which functioned to support a status hierarchy as well as to

generate knowledge. Within the hierarchy superintended by rhetoric, the adequacy of knowledge was not a function of its truth to nature. Instead, knowledge was considered adequate if it was universal (if it confirmed the commonplaces). In practice, the adequacy of knowledge in the ancient and medieval worlds was judged by an orator's ability to persuade; and the ability to persuade was largely a function of the speech act's conformity to rules—the rules that constituted rhetoric. As John Bender and David E. Wellbery have pointed out, the rules of rhetoric functioned both to reproduce the status hierarchy and to adjudicate among knowledge claims.[18] Even though the author of the first printed double-entry bookkeeping manual sought to borrow the prestige that rhetoric conferred by positioning this mode of mercantile writing within rhetoric's cultural field, however, some of the features of double-entry bookkeeping were implicitly at odds with the assumptions about hierarchy intrinsic to rhetoric. As a consequence, generalizing the rule-governed behavior that reproduced the orderly books, as merchant apologists tried to do, began to undermine the status and the knowledge hierarchies that rhetoric both needed and reinforced.

The pivotal position of double-entry bookkeeping—it simultaneously was situated within the status system upheld by rhetoric and tended to undermine this system—made mercantile writing particularly attractive to the natural philosophers of the seventeenth century, who sought models for the knowledge they claimed to produce. Mercantile writing was also attractive to natural philosophers like Robert Boyle because, even though merchants tried to use their stylistic affiliation with the old status system to enhance their prestige, they did not acquire the social authority they sought. As Boyle (and more explicitly, Sprat) contended, merchants and their writing could be appropriated as examples, but merchants posed no threat to natural philosophers, who could more easily command the king's attention because they seemed less interested in monopolizing his resources. Thus seventeenth-century natural philosophers not only used merchants (to gather information) and mercantile writing (as an example of "uninterested" knowledge production); they also helped keep merchants in their place, partly by effacing the importance of conventions of writing in the production of natural philosophical knowledge.

Examining in detail the workings of the double-entry system, as I do in chapter 2, shows us that merchants' ability to generate systematic knowledge and the effects that exceeded the demands of record keeping depended on constitutive fictions. But even if natural philosophers used merchants as an example, as I argue in chapter 3, does this mean that the knowledge natural philosophers generated was based on fictions too? Certainly Boyle's claims about the air pump did not depend on personifications like "money," nor did he

use the trope of balance to align natural philosophy with God's order. Nevertheless, as Steven Shapin and Simon Schaffer have argued, seventeenth-century
natural philosophy did derive its authority from claims that required a leap of
faith—even if these were not self-evident fictions. These claims included the
contention that the knowledge generated in the laboratory had nothing to do
with politics and the assertion that this knowledge, which was artificially contrived, could be confirmed by an audience of credible witnesses but held good
for the world at large.[19] In chapter 3, I discuss how natural philosophy depends
on such leaps of faith or canons of belief and show how the experimental
method and induction were reunited with what seventeenth-century philosophers argued were their opposites: the deduction and a priori reasoning associated with Aristotelian philosophy and Thomas Hobbes. This reunion occurred
in the work of William Petty, whose political arithmetic sought to use the experimentalists' claims *not* to rely on leaps of faith to present a variant of deductive logic as mere observation.

If both the experimentalists and Hobbes based their systems of knowledge
at least partly on belief, then why didn't systematic knowledge's dependence on
what one might call fictions undermine the experimentalists' claims? In chapters 3 and 4 I begin to develop two answers to this question. First, I demonstrate
that proponents of the new knowledge periodically repeated the gesture by
which Bacon distinguished the method of induction in the early seventeenth
century: they invoked *stylistic* difference to mark the novelty of a mode of
knowledge production that was supposedly closer to the particulars of the phenomenal world because it was more *transparent*. Emphasizing stylistic difference
where we might expect to find a description of method helped signal the supe-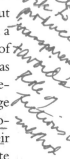
riority of the new mode of analysis without raising troubling questions about
the place of fictions in the method, because it focused readers' attention on a
mode of representation that seemed to be transparent, not on the stages of
analysis by which systematic knowledge was being produced. Thus Thomas
Mun, Thomas Sprat, William Petty, Daniel Defoe, David Hume, and early nineteenth-century champions of statistics all argued that the modes of knowledge
production they were trying to define jettisoned the ornamental excesses associated with rhetoric. For all these writers, number constituted the limit their
stylistic revisions aspired to reach, for number increasingly seemed to constitute
the most transparent—and thus the least "interested" or biased—form of representation.

My second answer to the question posed by the new knowledge's dependence on belief stresses the persistence of a strong theological strain in seven-
teenth- and eighteenth-century attempts to generate knowledge that was
simultaneously true to nature and systematic. What David Hume was to call a

fiction, in other words, seemed to more orthodox philosophers like God's plan; and the belief that Hume only reluctantly admitted into philosophy, as the necessary prop for system, other philosophers eagerly embraced. Because they understood the prop that supported systematic knowledge as providential design, most British philosophers of this period did not see a problem in the gaps that yawned in all systematic sciences, including the new sciences of wealth and society. Indeed, even after Hume refused to accept on faith *any* kind of explanation not based on particulars that could be observed, most moral and natural philosophers continued to refer what they failed to understand to God, and most continued to insist that the order of philosophy simply reflected the order God had written into the universe.

Even though the effect of Hume's skepticism was delayed, the issue he raised theorized the peculiarity inherent in the modern fact and made it visible *as a philosophical problem*. As what philosophers have dubbed the problem of induction, moreover, this epistemological conundrum was eventually to prove instrumental in undermining the authority of induction itself. Beginning in chapter 4, I assign the problem of induction a central place in my narrative, because I argue that the light this problem casts on providential narratives enables us to see what the authors of these narratives could not: that what some called God, others might interpret as (necessary) fictions. In the next section of this chapter I develop the theoretical assumptions implicit in the mode of analysis that leads me to read this way; for now I will simply note that, once Hume formulated the problem of induction—*even though he did not consider it particularly troubling*—the peculiarity inherent in the modern fact could be conceptualized as such. It was, to use Foucault's formulation, "in the true."

In emphasizing the problem of induction, I once more follow the lead of Peter Dear. In its modern usage, of course, induction is the method Bacon recommended for moving gradually from observed particulars to the generalizations that constituted systematic knowledge. As Dear points out, Aristotle also used the concept that Cicero translated as *inductio* (for the Greek *epagôge*), but Aristotle typically used this concept to refer to the production of *universal* statements that were capable of serving as premises in a demonstrative syllogism.[20] Because Aristotle believed that universals really existed apart from their appearance in phenomenal particulars, and because he assumed both that they could be demonstrated with logical procedures like the syllogism and that they were self-evident, he did not consider induction a problem. As Daston explains, moreover, because Aristotle believed that humans were "happily constructed so as to detect the unity of universals in the welter of particulars," he did not think it was necessary to elaborate—or interrogate—the stages by which general knowledge was forged from discrete particulars.[21]

When Bacon explicitly elevated the observed particular—especially the natural anomaly or the contrived event—to a new status, he also laid the groundwork for problematizing induction. If one had to resist premature generalization, after all, and if one could produce systematic knowledge only by reasoning from the phenomena one observed, then it was imperative to know how one moved from the particulars one saw to knowledge that was sufficiently general to explain things one had yet to see. As I suggest in chapter 3, Bacon was notoriously vague on this critical point, and as I have already suggested, he tended to emphasize stylistic difference precisely where methodological clarification seems called for. Nevertheless, and partly because Bacon was more interested in quelling the acrimonious controversy that characterized philosophical discourse in the early seventeenth century than in creating a blueprint for future science, he acted as if induction was an unproblematic method. David Hume's explicit formulation of the problem of induction in the 1740s constitutes what one might consider a *belated* effect of Bacon's empiricism, and the solutions to this epistemological problem that I examine in the concluding chapter constitute another belated effect—this time of the problem of induction itself.

Chapters 5 and 6 are devoted to the philosophical context in which Hume conceptualized the problem of induction and to various eighteenth- and early nineteenth-century attempts to address or neutralize this epistemological question. In these chapters I also describe eighteenth-century variants of the modern fact, because these variants were all developed in order to produce knowledge that was simultaneously true to nature (in some sense) and systematic; as such, all these variants of the modern fact also both incarnated and sought to manage the problem to which Hume gave a name. These variants appear in a group of related philosophical enterprises, all of which were descended from (and remained affiliated with) natural philosophy; they include experimental moral philosophy, conjectural history, political economy, and beginning in the nineteenth century, statistics. Although the participants in these practices took different positions on the central questions of providentialism, skepticism, and the ground of systematic knowledge, all the practices reveal that one effect of efforts to generate systematic knowledge was the production of a set of abstractions, which rapidly became the objects of these sciences. These abstractions, which include "society," "the market system" (then "the economy"), and "poverty," now constitute the characteristic objects of the modern social sciences, including the sciences of wealth and society.[22]

In chapters 6 and 7, I suggest that attempts to generate systematic knowledge about abstractions like society, the market, and poverty also created a new social position—that of the expert. In chapter 2, I argue that merchant apologists like Edward Misselden and Thomas Mun tried to create such a position in the early seventeenth century, and in chapter 3, I point out that William Petty attempted to convince the king that something like a numbers expert was essential to national well-being; but because of the priority that the developments I discuss in chapter 4 gave to a public (as opposed to sovereign) mode of knowledge production, expertise was not an essential component of eighteenth-century British conceptualizations of knowledge. Indeed, the emphasis that both natural and moral philosophers gave to a universal subject ("mankind" or "human nature") and to understanding the dynamics of subjectivity tended to work against the emergence of professional expertise as a necessary criterion of knowledge production. When David Ricardo recast political economy as a mathematical science, however, and when natural scientists and political economists began to form professional organizations early in the nineteenth century, they laid the groundwork for making the expert essential to (what was understood as) legitimate knowledge. In the concluding chapter I suggest that the protocols and systems of credentialing associated with such professional organizations offered one way to approach the problem of induction. As we see in the taxonomy of knowledge that J. R. McCulloch devised in 1825, collecting information about particulars could be separated from producing systematic knowledge if one was willing to hand the task of knowledge production over to experts and entrust them with settling what could then be seen as properly professional disputes.

METHODOLOGICAL CONSIDERATIONS

Given the range of secondary materials I have consulted for this book and the intensity of disciplinary debates, I need to locate my analytic method within the field of interpretative possibilities represented by the scholars I have already named. As everything I have said thus far should indicate, *A History of the Modern Fact* offers a *historical* analysis of certain texts and writing practices produced in Britain during what is loosely called the Enlightenment. Although I have drawn extensively on intellectual histories, however, the book is not simply an intellectual history. Indeed, for the most part I am less interested in the *content* of particular texts than in the *mode of argumentation* writers used, including whether or not they used numbers or references to mathematics and, in cases where numerical representation or mathematical tropes do appear, how these figures work in the semantic system constituted by the text.

Unlike most intellectual historians, in other words, I am not primarily interested in the influence one individual's ideas had on the ideas of others, nor am I primarily concerned with the development of particular abstractions (liberalism, for example), which intellectual historians typically detach from their original formulations for analysis. As I have already noted, these abstractions were a characteristic product of the kind of writing I examine here, and instead of simply taking them as given, I want to see how they acquired sufficient vitality to produce material effects. As I have also noted, one can identify epistemological effects like these abstractions only by looking at *the way an argument was conducted* and at how the mode of argumentation delimited *what could and could not be said*. In the chapters that follow I focus on how arguments were conducted, because I assume that questions of epistemology—what constitutes a unit of knowledge, how experience is sifted and arranged—underwrite what can be known. I am also interested in the way arguments were conducted because I assume that *how an argument is conducted constitutes the argument itself*: there are no ideas apart from their articulation. Because I believe that ideas are not separable from their articulation, I am also interested in the formal conventions of writing practices themselves. Like Stephanie H. Jed, whose treatment of Italian mercantile writing has been essential to my analysis of double-entry bookkeeping, I believe that the conventions encoded in features like the placement of words (and numbers) on the page help constitute—as well as delimit—how a text can mean.[23]

If I depart from intellectual historians in insisting that ideas cannot be separated from modes of representation, then I depart from Foucauldians in resisting any historical account that privileges ruptures or focuses only on discourses. In many ways *A History of the Modern Fact* has been influenced by Foucault's work, of course, especially by his early writing on epistemology (*The Order of Things*) and his late writing on governmentality ("Governmentality," in *The Foucault Effect*).[24] Indeed, some readers will recognize similarities between the account I offer here and the kind of history Foucault called genealogy. Like Foucault's genealogies, this book rejects the claim, advanced by historians as diverse as Marxists and liberals, that history reveals the unfolding of a single "logic" (the logic of capital, the logic of the liberal state); and like the variants of genealogy offered by Mitchell Dean and Ian Hunter among others, *A History of the Modern Fact* insists that texts and events generate effects in multiple domains, even those distant from the domain a writer intended to affect.[25] Even though they explicitly repudiate narratives that chronicle logics, however, most Foucauldians create the impression of origins and teleology when they insist that certain practices or events constitute ruptures. I have deliberately blurred the beginning and the end of the story I tell here because I want to insist that what Foucauldians identify as ruptures can also be interpreted as part of a continuous, if complex, process. When I suggest in chapter 3 that Bacon and Boyle proclaimed the novelty of induction in order to clear a space for new knowledge, for example, and when I argue in chapter 2 that a prototype for the basic epistemological unit of this new knowledge predated the new philosophy of the seventeenth century, I am rejecting the narrative of ruptures and revolutions for an account of processes whose continuities were effaced—most often for social (rather than strictly epistemological) reasons.

By his own account, Foucault emphasized narratives of rupture or discontinuity because he wanted to "substitute differentiated analyses for the theme of totalizing history ('the progress of reason,' 'the spirit of a century')." As he describes it, the history that results respects the differential rationality of various discourses: "The history of mathematics does not follow the same model as the history of biology, which itself does not share the same model as psychopathology."[26] While I share Foucault's suspicion of totalizing history, however, and while I wholeheartedly endorse his claim that discourses are neither structurally homologous nor likely to change in the same way (or at the same rate), I have not wanted to limit myself here to what he calls discourses. As he defined it in "Politics and the Study of Discourse," a discourse is a rule-governed practice: "What individualizes a discourse such as political economy is . . . the existence of a set of rules of formation for *all* its objects (however scattered they may be), *all* its operations (which can often neither be superposed nor serially connected), *all* its concepts (which may very well be incompatible), *all* its theoretical options (which are often mutually exclusive)."[27] I have chosen not to limit myself to such rule-governed practices partly because, by Foucault's own statement, a discourse is only a phase—albeit an important one—in the history of meaning-making practices. In *A History of the Modern Fact,* I begin with the epistemological unit of the fact because it exists prior to discourses (in both chronological and conceptual senses) and thus can help expose the affiliations among discourses even after the process of disaggregation has codified their differences. I have chosen not to limit this book to one or even a number of discourses, because doing so has led many Foucauldians to focus exclusively on one kind of rationality—say, political rationality—despite Foucault's insistence that one must trace the "divergence, the distances, the oppositions, the differences, the relations of various . . . discourses."[28] Limiting oneself to one rationality respects the outcome of that long (and uneven) process of disciplinary disaggregation that has created the modern, functionally differentiated domains; but respecting this outcome without understanding the historical process of disaggregation that has produced it tends to obscure the traces of likeness that linger in modern discourses. Simply respecting (repeating) the functional differentiation of modern domains tends to make numbers, for example, seem different in epistemological nature from metaphors and other kinds of figurative writing.

Instead of focusing on discourses, I have chosen to focus on epistemology because, as Shapin and Schaffer have argued, "questions of epistemology are also questions of social order."[29] Unlike discourses, which tend to (but do not necessarily) limit historical accounts to one domain, that is, questions of epistemology encourage the historian to consider how domains of knowledge production overlap (whether or not the modes of knowledge production have yet to be stabilized as discourses), how knowledge about nature resembles (as well as differs from) knowledge about society, and how knowledge practices are gradually differentiated, codified, and institutionalized *as different kinds of knowledge.* By focusing on the modern fact instead of political economy, for example, I have been able to identify the continuities between seventeenth-century natural philosophy and political philosophy (as Shapin and Schaffer have also done). By focusing on the peculiarity written into the modern fact, I have also been able to locate the knowledge practice that took up the project of generating knowledge about society after that abstraction was split off from nature in the seventeenth century (by the joint efforts of political theorists like Hobbes and experimentalists like Boyle). The eighteenth-century knowledge practice that pursued society as avidly as natural philosophy pursued nature was experimental moral philosophy. This discipline, *not* political arithmetic, was the disciplinary antecedent to political economy, which in turn was the disciplinary forebear of statistics—although, as we will eventually see, statistics was imported specifically to split off a task that Adam Smith had assigned to political economy: the task of *describing* the particulars of the social world.

As my references to Lorraine Daston, Peter Dear, Steven Shapin, and Simon Schaffer should make clear, the contemporary variant of historical analysis with which I want to affiliate *A History of the Modern Fact* is the branch of the history of science often called science studies. Recently, in fact, a French practitioner of science studies has offered a programmatic statement about the agenda of this subdiscipline that accords remarkably well with what I have tried to do in this book. In *We Have Never Been Modern,* Bruno Latour also rejects the historical model that privileges rupture in favor of what he calls a "sociology of criticism," which seeks to understand why historians (and historical agents) have *needed* a narrative of novelty and absolute breaks. Drawing heavily on Shapin and Schaffer's work, Latour argues that when modern historians identify ruptures between the premodern age and modernity and when they separate knowledge about nature (science) from knowledge about society (politics), they repeat the gestures first made by Bacon, Boyle, and Hobbes in the seventeenth century. Moderns have invented "the idea of Revolution" and a complex series of propositions about the differences between nature and society, according to Latour, because they have wanted simultaneously to believe that individuals could "influence their own fate" and to explain why volition so frequently fails.[30]

Following Shapin and Schaffer, Latour suggests that the disciplinary division introduced by Bacon, then by Boyle and Hobbes, installed what looks like (but is not) an insurmountable difference between nature and society. Nature, in Boyle's view, lies beyond all human institutions, including politics, but it can be studied (fabricated) in the laboratory; society, according to Hobbes, is an order contrived solely by human will, but it surpasses the humans who created it. This "invention of an absolute dichotomy between the production of knowledge of facts and politics," Latour argues, constitutes the fiction that grounds modernity.[31] As a fiction, it has had to be sustained by countless institutions and stories, including the story of historical ruptures; as a ground, this dichotomy has generated inestimable effects, including the production of the disciplinary difference with which my book is primarily concerned.

We Have Never Been Modern sets out a research agenda that Latour has yet to realize in his more detailed studies of knowledge production within the natural and medical sciences.[32] At times, moreover, it is difficult to see how Latour would go about implementing this agenda, since his book operates at such a level of abstraction that he seems to lose sight of the processes by which these abstractions were produced in the first place. Nevertheless, and even if it is only a programmatic statement, *We Have Never Been Modern* describes the kind of research agenda I have pursued in *A History of the Modern Fact*. Indeed, Latour recognizes the peculiarity I have identified in the modern fact, although he does not take this peculiarity as the object of those studies he forecasts. When he describes the kind of fact Boyle produced as "an object that is mute but endowed or entrusted with meaning," he animates the paradox that I have described as the modern fact's affiliation with both deracinated particulars and a priori or conjectural theories. When he argues that Hobbes's "naked and calculating citizen" constitutes the sociological counterpart to the natural philosophical matter of fact, Latour offers one way to understand why *A History of the Modern Fact* does not stay with the history of natural philosophy, as most examples of science studies (including Latour's own books) do. Instead of pursuing natural philosophy through its eighteenth-century incarnations, it veers onto the path less taken: the history of the sciences of wealth and society, which sought to produce systematic knowledge about the behaviors and subjectivity of the "naked and calculating citizen," the counterpart to the laboratory-generated or naturally occurring object of scientific analysis.[33]

The research agenda that Latour describes also accords with my conviction that it is high time for late twentieth-century critics to abandon the mission of

"unmasking" the benighted past that has become the rallying cry for so much of
what passes as Foucauldian critique. In the United States, the rigid and somewhat self-righteous sense of superiority implicit in the claim to unmask has itself been exposed in recent years by the epithet of "political correctness." In
France, Latour points out, criticisms of ideology critique or critical sociology
have focused on the scapegoating mechanism inherent in this practice.[34] In a
statement I can only wish were true, Latour writes that in the late 1990s "denunciation and revolution have both gone stale." Somewhat prematurely, I'm
afraid, he declares that "instead of really believing in it, we now experience the
work of denunciation as a 'historical modality' which certainly influences our
affairs but does not explain them any more than the revolutionary modality explained the process of the events of 1789."[35]

   In *A History of the Modern Fact,* I offer what I hope will be a model for moving beyond the historical modality of political correctness and denunciation.
To suggest how I propose we move beyond political correctness, let me describe
the analytic method I use in the following chapters. This method superimposes
two kinds of historical reading. The first will seem rather old-fashioned to post-
structuralists, for it focuses on the author's intentions and on the political, philosophical, and semantic contexts in which the text was produced. My desire here
is to identify the debates a text originally participated in, the semantic connotations and range of references available at the time it was written, and the terms
in which particular words and arguments would have made sense to the person
who wrote them and to the text's immediate audience.[36]

   If this first reading tries to approach recovery, then the second reading measures subsequent (mis)readings against two fields of connotations: the one that
had disappeared and the emergent field that was taking or had taken its predecessor's place. As fields of connotations change, generally in response to social,
economic, and political determinants that lie beyond the scope of this book, so
do configurations of knowledge. I think of these configurations as ensembles,
for they consist of hierarchies of kinds of knowledge, whose individual members may change definition and whose internal order may change too. As these
ensembles of knowledge change—as some ways of knowing acquire enhanced
prestige while others fall in status—methods and questions that once made
sense to an intellectual community become problematic. Indeed, they eventually become so problematic that a radical reconsideration seems necessary; this
reconsideration might alter the hierarchy of kinds of knowledge, or it might introduce a new mode of knowledge production or a new object of analysis. In
any case, the reconsideration often takes the form of a return to some existing
text, which once made sense but no longer does. In other words, it often takes

the form of a misreading that, when read retrospectively and in relation to the earlier text, looks like a solution to a problem that was never posed in the terms in which a solution is being offered.

I superimpose these two readings not because I think that reading backward (as one must inevitably do) can expose what "was really at stake" in some past text, but because I believe two things: that the only history we can construct is a history of (mis)readings; and that we cannot identify a reading's relation to its predecessor—we cannot identify it *as* a misreading—unless we also try to recover the field of connotations in which the first text was written. In the chapters that follow, I have generally focused on texts that, when read in relation to earlier texts, can be seen to realign the cultural configuration of knowledge practices or the configuration of individual disciplines in relation to which the earlier text was written. These texts are particularly significant for any historical epistemology because they make new theoretical or methodological problems visible, often (though not always) by introducing a new variant of some basic epistemological unit like the fact.

This double reading—first within a historical field of connotations, then back against its transformation—provides alternatives both to a narrative of rupture and to the imperative to unmask. Unlike narratives that privilege rupture, the mode of reading I practice here seeks the continuities that lurk within (what look like and can also be said to be) discontinuities. What I describe as a continuity looks to others like a rupture because the answer the later author gives doesn't even seem to address the issue the prior writer introduced; but it can be seen to do so *if one recognizes that the answer transforms what was initially a statement into a question, which the answer now addresses.* Thus Newton's recasting of induction at the very end of the seventeenth century transformed Bacon's confident statements about induction and Boyle's equally confident experimentalism into a question about how one linked one observation to another, which only a revision that emphasized mathematical laws could solve. By the same token, McCulloch's taxonomy of knowledge transformed the gap between discrete perceptions that Hume had noticed but not belabored into a question about how one could assume that what one had yet to see would resemble what one had already observed by offering a solution that separated the task of collecting data from the production of theoretical knowledge.^{37}

This method of reading offers an alternative to the imperative to unmask because it does not insist that the earlier work, which *becomes* problematic only as a result of subsequent changes and (mis)readings, was blind to (or duplicitous about) its "real" agenda. To put it in the simplest terms, this kind of reading does not argue that past writers overlooked the issues that interest late twentieth-century readers most—race and gender and class—because they were racist, sexist, or elitist. For the vast majority of writers I examine in this book, it would not have been possible to be racist, sexist, or elitist, because hierarchies of race, sex, and status simply organized their understandings of society and the world. Not until what has been called an anthropological perspective was generally available does it make any but the most anachronistic sense to think of white Western Europeans as racist; not until a feminist critique had been popularized (or vilified) in the work of Mary Wollstonecraft does it make sense to think of sexism; and not until the language of class and equality was common enough for privilege to look like *unnatural* inequality does it make sense to tar our ancestors with the brush of elitism.

Having said this, however, I also suggest that the mode of analysis I offer here enables us to see how issues that were not problems for past writers *became* problems for writers who returned to earlier texts. Once these issues become problems, moreover, it is almost impossible to read the earlier texts without reading through those issues—without, in other words, seeing the earlier texts as in some way blind, naive, or part of some "deeper logic" and therefore in need of unmasking. Indeed, that race, sex, and class have become such unavoidable problems for late twentieth-century readers helps explain why the imperative to unmask was so intensely felt in the 1970s and 1980s; this imperative, as Latour contends, was a way of scapegoating others for prejudices we had yet to escape. It also, I think, expressed the genuine need that Western academics felt (feel) to change a world in which academic work was (is) so routinely devalued and in which the abused minorities we sought (seek) to defend seemed (seem) so conveniently like us.

I include much of my own previous writing in the paradigm of denunciation that I now want to leave behind, although I like to think I never reached quite the pitch of self-righteousness that has so enraged conservative critics of the so-called tenured radicals. In leaving behind the paradigm of denunciation, I am not forsaking my political commitments to and in the present. I am still a feminist; to conservative critics, I probably still seem like a tenured radical. Certainly I still think that Robert Boyle's tendency to take his servants for granted violates the way I want to treat people less well-off than I am, and if I were creating new professions, I would not exclude women and peoples of color. Basically, I would rather live in the late twentieth century than in any of the centuries I write about here; but my historical position does not grant me diplomatic immunity or make it imperative for me to judge past writers. I realize that, inevitably, I am writing for my own contemporaries, but that does not mean that historical analysis is only for or about the present. To understand the past exactly and in its own terms may be impossible; but approaching these terms is a task worth undertaking, because the dynamics of epistemological change are clearer there than they can ever be for someone who writes about her own culture.

In the chapters that follow, I have devoted almost no attention to the issues that most trouble me and many of my contemporaries. One reason I do not discuss race, gender, and class in *A History of the Modern Fact* is that with a very few exceptions most of the writers I am concerned with were not primarily engaged with these issues. Even though women were excluded de facto from the ranks of economic experts by the codification of the rule-bound, credential-dependent practices adumbrated by double-entry bookkeeping, Luca Pacioli did not design the rules of double-entry bookkeeping in order to exclude women; even though peoples of color undeniably were adversely affected by the truisms about racial inferiority perpetuated by the conjectural historians, these Scotsmen were not trying to oppress the peoples they called barbarians— even though, as theorists of unanticipated consequences, they might be expected to have been more alert to the deleterious effects of their own philosophical discourse.

Another reason I do not address race, gender, and class here is that my earlier attempts to do so have convinced me that accounts of peoples hidden from history depend, at least to a certain extent, on the availability of histories of those who were neither hidden nor silent. I have published several articles in which I have sought to understand the role that gender and class played in the disciplinary division of knowledge. In this I have followed the work of John Barrell, Michael McKeon, and Clifford Haynes Siskin among others, and I still believe this is a critical story that has yet to be adequately told.[38] Nevertheless, before it will be possible to position accounts of the gendering of knowledge (for example) in relation to the kind of epistemological developments that enabled white men to divide up and discipline knowledge, we need more work on those epistemological developments themselves. By the same token, before we can fully appreciate the effects this disciplining had on women, peoples of color, and the poor, we need more historical studies like Susan Dwyer Amussen's *An Ordered Society* and Gretchen Gerzina's *Black England,* as well as more and more readily available primary texts, such as the poetry collected in Brian Maidment's *The Poorhouse Fugitives.*[39] I view *A History of the Modern Fact* as one contribution to a project that will require many other hands. My inability to deal adequately with race, gender, and class is a sign of how much work still needs to be done.[40]

The historical account I offer in the following chapters, then, is old-fashioned in some ways (in its inability to factor in the issues that now preoccupy us, in its intermittent attention to authorial intention). In other ways it may be almost unrecognizable, for my desire to use close textual reading to generate evidence about an extremely abstract object of analysis makes the book unusual, to say the least. To those who are suspicious of close reading—who argue, quite rightly, I think, that close reading typically illuminates the assumptions of the reader rather than meanings that could be said to reside (somehow) "in" the text—my response is that I most frequently focus on the way readers have read or how they have conducted their arguments instead of trying to infer what the texts meant (to contemporary readers) or mean (to me). To those who are suspicious of abstraction, my response is that the primary work of many of the texts I analyze here was to produce abstractions. Taken together, these texts produced the abstractions on which many of the models we now use to interpret our world are founded—including, inevitably, the one I use here. To understand how these abstractions work now, it seems to me we must understand the dynamics by which they were initially generated, the semantic worlds to which they originally belonged, and how they were reworked and used in ways their architects never intended.

Readers of my last book, *Making a Social Body,* will recognize that *A History of the Modern Fact* constitutes an earlier chapter of the history of British cultural formation that I offered there. Like *Making a Social Body,* this book seeks to describe the epistemological context that predates—but prepared for—the emergence of the identity categories that dominate so much late twentieth-century criticism; like *Making a Social Body,* it seeks to show how domains and disciplines were gradually disaggregated before the twentieth century; and like the earlier book, it seeks to show the unevenness of this process of disaggregation by giving some attention—especially in chapters 4 and 5—to kinds of writing that did not simply privilege numerical representation during the long eighteenth century. *A History of the Modern Fact* differs from *Making a Social Body* in at least two substantial ways, however, and it is important to point out these differences here. First, in *A History of the Modern Fact* I do not examine individual statutes like the New Poor Law, or voluntary societies like the statistical societies, or reform movements like the public health campaign. This means that much of the institutional mediation through which the ideas that were also expressed in texts were materialized or implemented is missing from the account I offer here. This is a grievous omission, and one I would *theoretically* like to rectify. While I believe in theory that one could—and should—describe the societies, movements, and institutions that mediated ideas in seventeenth- and eighteenth-century Britain too, it has proved much more difficult to isolate and specify these mediations. Although laws were certainly passed in the eighteenth century (too many, some would say), and though societies did sponsor reform initiatives, it was not until the kind of numerical data I describe here was valued and collected that gathering information became a central part of reform initiatives. This means that less information exists about seventeenth- and eighteenth-century initiatives than about their nineteenth-century counter-parts; but it also means that before the nineteenth century, power was enacted in a different way: it did not flow along chains of information and through bu-reaucratic hierarchies so much as it moved through conduits of affiliation and by reading and discussing the kind of texts I analyze here.

The second way *A History of the Modern Fact* differs from *Making a Social Body* is that, with the possible exception of Johnson's travelogue, this book does not discuss any texts that late twentieth-century readers would call literature. This is not because I consider seventeenth- and eighteenth-century imagina-tive texts immaterial, or because poets and novelists were uninterested in these debates. Instead, and apart from a few discussions of the fate of rhetoric in the eighteenth century, I leave literary texts aside because the (uneven) process of disciplinary disaggregation that I describe here authorized numerical represen-tation and the modern fact more generally precisely by distancing numbers from figurative language and (what we call) literature. My implicit argument throughout this book is that numerical representation has always had more in common with figurative language than the champions of the former tend to admit; but even if this is true, because the disciplines associated with the two modes of representation were conceptualized as ever more different from each other, it has proved impossible for me to do justice to their continuing affilia-tions and still construct a coherent narrative. In chapter 4 I do discuss the liter-ary genre of the essay, and in chapter 5 I devote some attention both to Adam Smith's theory of description and to Samuel Johnson's *Journey to the Western Is-lands.* For the most part, however, I have left the missing half of this story to oth-ers.[41] Certainly the disaggregation of domains that very gradually demoted classical rhetoric and elevated modes of knowledge production associated with numerical representation had as one of its effects the creation of a new set of values, which eventually came to be collected under the terms "aesthetic" and "literary." One of my greatest ambitions for *A History of the Modern Fact* is that it will encourage others to map the complex history of the *relationship* between numerical representation and figurative language within that epistemological unit I call the modern fact instead of simply asserting, as I too often do here, both that this relationship existed and that it has been obscured by the history of disciplinarity in whose shadow we work.

THEMATIC OVERVIEW

The chapters that follow contain several interlocking narratives, any one of which might organize a reader's experience of the book. In the introduction, I explained my rationale for choosing the particular texts and authors I examine here, but it might be helpful now to summarize briefly some of the themes that have emerged alongside and as an effect of these choices, so that readers engaged with any one of these themes might be assured that there is a reason for reading about authors whose work might be unfamiliar. The first theme that recurs in *A History of the Modern Fact* centers on the vicissitudes of the concepts of "interests," "interestedness," and "disinterestedness." In taking up these concepts, I am implicitly (and sometimes explicitly) entering into a dialogue with Albert O. Hirschman's influential *The Passions and the Interests,* but I have tried to provide more contexts for understanding these concepts than Hirschman did in 1977.[42] By connecting this cluster of concerns to both reason-of-state arguments about government and theoretical defenses of liberalism and by showing how what we think of as (scientific) impartiality entailed an argument about professional expertise, I have given interestedness and disinterestedness the prominent place I think these concepts deserve: at the intersection of political concerns about government and the epistemological stance we now call objectivity.

The second theme that recurs here emerges from another cluster of related terms: credit, credibility, credentials, and credulity. At some level each of these terms implies something about people's willingness (or need) to believe, but the range of the terms—from credit to credulity—also takes us from activities that seem to be merely economic to attitudes that seem exclusively psychological (and sometimes religious). One of the arguments of this book is that even behaviors that seem to be "merely" economic have always depended on mechanisms that solicited belief; another is that (what we call) psychological attitudes have always had material and political ramifications. One of my primary agendas has been to understand the structural place that belief occupies in any program of systematic knowledge; another has been to recover the relation between the instruments that assisted commerce (credit) and the systems of knowledge developed to describe and regulate trade (political arithmetic, political economy). Thus all the terms that circulate around "credit" call our attention to the leaps of faith that underwrite both the modern economic infrastructure and the knowledges with which we enable this infrastructure to support our lives.

As the cluster of terms around "credit" implies, a third theme that appears in *A History of the Modern Fact* involves the elaboration of a nontheological discourse about human motivations or subjectivity. I argue in chapter 4 that this discourse was elaborated only with the demise of sovereign government and the rise of liberal governmentality at the beginning of the eighteenth century, for only when it seemed important for legislators to understand how the mar-

ket system might encourage individuals to govern themselves did it begin to
seem politically expedient to understand why people acted as they did in the
market. In theorizing desire (for commodities), emulation, self-interest, and
virtue, eighteenth-century moral philosophers began to sketch a rudimentary
account of subjectivity, which constitutes the prototype of the nineteenth-
century psychoanalytic interpretation of subjectivity. In the early nineteenth
century, as I describe in chapter 6, theorists of wealth and society ceased to
be so interested in subjectivity, for as the ensemble of social sciences became
increasingly differentiated and specialized, the science of wealth adopted
new aggregative abstractions ("public opinion") in place of the old protopsychological abstractions that had engaged the moral philosophers ("human
nature," "the human mind"). Meanwhile, of course, and as part of this same disciplinary specialization, psychology was codified as the science of subjectivity
and its relation to political economy and philosophy became a matter of professional dispute.

   A fourth theme that dominates this book is the development of various
kinds of abstraction and their passage from epistemological concepts to institutional arrangements that have material effects. In the pages that follow I give
considerable attention to the relationships and differences among the most significant versions of abstraction: universals, generalizations, and aggregates. I
also discuss the reification of abstractions (like "trade" or "the market system")
in instruments (like bills of exchange) and institutions (like banks). In taking up
these subjects, I have tried to show both that abstraction has always played a critical role in the systematic knowledge projects that depend on the modern fact
and that the history of abstraction is inseparable from that philosophical
dilemma we call the problem of induction. As with all the large themes I treat
in this book, the history of abstraction calls for much more scholarly attention,
but I trust that readers interested in this history will be encouraged to think
about the subject and to see how our existing histories of the very long eighteenth century might be enhanced by including such topics.

2

Accommodating Merchants:
Double-Entry Bookkeeping, Mercantile Expertise,
and the Effect of Accuracy

S ince at least the seventeenth century, British efforts to formulate new "sciences" of wealth and society have been pervaded by the metaphor of bookkeeping. Sometimes this metaphor is explicit, as in Hobbes's *Leviathan* (1651),
where "reckoning" appears as the very type of rationality.[1] Sometimes the
trope is less obvious, as in Francis Hutcheson's 1728 reference to "computing
the *Quantities* of Good or Evil."[2] Occasionally, as in Hobbes's text, the allusion
to bookkeeping praises precise adherence to rules. More frequently, as in nineteenth-century political economists' references to "mere social bookkeeping,"
the allusion denigrates bookkeeping as "merely" precise in order to enhance the
prestige of some more sophisticated alternative, whether mathematical modeling or political or economic theorizing.[3]

In this chapter I contend that the metaphorical persistence of bookkeeping in descriptions of the emergent sciences of wealth and society preserves in
vestigial form the memory of a more profound relationship. This relationship
was not just between accounting and the embryonic sciences of society, but
between what counted as a fact and the way the systematic representation of
these facts enabled more general theoretical knowledge to be produced. Early
modern bookkeeping—especially the printed form of accounting known as
double-entry bookkeeping—was one of the earliest practices where a prototype of the modern fact was generated. In contrast to ancient facts, which referred to metaphysical essences, modern facts are assumed to reflect things that
actually exist, and they are recorded in a language that seems transparent. Since
the early nineteenth century, this transparent language has been epitomized by
numerical representation, although one of my purposes in this book is to argue that forging the relationship between the fact and numbers that seem
transparent entailed a long and uneven process. Double-entry bookkeeping
constituted one of the earliest systems to privilege both things in themselves
(the objects and money the merchant traded) and a formal system of writing
numbers that transformed representations of these things into usable facts.

The nature of the double-entry fact can be grasped by recognizing that this system of bookkeeping did not simply record the things merchants traded so that they could keep track of assets or calculate profits and losses. Instead, *as a system of writing,* double-entry bookkeeping produced effects that exceeded transcription and calculation. One of its *social* effects was to proclaim the honesty of merchants as a group. One of its *epistemological* effects was to make the formal precision of the double-entry system, which drew on the rule-bound system of arithmetic, *seem* to guarantee the accuracy of the details it recorded. While the social effect of displaying mercantile honesty derived in part from double-entry bookkeeping's original place within the field of rhetoric, it was also supported by this epistemological result, which I call the effect of accuracy. In other words, the rhetorical apology for merchants embedded in the double-entry form drew additional support from the epistemological claim built into this system of writing: the double-entry system seemed to guarantee that the details it recorded were accurate reflections of the goods that had changed hands *because* the system was formally precise. Because it aligned precision with accuracy, moreover, it also constituted the site at which a question essential to modern social scientific knowledge in general, and to the social scientific fact more specifically, was initially posed—although it was *not* yet formulated as a problem. This question was how to conceptualize the relation between the particular (quantifiable) details one could observe in the world and the general theories one could advance to explain them.

If double-entry bookkeeping was a site where both modern facts and what contemporary philosophers call the problem of induction first appeared, then why has the bookkeeper been assigned only a cameo role in histories of the sciences of wealth and society? And why have modern historians tended to separate the two facets that coexisted in double-entry bookkeeping, so that the modern use of numerical facts to generate theoretical knowledge seems to have two "roots"?[5] In the most general sense, the answers to these questions have to do with the relative—and changing—status of different kinds of knowledge—a subject I will address repeatedly in this book. As Stephanie H. Jed has argued, the erasure of accounting from histories of modern knowledge production and the division of knowledge into mere recording and higher theorizing have to do with the prestige attributed to abstraction over concrete knowledge in the early modern period.[6] As we will see in chapter 3, although this prestige had to be reworked, it persisted in seventeenth-century Britain *even as* the observation and recording of particular details acquired a new importance in the discipline of natural philosophy. In a narrower sense, the question of why the kind of modern knowledge epitomized by statistics came to be perceived as having two origins or roots can be answered by reference to two developments that profoundly affected the historical treatments of accounting and the social sciences. On the one hand, in the early nineteenth century accounting became an instrument designed to enhance calculation and administration; on the other hand, during the second half of the eighteenth century the relation between quantified particulars and theoretical knowledge came to be perceived *as a problem*. The first of these developments has led modern students of accounting to assimilate even early modern accounting to what Michel Foucault called governmentality; the second has encouraged historians of political economy to efface the problem of induction by disavowing political economy's relation both to political arithmetic and to accounting.[7]

In this chapter I offer a historical account of one early attempt to produce theoretical knowledge about wealth that seeks to avoid both these historiographical misrepresentations. By restoring accounting to the paradigmatic place suggested by the persistent figure of the bookkeeper, I want both to divorce early modern bookkeeping from governmentality, administration, and discipline and to recover the clear relation between particularized facts and systematic knowledge that characterized British "theory" before the second half of the eighteenth century. I do not pursue these goals by aligning a history of the rudimentary science of wealth with a history of accounting, however. Indeed, because accounting practices did not develop in the same way as did the production of theoretical knowledge about wealth or society, and because theoreticians of accounting never specifically addressed the epistemological innovation implicit in double-entry bookkeeping, accounting all but disappears from this book after this chapter. By taking up and then dismissing accounting, double-entry bookkeeping, and mercantile writing more generally, I do not mean simply to repeat the erasure I have just described. Instead, by restoring early modern accounting to rhetoric, the larger disciplinary field it occupied, and then tracing the relation between rhetoric and the emergence of some of the modern sciences, I want to show that the relation between accounting and theorizing about wealth and society constitutes a chapter in the uneven process by which the status hierarchy that characterized ancient society was gradually replaced by modern, functionally differentiated domains.[8] More narrowly, by restoring early modern accounting to the field of rhetoric, I want to demonstrate that what eventually emerged as the problem of induction, in both philosophy and natural science, initially appeared when rhetoric, which depended on a different kind of logic (place logic), began to lose its cultural authority to the new sciences formed in imitation of natural philosophy.

In the second section of this chapter I turn away from accounting per se to clarify the political context in which the fact-based knowledge production adumbrated by double-entry bookkeeping accrued additional social authority.

Even after the first double-entry manual was published in 1494, this kind of accounting did not immediately succeed either in transforming all Western merchants' bookkeeping practices or in achieving the rhetorical goal its early users
sought: to enhance the prestige of merchants as a group by proclaiming that the
profits they earned were just. Instead of being an effect of reformed accounting, the shift in the cultural value accorded merchants followed the creation of
mercantile expertise; and mercantile expertise began to be widely acknowledged only when late sixteenth-century political theorists started to grant
commerce a central role in preserving the power or "greatness" of a state. In the
second section of this chapter I argue that what has been called a "revolution in
politics" (in which the argument that governments should conform to absolute
principles of virtue and justice was replaced by reason-of-state arguments that
emphasized preserving and enlarging the state) constitutes the general context
in which we need to understand how the kind of fact produced by double-
entry bookkeeping began to seem politically valuable. Indeed, in the wake of
the revolution in politics such facts began to seem so valuable that monarchs
were increasingly persuaded to support or protect the merchants who claimed
they could produce them.   *BUT not Petty or 180*

This chapter therefore explores two episodes in the early history of what I
am calling the modern fact. The first episode features accounting. After a brief
glance at the architectural prehistory of modern accounting, I take up Luca Pacioli's publication in 1494 of the first manual on double-entry bookkeeping,
De Computis et Scripturis. I trace the ramifications of Pacioli's innovations by
placing the conventions of double-entry in the context of humanist rhetoric
and by examining these conventions as they were set out in John Mellis's *A
Briefe Instruction and Maner How to Keepe Bookes of Accompts after the Order of Debitor and Creditor* (1588), one of the first manuals to introduce Pacioli's method to
English readers. The second episode had its origins in Spain and Italy about
1570, but I take up the story once more in the early 1620s, when it began to
have a noticeable effect in England. By examining a critical debate about exchange, treasure, and the balance of trade, I show how merchant apologists like
Thomas Mun capitalized on contemporary reason-of-state arguments in order
to elevate mercantile experience to the status of expertise—and more precisely,
to represent experience as a form of expertise that had implications both
personal and national, both fiscal and political. Mun's representation of
the merchant's specialized experience, I propose, constituted a revision of
the Aristotelian idea that "experience" was commonplace—"what everyone
would know"—and his use of numerical representation to create a transparent
language reinforced the epistemological innovation foreshadowed by double-
entry bookkeeping. To enhance the prestige of mercantile experience, Mun

drew on the association that apologists for double-entry bookkeeping had forged between formal precision and moral rectitude or virtue, and he welded this association to the association forged by double-entry bookkeeping itself: the apparently indissoluble link between formal precision and accuracy.

Before I turn to early modern accounting, I need to make one methodological point about my own analysis: situating double-entry bookkeeping historically within the field of rhetoric, as I do below, means something very different from simply attributing a rhetorical function to accounting, as some other analysts have done.[10] Because rhetoric is no longer a vital discursive field, at least to the extent it was in the early modern period, many modern critics use the term *rhetorical* metaphorically, to mean simply "persuasive" or even "manipulative," without reference to the social functions that classical or Renaissance rhetoric once played. Thus many analysts who expose the "rhetoric" of accounting construct essentially ahistorical or formalist arguments, which describe the formal conventions of accounting and infer texts' reception from their formal properties. In contrast, by situating accounting within the *history* of rhetoric, I can begin to show both why apologists for early modes of accounting devised the particular formal conventions they did and why borrowing the tactics of rhetoric enabled bookkeepers to borrow some of rhetoric's cultural prestige as well. To demonstrate how accounting came to exercise social authority—how it came to seem persuasive—I need first to analyze the formal conventions of bookkeeping, then to restore to visibility rhetoric's now lost vitality—its function as *the* discursive field where status issues were negotiated and *the* instrument by which the status hierarchy was reinforced.

"THIS EXQUISITE DEEP-DIVING SCIENCE"[11]

Modern historians have offered two accounts of the origins of Western accounting, each holding important clues about the epistemological issues carried over into the sciences of wealth and society. The first account tends to imply that various parts of modern accounting were developed simultaneously in different Italian cities; and it attributes early accounting methods either to well-to-do householders' efforts to devise a system that would help in managing household economies or to Italian merchants' attempts to adapt household administrative techniques to extensive commercial enterprises.[12] The second account tends to focus on Luca Pacioli's codification of double-entry bookkeeping; it attributes the development of this kind of accounting to a university-trained elite and yokes the semantics of double-entry bookkeeping to rhetoric.[13] Although we will probably never have enough evidence to choose between these two accounts, it is possible to see that some of the issues addressed in household and mercantile accounting were preserved when accounting was codified by a Franciscan friar in 1494.[14]

The history that focuses on accounting's origin in the household and the countinghouse stresses the secrecy of this practice: as an instrument of "oeconomy," or the management of a household estate, written accounts were initially intended to be secret and to keep secret what was most private—knowledge about a man's family in the broadest sense of this term.[15] Such records belonged to a heterogeneous miscellany of documents and precious things, and they were probably heterogeneous in form as well. Written accounts were probably housed in locked strongboxes or chests along with bills of sale, IOUs, and family heirlooms; and they most likely consisted of interrelated financial and genealogical records, interspersed with commonplace sayings, prayers, and reminders that would have resembled modern diary entries more closely than account ledgers.[16]

In an essay on the fifteenth-century architectural innovations of Leon Battista Alberti, Mark Wigley has argued that the practice of early modern Italian accounting not only administered the family economies of merchants and other well-to-do householders but, in so doing, also participated in the creation of a kind of (class-specific) privacy that was new in the early Renaissance.[17] Effective administration, after all, depended both on the husband's knowing the extent and whereabouts of his "family" and on his ability to control access to this knowledge and to his family's resources. Indeed, although Wigley does not specify this point, when Alberti mandated the construction of a wall between the bedrooms of the husband and wife, he created the architectural conditions for *two* new kinds of privacy: the first, centered in one of the spouses' bedrooms, was sexual and functioned to protect the lineage of the family's resources by confining reproduction to the marriage bed; the second, situated in an even more hidden recess, functioned both to protect the records of the household estate from prying eyes and to reinforce the intellectual (that is, non[hetero]sexual) nature of the husband's authority by making sure that no one else had access to the family treasures. This second site of privacy probably began simply as a locked piece of furniture kept in the husband's bedroom.[18] Beginning in the fourteenth century and becoming a commonplace by the fifteenth, the "closet" or "study" (*studio*) became "the true center of the house"; it marked "the internal limits to the woman's authority in the house" and created "an intellectual space beyond that of sexuality."[19]

Although Wigley's argument about the extent of the privacy a fifteenth-century husband was likely to have experienced in the study has recently been challenged by reference to the omnipresence of the husband's steward or secretary, both of Wigley's points about the semantic role the study played in the residence of a well-to-do Renaissance family are relevant to the epistemological work of early modern accounting.[20] First, as the one part of the house that most other occupants, including the wife, were forbidden to enter, the study epitomized the spatial arrangement of power that was reproduced by double-entry bookkeeping (even though double-entry bookkeeping might not have been the kind of accounting used in every Renaissance study). The designation of one place as "private," in other words, even if this "privacy" constituted a "public" sign *of* privacy to the rest of the household, imposed a system of order on the movement of residents and the circulation of knowledge within the house. By this system, some kinds of movement and some kinds of knowledge (or some movements and some knowledge by some people) were rendered impermissible, and the "rule of the household" became synonymous with a "law of place." Echoing Xenophon, the Greek theorist of "oeconomy," Alberti described this law as putting "every thing in its place." Because everything in the household seemed to have its "natural place" within this spatial arrangement of power, Wigley continues, when something was missing, "the 'gaping space [would] cry out.' The structure [was] therefore a mechanism of detection. . . . The house [was] itself a way of looking, a surveillance device monitoring the possessions that occup[ied] it."[21]

As we will see in a moment, the spatial ordering of the double-entry ledger functioned as just this kind of surveillance system—with one critical exception. In the double-entry ledger, which was the most public book of the accounting system, *who* had access to the production or consumption of knowledge was not limited in the same way that it was in the early modern household. Because the double-entry system assigned a proper place to every entry, it saturated place with meaning and rendered place a mechanism of detection; but because several authorized persons were allowed to write in the ledger, place did not reflect the status of the writer but made writers interchangeable, so that place became a substitute for status, not its reflection. By the same token, the rigid conventions of the double-entry system functioned like the locked door of the study, in that these conventions rendered the recorded information inviolable. But—and again because these conventions allowed any authorized person to keep records as long as they did so according to rule—they did not restrict access to information so much as they stabilized it. In so doing, the conventions of double-entry made the information recorded in the ledger proclaim not just the honesty of a single merchant but the rectitude of a company, which included individuals of a variety of social ranks.

Wigley's second point, that the study created a space in the early modern household that was theoretically beyond—indeed, immune from—(hetero)-sexuality, also casts light on the epistemological work of written accounts. Recent historical treatments of the early modern secretary are surely correct in pointing out that anxieties about homosocial relations in the "closet" might well have circulated even if (or precisely because) women were not allowed to enter the husband's study.[22] However, Wigley's argument turns not on sexuality per se but, more narrowly, on Alberti's tendency to associate sexuality with excess. "Alberti [was] everywhere opposed to sensual pleasure, describing it as a 'vile appetite.'. . . Alberti condemn[ed] excess pleasure or, more precisely, pleasure understood as excess. Such pleasure [was] dangerous because it [made] men lose their reason and become the 'effeminate' servants of women."[23] Whether heterosexual or homosexual, such excess was potentially damaging to commercial success, which required self-control and privileged deferred gratification; the Renaissance closet was, theoretically at least, a site where such control could be enforced. Like the closet, the conventions of double-entry bookkeeping were intended to manage or contain excess. Most generally, its conventions were designed to contain all those kinds of excess that we might associate with risk. As we will shortly see, the conventions of double-entry writing simply excluded allusions to what no rules of writing could control: shipwrecks, storms at sea, and the wild fluctuations in currency rates that characterized the early modern economy.

The publication of the first double-entry bookkeeping manual constitutes the decisive episode in the histories that attribute accounting to the university elite. It also marks the moment when the relation between the spatial negotiations of power in the early modern household and the epistemological work of accounting can first be described as something more than metaphorical. That is, as an instrument for recording and monitoring the genealogical and financial secrets of a family or a family business, early accounting records can be said to have functioned *like* the study or the locked chest. Only when the rules of the double-entry system were codified, however, and published in what amounts to a textbook, does it become possible to speak of a *public system of accounting*.[24] This public system generated effects beyond individual accounting instances and did so even if merchants did not all immediately begin to keep their accounts in the double-entry form. Indeed, the formal features of the system of double-entry bookkeeping helped transform not only the claim that merchants were able to make about their status but the hegemony of the status system itself. The system was not simply *like* early modern studies, entry to which was no doubt regulated by rules that were explicit and informal but that must have varied with family circumstances. Instead, as a printed set of rules, which was specifically promoted as what we might call the industry standard, the double-entry bookkeeping system was an instrument designed to impose specific rules on a heterogeneous set of practices—to standardize bookkeeping. In promoting these standards, apologists for double-entry bookkeeping sought to make what had been a loose and class-specific set of rules governing the use of place into a generally adoptable and more easily enforced set of regulations governing writing.

More emphatically than earlier and more varied kinds of record keeping, then, double-entry bookkeeping transported the system of management unevenly realized in private households to the space of public writing. In so doing, it created a vehicle for producing public knowledge—that is, knowledge that was designed to function in public as a sign of something more than the information included in the books. Some merchants may have developed parts of the double-entry system and used them privately before textbooks began to appear (B. S. Yamey identifies the first traces of double-entry in the *massari* of Genoa as early 1340).[25] But the method could not work as a *system* of knowledge production until it could be imitated. The social function of double-entry bookkeeping—its role as an apologist for mercantile honesty—thus coincided with the appearance of printed books about it.

Because this codification involved the passage of published books into print and into the pedagogical apparatus, double-entry bookkeeping signaled another alteration in the early modern relations between privacy and the public. Whereas the architectural innovations Alberti described created a non(hetero)sexual kind of privacy associated with thinking, writing, and masculinity, the double-entry ledger introduced an interface between the company's "private" concerns and the "public" institutions of the government and the church. We will see that parts of the double-entry system continued to be private (in the sense of secret) even when this mode of writing was most elaborately devoted to playing a public (in the sense of juridical) role. In passing, it seems worth noting that this transformation of accounting from the earliest, and arguably most secluded, form of privacy into the arena where "private" business met the "public" state may also have contributed to the constitution of sexuality as even more "private" (in the sense of unmentionable) than business.[26]

Historians who have stressed the importance of Luca Pacioli's *De Computis et Scripturis* have also helped us see the connection between this kind of accounting and the primary instrument by which knowledge was produced in the medieval university: rhetoric. As early as 1985, James Aho pointed out that Pacioli, a Franciscan friar who was schooled in Scholastic rhetoric—and who was a cohabitant with Leon Battista Alberti late in the latter's life—modeled the parts of the double-entry system on the major elements of Ciceronian rhetoric: *inventio, dispositio,* and *elocutio.*[27] The formal conventions of double-entry bookkeeping, Aho argues, were devised to defend commerce against the church's ban on usury, which the church fathers castigated as a sin against justice. The "case" the double-entry ledger was designed to make, according to Aho, was that the business, the facts of which the ledger recorded, was honest and that its profits did no more than offset the risk its owners incurred. The ledger made this case by following certain stylistic conventions: its contents were concise, orderly, and systematic, and its details were (presumably) faithful to the facts. The convention that gives double-entry its name—the double transcription of each transaction, once in the debit section of the ledger and once in the credit section—was intended to demonstrate that the firm's profit was legitimate, Aho concludes: for every credit I am due, this double-entry declares, I owe just so much.

Grahame Thompson has recently extended Aho's argument by contending that Pacioli codified double-entry bookkeeping not simply to justify commerce but also to reemphasize belief in an order sanctioned by God. Pacioli's project was rhetorical, Thompson explains, not just because it made a "case" for merchants but "in so far as Pacioli was responding to a question or problem posed for religious belief more generally."[28] Double-entry bookkeeping addressed the question of why one should believe by reiterating in its very form the symmetry and proportion with which God invested the world. The double-entry ledger's balance was critical to the rhetorical work performed by this system of accounting, he claims, because the balance epitomized symmetry and fused proportion with the virtue of books well kept.

If we are to appreciate—and extend—these analyses, it is critical both to recapture the significance and history of early modern rhetoric and to reconstruct the workings of the double-entry system. Although historians of rhetoric offer stories at least as complex as the histories of accounting, all scholars agree that rhetoric began as a set of rules governing public speech. Aristotle attributed the beginning of rhetoric to a dispute over property. As John Bender and David E. Wellbery note, whether or not Aristotle's story is true, classical rhetoric was "a specialized system of knowledge, acquired, through formal education, in order to maintain property and negotiate social interaction."[29] Indeed, oratorical displays of rhetoric maintained property and reinforced the social hierarchy by discriminating among audiences based on rank, education, and social character. As an instrument of the educational system in Western Europe, rhetoric helped uphold the status levels that governed the distribution of power at least in part by regulating the production of knowledge.

Even if early modern accounting originated in individual households or in the practice of merchants, when the university-trained Luca Pacioli codified double-entry bookkeeping in accordance with the rules of rhetoric, he enhanced the status of accounting and—by extension—of all those merchants

who were willing to adopt it, because he aligned the form of mercantile knowledge production with the prestigious practice of rhetoric. In this sense, and
whether or not enhancing the status of commerce was Pacioli's primary goal,
double-entry bookkeeping could be used as an apology for the public character of merchants as a group. Paradoxically, however, using a form of writing derived from the rules of rhetoric to enhance the status of merchants—as
subsequent apologists for double-entry did—helped undermine the very status
ranking that rhetoric otherwise reinforced, because no matter how wealthy
they were, merchants were not considered the social equals of university men
or churchmen. Even though double-entry bookkeeping was initially positioned within the general field of rhetoric, then, merchants' appropriation of
rhetoric's cultural prestige helped erode the hierarchy that coexisted with the
ancient picture of a single kind of knowledge produced by like-minded men.

Of course, by 1494 the classical form of rhetoric had already been significantly altered, although it was still arguably central to maintaining a social hierarchy and though its formal method still monopolized what counted as
knowledge. Most obviously, by then rhetoric was no longer primarily an oral
practice. As the prominence of Erasmus suggests, by the early Renaissance
rhetoric had been extended to writing, especially the art of the letter. In the
elite Renaissance culture in which distance made communication difficult and
political intrigue rendered it mandatory, letter writing was *the* vehicle capable
of connecting and maintaining an international political network (largely
Western European). The epistolary decorum that was cultivated in letter writing was then repeated at other sites—such as the court—so that the rhetoric of
the letter became the model for possible speech situations. Thus Renaissance
rhetoric marked the transfer of knowledge production from speech to writing,
a transfer that, like the printing of rules for double-entry bookkeeping, was
eventually to have profound consequences for the status hierarchy. By the same
token, the publication of Erasmus's letters and his theoretical treatment of
rhetoric (*De Copia,* 1512) further complicated the distinction between "public" and "private" knowledge. Erasmus's letters did constitute part of a pedagogical apparatus that supposedly limited the production of knowledge to an
elite few. By printing these letters, however, Erasmus created a public model for
writing that, theoretically at least, helped enlarge the number of people who
could produce knowledge according to his rules.

The second point I need to make about Renaissance rhetoric is that, at least
until the 1570s, the classical writer whose style and ideas were considered exemplary was Cicero. Indeed, as Maurizio Viroli has pointed out, in fifteenth-
and sixteenth-century politics, "'reason' stands for the Ciceronian reason—the
recta ratio—which teaches us the universal principles of equity that must govern

our decisions in legislating, counselling, ruling, and administering justice."[30] Even though some Renaissance "restorers" of rhetoric, like Erasmus, felt compelled to resist extremist attempts to follow Ciceronian principles to the letter, most Renaissance humanists remained loyal to the central Ciceronian values, seeking only to adapt them to a modern Christian society.[31] To say that Renaissance rhetoric was Ciceronian is to call attention to both its content and its style. Substantively, "Cicero" stood for a combination of skepticism and Stoicism: Ciceronian skepticism was essentially epistemological, for he held that the frailty of the human senses renders certainty about the natural world impossible. Cicero's Stoicism, by contrast, appeared in the arena of morals, where he believed that reason *could* lead citizens to universal principles of equity through orderly debate. Cicero considered prudence, justice, temperance, and fortitude essential to a life that yielded both personal and social glory, and he maintained that displaying these virtues was a critical component of generating such glory, both for the individual man and for the state.[32]

In large part, a Renaissance humanist displayed Ciceronian virtues by adopting Cicero's style—in deportment, in dress, and, most important of all, in writing. The hallmark of Ciceronian style, especially as Erasmus described it, was *copia,* a concept that embraced the cognate subjects of material riches, natural plenty, and figurative abundance.[33] In rhetoric per se, *copia* was a particular resource or effect of writing; in practice, it mandated a method for producing variety in writing—for saying the same thing in an almost infinite number of ways. Variety, in turn, was valued both because it was thought to counteract the mind's tendency to grow sated or disgusted and because it allowed the writer to develop a range of nuances as elaborately differentiated as the status hierarchy itself.

In the early Renaissance, the cultural prestige accorded *copia* coexisted with the development of a number of "plain styles" of speaking and writing. Derived from Aristotle's *genus humile,* these plain styles were deployed in new kinds of preaching and theological writing, especially Puritan writing, which drew heavily on Protestant theology.[34] Such "plain" styles, to which I will return in the next section, were not completely antithetical to Erasmus's program. In *De Copia,* he paired *copia* with *brevitas,* which implied the "fitness" between words and meanings as well as figurative restraint. To Erasmus, *brevitas* was characterized by writing that so "compress[es] a subject that you can take nothing away."[35] Although Erasmus does not develop this concept in *De Copia, brevitas* was always available to Renaissance humanists as a rhetorical alternative to *copia* and was in fact intended to balance *copia* so as to prevent what Erasmus called "futile and amorphous loquacity" or "Asian exuberance."[36]

This is the point to return to double-entry bookkeeping, for creating the impression of a "fit" between a systematic arrangement of words and the things of the world constitutes one of the signal epistemological innovations of this kind of writing. Indeed, as we will see, in transposing this "fit" from the numbers recorded in the accounting ledger to the relation between the ledger and the merchant's activities, double-entry bookkeeping weakened the epistemological skepticism associated with Cicero. Far from arguing that the human senses were too frail to produce knowledge about the world, the double-entry system confidently showed how such knowledge could be created. It is necessary to recognize, however, that even though double-entry bookkeeping, by according new importance to observed particulars, can be said to have contributed to the epistemological upheaval that contemporaries call the probabilistic revolution, it was in the interest of merchants *not* to ask the question philosophers posed about this new kind of fact: whether this knowledge was certain or merely probable.³⁷

We can begin to understand how double-entry bookkeeping created this and other epistemological effects, which exceeded and helped undermine the epistemology and social hierarchy associated with rhetoric, by turning to John Mellis's *Briefe Instruction*. In the introduction to this text, Mellis aligns double-entry bookkeeping with rhetoric when he attributes a juridical function to the ledger. Like a public oath, which was the less prestigious equivalent of a gentleman's word of honor, the balances exhibited in the ledger constitute a public declaration of credibility. Some such formal declaration is necessary, Mellis implies, because merchants are no longer accorded the credit their oaths once procured and that they need in order to do business.

> Good Lord what a great and commendable thing is, the outward fayth or promise of a merchaunt that is just in dealing, the which in time past hath bene incomparable, In so much, that their othes [oaths] in confirming the treuth, in great common weales were made in this maner: Per fidem bonae & fidelis mercatoris, which is to say: By the faith of a good faithfull merchant. For without fayth and fidelitie betwixt man and man, it is not possible that our labours and travels can eyther be well maintained, continued or ended. (*Briefe Instruction*, 10)

Mellis conjures a fictitious "time past" when a merchant's honesty could be signified by using a single phrase both to establish historical precedent for the prestige he now claims and to designate the present as a fallen or debased age, whose "decay" is signaled by the distrust now generally directed against merchants. It is not clear what, if any, "time past" Mellis might have had in mind, but his claim that adopting double-entry bookkeeping will enhance the credibility of merchants implies that he thought English merchants could benefit from the prestige with which this system's affinity to rhetoric had originally invested double-entry. In the English context, in fact, where landownership was the basis of political power and commerce was often associated with Jews, sixteenth-century merchants might well have faced even more prejudice than had their counterparts of a century before in the Italian city-states, where commerce was considered a source of some social distinction.

Having placed double-entry bookkeeping within the field of rhetoric, however, Mellis immediately introduces one of the features that implicitly challenged the status hierarchy that rhetoric upheld. He points out that the double-entry ledger is merely one of a *system* of books, which must be taken together to understand what the all-important balances mean. In neglecting this system and focusing primarily on the accounting ledger, even historians who have recognized the relation between double-entry and rhetoric have failed to see the implicit challenge posed by this mode of accounting. The system of books that culminated in the double-entry ledger implicitly challenged the social and epistemological hierarchies for two reasons: first, because the form in which information was recorded *and reworked* in the system of accounting books both subjected writing to rules and *seemed* to make rule-governed writing a guarantee for the accuracy of the information recorded there (it seemed to make writing transparent instead of performative, as rhetoric so obviously was); and second, because the system of accounting books produced *writing positions* that made the individuals who wrote there interchangeable, regardless of their rank, instead of reiterating their social position as rhetoric was designed to do. These two effects—the transparency of writing, which followed from the effect of accuracy, and the creation of writing positions—become visible in Mellis's description of the system of accounting books.

Double-entry accounting, Mellis explains, consists of at least three account books, although it began with another book that was not part of the working system of accounting, and it could also include additional books.[38] This first book was the inventory (fig. 1). In it, Mellis tells his readers, the merchant was to list all his possessions—money, stock, jewels, household goods, lands, outstanding loans, and so on—and all his debts, in as much detail as possible and at the moment at which he initiated the bookkeeping. In the inventory, the merchant should be as expansive as possible—Mellis tells us that he "can in no wise make too large a declaration in writing" in this book (27)—and he should use both narrative description and numbers to convey the extent and kind of his possessions.

Mellis calls the second book (the first book of the system proper) the memorial; in it the master or his agent was to chronicle each day's business transactions as they occurred. Writing in the memorial was also prolix, and it also mixed narrative with numbers, for the contents were to include the names of the parties involved, the terms of payment, and all relevant details about "the marchandise, money, measure, weight, or number" (23). At some specified time (Mellis suggests every five or six days) an accountant or the master's agent was to transcribe the contents of the memorial into the journal, which constituted the second book of the double-entry system (fig. 2). Entries in the journal were to differ from those in the memorial in several significant ways. First, all the moneys were to be translated into a single currency, the money of account;[39] second, the placement of information on the page was to highlight certain critical relations between words and numbers by linking narrative information to monetary values (placed to the right and separated by a vertical line) and by creating the conditions for an index (through index numbers placed to the left of the words and again separated by vertical lines); and third, what narrative remained was to be written in "shorter sentence, without superfluous words" (30). The resulting abbreviation, to which number, placement, and verbal restraint all contributed, made it easier to transfer information to the system's final book, the ledger, for in the journal only those details considered essential remained. Each transaction now appeared as either a debit or a credit; all had been translated into a common money, which was written as a series of numbers; and each could be referenced through an index, whose existence was also signaled by numbers. The vertical lines that scored the pages of the journal to separate words from numbers had their counterpart in a horizontal rule that appeared at the top of each journal page; above this line the merchant was to place the date, written in an appropriately devout form that also mixed words and numbers ("Anno do 1587").

The final book of the system, the ledger, contained the characteristic double-entry account of each entry in the journal (fig. 3). These entries were arranged not chronologically but by kinds (cash transactions, transactions involving jewels, etc.) and by whether they were being entered as credits (placed on the right-hand pages of the book) or as debits (on the left-hand pages). This is the point at which virtue was made visible in the double-entry system, for entering the pertinent information about each transaction twice, once as a credit and once as a debit, enabled the accountant to add up the amounts, then to rectify or balance the sums of the entries on each set of facing pages. The accountant did not produce this balance simply by adding all the numbers on each page, however, for given the nature of commercial transactions, it almost never happened that expenditures and receipts concerning jewels, for example, actually equaled each other. This was true, first of all, because jewels constituted only one segment of a business's transactions, so that jewels were rarely exchanged simply for jewels, and second, because it was the aim of any merchant to turn a profit by selling jewels for more than he paid for them. Producing the

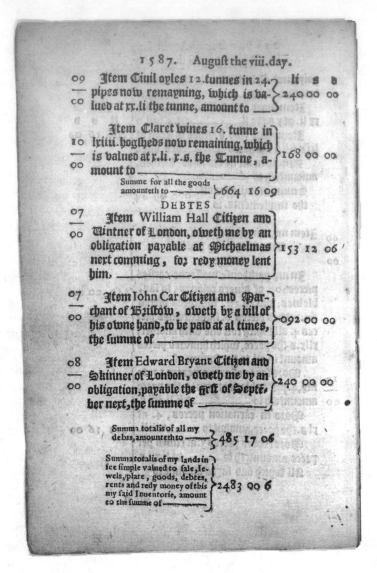

1587. Auguſt the viii.day.

09 Item Ciuil oyles 12.tunnes in 24.⎫
— pipes now remayning, which is va-⎬ 240 00 00
00 lued at xx.li the tunne, amount to __⎭

Item Claret wines 16. tunne in⎫
10 lxiiii.hogſheds now remaining, which⎬
— is valued at x.li. x.s. the Tunne, a-⎬ 168 00 00
00 mount to _____⎭

Summe for all the goods
amounteth to _____ ⎬ 664 16 09

DEBTES

07   Item William Hall Citizen and⎫
00 Wintner of London, oweth me by an⎪
obligation payable at Michaelmas⎬ 153 12 06
next comming, for redy money lent⎪
him. _____⎭

07   Item Iohn Car Citizen and Mar-⎫
00 chant of Briſtow, oweth by a bill of⎪
his owne hand,to be paid at al times,⎬ 092 00 00
the ſumme of _____⎭

08   Item Edward Bryant Citizen and⎫
— Skinner of London, oweth me by an⎪
00 obligation,payable the firſt of Septē-⎬ 240 00 00
ber next,the ſumme of _____⎭

Summa totalis of all my⎫
debts,amounteth to ____⎬ 485 17 06

Summa totalis of my lands in⎫
fee ſimple valued to ſale, Ie-⎪
wels,plate , goods, debtes,⎬ 2483 00 6
rents and redy money of this⎪
my ſaid Inuentorie, amount⎪
to the ſumme of _____⎭

FIGURE 1. A representative page of the inventory, the foundational book of the double-entry bookkeeping system. The second entry on the creditor page (the right-hand page) indicates the debt this merchant owes to Thomas Barton. It reads: "Item I owe unto Thomas Barton of Bristow Marchant, for the rest of an accompt betweene him and me for a partable viage into Spaine, due at his pleasure, the summe of——————053 17 06." Note also that at the bottom of the right-hand page (at the end of the inventory), Mellis begins to spell out the rules governing the transcription of entries and the transfer of entries from one book to another. From John Mellis, *A Briefe Instruction and Maner How to Keepe Bookes of Accompts after the Order of Debitor and Creditor* (London: John Windet, 1588). Courtesy, The Bancroft Library.

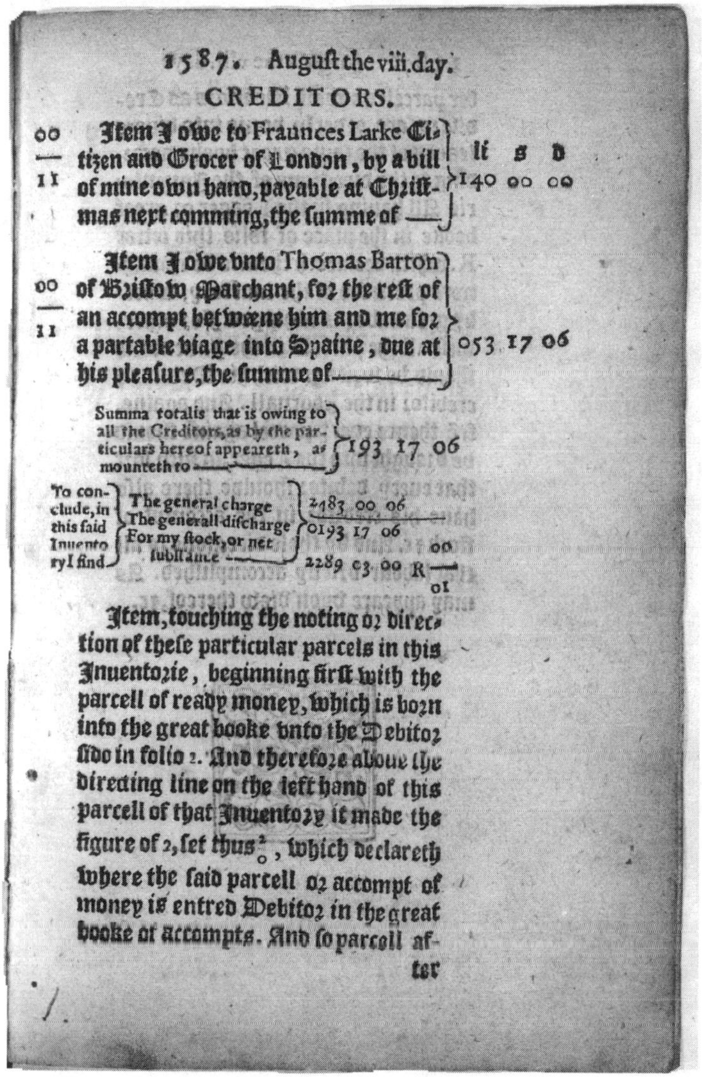

CREDITORS.

			lí	s	d
00	Item I owe to Fraunces Larke Ci-tizen and Grocer of London, by a bill of mine own hand, payable at Christ-mas next comming, the summe of ——		140	00	00
11					

00	Item I owe vnto Thomas Barton of Bristow Marchant, for the rest of an accompt betwéene him and me for a partable viage into Spaine, due at his pleasure, the summe of——		053	17	06
11					

Summa totalis that is owing to all the Creditors, as by the par-ticulars hereof appeareth, as mounteth to —————— 193 17 06

To con-clude, in this said Inuento-ry I find	The general charge	2483 00 06
	The generall discharge	0193 17 06
	For my stock, or net substance ——	2289 03 00 R —— 00 / 01

Item, touching the noting or direc-tion of these particular parcels in this Inuentorie, beginning first with the parcell of ready money, which is born into the great booke vnto the Debitor side in folio 2. And therefore aboue the directing line on the left hand of this parcell of that Inuentory it made the figure of 2, set thus $^2_0$, which declareth where the said parcell or accompt of money is entred Debitor in the great booke of accompts. And so parcell af-ter

FIGURE 1. Continued

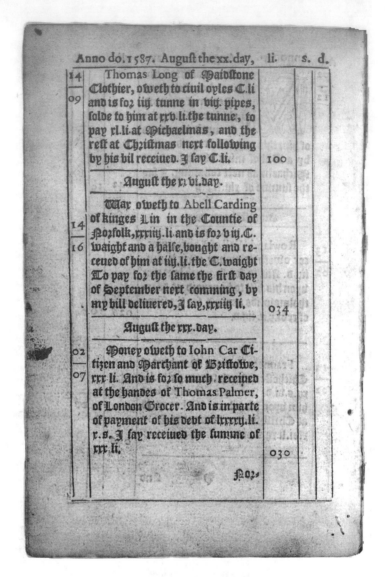

FIGURE 2. A representative page of the journal, the second book of the double-entry system proper. The last entry on the right-hand page records the transaction between the merchant's father-in-law and Barton as a debt that Barton owes to "money." "Money" is one of the personifications that grounds double-entry bookkeeping. Note also that money equivalents (book prices) have been graphically separated from the narrative by a vertical line to the right of the text. The numbers on the left side of each page index these entries to other books. From John Mellis, *A Briefe Instruction and Maner How to Keepe Bookes of Accompts after the Order of Debitor and Creditor* (London: John Windet, 1588). Courtesy, The Bancroft Library.

| 15 | Noꝛwich grogrames oweth to Robert Garſet of Noꝛwich Tay-ler, liiij.li, xiii s.ij.d.ob. and is foꝛ xx pæces bought of him, at liiij.s. vij.d the peece, to pay xxx.li, at hallontide next, and the reſt at Chꝛiſtmas fol-lowing, per my bill deliuered | 054 | 13 | 02 ob. |
| 16 | | | | |

September the firſt day.

| 02 | Money oweth to Edward Bri-ant Citizen and Skinner of Lon-don, CCxl. li. And is foꝛ ſo much receiued of him, in full payment of his debt, and deliuered his band, I ſay receiued CCxl.li. | 240 | | |
| 08 | | | | |

| 11 | Thomas Barton of Bꝛiſtowe Marchant, oweth to money xxx.li. foꝛ ſo much was paid him in parte of payment of his debt by the hāds of my father in law. I ſay xxx.li. | 030 | | |
| 02 | | | | |

H 2   Tho-

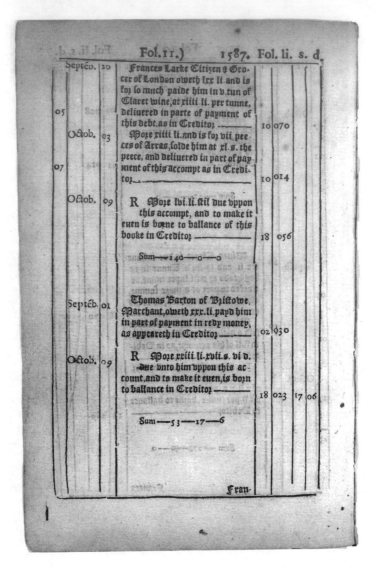

FIGURE 3. A representative set of pages from the ledger, the final book of the double-entry system. Debits appear on the left-hand page, and credits appear on the right-hand page. The first debit entry for Thomas Barton appears on the left, second from the end: "Thomas Barton of Bristowe, Marchant, oweth xxx.li. payd him in part of payment in redy money, as appeareth in Creditor———[folio] 2 030 [pounds]." Note that Barton "owes" this amount to "money"; that is, he has been paid from the account personified as "money" by the merchant's father-in-law. The credit entry for Barton appears at the end of the right-hand page and reads "Thomas Barton Marchant of Bristowe, is due to have liii.li. xvii.s vi[d] for the rest of an account with him in companie for a viage into Spaine due at pleasure without any specialtie, as in the Inventorie generall appeareth in folio 4 annexed to the fore part of the Journal booke———053 17 06." Because the account with

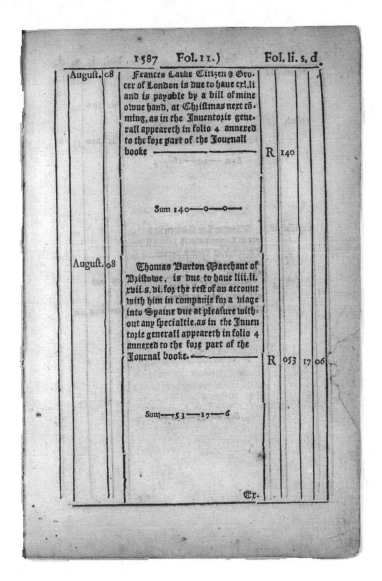

August. 08 — Frances Latke Citizen & Grocer of London is due to haue cxl.li and is payable by a bill of mine owne hand, at Christmas next coming, as in the Inuentorie generall appeareth in folio 4 annexed to the fore part of the Iournall booke —————————— R 140

Sum 140——0——0—

August. 08 — Thomas Barton Marchant of Bristowe, is due to haue liii.li. xvii.s. vi. for the rest of an account with him in companie for a viage into Spaine due at pleasure without any specialtie, as in the Inuentorie generall appeareth in folio 4 annexed to the fore part of the Iournal booke.—————— R 053 17 06

Sum——53——17——6

Cr.

FIGURE 3.   Continued

Barton does not balance, the merchant adds another entry to the end of the debit page: "More xxiii.li. xvii.s. vi d. due unto him uppon this account, and to make it even, is born to ballance in Creditor———[folio] 18 023 17 06." This number—£23 17s. 6d.—is the fictitious number, which is imported simply to balance the books. From John Mellis, *A Briefe Instruction and Maner How to Keepe Bookes of Accompts after the Order of Debitor and Creditor* (London: John Windet, 1588). Courtesy, The Bancroft Library.

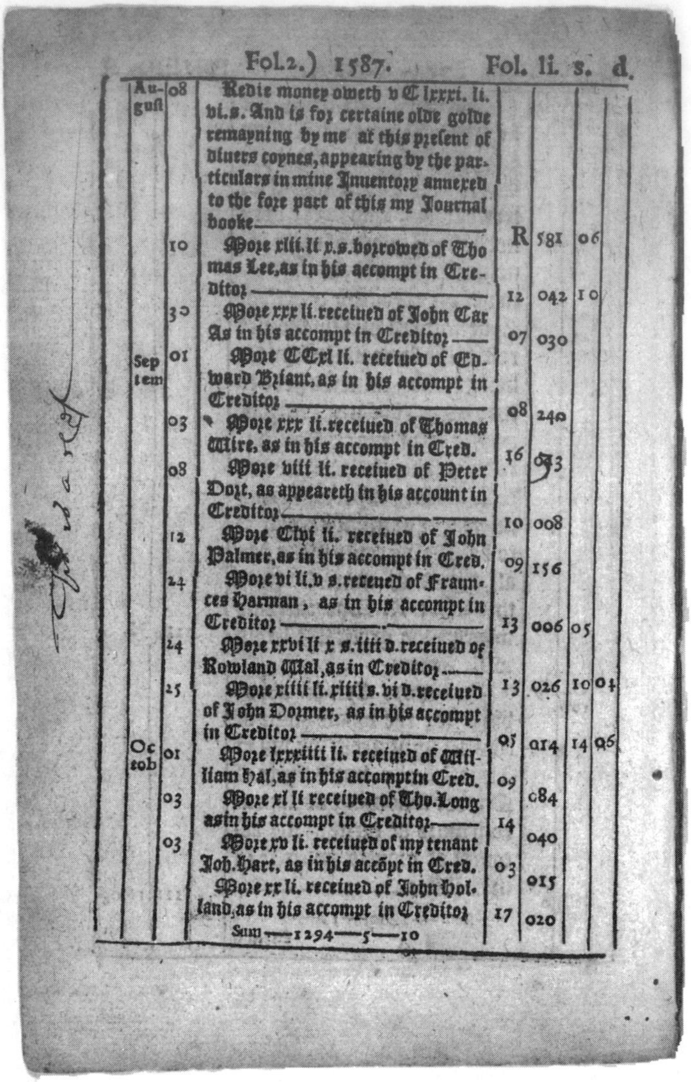

		Fol.2.) 1587.	Fol.	li.	s.	d.
August	08	Redie money oweth v C lxxxi. li. vi.s. And is for certaine olde golde remayning by me at this present of diuers coynes, appearing by the particulars in mine Inuentory annexed to the fore part of this my Journal booke	R	581	06	
	10	More xlii. li x.s. borrowed of Thomas Lee, as in his accompt in Creditor	12	042	10	
	30	More xxx li. receiued of John Car As in his accompt in Creditor	07	030		
Septem	01	More CCxl li. receiued of Edward Briant, as in his accompt in Creditor	08	240		
	03	More xxx li. receiued of Thomas Wire, as in his accompt in Cred.	16	033		
	08	More viii li. receiued of Peter Dort, as appeareth in his account in Creditor	10	008		
	12	More Clvi li. receiued of John Palmer, as in his accompt in Cred.	09	156		
	24	More vi li. v s. receued of Fraunces Harman, as in his accompt in Creditor	13	006	05	
	24	More xxvi li x s. iiii d. receiued of Rowland Wal, as in Creditor	13	026	10	04
	25	More xiiii li. xiiii s. vi d. receiued of John Dormer, as in his accompt in Creditor	05	014	14	06
October	01	More lxxxiiii li. receiued of William Hal, as in his accompt in Cred.	09	084		
	03	More xl li receiued of Tho. Long as in his accompt in Creditor	14	040		
	03	More xv li. receiued of my tenant Joh. Hart, as in his accöpt in Cred.	03	015		
		More xx li. receiued of John Holland, as in his accompt in Creditor	17	020		
		Sum — 1294 — 5 — 10				

FIGURE 4. This is another set of pages from the ledger. The sum that the merchant's father-in-law paid Thomas Barton on 30 August 1587 (as recorded in the journal) appears on the right-hand page in the third entry from the top. Note that the numbers on each of the facing pages have been added, and that the last entry on the credit page is the fictitious number imported to create the balance: "M lxxiii. li. xix.s. x.d. [£1,073 19s. 10d.] for so much ready money now at this present remayning, and to make this accompt even, is borne to ballance of this booke in Debitor——[folio] 18 1073 19 [effaced] 10." From John Mellis, *A Briefe Instruction and Maner How to Keepe Bookes of Accompts after the Order of Debitor and Creditor* (London: John Windet, 1588). Courtesy, The Bancroft Library.

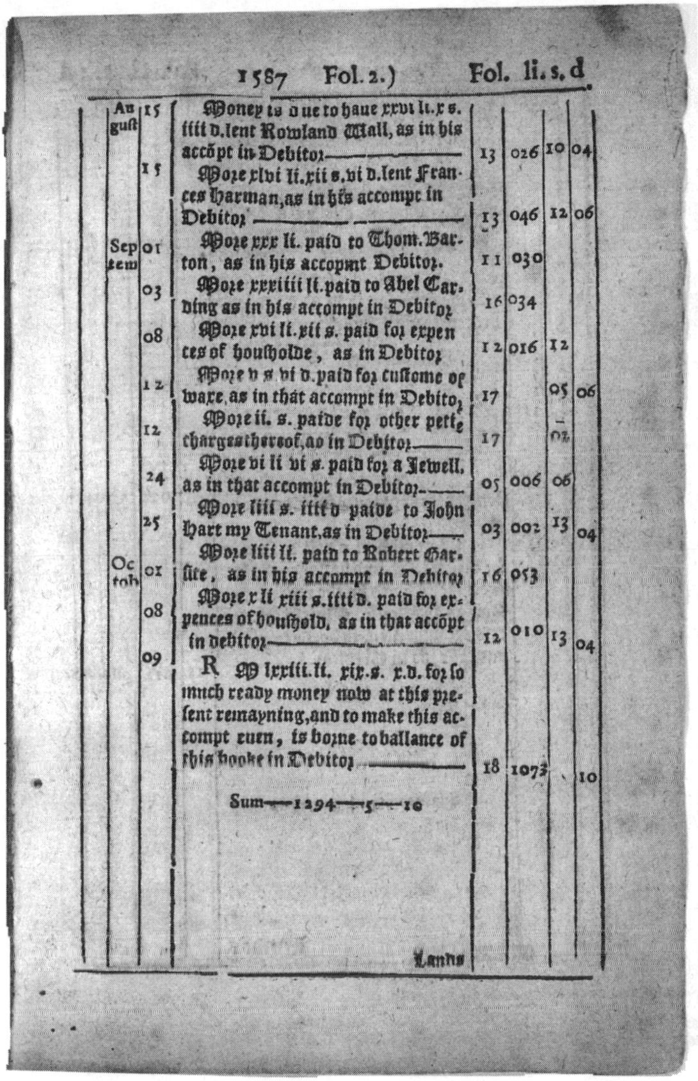

August	15	Money is due to haue xrvi li. x s. iiii d. lent Rowland Wall, as in his accompt in Debitor———	13	026	10	04
	15	More xlvi li. xii s. vi d. lent Frances harman, as in his accompt in Debitor———	13	046	12	06
September	01	More xxx li. paid to Thom. Barton, as in his accompt Debitor.	11	030		
	03	More xxxiiii li. paid to Abel Carding as in his accompt in Debitor	16	034		
	08	More xvi li. xii s. paid for expences of housholde, as in Debitor	12	016	12	
	12	More v s. vi d. paid for custome of ware, as in that accompt in Debitor	17		05	06
	12	More ii. s. paide for other pettie charges thereof, as in Debitor———	17		02	
	24	More vi li. vi s. paid for a Iewell, as in that accompt in Debitor———	05	006	06	
	25	More iiii s. iiii d. paide to Iohn hart my Tenant, as in Debitor———	03	002	13	04
October	01	More liii li. paid to Robert Garsice, as in his accompt in Debitor	16	053		
	08	More x li. xiii s. iiii d. paid for expences of houshold, as in that accompt in debitor———	12	010	13	04
	09	R M lxxiii. li. xix. s. x. d. for so much ready money now at this present remayning, and to make this accompt euen, is borne to ballance of this booke in Debitor———	18	1073		10

   Sum——1294——5——10

         Lands

FIGURE 4. Continued

R	Octob.	09					

Ballance of this booke oweth i. M lxxiii. li. xix. s. x d. for ready money remayning, as in Creditor ———— | 01 | 1073 | 19 | 10

R More CCC. li. for my farme, as in that accompt in Creditor | 03 | 300 | |

R More CClxxx. li. for my manfion houfe, as in that accompt in Creditor ———— | 04 | 280 | |

R L. li. xviii s. s. ob. for my Iewels, as in that accompt in Creditor | 05 | 050 | 18 | 01 ob

R More xxxiii. li. vi. s. viii. d. for my plate, as in that accompt in Creditor | 06 | 033 | 06 | 08

R More lviii. li. xix. s. iiii. d. for implements of houfholde, as in Creditor | 06 | 058 | 19 | 04

R More lxix. li. xii. s. vi d. owing by William Hall, as in Creditor | 07 | 069 | 12 | 06

R More lxii. li. owing by Iohn Car, as in his accompt in Creditor | 07 | 062 | |

R More lxxx. li. for iiii. Tun Ciuil oyles remaining, as in Creditor | 09 | 080 | |

R More Clxxxix. li. for xvii tun of wines claret remayning, as in Creditor | 10 | 189 | |

R More lx. li. owing by Thomas Long, as in his accompt in Creditor | 14 | 060 | |

R More xxx. li. owing by Iohn Bearden, as in his accompt in Creditor | 14 | 030 | |

R More xliiii. li. owing by william Harper, as in his accompt in Creditor | 14 | 044 | |

R More liiii. li. xiii. s. ii. d. ob. for xx. Norwich Grograms remaining, as in Creditor ——— | 15 | 054 | 13 | 02 ob

R More lx. li. owing by Thomas Vire, as in his accompt in creditor | 16 | 060 | |

R More xxxiiii. li. vii s. vi. d. owing by Spanish accompt, as in Creditor | 17 | 034 | 07 | 06

Sum——2480——17——2

Bal

FIGURE 5. This is a page of the balance at the end of the double-entry ledger. The entry for Barton appears at the end of the right-hand page: "More xxii li. [an error for xxiii. li.] xvii. s. vi. d owing to Thomas Barton, as in his accompt in debitor——[folio] 11 023 17 06." This number corresponds to, and offsets, the £23 17s. and 6d. that was imported into the debit page of the ledger (figure 3). From John Mellis, *A Briefe Instruction and Maner How to Keepe Bookes of Accompts after the Order of Debitor and Creditor* (London: John Windet, 1588). Courtesy, The Bancroft Library

			1587 Fol. 18.)	Fol.	li.	s.	d.
R	Octob.	09	Ballance of this booke is due to haue MM.CC.Lxxxix.li.iii.s. and is for so much being the very net rest of my estate at the beginning of this accompt, as appeareth in Debitor —	01	2289	03	
			R More Cxi.li.xvi.s. viii.d. for so much more gayned during the time of this accompt, as appeareth in Debitor folio ———	01	111	16	08
			R More Lvi.li. owing to Frances Larke, as in his accompt in Debitor ———	11	056		
			R More xxiiij.li.xvii.s.vi.d. owing to Thomas Barton, as in his accompt in debitor ———	11	023	17	06
			Sum —2480—17—2				

FIGURE 5. Continued

balance thus required something in addition to arithmetic. To balance the sums on the facing pages, the accountant had to supplement records of actual transactions with numbers that had no referent in the company's business. To make the sums on the pages tally, the bookkeeper added a number to the deficient side sufficient to offset their difference. (On the credit—right—side of the ledger in fig. 4, the accountant adds £1,073 19s. 10d. at the end of the entries "to makes this accompt even.")

Before returning to this vital number, which had no referent in the merchant's commercial transactions, let us examine the process of transcription and transfer that Mellis describes. It should be clear from my summary of Mellis's text that the transfers that constituted the double-entry system gradually reworked the form of the information recorded in the inventory and the memorial but (theoretically) not its content. In double-entry accounting, each transfer seeks not simply to express the same information in different words but to write this information in abbreviated and increasingly rule-governed form. The limit toward which this process of abbreviation moved was the number, for numbers allowed one to write in short form details considered pertinent to the initial transaction. The priority accorded numbers because of their brevity and the ease of calculation they afforded, of course, privileged quantification over qualitative descriptions; the priority accorded numbers tended to make details that could be quantified seem more pertinent than details that could not. At the same time, number also allowed the writer to translate quantity into prices, and the priority accorded prices privileged commodities over other numbered items such as dates and index references.

Even though numbers were (and are) crucial to the double-entry system, however, we should not assume that they were important simply because numbers were accorded universal respect in 1588 or because the numbers that appeared in the double-entry books were assumed to refer to things that could be (and had been) counted.[40] In the late sixteenth century, in fact, and despite mathematicians' efforts to enhance its reputation, number still carried the pejorative connotations associated with necromancy;[41] and some of the numbers recorded in the double-entry books—specifically the prices—never pretended to refer to prices in the actual world of commerce. Instead of gaining prestige from numbers, double-entry bookkeeping helped confer cultural authority on numbers. It did so by means of the balance, which depended, as we have begun to see, on that wholly fictitious number—the number imported not to refer to a transaction but simply to rectify the books.

For late sixteenth-century readers, the balance conjured up both the scales of justice and the symmetry of God's world. In the system of double-entry, this trope was represented visually, as the symmetry visible in the pair of numbers

that appeared on facing pages of the system's final book (figs. 3 and 4). As this illustrates, in the double-entry system, the image of the scales of justice was subordinated to the figure of God's order, as if the scales had been brought to
balance instead of indicating the weightier argument. Implicitly, this representation replaced the hierarchy of status, which was reinforced by rhetoric, with
an equivalence or even an identity, for unequal contestants have been superseded by identical figures. The balances signaled by such figures were thus
equated with justice not because one triumphed over the other but because
what they displayed—the identity of two numbers—could be easily verified,
first by simply comparing the two numbers and second by arithmetic, by
checking the addition that produced the numbers entered at the bottom of
each page. Even though number was not in itself the sign of virtue, *arithmetic*,
which followed its own formal rules, constituted a system in relation to which
one could judge right from wrong. In double-entry, then, the precision of
arithmetic replaced the eloquence of speech as the instrument that produced
both truth and virtue.

As I have already pointed out, the number added to create the ledger's balance had no referent in the actual world. It did not refer to any aspect of a commercial transaction, to a quantity of things, or to a price. Although this number
had no referent, however, it did have a counterpart in the books themselves: the
same number that the bookkeeper added to produce the balance on the ledger
pages was then entered again on the balance sheet that appeared at the end of
the ledger. On the balance sheet, the number was entered as either a credit or a
debit, depending on what was necessary to offset the form it had taken in its first
appearance. In figure 3 the end of the debit page records xxiii li xvii.s. vid. being "born to ballance in Creditor"; this same number appears again in figure 5,
on the credit page: "more xxiii[.]li. xvii.s. vi.d owing to Thomas Barton, as in his
accompt in debitor." This balance sheet served to rectify the entries in the
ledger, and, because balancing all the credits and debits also typically required
adding a fictitious sum (carried over into another book, which constituted the
starting point for a new year's or new cycle's accounts), it theoretically showed
whether the business was currently in debt or making a profit.

Because double-entry bookkeeping's sign of virtue—the balance—depended on a sum that had no referent—the number added simply to produce
the balance—the rectitude of the system as a whole was a matter of formal precision, not referential accuracy.[42] One could easily check whether entries had
been correctly transcribed from book to book; one could tell at a glance if the
sums on the ledger pages equaled each other and if the rectifying figure was
correctly entered on the balance sheet; and one could easily check the arithmetic. But it was less easy to follow the course of an individual's transactions

with the merchant, to tell exactly how a merchant stood with his creditors and debtors at any particular moment, or to tell whether the transactions initially recorded in the memorial were accurate. This precision, which was also a property of arithmetic, was an effect of the system of bookkeeping as a whole: only in relation to the other entries in the books could an individual entry be judged right or wrong; but in relation to those entries, the correctness of an entry could be judged absolutely, with no margin for error.

The formal precision of the accounting system made the figures recorded there seem accurate for two reasons. First, as I have just pointed out, the stages by which information was reworked from narrative to number did allow a reader to monitor the accuracy of the entries *in relation to other entries in the books.* Ensuring the accuracy of new entries, in fact, was aided by another convention that also equated both accuracy and virtue with writing according to rule. As Mellis explains, when an accountant transferred an entry from one book to another, he struck through the original entry so that he would not mistakenly enter it twice: thus, when he transferred information from the memorial to the journal, he crossed out the narrative entry in the memorial; and when he transferred a journal entry to the ledger, he struck through the former twice—once from left to right when he recorded it as a debit, and once from right to left when he recorded it as a credit.[43] This system of marks both ensured accuracy of transcription and transformed whatever errors were made in recording transactions into opportunities for displaying virtue, for if an entry was inadvertently recorded incorrectly, it was never erased, but merely struck through or marked with a cross and entered elsewhere correctly, with reference to the original, incorrect appearance.[44]

Thus the accuracy of *transcription,* which proclaimed the rectitude of the books, stood in for what could not be verified: the accuracy of the initial record of goods and transactions, which was recorded in the inventory and the memorial. Because so much more time and space were devoted to transcribing information from book to book than in making the initial record, the accuracy that could be verified assumed greater prominence than the writing that could not be checked. By the same token—and this is the second reason the system's formal precision made the figures recorded there seem accurate—the priority accorded to formal precision tended to *create* what it purported to describe. This complex effect of double-entry bookkeeping was a function of two related features of the system: its dependence on a series of personifications, and the way these personifications created writing positions that caused anyone who wrote in the books to subordinate personality (and status) to rules.

Even though the double-entry ledger's balance was the most obvious sign of the system's precision and virtue, to record the information from which that balance was derived as credits *and* debits, the bookkeeper had to create a set of personifications. Since commercial transactions were actually *either* expenditures *or* collections (or perhaps most typically, mixed transactions, as when jewels were purchased for a combination of credit, cash, and wheat), in order to write a transaction as *first* a credit and *then* a debit, the bookkeeper had to personify aspects of the business that did not necessarily correspond to actual sums of money, kinds of transactions, or even segments of the business.[45] Not only did they not refer to actual events or sums, but these fictitious entities were also not directly associated with or overseen by a single individual. "Stock" was one such personification; "Money" and "Profit and Loss" were others.

We can see how these personifications work if we follow a single entry from Mellis's sample inventory to the journal, then to the ledger. (Mellis does not include samples of the daily memorial, a point I will return to.) In Mellis's sample inventory, we discover that the merchant keeping the books owes Thomas Barton, a merchant from Bristow, £53 17s. 6d. for a "partable viage [voyage] into Spaine"; this sum is payable at Barton's pleasure—that is, due whenever Barton chooses to call it in (fig. 1). The journal records that on 1 September 1587 the merchant's father-in-law paid Barton £30; Barton is therefore described as owing "Money" £30, because he has received this amount from the cash of the firm (which obviously included the father-in-law; fig. 2). In the ledger, this £30 payment appears on the initial credit page for "Money" in this form: "[Money] xxx.li. [pounds] paid to Thom. Barton, as in his accompt [*sic*] Debitor" (fig. 4). Barton's name appears three more times in the ledger. The first two appearances constitute the "double entry" (fig. 3): on a debit page, we find "Thomas Barton, of Bristowe, Marchant, oweth xxx.li. payd him in part of payment in redy money, as appeareth in Creditor"; and on the facing, credit page we find "Thomas Barton Marchant of Bristow, is due to have liii.li. xvii.s. vi. for the rest of an account with him in companie for a viage into Spaine due at pleasure without any specialtie, as in the Inventorie generall appeareth in folio 4 annexed to the fore part of the Journal booke." At the bottom of the left-hand debit page, as I have already noted, we also see the fictitious sum imported to make the entries balance: "more xxiii.li. xvii.s vi.d. due unto him uppon this account, and to make it even, is born to ballance in Creditor." These entries indicate that no further payments have been made to Barton, so when Barton's name appears for the final time, in the ledger's balance, it is on the creditor page: "[Ballance of this booke] xxii.li. xvii.s. vi.d. owing to Thomas Barton, as in his accompt in debitor" (fig. 5). (Note that Mellis omits an "i" from the initial figure when he records the numbers the first time—that is, in the narrative account and as roman numerals. This error shows how easy it was to make a mistake, especially when using roman numerals; and since Mellis records the correct amount in arabic numerals, it suggests that the latter promoted accuracy of transcription.)

What we should notice about these entries is that even though it is clear from the journal that an individual (the merchant's father-in-law) paid Barton £30, when Barton's name appears in the ledger, all of the relationships seem to be with "Money." In the first of the paired double entries, Barton seems to owe £30 to "Money"; in the second, "Money" owes him £53 17s. 6d. The difference—£23 17s. 6d.—is what "The Ballance of this Booke" owes Barton at the end of the ledger. "Money" and "Ballance" were not individuals, of course, and they could not have been held responsible for the company's debts; Barton's ability to collect his debt would have depended not on the goodwill of "Money" but on the merchant's willingness to acknowledge that this personification had something to do with him.

Even if "Money" and "Ballance" were fictions, however, as conventions of an accounting system used to demonstrate honesty, such personifications also tended to hold real individuals responsible for the fictions the double-entry system required. In other words, the "personal-moral metaphors" of accounting tended to realize what they purported to describe—by encouraging the company's agents to act as responsibly as the books represented them as being.[46] This helps explain why the fictions that might have undermined the book's display of honesty did not necessarily do so—because preserving the precision of the system required anyone who wrote in the books to act as if these fictions were true and, in so doing, to help make them so. It also helps explain how the formal precision of the books created the effect of accuracy. Even though the information recorded in the books was not necessarily accurate, the combination of the system's precision and the normalizing effect that privileging precision tended to produce created the impression that the books were not only precise, but accurate as well. This in turn was critical to the social role the double-entry system was designed to play, for what social good were accounting books if what they recorded could not be taken as accurate accounts of transactions that had actually occurred?

The paradoxical phenomenon of self-actualizing fictions was repeated in the system as a whole, which purported to be, but was not, a system of total disclosure. In theory, the system of double-entry bookkeeping displayed the honesty of individual merchants and of merchants as a group by prominently featuring the easily monitored balances that signified virtue. In practice, however, not all the double-entry books were public, in the sense of being open to inspection.[47] Indeed, the differential publicity of the various books helped to generate the very complicated versions of "privacy" and "publicity" to which I have already alluded. Thus the ledger was open for all to see, and its balances were indeed the principal indexes to a merchant's honesty. The memorial was also public, in a rather more limited sense. As the book in which daily transactions were recorded, the memorial had to be accessible to all of a merchant's factors, even those who did not enjoy the master's complete confidence. By contrast, however, both the inventory and the journal were secret books, in the sense that access to them was strictly limited—to the merchant and his most trusted steward. The first was secret because it was commonly accepted that no man should know another's estate; the second was secret because circumspection about the current status of a company's transactions was as essential to establishing credibility and creditworthiness as was the appearance of honesty. Secrecy was critical because, given almost every early modern commercial concern's involvement with long-distance trade, at any given moment the actual state of a business's finances might not bear too much looking into. Ships not yet returned and debts not yet collected could mean that even an honest company might not actually possess the money it was theoretically worth; thus, even before the general acceptance of standardized instruments of credit, merchants engaged in a complex juggling act of money-to-hand and money-at-work. So unpredictable were the conditions of trade, Mellis points out, that a company's public books constituted the only place where a merchant could even *seem* to be in control (18–19).

Insofar as the fiction of total disclosure displayed the rectitude of a merchant, it underwrote his creditworthiness by proclaiming his credibility. Even though total disclosure was *only* a fiction, it contributed to the conditions that enabled the merchant to achieve the rectitude his books so prominently displayed. Beyond underwriting creditworthiness, moreover, the fictions essential to the double-entry system also tended to discipline any agent who wrote in the books, further enhancing the merchant's credibility and the reliability of his books. All early modern accounting manuals make it clear that it was critical that every agent must enter materials in the same form in the journal and ledger. Beyond this, every agent at home or in the field had to act as the merchant would act, and the merchant in turn had to act consistently—according to rule—so that his agents could predict and imitate what he would do. The priority given to writing to rule created writing positions, in other words, that simultaneously disciplined anyone who wrote in the books and made various writers effectively interchangeable.

To see how privileging precision at the level of writing tended to discipline agents in the field by creating writing positions, we have only to turn to one of the numerous commercial handbooks published in the early modern period. Such handbooks—I will take as my example *The Merchants Avizo* (1607)—were specifically designed to curtail the agent's initiative; more indirectly, they also bound the merchant–master to the rule of consistency. Thus, from the opening pages of *The Merchants Avizo,* the author enjoins the young agent to execute every transaction "according to [the merchant's] commission and direction. . . .See that at no time do you take any mans doings or dealings into your hands, without my leaue and counsel: because by ye trouble of other mens busines, you may neglect and frustrate mine own."[48] The injunction to the agent to act for the master is extended to the expectation that he will act *as* the master through a set of model letters, which the agent is asked to imitate or copy. Here is an obvious similarity with the letter imitation Erasmus extolled, but it is a similarity with a significant difference. Whereas Erasmus's elite readers were enjoined to imitate his style—that is, to write *in the way* Erasmus (or more precisely, Cicero) would write—the socially inferior reader of *The Merchants Avizo,* who was presumably an apprentice, was commanded to write *exactly* what appeared in the model. This is not a model of *copia,* in other words, but a model of literal reproduction intended to discipline the writer. To foster this discipline, the merchant–author supplies a template for almost every conceivable occasion—the arrival letter, the second letter after arrival, the first arrival letter in the second port, and so on. We see the author's relentless campaign against the agent's originality, personality, and judgment in the one non–business-related comment included in the only letter he allows the agent to send to a friend: "Little news I heare woorth the writing" (17).

The real object of such instruction, of course, was not to destroy the agent's initiative but to harness his energy to the master's designs, to make the agent respond voluntarily—or better still, automatically—as if he were the merchant or, better still again, a more predictable version of the merchant. For agent and master alike, bookkeeping contributed to this process in two ways. In the first place, the agent's books constituted a rule-governed arena where the merchant could confirm that the agent acted in the field as the merchant did at home. Because these "accounts current" were typically kept in double-entry format, the balance held the agent responsible for whatever profits the books showed while making the merchant liable for whatever debts the agent incurred in the merchant's name. In the second place, knowing that his records would be checked and that his own success depended on satisfying his master, the agent (ideally) began to monitor himself, to emulate the master whose consistency he assumed. To this end the author of *The Merchants Avizo* shrewdly offers to let his agent engage in some "small aduentures . . . for [his] priuate benefit," but he demands that the agent "do deliuer me an accompt of it, whereby from time to time I may see and know your estate" (5–6). The concluding homilies (or "Godly sentences") and the parable of a deferential young lion further supported the responsibilizing of the agent and the rectification of the merchant. As mnemonic devices rather than simply didactic instructions, these passages were designed to fix a more upright, more perfect version of the merchant in the agent's memory, where it could subtly influence both the agent, who loved and feared his master, and the merchant, who must have wanted to be what he told his agent he was.

By subjecting both merchant and agent to the rules of double-entry writing, this system of accounting functioned to discipline or control excess. In so doing, double-entry bookkeeping served the same function as the wall that Alberti directed be built between the husband's and the wife's bedrooms. More precisely, as we see in *The Merchants Avizo*'s reference to news that is not worth writing, both the wall and the rules of double-entry bookkeeping *constituted* some things as excessive—in this case excessive not just to the control or intellectual duty of the husband but to writing itself. In fact, the constitution of some things as excessive to a rule-governed system of writing is another of the fictions that support the precision of the double-entry system as a whole, for the entries transferred from book to book can be taken as equivalent to each other only if the details that are progressively omitted are constituted as excessive— that is, as inessential to the knowledge the system creates. Constituting some (primarily narrative) details as excessive, in turn, is the basis for the system's precision, for excising narrative details is necessary in order to privilege numbers, which can be added and balanced in a way that narrative descriptions cannot. Of course, constituting some things as excessive does not eliminate them. Instead, it subordinates those things to the writing that excludes them, and it renders writers who are willing to write according to rule more alike than different.

We can see this principle at work if we turn for a moment to Mellis's description of the book for which he provides no sample pages: the memorial. In his *Briefe Instruction,* Mellis notes that when a merchant and his agents were away, the initial record of a commercial transaction was often written in the memorial by a woman or a "young person." In the bookkeeping process, Mellis explains, these records were of fundamental importance; not only did they constitute the first (and in most cases only) account of a transaction, but if the ledger was lost or destroyed, all the books of the system could be reconstructed from the memorial. When the information contained in the memorial was subjected to the rules of the accounting system so that it could be transferred, in largely numerical form, to the journal, the contributions of women and young persons were effaced. Indeed, Mellis provides no sample entries from the memorial *precisely because* the writings of women and youths do not conform to rules: "For as much as . . . the seruantes learned and unlearned . . . may enter & write [in] the Memoriall, after the capacitie of their mindes, wherefore may no perfitte doctrine be giuen in ordering of the same" (26–27).

Because women and young persons wrote "after the capacitie of their mindes" and not according to rule, no "perfitte doctrine" (no rules) could be given for their writing. Women and youths were represented as exceeding this kind of representation, in other words, and this kind of writing could be represented as "perfitte" because it excluded that which was, by definition, excessive to it. In one sense, of course, what was excessive to the "perfitte" order of double-entry accounting was not the writing of women or youths per se but the detailed narrative of the initial business transactions, which could be written by anyone. These details were not called excessive, however, because they were represented by the numbers that had replaced them. In another sense, what was excessive to the "perfitte" order of double-entry accounting were all the factors that accounting could not possibly order, and that it did not even try to represent. These other factors, mentioned explicitly in none of the early modern accounting books, can be assimilated to a term already in use in the early modern period: risk.[49]

Even though manuals on double-entry bookkeeping do not catalog the kinds of risk an early modern merchant would typically have faced, we can get some idea of them in a text like Lewis Roberts's *Merchants Mappe of Commerce*. Roberts's *Mappe* was designed to be an all-purpose reference guide to commerce, with details on regional currencies and trading conventions, national laws and geographical longitudes, local customs and dates of major fairs. Beyond simply providing information, however, the opening chapters of the *Mappe* allude to circumstances more sinister (and insurmountable) than ignorance. When he notes that a merchant must understand how ships are built, for example, Roberts alludes to unsafe vessels; the caution that a merchant must learn to navigate hints at ships captainless and lost; and the argument that a merchant must understand insurance implies risks both financial and bodily.[50] At one point Roberts even refers directly to "the incommodity and danger" of a man's carrying money from place to place, a phrase that raises the specters of pirates, highwaymen, and banditti (12).

In the world idealized in double-entry accounts, there are no such dangers: all the ships return safely, all the moneys are realized; the haggling and bargaining of the marketplace are long past, and the future has already arrived. Like uncertainty, risk and human labor have disappeared from view, and the only threat worth noting is the one an error poses. Whereas the heterogeneous details of the business transaction could theoretically be represented, then, by the numbers that (male) agents wrote in their books, no writing could represent all these forms of risk in such a way as to neutralize its power. Indeed, early modern commerce *depended* on the excessiveness of risk to writing, for fluctuations in prices, production, and demand were the source of the profit that made commerce worth pursuing in the first place. In the next section I describe the larger system of mercantile accommodation in which such risk was rendered manageable by being subjected to another convention of mercantile writing; but risk was *not* subjected to writing in the pages of the double-entry accounts. Instead, in the double-entry system, risk was simply omitted—not struck out, like the errors whose inclusion was an integral sign of honesty, but effaced, like the women and young persons who wrote in the memorial but whose (narrative and unruly) writing was considered unfit.

The effacing of whatever contributions women and youths made to early modern commercial transactions illustrates how the formalizing of accounting—its transformation into a codified system of public accountability—tended to privilege not just a rule-governed kind of writing but also the system of education and credentialing by which particular individuals (almost always men) were rendered obedient to such rules. Although it is undoubtedly true that women kept informal household account books in this period and that, as Mellis admits, they wrote in the memorials of companies that were no longer simply extensions of the household, women do not figure in accounts of double-entry accounting because what was at issue in such apologies was the professionalizing of bookkeeping. Such professionalizing required double-entry's apologists to differentiate this system from other kinds of record keeping that resembled it in every way but double-entry's adherence to its own rules. Such professionalizing, in other words, which was necessary if double-entry bookkeeping was to become the industry standard and so realize the credibility of those merchants who adopted it, depended on its apologists' defining as excessive those kinds of writing that any literate person could perform. Apologists for double-entry bookkeeping could claim that the records were accurate because they could demonstrate that the rules were precise, but they could defend this mode of writing as the one legitimate sign of mercantile credibility only by opposing such rule-governed writing to the unruly writing associated with women, who were increasingly excluded from the apprenticeship system in the early modern period.

Before I turn to the other systems of mercantile writing that supported the claims double-entry bookkeeping made on behalf of merchants, I need to underscore the challenges this kind of accounting posed to the status and epistemological hierarchies superintended by rhetoric. The first challenge came, albeit indirectly, from what I have called the effect of accuracy. Here we need to remember that the effect of accuracy was just that: an effect, not a verifiable reflection of the fit between words or numbers and measurements or counts. In part, as I have already pointed out, a degree of *in*accuracy was necessary to the system: because all early modern merchants depended for their profits on some long-distance trade or credit transactions, the ledgers could never be temporally aligned with the company's actual money; some of the debts owing to the merchant would only eventually be paid, for example (and some would never be), even though the books recorded these debts as if repayment could be taken for granted. By the same token, the rhetorical function of the ledger—to display the merchant's honesty and thus his creditworthiness—always tended to surpass the ledger's referential function. It was necessary, in other words, for the merchant to represent himself as solvent even if he was not *in order* to establish the credit necessary to make himself so.

In part, however, the inaccuracy of the double-entry system was an effect of the priority the system gave to precision. If the entries were to balance, and if entries were to be transferable from book to book, then the most significant accounting numbers—the prices—had to be consistent with the prices that obtained at the moment of inventory *rather than* reflecting the market value current at the moment of transfer. Thus, even though English merchants participated in an international system where variations in amounts of precious metals and other local conditions meant wildly fluctuating prices, the bookkeeper did not adjust the prices he recorded to reflect current prices (to take account of what we would call appreciation or depreciation). This is why I said that the final balance of the ledger could only *theoretically* show whether the business was making a profit and why it is difficult to make the case, as some modern historians try to do, that double-entry bookkeeping was a prototype of managerial accounting. If the double-entry system was to work as a sign of mercantile virtue, then the numbers could be only *nominal* prices—numbers that referred to, by repeating, other numbers in the books—not *real* prices, which might have been realized in an actual trade.

Having said that double-entry bookkeeping assigned priority to formal precision over accuracy, however, I must reiterate the point I have repeatedly stressed: that the formal precision of the books created an effect of accuracy. Even though the accuracy of the initial records could not be verified, the formal precision of the books made the records function *as if* they were not only precise but accurate as well. Paradoxically, this effect of accuracy tended to represent writing as transparent to its object, *even though* it both created and depended on the fictions I have described. Because these were account books and not simply rhetorical arguments, the double-entry books seemed to privilege transactions in the world instead of writing; because they recorded (however inaccurately) specific exchanges and quantities and prices, they seemed to privilege those empirical particulars that have become our modern facts. Rendering writing transparent to such particulars—even if such transparency was merely a result of the effect of accuracy—challenged the epistemology of rhetoric because it made language seem to point to the natural world instead of to the speaker or to rhetoric's rules.

The second challenge that the epistemological effects of double-entry bookkeeping posed to the hierarchies superintended by rhetoric follows from the system's tendency to create writing positions. Whereas the classical system of rhetoric, and even its revived, humanist incarnation, tended to reinforce the status hierarchy of society, double-entry bookkeeping's writing positions weakened status differences by making every writer who was willing to write to rule equivalent, even (as with the agents in the field) interchangeable. Most analysts who have noticed this effect have stressed its disciplinary quality, as I have done. Equally to the point, however—indeed, more significant from a historical perspective—is that the generalized subject positions created by double-entry's preference for writing to rule anticipate the universal human subject so critical to the claims of seventeenth-century natural philosophy. Whereas rhetoric presupposed a social contest where contestants were unequal, modern science takes as its opposite nature itself; and as Bender and Wellbery maintain, the human who addresses—and seeks to conquer—nature is no longer a particular individual but becomes a "neutral or abstracted subject" whose rank has been subordinated to a claimed neutrality.[51]

Although a few historians of accounting pay attention to some of the epistemological effects I have described, almost no one who writes about either the production of economic *theory* or the sciences of wealth and society more generally has done so. As I pointed out at the beginning of this chapter, modern historians of accounting and the social sciences have tended to separate the two kinds of knowledge production because of developments that occurred in the late eighteenth and early nineteenth centuries. This separation has made it difficult to see how the kind of fact produced by the double-entry system adumbrated the modern, empirical facts celebrated by Bacon in the seventeenth century. More important, the separation of double-entry bookkeeping from the history of efforts to produce knowledge about the natural world has made it difficult to recognize the critical role of the epistemological formula essential to seventeenth-century natural philosophy: precise, and therefore accurate. Before I turn to the natural philosophical adaptation of this formula in chapter 3, I should explain how the kind of writing associated with double-entry bookkeeping helped generate an image of a law-governed commercial system that could be abstracted from accounting per se. By the second decade of the seventeenth century, in English debates about the shortage of coin, we can begin to see how a specialized mode of knowledge production associated with the profession of trade or commercial exchange was beginning to be defended as the proper way to understand the relation between private merchants' transactions and national security. The defense of professional merchants, in turn, proved to be a critical contribution to the eventual abstraction of (what we would call) a national economy.

FROM RHETORIC TO REASON OF STATE

In England, the first theoretical model of a commercial *system,* which bore a complex relation to the rule-governed system embodied in double-entry bookkeeping, began to be elaborated in debates about why money seemed in short supply in the early 1620s. For my purposes these debates are also relevant because they constituted the occasion for codifying a mode of writing about commerce that eventually came to seem authoritative for producing economic matters of fact.[52] The initiation of a law-abiding system of commerce, in other words, was closely related to the creation of a rule-governed mode of writing about that system; and the constitution of economic "expertise" as a certain style of writing was inextricably—but complexly—linked to the developments within the discourse of rhetoric to which I have already alluded. Although professional or expert writing about commerce in general was not derived directly from the conventions of writing codified as double-entry bookkeeping, moreover, the two kinds of writing were tightly bound up with one another, as their complicated legacy demonstrates.

Three contributions to the debate about money—Gerald de Malynes's *Center of the Circle of Commerce* (1623), Edward Misselden's *Circle of Commerce* (1623), and Thomas Mun's *Englands Treasure by Forraign Trade* (1622 or 1623)— constitute the basis for the analysis I offer here.[53] If the explicit subject of these texts was how to understand the current and local (that is, English) "decay of trade," however, in trying to account for the shortage of both treasure and trade Malynes, Misselden, and Mun also raised questions about whether commerce belonged to the earthly kingdom whose stewardship had been entrusted by God to the monarch or whether, as a self-regulating domain, it should be overseen by the prince only as long as this control served the best interests of the state. In so doing, Misselden and Mun also both borrowed from and contributed to a theory of politics that had begun to acquire considerable currency throughout northern Europe in the last quarter of the sixteenth century: *ragion di stato, raison d'état,* reason of state.

Before I take up the 1620s debate, let me make three methodological and theoretical points. The first is that even though Malynes, Misselden, and Mun can be seen in retrospect as having contributed to the theoretical abstraction of what would first be called "the market system," then "the economy,"[54] none of these writers were specifically trying to devise an image for this system of international and domestic trade. The terms they used were the terms most common in the period: *trade* and *commerce.* Malynes occasionally refers to "generall Trade" or the "Politike Body of Trafficque," but insofar as he saw a system, it consisted of a more or less linear chain of causes and effects, and his goal was to

Cambio

find the "efficient" or final cause of the development that interested him: the current depression in the English cloth trade.[55] Misselden also seems uncertain whether it was necessary to find a way to describe a system of commerce. At one point, for example, he worries the problem of definition and even tries to trace the etymology of *bourse* and *cambrio* so to understand how a place (the literal markets where money and goods were exchanged) and an activity (exchange) bred such general terms as *purse* and *exchange*. When he attempts to understand "commerce" or "trade" in an abstract sense, however, Misselden gets embroiled in an Aristotelian inquiry about what features of trade constitute its "essence" (as opposed to its "form" or its "materia"). Candidates include exchange, gain, treasure, money, and commodities, but Misselden's exercise does not yield a consistent analysis about the relations among these features, much less a single set of terms for economic activity.[56] Mun seems not even to participate in the enterprise of generalization, for he explicitly delineates his subject—"forraign Trade"—as only one part of a larger subject, whose extent does not particularly concern him.[57] Indeed, it is generally accurate to say that the economic world of these merchant writers consisted exclusively of imports, exports, treasure, and exchange. Because they extracted surplus value from *circulation,* the merchant analysts generally did not look beyond the factors that directly affected trade, prices, and demand on the international market.[58]

The contributions these writers made to conceptualizing and naming a modern "market system," then, were indirect and incidental to their specific intentions. In fact, the most prominent innovation by Misselden and Mun was their argument that merchants possessed critical expertise about "trade" and that only they knew how to read existing records so as to generalize knowledge about "commerce." This was not a trivial argument, because it entailed representing "trade" as a semiautonomous, law-abiding domain, because it significantly reworked the concept of "experience," and because it helped recast the idea of a prince's "greatness" so that it rested on the twin supports of treasure (money) and population.

This leads to my second point. Historians of economic thought, like D. C. Coleman, have traditionally been skeptical about whether these and other "mercantilist" writers influenced government policy in the seventeenth century. Indeed, ever since Adam Smith scornfully christened Misselden and Mun "mercantilists," seventeenth-century writers who focused on England's balance of trade have received scant respect.[59] With the historiographical revision inaugurated by Maurizio Viroli and Richard Tuck, however, it has become possible to identify a group of early modern theorists whose political writings made a variant of the same case Misselden and Mun made for the critical role that merchants played in the modern state.[60] Beginning about 1570 and drawing on the classical work of Tacitus, these historians have argued, various writers across Europe (including Francis Bacon in England) identified money and population as the critical bulwarks of national strength. So telling is the overlap between the work of these theorists and the writing of the mercantilists that Tuck has claimed that "works on 'economics' were parallel to works on military organisation, as textbooks for the princes of the post–constitutional states."[61] In what follows, I generally follow the argument advanced by Viroli and Tuck, although, in giving more attention to the mercantilists than to the political theorists, I focus on the contributions the former made to economic writing *in the context of* another body of writing, which specifically retheorized political power.

This, finally, leads to my third introductory observation. As we will see, a complex relation existed between the writing of Malynes, Misselden, and Mun and both the revision of rhetoric associated with Tacitus and double-entry bookkeeping more narrowly. Whereas Misselden and Mun in particular implicitly supported the Tacitists, who were advising princes to pay attention to their treasuries and to adjust means to ends, only Mun can be said to have adopted the plain style associated with Tacitus's actual writing. Because Mun adopted this style and Malynes and Misselden did not, we see in the debates between Malynes and Misselden a contest between an advocate for the traditional approach to sovereign power and a champion of the more modern *ragion di stato* entirely in a style of writing associated with Ciceronian rhetoric. Only in Mun's writing do we see a coalescence of argument and style, and this, I argue, marks the consolidation of an authoritative manner of generating economic matters of fact. We do not have sufficient evidence to argue that Mun did this because he sought to endorse Tacitus's style (or the Puritan plain style) as an epistemological innovation. The most we can say is that Mun's adopting the plain style, as he undeniably did, reveals that this stylistic option was available to him, even if its epistemological implications had yet to be theorized.

Mun's style, in turn, bears a complex relation to double-entry bookkeeping. I argue below that the critical component of the argument that Misselden and Mun made on behalf of merchants turned on the system of mercantile accommodation, that informal network of agreements to honor the bills of exchange drawn by creditworthy merchants. The creditworthiness of these merchants, of course, was demonstrated in large part by their ability to keep rule-governed, balanced account books. Because individual merchants' use of double-entry bookkeeping took place within the larger system of mercantile accommodation, the former actively contributed to maintaining the latter, just as the latter depended on the displays of honesty manifest in double-entry ledgers. Thus when Misselden, then Mun, drew on the image of "balance" to identify the desirable goal of trade, he mobilized a figure that captured the rectitude made visible by double-entry bookkeeping and that supported the all-important system of mercantile accommodation, even though the "balance" the nation sought was not one in which credits equaled debits but one in which incoming treasure exceeded outgoing money.

The debate that took place among Gerald de Malynes, Edward Misselden, and Thomas Mun was indirectly occasioned by the social disruption and unrest caused by the depression that afflicted England's cloth trade in the early 1620s.[62] Modern historians have helped recover some of the details of this massive dislocation. B. S. Supple explains that the dramatic decline in foreign demand for English cloth beginning early in 1622 led almost immediately to widespread social unrest. Wiltshire and Glocestershire were the hardest hit by rioting, but by May Somerset, Devon, Dorset, Berkshire, Northamptonshire, and Hampshire were all seeing "tumultuous assemblies."[63] Historians of early modern poverty have placed these uprisings in a wider context. J. Thomas Kelly and A. L. Beier have argued, for example, that the fury that ignited in 1622 was simply a particularly flagrant protest against conditions that had been making the poor poorer since Elizabeth's reign.[64] These conditions, Richard Halpern proposes, were simultaneously economic and social; they were, moreover, the same conditions that, in fueling the English cloth industry, had enabled merchants like Misselden to grow rich through trade. As the source of profit from land shifted in the Tudor period from tenants' agricultural labor to wool production, landowners had driven workers from their customary homes so that the well-to-do could raise sheep on unoccupied pastureland; the result, all these historians agree, was a huge increase in the vagrant population, which did not begin to abate until 1660.[65]

The more immediate occasion for the publication of the three pamphlets I am concerned with was a series of official debates about treasure and trade, organized by James I in 1622 partly in response to the widespread social unrest.[66] In April of that year the government appointed a twelve-man committee of investigation to analyze the trade depression, and in October a permanent commission was established to oversee trade; the latter was the precursor to the modern Board of Trade.[67] Even though contemporaries, like modern historians, tended to view the decay of trade and the scarcity of money as related problems,[68] the English government initially responded to them as if they were separate: alongside the committee and commissions charged with investigating the decay of trade, James I appointed a select committee in the spring of 1622 to report on the alleged abuses of exchange, which emphasized the problem of treasure.[69] This committee's report, issued in May, was so controversial that the

king appointed a rival committee staffed largely by merchants, including
Thomas Mun. The second committee on exchange offered an explanation for
the fiscal crisis that differed sharply from the first committee's. This rivalry was
intensified in June, when the committee on the decay of trade, which also included Mun, issued a report endorsing the analysis of the second exchange
committee—that is, subsuming monetary scarcity into the problem of the balance of trade.[70]

Historians of economic thought have long complained that the participants in the debates about treasure and trade were financially and professionally
invested or "interested" in the positions they took; judged by the modern criteria of "disinterestedness," such investment has seemed to preclude objectivity
and therefore to disqualify these writers from the pantheon of "economic theorists."[71] But two things about this debate considerably complicate this modern judgment. The first is that James I selected various committees' members
because their occupations gave them knowledge about trade otherwise unavailable to the monarch; as a consequence these writers were inevitably "interested"—in the older and nonpejorative sense of having a legal interest—in the
practical implications of their theoretical positions. The second important fact
about this debate follows from the new priority assigned to firsthand experience, which was very gradually being reconceptualized as expertise.

As we will see, the campaigns waged on behalf of mercantile experience by
merchant apologists like Misselden and Mun took place in a period when the
concepts of interest and experience were undergoing dramatic revision. As part
of the language of the new reason-of-state arguments, interest was gradually
being removed from its older, juridical context and recast as a political *and* economic term. Gradually, and in the context of subsequent developments in the
domains of politics and religion, two new terms were introduced that carried
the evaluative connotations familiar to modern readers: "disinterested" (in the
sense of being unbiased by personal interest), which came into use about 1659,
and "interested" (in the narrower sense of self-serving), which was in use by
1705.[72] The appearance of these new terms suggests that within the domain of
political theory and in the course of the seventeenth century, it gradually became possible to think that the producers of knowledge had—or were superior
to—a personal, self-serving investment in the knowledge they generated, instead of simply seeking knowledge (or "truth") for its own sake.[73]

This alteration in the concept of the adviser's relation to policies coincided
with gradual changes in understandings of the knower's relation to knowledge.
Critical to this epistemological development was a reconceptualizing of "experience." As Peter Dear has argued, the Aristotelian concept of experience emphasized both commonplaces and the communities in which what constituted

a commonplace was adjudicated. "An 'experience' in the Aristotelian sense was a statement about *how things happen* in nature, rather than a statement of *how something had happened on a particular occasion,*" Dear writes. "For Aristotle, the nature of experience depended on its embeddedness in the community; the world was construed through communal eyes."[74] By the early seventeenth century, a new concept was beginning to rival the Aristotelian truisms. This new way of understanding experience stressed the particularity of individual events and individual observers. Although it was not yet fully individualized (even Baconian rarities were assimilated to generalizations about nature), this way of conceptualizing the production of knowledge began to shift the emphasis away from commonplaces and communities toward specific observations of particular events and specific—and eventually expert—observers.

These developments indicate that the seventeenth century witnessed a realignment of the relations among politics, religion, economic activity, and the production of knowledge. This realignment created the idea that abstract knowledge (theory) could be value-free *because* it was based on specific experience and because it differed from another kind of knowledge, which was "biased" because "self-interested," usually (though not always) in an economic sense. These changes in contemporary understanding of the nature of knowledge and of knowledge production more generally are central to the concerns of this book, and I will return to their legacy repeatedly in the chapters that follow. For now let me simply note that the modern claim that writers like Misselden and Mun were interested and therefore not theorists is one outcome of changes in the very concept of interest, which was initially elevated to an honorific position when reason-of-state theorists revalued specific kinds of experience, then devalued when the consensus about what constituted a state's interest shattered in the wake of the religious and political wars. The constitution of a new form of disinterestedness entailed a change in the symbolic meanings of commerce, merchants, mercantile expertise, national greatness, and public service more generally. The enhanced prestige of numbers, to which double-entry bookkeeping had already contributed, played a complex role in this transvaluation of values: to the extent that numbers were considered disinterested because transparent to their object, so too were those who produced numerical knowledge.

For analysis, I divide the 1622–23 debates into two phases. Generally speaking, in the first, exchange was identified as the source of current problems; in the second, the culprit seemed to be the balance of trade. In the initial part of my analysis, I focus on Gerald de Malynes's *Center of the Circle of Commerce* and Edward Misselden's *Circle of Commerce,* both published in 1623. I consider these two texts in what may seem like reverse order, for even though Malynes's *Cen-*

ter was a direct response to Misselden's *Circle,* the latter was written in response to two texts that Malynes had published in 1622, *Lex Mercatoria* and *The Maintenance of Free Trade.* The second constituted a rejoinder to Misselden's 1622 *Free Trade, or Means to Make Trade Flourish,* but the first gauntlet in what had become a fierce dispute by the 1620s had been thrown down by Malynes in 1602 with the publication of *A Treatise of the Canker of England's Commonwealth.* In the second part of my analysis I turn to Thomas Mun, who is now universally considered the most sophisticated of these three writers.[75] My analysis of Mun focuses on the style of *Englands Treasure,* for I argue that the "advance" his work represented over that of Malynes and Misselden was as much a matter of style as content.

In 1623 Gerald de Malynes, an assay master of the English mint (and thus a royal appointee), and Edward Misselden, a prominent Hackney merchant, traded blows in print about the nature and function of the exchange rate, whose unfavorable nature Malynes held responsible for the shortage of English coin.[76] According to Malynes, who served on (and possible chaired) the first committee on exchange in 1622, the rate of exchange should be fixed by the king so as to reflect the weight and fineness of the metal contained in various coins. Malynes argued that because this would enable merchants to trade at par with other countries, it would stop the outflow of English coins. Misselden countered that exchange was like every other feature of a market economy—uncertain—and that it did not reflect metal but rather responded to variables like the supply of commodities, which money was used to purchase. At stake in this debate were three points central to the reconceptualizing of knowledge and power in this period: What, if anything, was the ground (or "center") of value? On what basis should knowledge about economic matters be considered authoritative? And what should be the relation between a prince, the nation's treasure, and the "greatness" of the country?[77]

Malynes's answers to these questions all emerged from his unqualified support of the sovereign power of the monarch as the representative of God on earth. For him, "the inward value of Siluer and Gold by weight" should be the true basis of exchange; the "goodnes of Gold in value" should be fixed by the mint; and both the mint standards and the rate of exchange should be determined by the monarch and enforced by royal proclamation (*Circle,* 11, 21). In this sovereign system, the intrinsic worth of precious metals was to be the ground of value, and this worth both reflected and reflected upon the authority of the king, whose power, in turn, derived from God and was the foundation for the nation's greatness. Malynes was not claiming that gold and silver had a natural value that derived from their scarcity or durability. Instead, he was arguing that the value was whatever the king, as God's representative, said it should be.

In contemporary parlance, the value at which money traded was considered "extrinsic" (*valor extrinsecus*), and once its nominal price had been set, this price was to be as sacred as the monarchy that guaranteed it. The power to make the extrinsic value of gold and silver correspond to their intrinsic value, then, was a sign of the monarch's authority, and the king's willingness to honor this value (that is, not to debase or enhance the currency) was considered an "essential Mark" of the monarch's embodiment of justice. Sir Robert Cotton formulated the reciprocal relation between good money and good stewardship thus: "Princes must not suffer their Faces [as stamped on the coins] to warrant Falshood."[78]

For traditionalists like Malynes, external value was not the opposite of intrinsic value (*valor intrinsecus*) but its realization.[79] By 1623, however, this conventional wisdom had become harder to defend. This was true partly because individuals within England (including nearly every monarch since Henry VIII) had manipulated the relation between the amount of metal a coin contained and its value.[80] And partly it reflected a growing sense that even if he were willing to do so, the monarch's ability to set the extrinsic value of money in a domestic context could not dictate the fate of English money in an international system of finance and exchange.

Recognizing that the seventeenth-century fiscal system differed in significant ways from the modern system helps clarify some of the points in this otherwise confusing debate. In the first place, because England lacked all but the most rudimentary facilities for credit, gold and silver served both as the repository of the nation's treasure and as the medium of exchange. This helps account for the attention contemporaries devoted to money, and it also explains why they assumed that the international balance of trade was connected to the domestic shortage of coin. England had no gold or silver mines, so to obtain the treasure whose liquidity was essential to the monarch's power, the English had to trade goods for money; and the money that simplified all exchanges was inseparable from the bullion that also constituted treasure.[81] Thus, if there was a slump in international demand for English goods, as there was in 1622, the amount of precious metals entering England—and therefore the amount of currency in circulation—declined.

The problems caused by the conflation of treasure (bullion) with the medium of exchange (currency) were compounded by another feature of the seventeenth-century fiscal system, the susceptibility of the English coinage to clipping, sweating, and other forms of mutilation. The official (and therefore legal) counterparts to this illegal manipulation of the relation between the amount of metal a coin contained and its value were the king's enhancement and devaluation of the currency, which were often responses to the interna-

tional monetary conditions that made the chronic illegal manipulation especially attractive. Beginning in 1545, for example, with the opening of the Po-
tosí mines in America, silver began to flow into the European market. It did so
differentially, however, and in such a way as to change both the relative fiscal
power of various countries (initially in favor of Spain) and the value of silver in
relation to gold. Given the adamant resistance in England to enhancing silver,
whose use as a domestic medium of exchange made its stability important, it
became increasingly profitable for unscrupulous Englishmen to clip or sweat
English silver coins, then export the melted clippings for the higher prices available in foreign markets. Following the logic known as Gresham's law, bad coins
tended to drive out good, until the "silver" coins circulating in England hardly
deserved that name.[82]

In the context of such variables, it seemed increasingly unlikely that the
monarch would set and maintain the extrinsic value of the currency in relation
to the intrinsic value of the metal; even if he did so, it seemed more and more
implausible that this could rectify the unfavorable balance of trade that saw silver flowing out of England between 1615 and 1622. Unlike Malynes, then, Edward Misselden jettisoned altogether the notion that currency should have a
fixed ground of value. His even more audacious claim was that asking the king
to set the rate of exchange would be ruinous to England's fiscal health. Implicit
in Misselden's position were the critical ideas that the king was not the guarantor of value and that trade, although essential to the prince's—and thus the nation's—power, should not be debated as a *political* matter by legalists who were
interested in defending absolute values, but should be addressed by people who
knew enough about commerce to discuss it in new terms, which would make
trade a fit instrument for political use. Implicit in Misselden's position were the
novel ideas that trade had its own dynamic and that, if left alone, merchants
would circulate both money and goods in such a way as to enhance the state's
interest along with their own. "If you should so limit or restraine *Exchanges,* that
no man should take or deliuer any mony, but according to the iust finenes," Misselden wrote, "then the vse of *Exchanges* in all places would bee taken away. For
then there would be no aduantage left neither to him that deliuereth, nor him
that taketh, when mony must bee answered with mony in the same *Intrinsique*
value" (*Circle,* 97). What drives trade, in Misselden's view, is not the assurance
that the monarch stands behind value or that one can exchange at par, but the
uncertainty that follows rates that vary "according to the circumstances of *time,*
and *place,* and *persons,* because this . . . allows for—although it does not guarantee—profit to him who is willing to risk" (*Circle,* 98).

With Misselden's claim that uncertainty, risk, and a purely circumstantial
rate of exchange underwrite the vigor of trade, we see the rudiments of an

emergent concept of an autonomous economic domain, although for him the abstract system of trade was still limited to international commerce.[83] Misselden defended his challenge to sovereign power by reference to "commutative" justice, the justice of the market. He claimed that the "natural liberty" of the market had created a new sphere of activity, where freedom had replaced obedience and justice inhered in commercial success. As a new application of Aquinas's typology of justice, Misselden's remarks bear quoting at some length:[84]

> Which *Taking* and *Deliuering,* as it is *A voluntary Contract, made by the mutuall consent of both parties,* so are both alike free to *Take* and *Deliuer* at their owne pleasure, as in all other contracts and bargains of buying and selling. And trade hath in it such a kinde of naturall liberty in the course and vse therof, as it will not indure to be fors't by any. If you attempt it, it is a thousand to one, that you will leaue it not worse than you found it. . . . Naturall liberty is such a thing, as the will being by nature rightly informed, will not endure the command of any, but of God alone. Which must be vnderstood of naturall liberty in the vse of things indifferent; and not of Regall authority in the exercise of gouernment. . . .
>
> *Iustice* is said to be *Distributiue* or *Commutatiue. Distributiue Iustice* is so called *a Distribuendo,* because it giueth euery man his owne, by a *Geometricall proportion,* as the *Ciuilians* speake: that is, with respect to the quality of the *Person,* not the *Thing. Commutatiue Iustice a Commutando,* because it giueth to euery man his own, by an *Arithmeticall proportion,* that is, with respect to the quality of the *Thing,* not the *Person.* This last is placed in Commerce and Contracts, because by the rule of Iustice there ought to be an equality in buying and selling. (*Circle,* 112–13)

As an equality of things freely exchanged in the market, this more equitable balance ignores the hierarchy of rank (and thus both the old status hierarchy and the traditional basis of disinterestedness), and though Misselden carefully limits the sphere of its influence to the "vse of things indifferent" (that is, not pertaining directly to the monarch), he clearly accords it a degree of authority that assumes a reconception of the prince's relation to treasure. In a domain separate from the political sphere, says Misselden, commerce and contracts rule, "naturall liberty" flourishes, and trade runs merrily and without restraint.

Misselden's vision of the "naturall liberty" of trade did not quite make trade a natural entity whose dynamics were understood to partake of or even resemble the natural laws of motion. For Misselden, the "liberty" that trade "hath in it" was a function of the freedom with which buyer and seller entered into the contract to exchange, not a sign that trade was governed by the "laws of nature." As we will see in later chapters, such laws were eventually invoked to explain—or justify—the autonomy of "the market system," but in the 1620s, with the concept of the laws of nature still being stabilized, Misselden's reference is to earthly law, not to the natural articulation of divine will.[85]

We can see that Misselden's frame of reference was juridical, not natural, by his reference to the scales. As the emblem of justice, moreover, the scales also supported Misselden's specific argument, for scales could signify a *relational* form of justice, not one that depends on a fixed equivalent or ground. Predictably, Misselden's preference for this figure provoked Gerald de Malynes's scorn precisely because the scale has no referent, no "center." To Malynes, a scale can produce no certain knowledge, because all a balance can do is tell whether one has lost or gained (not what value is) (*Center*, 55–56). Malynes's assaults on the image of the scale implicitly questioned Misselden's promotion of commutative justice, his endorsement of commercial freedom, and his effort to recast the relation between political power and economic expertise.

Malynes's ridicule of the scales also indirectly targeted double-entry bookkeeping, which by 1623 was being actively promoted as the best mode of mercantile accounting. Indeed, Malynes's fierce critique of Misselden constitutes one of the most telling early modern assaults on the mode of knowledge production inherent in double-entry bookkeeping, for Malynes recognized, as its apologists did not, that the systematic knowledge it represented could present its precision as an accurate representation of commerce only by effacing the work performed by the system itself. Malynes recognized, for example, that in both measuring by scale and keeping double-entry books, one could reach a conclusion only if one artificially stopped time. Stopping time, in turn, erases the uncertainty that is inherent in every aspect of trade. This uncertainty may be effaced for the purposes of reckoning, Malynes argues, but it continues to operate in the real world of exchange, in such a way as to make all the figures written in the books merely conjectural. Thus, Malynes charges, the uncertainty of foreign currency rates makes any estimate of a merchant's profit "meerely coniecturall"; the variation in customs charges makes projections of these costs "most incertaine"; and even the customs records that do exist are "but a supposition" (*Center*, 58–59). "What *Center* is there in this *Ballance?*" Maylnes asks scornfully. "Is it not like vnto those great *Balloons,* that men play withall fild with wind? for there is not any sollid substance, but all is coniecturall and immaginary" (*Center*, 59).

"Coniecturall" (or conjectural) is a word that will reappear in this book, for the relation between systematic *theories* about fiscal matters and the (often numerical) data these theories were supposedly based on was a chronic site of debate, even before induction was cast as a problem in the second half of the eighteenth century. We can grasp the reason for this debate by imagining the double-entry fact—the numbers that appear in the ledger—as an intermediary between empirical events and the theoretical system that constitutes the site of general knowledge. As intermediaries, the double-entry numbers seem both to record transactions that actually occurred *and* to belong to (indeed, to emanate from) the double-entry system itself. Yet either of these claims—the accuracy of the numbers or the capacity of the system to generate truth—was open to dispute. Since the relation between numerically rendered data and systematic knowledge had yet to be theorized, early modern efforts to assign cultural authority to numerical representation always had to negotiate the animus directed against whatever representational practice went under the epithet of "mere conjecture" or "sheer speculation." By the middle of the eighteenth century, as we will see in chapter 4, a kind of empiricism that could be (but not always was) linked to numerical specificity was frequently *opposed* to conjecture; "fact" became the antidote to "mere conjecture."

At this early stage of the campaign against granting merchants undue power and numbers unwarranted cultural authority, however, Malynes specifically yoked numerical representation to conjecture by arguing that numerical facts were not the antidotes to conjecture but its symptoms. In subsuming numbers into conjecture, he also called into question the formula "precise, and *therefore* accurate." Malynes argued both that the precision so prominently displayed in account books and in those theoretical systems that emulated them was only a property of the system (not a property of what the system purported to represent) and also that no system could claim to be better—in the sense of more accurate—than the numbers that were fed into it. Despite their spurious precision, Malynes charged, the accounting methods by which numbers were currently being manipulated to determine the balance of trade were imperfect because the system was only a model and because the available numbers were simply wrong. At one point he goes even further, charging that all numerical accounts that purport to ground fiscal policy are useless, precisely because no matter how precise *or* accurate the numbers, such accounts are *only writing,* which has no necessary relation to reality or to policy. "Can the making of a Ballance cause moneys and Bullion to be brought in, or hinder the transportation of moneys?" he asks. The merchant champions such writing, Malynes charges, because privileging the balance of trade privileges numbers and accounting; the merchant champions numbers and accounting because only he knows how to keep the books. As secret books, moreover, the merchant's accounts can mask the extent to which policies he represents as underwriting national well-being actually favor the merchant class. "*Albeit the generall is composed of the particular, yet it may fall out, that the generall shall receive an intollerable preiudice and losse by the particular benefit of some.* . . . Kings and Princes are to sit at the sterne of Trade, which caused the wise man to say: *Consult not with a Merchant concerning Exchanges*" (*Center,* 45).

Writing in opposition to Malynes, Misselden claimed that the scale does

GIGO

represent the best image for capturing how a nation's strength should be cali-
brated, because in order to work all systematic knowledge has to do is be internally consistent and convey the relations among the parts of whatever it stands
for. In Misselden's defense of the scale, then, we can identify both the double-
entry bookkeeper's preference for precision over accuracy and an explicit defense of what I have called the effect of accuracy. For Misselden it does not
matter if the numbers recorded in the customs records are accurate or complete, or even if conjecture plays a prominent role in projecting profit. What
matters is that the accounting method be precise—that is, consistent—and that
the numbers entered in the book show the value of England's incoming treasure *relative* to outgoing bullion. Such accounts, he argues, will also provide a
basis for determining the value of England's resources relative to those of other
countries (*Circle,* 126). Even if the knowledge produced by such accounts is not
accurate, Misselden maintains, accounting is a "science" that produces usable
knowledge, because it enables us to visualize what was previously hidden (*Circle,* 130).

For Misselden, what underwrites the mode of knowledge production
epitomized by the scale is not sovereign authority but mercantile expertise.
Thus Malynes was right to suggest that Misselden wanted to shift power to the
merchants. According to Misselden, casting and verifying the vital accounts of
national wealth should be the work of the true experts in this age of trade:
"Who can enter into consideration of the quantitie or qualitie of Commodities, whether native or forreigne, exported or imported, deare or cheape, comparable to Merchants?" he asks rhetorically. "If the Ballancing or ouer
Ballancing of trade by the disproportion therof, can be said to be evident to any,
surely it can be evident to none more than to expert Merchants" (*Circle,* 16).

To Misselden the new instruments of accounting—especially double-
entry bookkeeping—make the general system of commerce visible, but only to
those who know how to use them. Indeed, one could say that privileging accounting as the best instrument for assessing the nation's treasure, as Misselden
does (*Circle,* 127–29), begins to *create* a system of commerce as a set of written
accounts that can be seen, judged, and balanced—if not by setting credits
against debits (as in double-entry bookkeeping), then by setting incoming treasure against outgoing money. In the form of such accounts, the prince will then
be able to see the nation's wealth, and with the help of merchant experts, he will
be able to evaluate and enhance the nation's greatness. The ability to see the nation's treasure not as an absolute number but as a function of the balance of
trade, Misselden claims, is the necessary first step in making judicious policy.
"This is the first *End* of our *Ballance of Trade,*" he declares. "It shewe vs our Case
in what Estate we stand: It shew's the Causes of our Decay of trade. It represents

those causes in Capitall Characters, that he that run's may reade *Excess* and *Idleness*" (*Circle,* 134). In Misselden's climactic image of the balance of trade, a system of commerce is literally realized, in the form of a giant glass globe that the king can enter and consult at will.

> It is said of *Sapor* the King of *Persia,* that he caused a great globe to bee made of Glasse, of such curiosity and excellency, that himselfe might sit in his throne, and he and it, in the *Center* thereof, and behold the motions and reuolutions of the Starres, rising and falling vnder his feet: as if he that was a mortall man, would seeme Immortall. And surely if a King would desire to behold, from his throne, the various reuolutions of Commerce, within and without his Kingdome; he may behold them all at once in in [*sic*] this Globe of glasse, *The Ballance of Trade.* (*Circle,* 142)

Misselden's image of the exotic globe is one figure for what would eventually be theorized and named as the semiautonomous market system. Even though it metaphorically embodies the knowledge made possible at a national level by a mode of representation and accounting affiliated with double-entry bookkeeping, however, this figure does not work exactly like the double-entry system. Not only is the model of the globe a figure that sets commerce in motion for the king, rather than stopping or spatializing it (as both the double-entry ledger and subsequent variants of economic tables and graphs do), but the glass globe also produces knowledge through metaphor, not by means of numbers. As a figure of speech, the image of the globe belongs to the traditional arsenal of rhetoric or, more specifically, to that variant of Ciceronian rhetoric in which disputation and elaborately ornamented language were considered the principal instruments of knowledge production. In fact, in style—although not in content—Misselden's *Circle* resembles Malynes's *Center.* Both texts embody the conventions of Renaissance writing associated with Cicero and Erasmus: all their texts are peppered with Latin quotations and references to Aristotle; at times both Malynes and Misselden proceed by a strict application of Scholastic logic; and ornamental language dominates every page of these texts. I will return presently to the social meanings that these conventions of writing had begun to acquire by the second decade of the seventeenth century. First, however, I turn to Thomas Mun, who elaborated the concept of the economic domain not by using ornate figures like the glass globe, but by employing a version of the "plain style" that Puritans in particular had begun to use to counter the stylistic and epistemological claims of classical rhetoric.

Whereas the rhetoric of Malynes and Misselden bristles with elaborate analogies, extended allegories, proliferating similes, and allusions to texts both classical and popular, Thomas Mun's style is declarative and relatively un-

adorned. Consistently, Mun nudges those tropes he does use toward what, in relation to the copiously elaborated metaphors of Malynes and Misselden, we
might call literalization or deflation.[86] *Englands Treasure by Forraign Trade* opens
with a familiar simile, but instead of elaborating the metaphorical capacity of
this figure, Mun diminishes its figurative valence. One way he accomplishes this
is by using numbers to set off the parts of his argument. "The Merchant is
worthily called *The Steward of the Kingdomes Stock,*" Mun explains to his reader;
"and because the nobleness of this Profession may the better stir up thy desires
. . . I will briefly set down the excellent qualities which are required to make a
perfect Merchant." He continues:

> 1. He ought to be a good Penman, a good Arithmetician, and a good Accoump-
> tant, by that noble order of *Debtor* and *Creditor*. . . .
> 2. He ought to know the Measures, Weights, and Monies of all forraign Coun-
> tries. . . .
> 3. He ought to know the Customs, Tolls, Impositions, Conducts, and other
> charges. (*Treasure,* 2)

This use of numbers reinforces their capacity for semantic transparency;
using numerals to separate one part of the argument from another, as if they
were asterisks, presents numbers as pointers, not meaning-generating figures.
Note, however, that Mun uses numbers not as a transparent window onto empirical objects, but only as transparent within his system. When he offers a concrete example to illustrate his theoretical positions, for example, the numbers
do not refer to actual money but simply hold the place of real money: the numbers are exemplary, not referential. Thus, to demonstrate that monarchs should
not hoard the income from foreign trade, Mun summons numbers "to make
this plain." "To make this plain, suppose a kingdom to be so rich by nature and
art, that it may supply it self of forraign wares by trade, and yet advance yearly
100000l. in ready mony: Next suppose all the King's revenues to be 900000l.
and his expenses but 400000l. whereby he may lay up 300000l. more in his
Coffers yearly than the whole kingdom gains from strangers by forraign trade;
who see not then that all the mony in such a State, would suddenly be drawn
into the Princes treasure, whereby the life of lands and arts must fail and fall to
the ruin both of the publick and private wealth?" (68–69). We know that these
are not referential numbers because they are round figures; yet as round figures,
they make Mun's point as well as accurate numbers would have done.

The exemplary nature of Mun's numbers is of a piece with his defense of
mercantile experience—a topic I will return to shortly. For now, it is significant
that Mun never theorized his preference for numbers over figures of speech or
his theory of style more generally. Clearly, plain style was available to him in the

1620s. Indeed, as we will see in the next chapter, by that time Francis Bacon had already begun to elaborate the relation between such a plain style and the new epistemology that privileged observed particulars over rhetorical display. In *Englands Treasure by Forraign Trade* Mun never cites Bacon, however, and, because he does not self-consciously comment on the connection between style and epistemology, we can only regard Mun as part of a transition—from arguments that focused on style to arguments that invoked epistemological premises—not as a stylistic theorist.

Mun's writing does remind us that the transition from arguments that deployed rhetoric to positions articulated in epistemological terms took place within a debate that also focused on (what we would call) economic issues, for at one level his antipathy to figurative language seems to echo his antibullionist position. Figurative language might have seemed like money to Mun: a medium of exchange that too frequently seemed valuable for its own sake rather than for what it could signify (or purchase). At another level—and here our analysis does not have to rely on homology—Mun's tendency to avoid figurative language expressed the irritation he seems to have felt toward fellow merchants like Misselden, who used such language to enhance their own status by presenting finance as an obscure art. Mun was not opposed to Misselden's effort to authorize professional merchants, of course. Indeed, as we have already seen, he served on the critical royal commissions charged with advising the king on how to respond to the current crisis in trade, and he was committed to making mercantile expertise *the* knowledge of record. Even though he supported Misselden's defense of merchants, however, Mun repudiated the copious rhetoric with which some merchant apologists mounted this defense. In general, Mun does not target Misselden, no doubt because the two agreed about most substantive issues. Instead he directs his explicit attention to Malynes, whom he takes on in chapter 14 of *Englands Treasure.* In this chapter we see the magnitude of Mun's stylistic innovation, for whereas Malynes and Misselden outdo each other in the venom of their invective, Mun treats his rival respectfully: "I find him skilful in many things. . . .his Works . . . deserve much praise" (44).

Mun's disagreement with Malynes turns on the place that exchange rates occupy in the dynamic of trade. Whereas Malynes had argued that unfair exchange rates created an undervaluation of English currency, which the king could remedy, Mun counters that the rate of exchange is a function of other factors, most prominently the difference between the money brought into a country through foreign trade and the money expended for foreign goods. "It is not the *power of Exchange* that doth enforce treasure where the rich Prince will have it," Mun concludes his point-by-point refutation of Malynes's argument, "but it is the money proceeding of wares in Forraigne trade that doth enforce the exchange, and rules the price therof high or low, according to the plenty or scarcity of the said money" (55). The logical extension of Mun's position, of course, is that no royal statute will make a country amass treasure. Like Misselden, then, and unlike Malynes, Mun sees commerce as a semiautonomous domain. Unlike Misselden, however, Mun moves this domain one step closer to the domain of nature, which is governed by its own (God-given) laws. When he refers to "the plenty or scarcity of the said money" ruling price and to "a Necessity beyond all resistance," Mun removes the "natural liberty" of trade from the domain of justice and attributes to it a mode of lawfulness that seems innate. "For so much Treasure onely will be brought in or carried out of a Commonwealth," Mun concludes his disagreement with Malynes, "as the Forragn [*sic*] Trade doth over or under ballance in value. And this must come to pass by a Necessity beyond all resistance" (88).

In Mun's work, because these laws are innate—that is, not the effect or sign of contracts—they are never manifest in the entities themselves. Instead, Mun insists even more emphatically than Misselden that identifying these laws requires a special expertise, which includes knowledge about how to use various mercantile instruments—double-entry bookkeeping, for example—but which he finally refers to mercantile experience. This commitment to firsthand experience helps explain Mun's repeated references to his personal observations. In this respect, as in his repudiation of disputation, Mun's epistemological assumptions depart from the assumptions implicit in Aristotelian logic and classical rhetoric, which (to varying degrees) both Malynes and Misselden endorsed. Like Scholastics more generally, Malynes and Misselden derived their authority partly from the bodies of classical literature to which they alluded and partly from the power of their own rhetoric, which they used to demonstrate universal principles and to dismantle and discredit their opponents. By contrast, Mun almost completely ignores textual authority (unless the texts in question are customs records), and he scorns elaborately ornamented and vituperative language for simple summaries of eyewitness accounts. The knowledge Mun produced, in other words, was generated neither in an intertextual context nor by means of a rhetorical contest. Instead, he referred his claims to his own firsthand experience. To counter Malynes's claim that merchants have conspired to manipulate the exchange rate, for example, Mun cites his personal knowledge: "I have lived long in *Italy,* where the greatest Banks and Bankers of Christendom do trade, yet could I never see nor hear, that they did, or were able to rule the price of Exchange by confederacie, but still the plenty or scarcity of money in the course of trade did always overrule them and made the Exchanges to run at high or low rates" (51).

In such passages, Mun seems to derive the principles he associates with the laws of trade from his personal experience. This is not the case, however. Although Mun was rejecting classical rhetoric as the best instrument for producing knowledge about trade, and although he was turning away from the Aristotelian notion of experience as commonplace knowledge, he was not claiming to derive regularities from particular empirical observations, as nineteenth-century theorists of the economy were to do. Mun's numbers, we remember, did not even pretend to reflect actual money or transactions but simply constituted examples of representative or even hypothetical cases. Like early seventeenth-century natural philosophers, Mun simply *assumed* that his object of analysis was lawful. References to his own experience thus constituted a form of personal testimony—a *social* guarantee for the credibility of his statements—not allusions to experiments or observations from which he had derived his conclusions.[87]

Because he assumed that trade obeyed laws, it did not even matter to Mun that the numbers recorded in the customs records did not always support his theories. Indeed, the inadequacy of these numbers—which constituted one of the few points these three writers all agreed on—became the basis of Mun's claims about mercantile expertise. For him, mercantile expertise consisted of the ability to *interpret* numbers, not to gather data through personal experience. Mun maintained, for example, that one should consult the customs house records to see the overall balance of trade; but he also argued that customs records were inevitably incomplete, that they were more often estimates than accurate reflections of the money and goods that passed through the customs houses, and that as a consequence they should be used only as a starting point for achieving the most useful numbers (84). These genuinely useful numbers—the numbers that revealed the laws of the market and therefore the balance of trade—ultimately were to come not from the customs records but from the manipulations knowledgeable merchants performed on these records. Thus Mun tells us to value imports at 25 percent less than they are rated (85), to multiply the value of exports "by twenty, or rather by twenty-five" (86), and to offset the money he imagines Catholics to send out of England every year with the amount he guesses foreign princes surreptitiously send into England to pay their spies (87).

Eighteenth-century theorists of the market system, like Adam Smith and David Hume, ridiculed the group they called the mercantilists for privileging a concept whose exact numerical value they could never establish. As early as 1696, Nicholas Barbon admitted that it was impossible to know a country's imports and exports and therefore to assess its balance of trade. "There is nothing so difficult, as to find out the balance of trade in any nation," Barbon acknowledged in his *Discourse concerning Coining the New Lighter Money;* "or to prove, if it could be found out, that there is any thing got or lost by the balance."[88] In the scorn of these writers, we see the extent to which the formal precision that did characterize Mun's accounting had come to be taken for (imperfect) accuracy by the end of the century; Barbon and Smith expected Mun's calculations to be based on accurate data, not simply to be precise in relation to the model he had devised. For Thomas Mun, who wanted to authorize merchants and not the numbers themselves, the accuracy of the numbers was not the decisive issue; his primary goal was to develop an analytic model that demonstrated his thesis that encouraging trade was more important than fixing the rate of exchange and simultaneously used this demonstration to promote expert interpretation. To do so, he used a form of argumentation, a method of demonstration, and a mode of representation that proved far more effective than the particular doctrines he advocated. That the *content* of Mun's theory would not survive the seventeenth century was clear as early as 1688, when Sir Joshua Child dismissed as quixotic the campaign to determine the balance of trade.[89] That Mun's *style* of producing economic knowledge would triumph is clear in the Royal Society's campaign against rhetoric and in the increasing importance assigned to both numerical representation and expertise by those who wanted to generate knowledge about politics or trade. To begin to understand why the mode of producing knowledge that Mun practiced became authoritative in both science and economics, it is necessary to place the subjects of civility, style, and mercantile accommodation in the context of arguments about reason of state and to position all these ideas in relation to the changes that altered the practice and status of rhetoric in the early seventeenth century.

Thomas Mun's relatively respectful treatment of Gerald de Malynes indicates not simply that he possessed good manners but that he practiced one form of civility rather than another. By Ciceronian, or even Erasmian standards, Malynes and Misselden would not have been considered uncivil; indeed, insofar as their rhetoric conformed to the conventions associated with Erasmian humanism, the style they shared also constituted a form of civility, and the method they used—a written variant of rhetorical debate—constituted the mode of knowledge production that had prevailed for centuries. During the course of the sixteenth century, and largely through the English grammar school curriculum, the style associated with Erasmus and Cicero had come to be considered the natural complement to "lyberall science." According to Richard Halpern, "Social manners and literary style thus cooperated to produce a subject 'well fasshyoned in soule, in body, in gesture, and apparayle.' . . . The 'well-fashioned' or civil subject [was] an aesthetic ideal that expand[ed] the concept of 'style' to cover the whole range of social bearing. To produce a civil subject [was] to produce a 'style'—of manners, dress, and discourse."[90]

By the second decade of the seventeenth century, however, this part of the humanist curriculum had come under fire, at least partly because, as Bacon charged, the study of "eloquence and copie of speech" led scholars to "hunt more after words than matter."[91] Bacon's accusation, which seems to be directed against a specific effect of rhetoric—ornament—reflects the extent to which rhetoric had been *reduced to* ornament by the early seventeenth century. The historical stages by which this occurred constitute a much larger story than I can summarize here; this story involves both logic's appropriation of some of the functions of classical rhetoric and the displacement of *copia* as the proper end of formal language use.[92] It is also the case that the development I want to illuminate here—the rise of theories of reason of state—was only one of several factors that exerted pressure on contemporary understandings of discourse; others include the critique of Scholasticism and the emergence of radical Protestant theology. Whatever the relationship among these factors—and this remains for other scholars to sort out—we can identify as their overall effect a transformation in available models for civility, government, and knowledge.

Because it produced ornament, or *copia,* not only as an instrument for discovering timeless principles but also as an end in itself—as a matter of style—rhetoric was open to the charge that what had once seemed like a rule-governed system of argument designed to settle disputes was simply a machine for producing meaningless variations, which by definition could know no limit. As early as the 1530s, then with increasing vehemence during the last decades of the century, critics of Erasmus's practice charged that the premium rhetoric placed on verbal acrobatics and pyrotechnic invective tended to separate style from content, to privilege elocutionary skill over integrity or political commitment, and therefore to undermine rather than encourage discrimination between falsehood and truth.[93]

Such charges, which constituted first part of and then a reaction to rhetoric's gradual reduction to ornament and its cultural demotion more generally, must be understood in relation to the history of political theory, for the denigration of Ciceronian style coincided with the revival of Tacitus and the elaboration of a new political theory that differed from Ciceronian politics in both style and substance. This political theory, which the papal diplomat Giovanni Botero dubbed *ragion di stato,* was developed in various parts of Europe beginning about 1570, specifically in response to the demise of the free city-republics in Italy and their replacement by principalities and tyrannies.[94] In these new states, political knowledge was not produced in public by the debate of citizens or legalists but was formulated secretly, by a set of councilors versed in political philosophy and history.[95] Its production was guided not by Ciceronian reason, understood as a faculty capable of discovering universal principles of equity, but by instrumental reason, which could adapt means to ends.[96] Its aim, finally, was not to preserve a just society but to create a set of practices that would enable the prince to control his people. To this end, reason-of-state arguments urged princes to exercise prudence, which in turn required them to recognize the cardinal rule of human nature and government: that because all men are motivated by self-interest, interest is the best instrument of rule.[97]

Reason-of-state arguments elaborated the juridical meaning of *interest* in such a way as to take account of—or produce an image of—a society splintered by competing interests, both political and economic. Indeed, whereas those political philosophers who followed Cicero tended to define politics as a practice different in kind from economic activity, reason-of-state theorists canceled the difference between politics and economic activity, and they represented political machinations and commerce alike as contests for personal gain.[98] It was against this backdrop that the modern concept of "disinterestedness" arose in the second half of the seventeenth century, for not until society was conceptualized as a congeries of competing interests that *lacked an institution capable of negotiating those interests* was it possible to imagine a state of mind that might be called disinterestedness, much less a new method to take the place of the old rhetoric in producing a kind of truth that met the emergent criteria of adequacy to the natural world.

In the description of reason-of-state arguments I have offered thus far, the virtue that I argued was written into double-entry bookkeeping seems to have no place. To a certain extent this is true, for what recommended one practice over another, especially to early theorists of reason of state, had nothing to do with virtue as Cicero and even Erasmus would have understood that term. Instead, what elevated commerce to a status of unrivaled importance to these theorists was the assumption that wealth was necessary to underwrite a prince's greatness; and for most of the countries of Western Europe, which had no indigenous supplies of precious metals, wealth could be obtained only by international trade. As the primary means of increasing national treasure, commerce was called by one theorist of reason of state "the nerves of the commonwealth."[99] To a certain extent, however, the claim to being virtuous advanced on behalf of early modern merchants, partly by reference to the virtue displayed in their well-kept books, supported the *revision* of reason-of-state theories that occurred in the half century following the publication of Botero's book. In an attempt to make reason-of-state arguments compatible with emergent theories about natural and divine laws, late sixteenth- and early seventeenth-century theorists increasingly discriminated between a form of reason of state that benefited the collective interests of the people (*salus populi*) and a form of bad government that simply served the prince's private interests.[100]

In elaborating this distinction, political theorists had before them an example of civility that was epitomized by merchants. This mercantile civility did not emphasize style, fashion, or manners so much as it stressed conventions of trust, agreement, and honor. Contemporaries called this form of civility, which had been developed to ease financial transfers between currencies and across long distances, the system of accommodation.

I suggest that merchant apologists like Thomas Mun were able to enhance the status of their own profession even above that granted by reason-of-state theorists by representing the conventions that prevailed among merchants as providing the necessary *theoretical* support for modern government. They could so enhance their own prestige for two reasons. In the first place, merchant apologists differed from reason-of-state theorists in stressing the difference between the domains of commerce and politics. While commercial activity was obviously critical in the most practical sense to national reputation and strength, they insisted, the domain of commerce was also a counterweight to the domain of politics; for commerce, unlike politics, was governed by its own imminent laws. This meant that whereas political decisions had to be made piecemeal and according to contingencies, commercial decisions could be made in the light of universal laws, whose operations could be generalized into theories and applied in an infinite variety of circumstances.[101] Political decisions could then be made against the backdrop of what was increasingly represented as a relatively stable domain, for if they knew the laws of trade princes would be able to judge at least some of the effects of local decisions. In 1601 John Wheeler, an apologist for the Company of Merchant Adventurers, located the "root or fountain" of these laws of trade in a single law of human nature: the universal propensity to truck, barter, and exchange.

There is nothing in the world so ordinarie, and naturall vnto man, as to contract, truck, merchandise, and traffike one with another, so that it is almost vnpossible for three persons to conuerie together for two houres, but they wil fal into talk of one bargaine or another, chopping, changing, or some other kinde of contract. Children, as soone as euer their tongues are at libertie, doe season their sportes with some merchandise, or other: and when they goe to schoole, nothing is so common among them as to change, and rechange, buy and sell of that, which they bring home with them. The Prince with his subjects, the Maister with his seruants, one friend and acquaintance with another, the Captaine with his souldiers, the Husband with his wife, Women with and among themselves, and in a word, all the world choppeth and changeth, runneth & raueth after Marts, markets and Merchandising, so that all things come into Commerce, and passe into traffique (in a maner) in all times, and in all places.[102]

Wheeler's apology for commerce, and for merchants in particular, did not serve the same function as did the reason-of-state theorists' claim that every human being is motivated by self-interest. Whereas early reason-of-state theorists used the idea of interest to justify the prince's suspending his country's laws, Wheeler (like Mun) insists that merchants advance their interests by obeying laws—not the laws of their state but the quasi-natural laws of commerce. Implicitly, Wheeler argues that the laws of commerce—or at least the merchants' willingness to behave as if there were such laws—provide (as well as support) a model of good government. This claim constitutes the second reason that merchant apologists were able to enhance the status of their profession at the very time when the morality of both rhetoric and reason-of-state arguments was being questioned. Despite their individual self-interest, these writers argued, merchants had voluntarily agreed to be governed by certain conventions *as a professional group* in order to advance the collective interests of that group for the well-being of all. In so doing, Wheeler asserted, professional organizations like the Merchant Adventurers both supported the welfare of the state and provided a model for good government in general. "I take it as granted," Wheeler explained, "that the State and Common wealth hereby reapeth more profit, then if men were suffered to run a loose, & irregular course without order, command, or ouersight of any: whereby many griefes, hurtes, dissentions, and inconueniences, besides no smal dishonour [f]or the Prince, and State would in short time arise, as heretofore they haue done for want of sage and discreet gouernment, of which remedie seeing the aforesaid Companie is sufficiently prouided."[103]

Although Wheeler does not discuss this in detail, the centerpiece of the merchants' model of good government was the system of accommodation. When one merchant "accommodated" another, he accepted or extended a bill of exchange for money due or promised; in so doing, one merchant assumed that the other was creditworthy, that he could be trusted to make good on his promise to redeem the bill. Bills of exchange were critical to the early modern international financial system because they simplified the transfer of money, thereby allowing a merchant from one country to purchase goods in another country and another currency without carrying large sums in cash.

As the cornerstone of mercantile credibility, bills of exchange guaranteed that trade would go on as an institution greater than the sum total of merchants and whether an individual merchant was honest or not. The creation of a *system* of mercantile virtue, in which merchants participated simply by being merchants, constitutes the critical link between mercantile accommodation and reason-of-state arguments, as these theories were reformulated to offset the charge of amorality. Theorists who reworked Botero's original emphasis on the prince's self-interest did so not by returning to timeless principles of justice but by elevating the idea that a people or a state could have a common interest, which the prince could serve. Mercantile accommodation, anchored in the specific instrument of the bill of exchange, modeled just such a shared and *positional* interest, and it also demonstrated that, while the government of this commonality assumed that every individual who participated in the polity shared the same interests, they had to do so in only one sense: every merchant had to have an interest in commerce in general, but each merchant could still pursue his individual interest even when it competed with his rivals'. Thus government could be effective by simply alluding to the common interest it claimed to represent; princes did not have to create this common interest but could merely gesture toward it.

In the complex relations among double-entry bookkeeping, the development of mercantile expertise, the epistemological developments by which rhetoric was eventually demoted, and the reworking of political theories that advocated reason of state, we see not only the underpinnings for this revolution in politics but also an intricate series of exchanges between what we might call high and low cultures. Thus apologists for double-entry bookkeeping, which was an instrument of socially inferior merchants, initially derived social credibility by positioning this kind of writing within the elite practice of rhetoric. In doing so, early modern bookkeepers enhanced the status of both commerce and numerical representation in general. Once virtue had been welded to a particular form of mercantile writing, however, this kind of writing could protect the association between writing whose precision seemed to guarantee accuracy and virtue *even after* the high cultural practice in relation to which this virtue had originally been developed was reworked and demoted in the early seventeenth century. By a similar historical process, apologists for mercantile expertise, who based their claims about merchants' credibility on the manifest rectitude of their books, found a supportive context for their arguments in the high cultural arena of political theories about reason of state. Then, completing this cultural exchange, one part of the claim made in the low cultural practice of trade—that merchants' self-government constituted a model government—provided political theorists with an example of the kind of common interest that they argued a prince could recognize and represent.

Mercantile accommodation is critical to this series of cultural exchanges not only because it modeled good government for theorists anxious to exonerate reason of state, but also because it provided a model of social behavior that was an alternative to that dictated by the Erasmian model of civility. Indeed, we can call accommodation a kind of civility, both because it entailed a certain style of writing—the abbreviation, conciseness, and numerical representation epitomized by double-entry bookkeeping—and because, in the absence of any enforcement agency or laws that recognized interest, it required a kind of compliance that simultaneously reproduced a social hierarchy and encouraged social ambition. Indeed, mercantile accommodation permitted individual merchants to improve their social standing precisely by stabilizing a system of rewards and risks and by codifying the behaviors that would be rewarded within this system. Like double-entry bookkeeping, the system of mercantile accommodation created status *positions,* which were linked to both financial success and manners, but it did not dictate who occupied these positions, nor did it equate manners with rank or birth. Like double-entry bookkeeping, mercantile accommodation implicitly challenged the old hierarchy where status was given, not earned, and it did so by means of a style that valued conciseness and precision, not eloquence or *copia.*

The tendency of mercantile accommodation to create a positional status system instead of referring status to birth was the implicit issue in the debate between Misselden and Malynes: Malynes wanted to preserve a social order in which value was an expression of identity, whereas Misselden insisted that value was a function of various relationships within the system of commerce as a whole. Malynes believed the king should govern commerce as he did everything else, because the king was the center of value; Misselden thought the king should learn from trade rather than trying to govern it, because mercantile civility drew on common interests to create the kind of government appropriate to a group willing to emulate a lawful domain. Mun elaborated this point when he stressed both the lawful nature of commerce and the importance of a favorable balance of trade. When he rejected the Scholastic mode of disputation for the mercantile mode of modeling with numbers, however, Mun moved the debate onto new terrain. In adopting the style of double-entry bookkeeping for writing about commerce, he simultaneously aligned the production of economic knowledge with a form of civility that seemed constructive, consolidated the relation between numbers and impersonality, and reinforced the impression that figurative language was irresponsible, capricious, and possibly seditious as well. By constructing concise, orderly models in abbreviated prose that relied heavily on numbers, Mun set a new standard for civility in writing about money matters. Like the double-entry conventions it drew from, this style created the effect of accuracy by foregrounding arithmetic and personal observation; it promoted dispassion while naturalizing the mania to trade; and it presented as irrelevant transactions and desires that did not register in the arena of exchange.

In adopting this style, Mun also helped enhance the reputation of merchants as a group. In celebrating accommodation as an alternative to humanist style, merchant apologists found a basis for claiming some nobility for their calling. In 1601 John Wheeler declared the "estate" to be "honorable" because the good government essential to profitable trade set an example for the government of the realm.[104] When Mun called the merchant the "*Steward of the Kingdoms Stock*" in 1623, he elaborated this claim in language that was persuasive because concise. The merchant's work, Mun continued, is "a work of no less *Reputation* than *Trust,* which ought to be performed with great skill and conscience, so the private gain may ever acompany the publique good."[105]

In this chapter I have argued that mercantile writing—both double-entry bookkeeping and mercantile accommodation more generally—played a greater role in the transition from the old status hierarchy to modern, functionally differentiated domains than historians have typically acknowledged. As the mode of writing that produced an early version of the modern fact, double-entry bookkeeping contributed substantially to the epistemological revolution we associate with the seventeenth century. As a system of writing that merchants could appropriate to figure their own virtue, the first codified method of accounting initially borrowed from, then helped vitiate, the system of rhetoric that upheld the status and epistemological hierarchies of the classical world. In enhancing the prestige of merchants—even as rhetoric was being demoted and reason of state was being reworked—mercantile writing also helped make a case that numbers were useful not just to commercial men but to states as well. As English monarchs were gradually convinced that numbers could be useful, both because they afforded precise records of money and therefore indexes to national wealth and because they could be treated as accurate records and therefore excuses for policy, they slowly became willing to support the individuals who made the numbers make sense. To understand how double-entry bookkeeping's apology for mercantile virtue was displaced by an argument for numerical expertise, it is necessary to turn from mercantile writing and address the epistemological revolution that elevated the particulars accountants professed to record to the status of natural matters of fact.

The Political Anatomy of the Economy:
English Science and Irish Land

In describing even a rudimentary *system* of commerce and trade, Thomas Mun elaborated one of the two contributions that double-entry bookkeeping made to the constitution of the modern fact: the role of the formal and theoretically coherent system in generating (what count as) facts. Despite his frequent recourse to numerical representation, however, he did not elaborate double-entry's other contribution. As we have seen, the particulars Mun cited—especially the numerical ones—were not observed particulars or actual treasury reserves. Indeed, even his references to personal observation did not privilege firsthand data collection, historical events, or the details themselves so much as they supported a social or even juridical argument about his own credibility. Thus, despite the relative sophistication of Mun's conceptualization of the commercial system, he did not help elevate the observed particular or the historical event into the preeminent position they enjoy in that peculiar epistemological unit, the modern fact.

The modern fact finally emerged as a theorizable component of knowledge production only as an effect of two related developments in the history of epistemology: what looked like or could be presented as the complete separation of observed particulars from theories, and the elevation of particulars to the status of *evidence* capable of proving or disproving theories. These developments, which effectively inverted the priorities of the Aristotelian concept of knowledge, must be considered in the light of a larger revolution that was not only epistemological but political and religious as well. Indeed, one could say that what has been called the epistemological revolution of the seventeenth century was largely an effect of political and religious contests that began in the sixteenth century: because of the challenge that Protestant reformers posed to the Catholic Church's traditional monopoly over both social and spiritual authority, the medieval idea that knowledge is centered in theology and based on received authority gave way to a variety of defenses of different kinds of knowledge. After the Reformation, theology no longer reigned as the exclusive site of knowledge production, and knowledge itself had fragmented into various kinds—and degrees—of certainty, only one of which was considered capable of compelling assent.[1]

In this chapter I initially focus on two responses to the fallout of the religious wars: Thomas Hobbes's effort to expand reason-of-state arguments beyond the domain of politics and into the domain of philosophy more generally, and the Royal Society's decision to sidestep both theological and political questions in favor of a new kind of knowledge that theorized and depended on the modern fact. In my discussions of Hobbes and the Royal Society I investigate how far they conceptualized the modern fact as such, and I consider how—if at all—they associated numerical representation with either observed particulars or theoretical generalizations. In this section I also pay special attention to the concept of "interest," for a profound disagreement over the nature of interest and the role it should play in the production of knowledge was central to the differences that made the contest between Hobbes and the members of the Royal Society so bitter. Whereas Hobbes tried to adopt the neutral definition of interest that political theorists like Botero had endorsed, Robert Boyle and spokesmen for the Society like Thomas Sprat insisted that production of natural knowledge could (and should) be "uninterested," because interested knowledge was invariably distorted by self-serving motives. One unintended effect of this debate about interest was a cultural revaluation of the contribution merchants could make to modern knowledge. Because merchants were thought to have no investment in proving or disproving natural philosophical hypotheses, Boyle identified this professional group as particularly suited to gathering the kind of information—the facts—from which natural philosophical knowledge could be produced. Boyle's endorsement of merchants as fact-gathering instruments inadvertently reinforced the case that Misselden and Mun had made for enhancing the reputation of merchants as a group—even though they generally remained objects of considerable suspicion in Britain right through the seventeenth century.[2]

I devote the rest of this chapter to the complex amalgamation that William Petty forged in the late seventeenth century from the two kinds of knowledge production epitomized by Hobbes and by Boyle. Petty's theoretical and practical amalgamation, which joined Hobbesian deduction to the experimentalism championed by the Royal Society, begins to explain why the modern fact eventually proved attractive to governments seeking to strengthen their national and international positions. Petty advanced his controversial theories, which he increasingly claimed to derive from observed particulars, in a set of arguments on why the king should sponsor projects to generate knowledge about the economy. Like Misselden's and Mun's campaigns to promote the balance of trade as the index to national well-being, Petty's projects focused on the relation between knowledge about fiscal matters and state government. Like Misselden and Mun again, Petty reinforced the semantic connection that was critical both to the eventual elevation of merchants' reputation and to the creation of economic expertise more generally: the connection between numerical representation and a kind of knowledge considered authoritative because it was disinterested. We will see, however, that Petty's knowledge projects also differed in important ways from these early apologies for merchants. These differences result both from Petty's new focus on domestic production and circulation and from his frank skepticism about merchants' willingness to provide any information that did not serve their profession's interests.

Although his position in the pantheon of "economic theorists" has been vehemently disputed over the past two centuries, William Petty thus constitutes the critical link between natural philosophy and the promotion of systematic sciences of wealth and society in the seventeenth century.[3] As both a founding member of the Royal Society and the architect of political arithmetic, Petty served as a literal go-between, carrying natural philosophical protocols into the exploration of social and economic issues and social and economic concerns into the discussions of the Royal Society. As both a student of Thomas Hobbes and a professional associate of Robert Boyle, Petty was in the unusual position of being trained in—and committed to—both the deductive method associated with Descartes and the French Mersenne circle and the inductive method enshrined in the English Royal Society.[4] Being so trained, Petty illuminates many of the complexities of the modern fact—its peculiar claims to exist only in and of itself *and* to lay the foundations for general knowledge, the so-called axioms of nature. Finally, Petty had as much—perhaps more—at stake in establishing the disinterestedness of the new fact-based knowledge as did most of his colleagues in the Royal Society, because the extensive holdings he acquired in Ireland were the target of legal proceedings that specifically questioned his personal investment in the categories he helped write into law.

Before turning to Bacon, Hobbes, and the Royal Society, I need to place my understanding of the modern fact in relation to the two most sustained treatments of this subject in recent scholarship, which are critical to any consideration of modern facticity, including my own. In 1991 Lorraine Daston directed our attention to facts in particular when she pointed out that the modern meaning of facts, "in the sense of nuggets of experience detached from theory," entered the English language in the early seventeenth century, at the same time that Bacon identified *"Deviating Instances"* as the proper objects of a reformed natural philosophy.[5] According to Daston, the modern fact is a "deracinated particular" raised to the status of "the indubitable core of knowledge, more 'certain and immutable' than axioms and syllogistic demonstrations."[6]

In Daston's account, such facts were accorded cultural authority in the academic communities of Western Europe not because they were reliable or even particularly plausible but simply because they were detached from theory. Since rival theories were considered the primary source of conflict both within these academic communities and in society at large, natural philosophers embraced deracinated particulars because the supposed independence from theory that facts enjoyed enhanced academic civility and thus promoted cooperative knowledge-making ventures. One of the ways that champions of "courteous science" tried to lower the temperature of Scholastic debate, Daston points out, was to persuade disputants that knowledge was common property, not a matter of personal interest. In this regard she quotes the French publicist and social visionary Théophraste Renaudot, from 1656. "One of the primary means to which we were driven in order to prevent these [riots and pedantic insults] was to persuade each participant that he had no interest in defending what he had initially proposed, and that an opinion once put forward was a fruit exposed to the [whole] company, and that no one should fret about making it his property."[7]

In an essay published in 1987, then more elaborately in his 1995 *Discipline and Experience: The Mathematical Way in the Scientific Revolution,* Peter Dear has argued that the emergence of the modern fact has to be understood as part of the reworking of the more general category of experience. What was new in the seventeenth century, Dear contends, is the idea that one should deliberately acquire experience in order to test theoretical propositions. By extension, the value accorded deliberately acquired (or experimentally devised) experience also tended to privilege singular events, for to claim that one should set out to obtain some experience was to acknowledge that commonplaces were not decisive. Thus "natural philosophers and, especially, mathematical scientists increasingly used reports of singular events, explicitly or implicitly located in a specific time and place, as a way of constructing scientifically meaningful experiential statements."[8] For my purposes, Dear's most telling observation is that such singular events gradually gained the status of *evidence.* "The singular event could not be *evident,*" he notes, "but it could provide *evidence.*"[9]

Taken together, Daston's and Dear's analyses illuminate the paradoxical character of the modern fact. On the one hand, and for all the reasons Daston advances, beginning in the seventeenth century facts could be represented as separate from and prior to all theory. In her discussion of these facts, however, Daston accepts too readily the claim of the seventeenth-century writers. She does not note the other side of the paradox, which Dear emphasizes: that out-side of natural history, singular events were considered important, in the sense of being meaningful, *only* when they were held to constitute evidence—most generally, evidence of what Bacon called "true axioms," but more specifically, evidence of what counted as a scientifically meaningful experiential statement. The modern fact, in other words, could be represented either as mere data, gathered at random, or as data gathered in the light of a social or theoretical context that made them seem worth gathering.

This peculiarity of the modern fact may well stem from the double mean-ing that *factum* had in Latin. As the *Oxford English Dictionary* records, in classical Latin *factum* occasionally had the extended sense of "event, occurrence"; in Scholastic Latin the modern sense of *fact*—as "something that has really oc-curred or is actually the case"—was therefore developed *alongside* the more tra-ditional sense of commonplace. The more controversial meaning of the modern fact, which the *OED* places last in its principal definition—the fact as "a datum of experience, as distinguished from the conclusion that may be based upon it"—may have derived from its seventeenth-century proponents' desire to differentiate the knowledge they produced from the language- and text-based mode of knowledge production associated with Scholasticism, a distinc-tion it would have seemed all the more necessary to establish *because* Scholastic Latin allowed for this critical ambiguity in the meaning of *factum*.

Whatever its origin, the paradoxical nature of what I am calling the mod-ern fact has made it possible for generations of commentators to emphasize one side of the concept or the other, depending on the kind of argument they have wanted to make. Thus some insist, as Bacon did, that fact collection is separate from and prior to interpretation and theory, whereas others argue, as G. Robertson did in 1838, that facts cannot exist—in the sense of being meaning-ful—unless they speak to some relation, which is always implicitly theoretical: "The fact, the thing as it is without any relation to anything else, is a matter of no importance or concern whatever: its relation to what it evinces, the fact viewed as evidence, is alone important."[10] The paradoxical sense of the modern fact is perfectly captured by one of the preeminent late twentieth-century philosophers of science, Thomas Kuhn, in his attempt to explain how scientific paradigms change.

> Discovery commences with the awareness of anomaly, i.e. with the recognition that nature has somehow violated the paradigm-induced expectations that govern normal science. It then continues with a more or less extended exploration of the area of anomaly. And it closes only when the paradigm theory has been adjusted so that the anomalous becomes the expected. Assimilating a new sort of fact demands

a more than additive adjustment of the theory, and until that adjustment is completed—until the scientist has learned to see nature in a different way—*the new fact
is not quite a scientific fact at all.*[11]

As we will see in a moment, the ambiguity implicit in the modern fact was
present even in Bacon's writing; and as we will see at the end of this book, this
ambiguity tended to attract attention whenever an emergent science became
embroiled in political or social controversies. The claim that facts were not theoretical, in other words, tended to be emphasized whenever theoretical disputes raged around a new science and as a means of removing "scientific"
knowledge production from the arena of controversy. This, of course, exactly
describes the situation and the agenda of the Royal Society in war-torn seventeenth-century Britain. It is in this context that we can begin to see how the
modern fact was initially theorized as instrumental to general knowledge about
nature and, by extension, to theories about society and wealth as well.

THE CRISIS IN KNOWLEDGE AND THE QUESTION OF METHOD

Any account of the complex British responses to the sixteenth-century crisis in
knowledge must begin, however briefly, with Francis Bacon, for Bacon's "Great
Instauration" constituted an influential attempt to reorder the priorities of
modern knowledge production in such a way as to elevate the "discovery" and
use of observed or historical particulars and to demote the Scholastic "cultivation" or contemplation of accepted commonplaces.[12] Although a detailed
study of Bacon is beyond the scope of this book, I do need to make two points
about Bacon's new science. The first point concerns the relation between his
new method and the facts it privileged and produced: Bacon's innovative instrument for producing knowledge about the natural world—his *Novum Organum*—was "experiment," which bore a complex and much-discussed
relation to "experience" and which helped codify the paradoxical nature of the
modern fact. Bacon's project as a whole, moreover, was also intended to reform
language use by elevating "plain style" over ornament, because he considered
"unadorned brevity" more appropriate to the cataloging of facts than "a treasury of eloquence."[13] I will presently take up the related subjects of experience,
experiment, facts, and style.

My second point concerns Bacon's understanding of the uses to which
knowledge should be put and, by extension, the kinds and limits of desirable
knowledge. For Bacon, who was the consummate statesman, all knowledge
should be formulated in such a way as to serve the state, which, in a period in
which both religious and epistemological heterodoxies proliferated, meant

supporting the monarch's ability to adjudicate and use what counted as truth. As Julian Martin has argued, Bacon wanted to reform natural philosophy because he wanted to reorient and strengthen the Tudor monarchy; to this end, he responded to the contemporary crisis of knowledge by promoting his new approach to knowledge production as the preeminent instrument of reason of state.[14] His famous aphorism "knowledge is power" should be understood in this context: as a statement about the instrumental value that a certain kind of knowledge could have for a specific form of political power.

As we will see later in this chapter and again in chapter 4, *experience* and *experiment* became increasingly central terms, first in mid-seventeenth-century efforts to define the relation between new kinds of instrumentally produced knowledge and the received, Aristotelian mode of knowledge production, then again in the eighteenth-century revision of moral philosophy. For both groups of epistemological reformers, Bacon's name served as a convenient shorthand for the "inductive" or "experimental" method, although what later theorists meant by "induction" and "experiment" (not to mention the extent of their familiarity with Bacon) varied considerably.[15] The variety of interpretations his writings authorized stems both from the novelty of his enterprise and from a certain imprecision in his descriptions of method. On the one hand, the aim of Bacon's enterprise was clear and explicitly revolutionary: he wanted to completely restart "the whole operation of the mind . . . so that from the very beginning it is not left to itself, but is always subject to rule" (*Novum Organum*, 38 [part 1, preface]). To accomplish this, Bacon rejected those "axioms now in use," because "they have been derived from a meagre and narrow experience and from a few particulars of most common occurrence"; he also rejected the mode of Scholastic debate by which truth was demonstrated about them (50 [part 1, aphorism 25]). Instead of axioms designed both to draw on and illuminate common occurrences by rhetorical demonstration, Bacon advocated "experiments," which could elicit from nature those "singular instances" capable of disclosing truths about nature that were otherwise hidden (195–96 [part 2, aphorism 28]).

One effect of Bacon's elevation of experiment was thus to elaborate the contribution to the modern fact represented by double-entry bookkeeping: whereas Aristotle and his followers sought universals and thus tended to summon generalizations as facts—"that which is always or that which is for the most part," in Aristotle's words—Bacon focused on particulars, on specific events, that he and his followers described in first-person, eyewitness, historical accounts.[16] For the most part, Bacon seems clear that the particulars he sought were valuable precisely to the extent that they departed from commonplaces, to the extent, that is, that they did not follow from received opinion or theory.

Thus Bacon especially valued "*Singular Instances,*" which "show bodies in the concrete, which seem to be outlandish and broken off in Nature," and "*Deviating Instances,*" which are "errors of Nature, sports and monsters, where Nature deviates and turns from her ordinary course." Bacon valued such rarities because "they correct the understanding in regard to ordinary things, and reveal general forms" (195, 196 [part 2, aphorisms 28, 29]). We see that Bacon believed such singular facts could be separated from theory by the division of labor he sets up in his "Preparation towards a Natural and Experimental History." In this text Bacon insists that collecting the materials for the science he proposes is work different from and inferior to the "actual intellectual work" of producing general knowledge. The former, he notes, may be left to "agents and merchants"; the latter should be the business of learned men like Bacon himself ("Preparation," 298).

Even though Bacon privileges rarities because they depart from both commonplaces and theory, and even though he distinguishes the mere collection of facts from interpretation or induction, two features of his account of method suggest that even Bacon was unwilling to completely sever the link between these new facts and theory. In the first place, he repeatedly rejected the kind of data he called "individuals," which he associated with natural history. "Those copious natural histories that offer a wealth of graphic descriptions of species and their curious variety are hardly what is needed," Bacon explained. "For such trifling variations are only the sports and wanton freaks of Nature, and come near to being the nature of individuals; they provide a kind of pleasant ramble among the things they describe, but only slight and often superfluous information to the sciences" ("Aphorisms on the Composition of the Primary History," 303). Instead of "individuals," which would be worthless to the philosophical project of discovering "true axioms," Bacon wanted his agents to catalog "the more outstanding instances in each and every kind," because these would aid the production of general knowledge ("Aphorisms," 307).

The distinction between "individuals" and "outstanding instances" leads to my second point: the *reason* for observing and recording particulars was to lay the foundation for science, which, for Bacon as for Aristotle, was general knowledge about causes. To a certain extent this general knowledge was *always* theoretical, because one always had to assume that the observed particulars were types of other, similar particulars that one had not observed. As we will see in a moment, then at greater length in chapters 4 and 5, this foundational theoretical assumption would eventually be formulated as the problem of induction. In his defense of induction, however, Bacon presents his method not as a problem but as a solution. Specifically, he imagined his method as a solution to the problem that the Aristotelian understanding of experience had bequeathed to knowledge production. This problem was that, as long as one worked only by "a meagre and narrow experience and from a few particulars of most common occurrence," one could never expand what was known, because if "fresh particulars" contradicted the received axiom, "the axiom is rescued by some frivolous distinction, when the more correct course would be for the axiom itself to be corrected" (*Novum Organum,* 50 [part 1, aphorism 25]).

To a certain extent, then, because they were both sought for and conceptualized as part of the enterprise of philosophy *as distinguished from "copious" natural history,* Baconian facts always carried an aura of theory. Bacon's agenda was to distinguish between the received notion of experience, which involved looking to nature only to find what one already knew, and a new concept of experience that elevated natural particulars to a new importance. He did not explore whether one *could* look at nature without some theory to organize what one saw, although his elaborate discussion of the "idols" that distort perception shows he was aware that, if this was possible, it was very difficult to do. Instead of distinguishing between a mode of observing that is blinkered by theoretical presuppositions and one that is absolutely theory-free, Bacon emphasized other distinctions that hold the place of the distinction he does not make. Thus he distinguishes between jumping too quickly to generalizations and moving at an appropriate pace through intermediate levels of generalization (70 [part 1, aphorism 64]), and he discriminates "what is loosely and vaguely observed" from what is "verified, counted, weighed or measured" (107 [part 1, aphorism 98]).

The method Bacon nominated to solve the problems bequeathed by the Aristotelian sense of experience, of course, was experiment. Because this method conformed to rules, it theoretically counteracted the tendencies simply to project the mind's internal order onto the external world and to discover about nature what one already knew about the self. According to Bacon, experiment should have enabled readers to discriminate between the old and new senses of experience, to distinguish "true" experience from "silly" experience, which at various points in the *Novum Organum* seems like either experience that is too limited by theoretical preconceptions or experience that is not disciplined at all.[17] These two kinds of silly experience correspond to the two practices he was attempting to avoid: Scholastic philosophy, which was too theory-bound, and natural history, which consisted simply of catalogs of particulars. Baconian experiment obviously differed from both of these inferior kinds of experience, but it is not always absolutely clear how. This ambiguity stems in part from the fact that Bacon also used *experiment* in a variety of senses, not all of them involving the use of instruments or even the formulation of a specific question designed to provide a specific answer about nature.[18] For example, he acknowledged that members of the "Empirical school of philosophy" (which included alchemists) performed experiments of a (an inferior) kind, and he also claimed that effective experiments could be carried out by reading books. In general, he tended to emphasize what a proper "experiment" was not rather than specifying what it was.[19]

Even if he did not—or could not—create absolute distinctions between the versions of experience and experiment he advocated and those of the Scholastics, natural historians, and empirics he opposed, Bacon could and did distinguish between the *style* common to all these writers and his own. Momentarily conflating Scholastics and the writers of "copious natural histories," he decries stylistic excess in favor of brevity, succinctness, and what looks like a transparent kind of language.

> Things that only serve to adorn a speech and provide similes and a treasury of eloquence and inanities of that kind should be utterly avoided. And each and every thing that is included should be set forth briefly and concisely, so that they may be nothing less than words. For no one collecting and storing materials for building a ship or other such structures thinks of arranging and displaying them prettily to please the eye (as in shops), but is careful only to see that they are sound and of good quality and take up the least room in his stores. And this is exactly what should be done. ("Preparation," 302)

By opposing the "treasury of eloquence" to another kind of warehouse, where materials are stored for use, Bacon forges an opposition that stands in for oppositions that are harder to make: between Scholastic experience and modern experience, between the ancient fact and the modern fact. Even if it was difficult for Bacon to define exactly what he meant by experiment in the 1620s, it was easy for him to distinguish between copious rhetoric and the plain style of the new science. Thus stylistic difference could hold the place for more telling kinds of distinctions, at least until the new mode of knowledge production associated with the modern fact could be theorized in more detail.

In the 1660s, Robert Boyle and his colleagues would begin to define some of the protocols for the kind of experimental natural philosophy they endorsed, although when apologists like Thomas Sprat began to describe these protocols, it proved easier once more to focus on style rather than content. In his *History of the Royal Society,* Sprat justified his imprecision about method with an elaborate defense of the antimethodological nature of the method itself; for Bacon, imprecision seems more an effect of the novelty of his enterprise than part of a theoretical position. Wanting to clear away those modes of knowledge production that obstructed an interrogation of nature and bred civil conflict *without* succumbing to skepticism, he focused more on differentiating himself from—and relating his new practice to—what had come before than on working out the new practice in all its details.

To a certain extent, moreover, Bacon did not have to decide exactly what experimental natural philosophy would be, because once he had laid out some of its features—that, for example, it would call upon as well as assemble giant natural histories that would provide compendiums of all available knowledge—it became clear that no individual natural philosopher or group of philosophers would be able to create all the elements of a workable experimental practice. Unlike Boyle and his colleagues, Bacon did not try to specify either the protocols of experimentalism or a set of institutional manners capable of discriminating between reliable and unreliable philosophers, because the enterprise he described existed on too large a scale for a single society to manage. By his own account, the natural philosophical project was "a matter of very great magnitude, and cannot be accomplished without great labour and expense." Calling it "a kind of royal work," Bacon referred responsibility for it to the only authority capable of financing and protecting this magnificent project: the king ("Preparation," 297). Whereas the members of the Royal Society specifically limited the kind of knowledge they sought to produce in order to avoid the religious and political controversies that had erupted in civil war in the 1640s, moreover, Bacon wanted to extend the reach of natural philosophy because, in 1620, he considered natural philosophical knowledge an instrument capable of reinforcing the king's power so that heterodoxy could be ruled out of court.

Bear in mind that the knowledge-producing instrument Bacon described both supported *and required* an entire set of *state-sponsored* institutions. In keeping with its imbrication with the monarchy, to Bacon what counted as "useful" knowledge was also a matter of state. For him it was critical to subordinate those personal prejudices or interests that might distort the search for knowledge to some external rule, because the individual's first responsibility was to the state, not either to himself *or* to knowledge, abstractly considered. In early seventeenth-century England, a commitment to knowledge that could serve the state meant a commitment to policies that could curtail religious, political, and epistemological controversies while permitting imperial expansion. Bacon's natural philosophical project was a royal undertaking not only because it required massive funding but also because it was intended to supply knowledge that would reinforce the king's power to decide what measures would serve his people's interests. Similarly, in Bacon's view the king had to sponsor the natural philosophical project because, to ensure peace where heterodoxy and controversy prevailed, knowledge had to be backed by royal authority.   *of STS*

Just as the knowledge Bacon sought to produce both required and sup-   *or Hobbes* ported the monarchy, so the rationality of quasi absolutism defined the limits of his natural philosophical project. Reason of state, in other words, not only provided positive guidelines for natural philosophy, which dictated the kinds of knowledge that would serve the interests of the ruler; it also declared some things off-limits to the philosopher because they were state secrets. Bacon's insistence that some kinds of knowledge should not be sought *for political reasons* is clearest at the conclusion to the *Advancement of Learning,* where he identifies "government," along with theology, as one of the few arenas into which the natural philosopher must not venture. "Concerning Government, it is a part of knowledge secret and retired, in both these respects in which things are deemed secret; for some things are secrets because they are hard to know, and some because they are not fit to utter. We see all governments are obscure and invisible."[20] If he considered knowledge to be power, then, Bacon's respect for power exceeded—or at least qualified—his respect for knowledge and dictated the limit of that otherwise enormous arena in which the natural philosopher would be allowed to work.

Despite his efforts to limit, and so neutralize, controversies about his natural philosophical project, Bacon's campaign to reconstitute knowledge on the ground of experiment did not foreclose contemporary battles over religion, the nature of knowledge, or the role the state should play in sponsoring knowledge. Nor did his effort to harness the production of knowledge to reason-of-state political theories protect those theories from a series of political controversies that erupted almost immediately upon Bacon's leaving office in 1621. In 1626 Charles I tried to force a reluctant Parliament to lend him money for policies that his ministers defended as "necessary" to the "public good."[21] In the ensuing furor, then again between 1634 and 1638 in the notorious wrangle over ship money, it became clear that once one admitted the argument that policy should be made according to the interests of the state and not (invariably) according to law, it became necessary to decide who had the authority to define those interests and the "emergencies" in which laws could be abrogated.[22] With both Royalists and Charles's opponents using the reason-of-state language of *salus populi,* "necessity," and state "interests," the Baconian argument that knowledge production should serve the king seemed less obvious than it had a decade before. As Clarendon pointed out in the 1630s, the same arguments about interest and necessity that Charles had used to extend his authority were used against him by his adversaries; this was possible because, once the unity of the commonwealth had been shattered, everyone could be said to have interests, and (nearly) everyone could claim to be able to define the common interests of the state.[23] In the civil war in which these contending claims were played out, the language of *self*-interest began to appear for the first time in British political discourse,[24] and this created the context in which the continuing knowledge crisis was addressed again, first by Thomas Hobbes in his *Leviathan*, then by the members of the Royal Society, beginning in the 1660s.

When Thomas Hobbes published *Leviathan* in 1651, he was an exile in Paris, looking across the Channel at a state that lacked a monarch, whose universities were in disarray, and where alchemists and Paracelsian physicians competed with university-bred Scholastics for the authority to generate knowledge.[25] In this context of political, religious, and epistemological turmoil, Hobbes deemed it politically necessary to use a mode of analysis capable of *compelling* assent, because only a government to which everyone assented could put an end to what Hobbes viewed as universal chaos. In these times of ceaseless dissension, he explained, "there is no place for Industry; because the fruit thereof is uncertain: and consequently no Culture of the Earth; no Navigation, nor use of the commodities that may be imported by Sea; no commodious Building; no Instruments of moving, and removing such things as require much force; no Knowledge of the face of the Earth; no account of Time; no Arts; no Letters; no Society; and which is worst of all, continuall feare, and danger of violent death; And the life of man, solitary, poore, nasty, brutish, and short."[26]

In the mid-seventeenth century, the only analytic method considered capable of compelling assent was mathematical demonstration. Following Galileo, Hobbes contended that mathematical demonstration had the potential to compel assent because it produced *certain* knowledge; thus mathematical demonstration was superior to experimental demonstration, which he considered capable of producing only *probable* knowledge. Ideally, he believed, the process of reasoning used in mathematics could be generalized to knowledge production, so that philosophy could become a kind of arithmetic. In *Leviathan*, Hobbes called this kind of knowledge production "reckoning": "REASON, in this sense, is nothing but *Reckoning*."

> When a man *Reasoneth,* hee does nothing else but conceive a summe totall, from *Addition* of parcels; or conceive a Remainder, from *Substraction* of one summe from another: which (if it be done by Words,) is conceiving of the consequence of the names of all the parts, to the name of the whole; or from the names of the whole and one part, to the name of the other part. . . . These operations are not incident to Numbers onely, but to all manner of things that can be added together, and taken one out of another. . . . The Logicians teach the same in *Consequences of words;* adding together *two Names,* to make an *Affirmation;* and *two Affirmations,* to make a *Demonstration;* and from the *summe,* or *Conclusion* of a *Syllogism,* they substract one *Proposition,* to finde the other. Writers of Politiques, adde together

Pactions, to find mens *duties;* and Lawyers, *Lawes,* and *facts,* to find what is *right* and *wrong* in the actions of private men. In summe, in what matter soever there is place for *addition* and *substraction,* there also is place for *Reason;* and where these have no place, there *Reason* has nothing at all to do. (*Leviathan,* 110–11)

Despite Hobbes's professed confidence in reckoning, by 1651 he had concluded that one could not rely on the method of mathematics to settle disputes. Mathematics could no longer compel assent, Hobbes lamented, because reason no longer triumphed over the passions; and reason no longer triumphed over the passions because the English Civil War had so fragmented and inflamed individual interests that people were willing to quarrel over even the laws of mathematics, if they might profit by doing so.[27] In a world where people were ready to suppress mathematical rules if it served their interests, Hobbes realized that appeals to reason had to be supplemented by an appeal to the passions. In the middle of the seventeenth century, the one device considered capable of engaging the passions was rhetoric.

Critics have long puzzled over Hobbes's obvious reliance on rhetorical figures, because he was such an outspoken critic of figures of speech.[28] Recently, however, Quentin Skinner has argued persuasively that Hobbes changed his mind about the relative power of reason and rhetoric. Skinner contends that in the works written before the late 1640s, Hobbes positioned himself firmly against the skepticism that the humanist method of rhetorical argument encouraged. In *The Elements of Law* (1640) and in *De Cive* (1642), Skinner explains, Hobbes maintained that "so long as we reason aright from premises based in experience, we shall be able not merely to arrive at scientific truths, but to teach and beget in others exactly the same conceptions as we possess ourselves." In 1651, by contrast, Hobbes endorsed the very humanist analysis that he had previously rejected—although he still repudiated skepticism—now agreeing with Cicero that reason lacks the capacity to persuade and that eloquence is necessary to make reason's truths compelling.[29]

While Skinner is right to point out that Hobbes seemed to think he could sever the connection between rhetoric and skepticism, so as to use the former without mobilizing the latter, he does not explain precisely how Hobbes sought to exempt rhetoric from the criticism increasingly leveled against eloquence and ornament. Nor does he explain how Hobbes used rhetoric to counteract the quarrels over interests that had torn the commonwealth apart. By placing *Leviathan* in the context of the constitution of the modern fact, we can see that Hobbes pursued both ends through a two-part strategy. On the one hand, he appealed to premises that he claimed to derive from observation. Appealing to observation instead of language seemed to remove his rhetoric from the increasingly dubious realm of language games and thus to disarm contemporary complaints against rhetoric. Hobbes's appeals to observation did not render his method inductive, however, nor did his observation generate modern facts.[30] For on the other hand, Hobbes represented the conclusions he drew from these observations as *universally valid,* not because they reflected axioms inductively reached, but because the *position* from which he made the observations constituted a representative stance that, once assumed, rendered the observer the *type* of universal human nature. By creating and occupying this position, Hobbes encouraged his readers to imagine that he represented them; by extension, if every man occupied the same position, then their interests would be identical to each other's—and to his.

We can see this two-part strategy in Hobbes's introductory invocation of the adage "*Nosce teipsum, Read thyself.*" By way of invitation into the text of *Leviathan,* Hobbes explains that if every reader follows the author's example, he will read in himself exactly what Hobbes has read in himself. Thus Hobbes's observations about himself, which seem to be based on actual introspection, get raised to the level of common experience, common sense, or self-evident truth.

> There is another saying not of late understood, by which [men] might learn truly to read one another, if they would take the pains; and that is, *Nosce teipsum, Read thyself:* which was not meant, as it is now used, to countenance, either the barbarous state of men in power, towards their inferiors; or to encourage men of low degree, to a sawcie behaviour towards their betters; But to teach us, that for the similitude of the thoughts, and Passions of one man, to the thoughts, and Passions of whosoever looketh into himself, and considereth what he doth, when he does *think, opine, reason, hope, feare,* &c, and upon what grounds; he shall thereby read and know, what are the thoughts, and Passions of all other men, upon the like occasions. . . .He that is to govern a whole Nation, must read in himself, not this, or that particular man; but Man-kind: which though it be hard to do, harder than to learn any Language, or Science; yet, when I shall have set down my own reading orderly, and perspicuously, the pains left another, will be onely to consider, if he also find not the same in himself. For this kind of Doctrine, admitteth no other Demonstration. (*Leviathan,* 82–83)

As Quentin Skinner has argued, phrases that alluded to common experience were staples of classical rhetoric; Aristotle had pointed out in book 3 of *The Art of Rhetoric* that "it conferres also to perswasion very much to use these ordinary formes of speaking, *All men know; 'Tis confessed by all; No man will deny* and the like."[31] What is noteworthy about Hobbes's adaptation of this device, however, is that whereas Aristotle could simply *assume* that "all men" would know, confess to, or agree on certain things, Hobbes had to create a position that would *confer* commonality and that would do so by appealing to the *self.* Hobbes conferred commonality on his readers by identifying himself as the representative of "all men"; he appealed to self-interest by identifying the interests of the commonwealth with the interests of every (identical) reader.

The common position Hobbes creates in *Leviathan* becomes the basis for a set of universal principles, which he refers to with the terms *laws* and *nature:* "human nature," the "laws of nature," and the "dictates of nature." These "laws" are what Hobbes claims his science can discover; but he also presents them as the premises from which his reckoning sets forth. This claim clearly distinguishes Hobbes's method from Baconian induction, just as his use of a common position to create a universal subject distinguishes the knowledge he produces from the modern fact. In this sense as in others, Hobbes's method constitutes a revival of Aristotelian deduction. Keep in mind, however, that the problem of interests forced Hobbes to adapt Aristotle's method. Claiming that a demonstration could produce certain knowledge by setting out from certain premises was the touchstone of Aristotelian scientific demonstration, but because of the problem of interests, Hobbes could not simply assume that the premises he asserted were certain—that is, universally agreed upon—principles.[32] The problem of interests, in other words, the recognition of which forced Hobbes to supplement reason with rhetoric, also required him to find a way to make the premises he claimed were universal *seem* universal to his readers, for the problem of interests had rendered the method of Aristotelian Scholasticism questionable as it had not been as recently as two decades before.

Hobbes supplemented the appeal implicit in the universal position he created and assumed by invoking arithmetic, for even though individuals *might* dispute mathematical laws if it would serve their interests to do so, Hobbes assumed that mathematical operations did not yet serve anyone's interests. Yoking his assertions about the "self-evident" truths of human nature to arithmetical operations thus enabled him to use the formal precision of the mathematical method to create the effect of certainty in a society where the consensual grounds Scholastic certainty had been based on no longer obtained. In *Leviathan* arithmetic merely served an illustrative function; the figure of mathematics or arithmetic was holding the *place* of certainty in a world in which certainty was no longer available because consensus was no longer possible.[33]

The role played by the figures of mathematics and arithmetic is clearest in Hobbes's frequent use of "reckoning" to name and illustrate his method, but the image of quantification also appears in a more subtle guise—as, for example, in his discussion of the "state of nature." In this state, he explains, we find three sources of dissension, the virulence of which ultimately forces human beings to accept the rule of a powerful monarch. Hobbes simply states that these sources

but wasn't H trying to reestablish consensus by this strategy?

of dissension (competition, diffidence, and glory) are rooted in "human na-ture," but he demonstrates (or "proves") this conclusion by translating what his theory of competing interests might have presented as a heterogeneous range of inclinations and preferences into a single desire, *which can be measured proportion-ately* (or numerically). Representing all individuals as motivated by the same de-sire symbolically solves the problem of divergent interests by translating difference into degrees of the same interest; it solves the problem of difference because it makes arithmetic, whose method is capable of producing certain knowledge, an integral part of social and ethical analysis. In demonstrating the third source of dissension, for example, Hobbes assumes that all men desire glory and that they measure glory quantitatively, in terms of the "rate" of a sin-gle kind of respect. "Men have no pleasure, . . . in keeping company, where there is no power able to over-awe them all. For every man looketh that his companion should value him, at the same rate he sets upon himselfe: And upon all signes of contempt, or undervaluing, naturally endeavours, as far as he dares . . . to extort a greater value from his contemners, by dommage; and from oth-ers, by the example" (*Leviathan,* 185).

In this illustration, quantification functions as an implicit metaphor for so-cial relations: people interact with each other as if there were a fixed amount of respect for which everyone vies. But quantification also functions as an inter-pretive instrument: by translating incommensurate differences or qualities into something that can be quantified, Hobbes renders social relations amenable to arithmetic. There are obvious affinities between Hobbes's use of quantification in examples like this and the translation the double-entry system performs upon the social interactions that occurred around trade in the warehouse or shop. In both cases, translating sociality into numbers allows for a level of preci-sion and certainty that could not be attained through narrative descriptions or comparisons of incommensurate things. In both cases, moreover, it is the pre-cision of the method that creates the effect—in Hobbes's case the effect is of certainty; in the case of double-entry bookkeeping, accuracy is the effect.

Because the conditions necessary to produce certain knowledge *in the do-main of politics* did not obtain in 1651, the figures of arithmetic in *Leviathan* could be only instruments of persuasion, not parts of a logical or mathematical demonstration. With the example of accounting before him, Hobbes could easily imagine numbers serving this function; so vivid is bookkeeping's poten-tial to persuade, in fact, whether or not the ledger's precision is accurate, that Hobbes specifically warned against taking accountants at their word.[34] In *Leviathan,* Hobbes sought to use arithmetic's ability to persuade in order to convince the prince in exile, Charles Stuart, that he should return, reclaim the throne, and produce the social conditions necessary to restore civil, religious, and epistemological accord—that is, to prove that what Hobbes claimed in *Leviathan* was true: that a monarch could create the conditions in which certainty would be possible by enforcing social accord. Hobbes's argument might have been self-interested in a narrow sense—eventually, when Charles did return to England, he awarded Hobbes a pension—but this is not obvious in *Leviathan*. Instead, the argument seems simply logical—like an appeal to common experience, of which Hobbes's experience seems merely to be representative.

Hobbes's turn to rhetoric in 1651 should be read in relation to the epistemological revolution of the seventeenth century, if only because what looks like an appeal to actual observation in *Leviathan* emphasized one facet of the modern fact while setting aside the other. Hobbes used the methodological precision perfected in double-entry bookkeeping to create the effect of certainty; but he had no use for observed particulars, for the knowledge he sought to create concerned not what actually existed but what might be brought about. His turn to rhetoric should also be read in relation to the reason-of-state political arguments that proliferated in Britain after the 1620s, because by 1651 Hobbes had become convinced that individuals did act out of a self-interest so powerful that it could overwhelm reason. Unlike theorists of reason of state, however, who were concerned not primarily with *self*-interest but with national interest and who typically appealed to nothing more substantial than expediency and immediate necessity to justify their policy recommendations, Hobbes *was* worried that self-interest would overpower reason or disguise itself as concern for the state. Largely as a consequence, he cast his suggestions as scientific—and as certain, because demonstrable—truths. Ironically, it was this claim to absolute veracity, which Hobbes made by analogy to mathematical precision, that helped make *Hobbes* a watchword for the very heterodoxy he sought to foreclose. As reason-of-state arguments continued to be reworked in the second half of the seventeenth century, it became unacceptable to represent human nature as driven by interests that were essentially and exclusively antisocial. As we will see in the next chapter, theorists from Pufendorf to Hutcheson to Adam Smith gradually reshaped the sixteenth-century theory of interests until it came to seem that self-interest, *especially when expressed as the love of monetary gain,* was the necessary building block for civil society, not its unassimilable antagonist.[35] Hobbes's depiction of self-interested human nature was not the only reason he was so roundly vilified, of course. As we will see in chapters 5 and 6, his deductive method, which smacked ominously of French theory, was also a target for subsequent generations of British writers. Yet because Hobbes incorporated the figure of arithmetic so effectively, the rhetorical function of this trope went virtually unnoticed by his readers. Indeed, at least one of his followers—William Petty—took his figurative use of arithmetic literally, so as to develop from Hobbes's Aristotelian Scholasticism a method of quantifying *and commodifying* the "value of a man."

Even such brief surveys of Bacon and Hobbes enable us to see that many of the assumptions basic to Aristotelian Scholasticism had come to seem questionable by 1650. In the early 1660s, in the wake of the Restoration, another response to the knowledge crisis of the sixteenth and seventeenth centuries was institutionalized. This response, which was formalized in 1662 with the founding of the Royal Society, claimed to reject Aristotelianism even more emphatically than Bacon had done *and* to create a domain of knowledge production *outside* of political and theological discussion. Instead of producing certain knowledge through formal and rhetorical demonstrations of commonly accepted premises, as Hobbes had tried to do, the members of the Royal Society sought to generate probable knowledge by means of experiments and with the help of instruments. Instead of asserting that the knowledge they produced had theological or even political implications, as Bacon had claimed, the Royal Society proudly limited the kind of knowledge to which its members aspired to "Matters Philosophical, Mathematical and Mechanical." Accepting this limitation in the kind of knowledge it produced earned the Society the right to assemble and to publish and communicate knowledge in the repressive context of the Restoration—rights that were expressly forbidden to other contemporary groups, which also wanted to generate knowledge—albeit of a distinctly more egalitarian (and thus politically subversive) kind.[36] The royal charter granted by the king in 1662 gave the Society "full Power and Authority . . . to print such things, matters and businesses concerning the said Society" and "to hold Correspondence and Intelligence with any Strangers . . . without any Interruption or Molestation whatsoever: Provided that this Indulgence or Grant be extended to no further use than the particular Benefit and Interest of the Society, in Matters Philosophical, Mathematical and Mechanical."[37]

Such royal privileges helped create the institutional conditions by which the Royal Society eventually rendered its preferred mode of knowledge production authoritative. To understand how the Society achieved this goal, however, we also have to see how its variant of Baconian experimentalism addressed the concerns that bedeviled Hobbes, for like Hobbes, the members of the Royal Society constructed their method in the context of political theories about reason of state and the priority of interests. Unlike Hobbes, however, members of the Royal Society did not address the problem that the idea of interest introduced into the production of knowledge by resuscitating Scholastic demonstration. Instead, members sought to forge a new relationship between producers of knowledge and the knowledge they generated. This stance, which Thomas Sprat, the Society's historian and apologist, called "uninterested," opened the possibility that personal interests could be set aside in favor of another kind of ambition: the desire to produce (or discover) universal knowledge about the natural world instead of either the contingent knowledge associated with reason of state or the instrumental knowledge associated with an individual's monetary gain.

The pathbreaking work of Steven Shapin and Simon Schaffer has established that the modern fact was theorized and produced by members of the Royal Society.[38] Recently, moreover, Steven Shapin has demonstrated in some detail how canons of civility supported the constitution of the natural philosophical matter of fact.[39] Neither Shapin nor Schaffer makes the point that seems critical to me, however: it was the specific emphasis that members of the Royal Society gave to the facts they produced that *necessitated* the invocation of civility. Because they argued that facts were separable from both theory and method *in order to decrease the likelihood of civil dispute,* the experimentalists had to invoke some other rule-bound practice so as to stabilize facts—to place what counted as a fact beyond dispute and, by doing so, to make it meaningful. As we saw in the case of double-entry bookkeeping, specific details become meaningful only when they are rendered subservient to the system, because the system—which is the method embodied—constitutes the theoretical apparatus that makes the particulars seem related to each other and, by extension, to some general principle. In the experimentalism of the seventeenth-century natural philosophers, the rule-bound practice that took the place of systematic method was the social system of civility, which, in late seventeenth-century Britain, preserved without rigidifying a status hierarchy shaken by the Civil War.

As he describes the process by which he generated modern natural facts, Robert Boyle acknowledges that the experimental method is less methodical than Scholastic demonstration. Boyle's records of his experiments are detailed, they are filled with references to concrete particulars, and they are punctuated with the first-person pronouns and historical specificity that have become the marks of eyewitness reporting. These narratives also report accidents, interruptions, and breakages, however, any one of which would have imperiled both replication and the reduction of experimental procedures to a single method. In Boyle's description of his attempt to dissipate and reunite the parts of common amber, for example, he reports that when he "stepped aside to receive a visit," an assistant turned the heat up under the retort, causing it to break. Boyle tried the experiment again, "but, having been by intervening accidents hindered from finishing the experiment, we missed the satisfaction of knowing to what it may be brought at last."[40]

The methodological obduracy of experimental science was partly a function of the fragility of seventeenth-century instruments, the absence of anything resembling stable laboratory conditions, and the assistants' relative lack of
training in how to use the delicate equipment. Yet seventeenth-century natural
philosophy also lacked methodological rigor because there were theoretical
reasons for resisting it. In their desire to repudiate Aristotelian science even
more decisively than Bacon had done, the experimentalists rejected any effort
to limit in advance what they might find; and once one threw out all hypotheses, who knew what detail might prove relevant? In his *History of the Royal-
Society,* Thomas Sprat theorized the antimethodological stance of the experimentalists as critical to the knowledge they produced.

> In the order of their *Inquisitions,* they [the experimentalists] have been so free; that
> they have sometimes committed themselves to be guided, according to the seasons
> of the year: sometimes, according to what any foreiner, or English Artificer, being
> present, has suggested: sometimes, according to any extraordinary accident in the
> Nation, or any other casualty, which has hapned in their way. By which roving, and
> unsettled course, there being seldome any reference of one matter to the next; they
> have prevented others, nay even their own hands, from corrupting, or contracting
> the work: they have made the raising of *Rules,* and *Propositions,* to be a far more
> difficult *task,* than it would have been, if their *Registers* had been more *Methodical.*
> Nor ought this neglect of consequence, and order, to be only thought to proceed
> from their *carelessness;* but from a mature, and well grounded *premeditation.* For it is
> certain, that a too sudden striving to reduce the *Sciences,* in their beginnings, into
> Method, and Shape, and Beauty; has very much retarded their increase. . . .By
> their fair, and equal, and submissive way of *Registering* nothing, but *Histories,* and
> *Relations;* they have left room for others, that shall succeed, to *change,* to *augument,*
> to *approve,* to *contradict* them, at their discretion. By this, they have given *posterity* a
> far greater power of judging them. . . .By this, they have made a firm *confederacy,*
> between their own *present labours,* and the Industry of *Future Ages.* (*History,*
> 115–16)

As Sprat described it, such resistance to method constituted a desirable
form of flexibility, or decorum, critical to the *social* agenda that underwrote the
natural philosophical enterprise: the founding of an amicable society, in which
differences did not foment sectarian dissent. Viewed from another perspective,
the resistance to method, which was a *political* stance, made some *social* arrangement necessary, because once the Society rejected the epistemological grounds
for adjudicating truth claims (the argument from method), only social grounds
remained. The members of the Royal Society staked their *political* claim on an
appeal to nature: if nature, not the philosopher, was the ultimate arbiter of

truth, then whatever political or religious beliefs individual philosophers held would be rendered irrelevant by orderly investigations of natural phenomena. Members of the Royal Society staked their *social* claim on a double appeal—to the juridical domain and to the status hierarchy whose remnants were still upheld by law. Thus the Society insisted that natural philosophical investigations take place in the presence of reliable witnesses: collective witnessing made the production of truth a public act, and if numerous individuals observed the same experiment at the same time (or replicated it elsewhere and later), then collective witnessing would convert self-serving disputes into mutually accepted knowledge.[41]

For members of the Royal Society, of course, reliable witnesses were most likely to be gentlemen like themselves. Even though Sprat congratulated the Society for "freely admitt[ing] Men of different Religions, Countries, and Professions of Life" and for founding "a Philosophy of *Mankind*," once he began to describe these men, he had to acknowledge that one class of individuals was more suited to the Society's ideals than others, because one class was more credible than others.

> But, though the *Society* entertains very many men of *particular Professions;* yet the farr greater Number are *Gentlemen, free, and unconfin'd.* By the help of this, there was hopefull Provision made against *two corruptions* of Learning, which have been long complain'd of, but never remov'd: The *one,* that *Knowledge* still degenerates, to consult *present profit* too soon; the *other,* that *Philosophers* have bin always *Masters, & Scholars;* some imposing, & all the other submitting; and not as equal observers without dependence.[42]

By moving rapidly from the fact of exclusion to the benefits such exclusion conferred upon knowledge, Sprat downplays the extent to which the Society's much heralded consensus depended on the like-mindedness of its members *and* on the social fact that underwrote it: with a very few exceptions, the fifty-five members of the Society were all gentlemen—that is, "free" from the necessity to earn a living with their hands.[43]

Apologists for the Royal Society muted the class—as well as the gender and racial—biases intrinsic to their construction of credibility by insisting that the knowledge they produced was general knowledge, which equally concerned (if it did not equally emanate from) every individual who inhabited the natural world. Assumptions about the universality of "human nature" and "the nature of mankind," in other words, underwrote apologists' claim that the subject of science was a universal subject, differentiated only incidentally by class, gender, race, nationality, or religion. The strategy by which Sprat creates this universal subject resembles but is not identical to the strategy by which Hobbes creates a universal subject in *Leviathan*. Whereas Hobbes created a *representative subject position* that implies an abstract but universal nature, Sprat implies that the human nature that counts is the *material body,* which, like other parts of the natural world, is the proper subject of science. We might note in passing that neither Hobbes nor Sprat sought to understand the subjectivity of the universal subject. As we will see in the next chapter, conceptualizing subjectivity did not seem politically important until the demise of the sovereign mode of government, for only when individuals were allowed to govern themselves did it seem necessary to theorize how they did so.

Sprat's elevation of bodily nature over class, gender, race, and nationality, then, is of a piece with his desire to displace those differences that mattered most in seventeenth-century England: religious and political affiliations. To register the magnitude of Sprat's claim, we do not have to marshal our own sensitivity to differences like race and gender; we simply need to remember that he and his colleagues produced the universal subject of science in the wake of historical events in which some kinds of difference—religious beliefs and political affiliations—mattered so much that people killed each other, fragmented a church, and beheaded a king. In the context of the Restoration, and with the memory of the Civil War as immediate as one's youth, to proclaim that natural philosophy produced noncontroversial knowledge about a universal subject whose differences did not count was an explicitly polemical gesture. This gesture was fraught with political and religious implications; it staked its future on the utopian belief that contentious individuals could be persuaded that knowledge about nature was <u>more important</u> than beliefs about the proper forms of worship or the right way to govern a state.

To make the Royal Society seem like a model for such harmonious social relations, Sprat had to revise some of the details of its actual origins.[44] According to Sprat, many of the individuals who eventually constituted the Society first began to meet after the Civil War, in John Wilkins's rooms in Wadham College, Oxford. Although Sprat may have been correct as to time, place, and personnel, his descriptions of the philosophers' shared desire for some alternative to "the passions, and madness of that dismal Age," as well as his representation of natural philosophy as diversion or entertainment, simultaneously efface the theological radicalism that Wilkins and his friends had embraced in the 1640s and present natural philosophy as an alternative to "serious" knowledge—in other words, to theology and political theory.

> For such a candid, and unpassionate company, as that was, and for such a gloomy season, what could have been a fitter Subject to pitch upon, than *Natural Philosophy?* To have been always tossing about some *Theological question,* would have been,

to have made that their private diversion, the excess of which they themselves dis-lik'd in the publick: To have been eternally musing on *Civil business,* and the dis-tresses of their Country, was too melancholy a reflexion: It was *Nature* alone, which could pleasantly entertain them, in that estate. The contemplation of that, draws our minds off from past, or present misfortunes, and makes them conquerers over things, in the greatest publick unhapiness: while the consideration of *Men,* and *humane affairs,* may affect us, with a thousand disquiets; *that* never separates us into mortal Factions; *that* gives us room to differ, without animosity; and permits us, to raise contrary imaginations upon it, without any danger of a *Civil War.* (*History,* 55–56)

By ascribing civilizing and psychological functions to natural philosophy, and by contrasting natural philosophical knowledge to both theology and pol-itics, Sprat created a space for an alternative kind of knowledge—one that did not precipitate factionalism or sectarianism but instead answered what he rep-resented as a universal human need—the need for quiet, pleasure, and a sense of power over natural things, which, unlike beliefs or ideas, would submit to con-trol. In so arguing, Sprat set aside both the Aristotelian assumption that com-mon experiences and rhetorical demonstration automatically generated a consensus about self-evident truths and the deadly squabbling that had broken out in the wake of that paradigm's collapse. By the same token, when he iden-tified the meetings at Wadham as a setting in which participants had "room to differ, without animosity," he created a model for a kind of knowledge produc-tion that rejected the contests intrinsic to rhetorical debate and that did so specifically to foreclose the chance of civil war.

For the most part, then, what seems to late twentieth-century readers like a controversial claim—that the members of the Royal Society could speak for and represent a universal subject of science—seemed unequivocal to some late seventeenth-century readers. This is true not only because like-minded men claimed a virtual monopoly over the production of natural philosophical mat-ters of fact but also because it seemed so pressing to so many of Sprat's contem-poraries that some model be found for a form of sociality capable of producing consensus instead of war. Sprat's claim that "free and unconfin'd" gentlemen should be entrusted with the production of natural philosophical knowledge may also have attained the authority he sought because, in practice, members of the Royal Society did *not* limit knowledge production to gentlemen.

In 1673 Robert Boyle explicitly identified at least one other group whose members should be taken at their word—merchants. Recommending a paper on ambergris to the Royal Society, Boyle advised that his colleagues "look on this account, though not as compleat, yet as very sincere, and on that score

Boyle quoted in Dear. "Totius in verbe" 156

116   CHAPTER THREE

Credible, if you consider, that this was not written by a Philosopher to broach a *Paradox,* or serve an *Hypothesis,* but by a Merchant or Factor for his Superiors, to give them an account of a matter of fact."[45] While this statement obviously uses the merchant's relative lack of learning to criticize Aristotelian Scholastics, it also hints at the complex relationship that always obtained between claims that the Society's apologists made about themselves and statements they made about merchants. Examining Sprat's allusions to merchants in his *History of the Royal-Society* will enable us to see how natural philosophers underpinned their own authority not by the infallibility of the experimental method but, to a surprising degree, by the credibility of mercantile practices. This in turn will prepare us to see how theorists of the market system—that domain that was just becoming visible as another rival to politics and religion—enhanced *their* authority by modeling economic matters of fact on natural philosophical knowledge.

Like Boyle, Sprat identifies merchants as reliable witnesses, and like Boyle again, he initially associates merchants' credibility with their lack of knowledge (relative to philosophers or Schoolmen). Indeed, in Sprat's comment we can see how separating facts from theory, as Bacon also (intermittently) did, could serve a social function: drawing a distinction between facts and theory could permit social inferiors to help make natural knowledge without imperiling the status hierarchy that made some kinds of knowledge more authoritative than others. "Though they bring not much knowledg, yet bring their hands, and their eyes uncorrupted," he writes of merchants, and he recommends to his colleagues "such as have not their Brains infected by false Images; and can honestly assist in the *examining,* and *Registring* what the others represent to their view" (*History,* 72–73). Peter Dear has argued that Sprat associated the relatively low status of merchants with a form of disinterestedness,[46] but he does not notice that he links this disinterestedness to the transparency of mercantile instruments. Indeed, Sprat could represent the merchant as a reliable and disinterested instrument for recording natural matters of fact because, in numerical representation and double-entry bookkeeping, the merchant possessed devices designed to record economic transactions while making the writer seem to disappear. For this reason, the ability to "register" natural matters of fact "honestly"—that is, uncorrupted by preconceptions or theories—could be equated with the use of instruments that could generate the effect of transparency and therefore of impartiality, whether those instruments were merchants or the conventions of representation that merchants used. Thus disinterestedness, as a form of impartiality or superiority to personal interests, emerged not just in relation to landowners' freedom from manual labor, where we have been taught to find it, but also at the nexus of natural philosophy and trade. In this conjunction, disinterestedness was a concept in which an unequal social relationship—the natural philosopher's ability to adjudicate which facts were credible—was masked by the use of representational instruments that were borrowed from merchants and that seemed to efface the individual who reported the facts.

As recording instruments, Sprat claims, *English* merchants are particularly valuable to natural philosophers because, *despite* their low social status, in foreign countries they behave like gentlemen—that is, they act as if they are superior to their financial interests. To emphasize "the *Noble, and Inquisitive Genius* of our *Merchants,*" Sprat contrasts the demeanor of English merchants to that of their Dutch rivals: "The *Merchants* of *England* live honourably in forein parts; those of *Holland* meanly, minding their gain alone: ours converse freely, and learn from all; having in their behaviour, very much of the *Gentility* of the Families, from which so many of them are descended. . . .Of the *English Merchants* I will affirm, that in all sorts of Politeness, and skill in the *World*, and *humane affairs*, they do not onely excel them, but are equal to any other sort of men amongst us" (*History*, 88). This ability to ape gentility *when abroad* makes merchants the perfect vehicles for that international correspondence on which the Royal Society based its claim to produce universal knowledge: "There will scarce a Ship come up the *Thames,*" Sprat confidently asserts, "that does not make some return of *Experiments*, as well as of *Merchandize.*"

Sprat's acknowledgment that the authority—or at least the reach—of the Royal Society depended on the willingness of merchants to ferry information and experiments is always qualified by his ability to put merchants in their place. After all, while it may have been important to acknowledge the contribution of others so as to minimize class resentment, it was also crucial not to cede too much authority to a profession that, within England, was waging its own campaign for the monarch's patronage. Thus he repeatedly reminds his readers that merchants are gentlemen only when abroad (and then only pseudogentlemen), that they are really only the philosophers' instruments, and most significant, that they have yet to realize that their mercantile instruments have the potential to register something more meaningful than trade.[48] When Sprat describes the ecumenical agenda of the Royal Society, he offers just such a backhanded compliment to merchants by predicting that members of the Royal Society will create the institutions that merchants can only squabble over. In applauding the Society's desire to bring "Men of different Religions, Countries, and Professions of Life" into a community devoted to consensus, he specifically compares the outcome with mercantile institutions that remain blocked by merchants' inability to cooperate: a national bank and a port that levies no tariffs on trade. "By their *naturalizing* Men of all Countries, they [the members of the Royal Society] have laid the beginnings of many great advantages for the future. For by this means, they will be able, to settle a *constant Intelligence,* throughout all civil

Nations; and make the *Royal Society* the general *Banck,* and Free-port of the World." "A policy, which whether it would hold good, in the *Trade* of *England,* I know not," Sprat adds; "but sure it will in the *Philosophy*" (*History,* 64).

Sprat's ambivalence toward merchants and their instruments surfaces again in the *History of the Royal-Society* when he addresses the question of *style.* Indeed, his much-analyzed discussion of the Society's writing and speaking style appears at a critical point in the *History,* precisely where he leads the reader to expect a description of the experimental method. Instead of offering an account of method, however ("of . . . the Method . . . I shall shortly speak in another place"), he discusses at length, and with uncharacteristic passion ("for now I am warm'd with the just Anger, I cannot with-hold my self") the Society's repudiation of copious rhetoric. Sprat associates "vicious abundance of *Phrase,* this trick of *Metaphors,* this volubility of *Tongue*" with the worst abuses of his time; so heinous is "the luxury and redundance of *speech*" in his view that "it may be plac'd amongst those *general mischiefs;* such, as the *dissention* of Christian Princes, the *want of practice* in Religion, and the like" (*History,* 109, 111, 112, 113). To preclude the atrocities that follow from this immoral style, Sprat continues, the Royal Society has willingly adopted the self-restraint that he associates with merchants.

> They [the members] have therefore been most rigorous in putting in execution, the only Remedy, that can be found for this *extravagance:* and that has been, a constant Resolution, to reject all the amplifications, digressions, and swellings of style: to return back to the primitive purity, and shortness, when men deliver'd so many *things* almost in an equal number of *words.* They have exacted from all their members, a close, naked, natural way of speaking; positive expressions; clear senses; a native easiness: bringing all things as near the Mathematical plainness, as they can: and preferring the language of Artizans, Countrymen, and Merchants, before that, of Wits, or Scholars. (*History,* 113)

This famous passage has typically been interpreted in the context of other seventeenth-century projects to reform language or create a universal character.[49] Certainly the *History* was a contribution to the program of language reform, for Sprat, like contemporaries such as John Wilkins, wanted to reduce the political and religious conflicts that had provoked a civil war to the more manageable problem of language use. While this is a meaningful way to read Sprat's *History,* however, for my purposes, it is equally crucial to see how he enlisted merchants in this project. In his account, merchants somehow operate outside, or even before, political and religious controversy; they are "primitive" and trade words parsimoniously, like things or money; their speech is "naked," "natural," "native," and most important, it is characterized by "Mathematical plainness." In this passage Sprat carefully qualifies the status he grants merchants (he ranks them with "Artizans" and "Countrymen"), but he uses their relative indifference to politics and religion as a *positive* trait, which makes their practice the model not only for the kind of transparency he associated with numerical representation but also for the consensual society he sought to create.

In the late seventeenth century as before, many landowning gentlemen reviled merchants, precisely because their investment in international commerce and liquid capital did *not* give them more stake in England's well-being than in their own. In this context, it seems paradoxical that Sprat celebrated merchants as the ideal citizens of a new civil society. There are two ways to interpret this paradox. First, merchants did enjoy at least some of the professional advantages the Royal Society was intended to institutionalize. Like the Royal Society, many mercantile organizations were granted monopolies by the king, and as with the Society's knowledge, the benefits merchants conferred on their countrymen were considered to be indirect—the consequence of a gain that was not conceptualized specifically in political terms or generated for the immediate use of the king. Indeed, even though Sprat mentions it nowhere in his *History,* mercantile accommodation might well have been the prototype for the conventions to which members of the Royal Society voluntarily submitted, just as it was explicitly the forerunner of the "general Banck" that Sprat wanted the Society to become. Certainly, numerical representation—the linchpin of the kind of knowledge merchants both produced and promoted—was the form of representation that he idealized for natural philosophers too. To so idealize numerical representation, of course, enhanced the impression already created by mercantile instruments like double-entry bookkeeping: that numbers were a transparent window onto the world of things, that this "naked" way of writing averted all the problems associated with that copious practice to which some critics had reduced rhetoric by 1667, and that numbers transcended politics and "interest" altogether.

The second way to interpret the begrudging respect Sprat accords merchants is to place it in the context of another vicissitude in the complex genealogy of interests and disinterestedness. Even though the members of the Royal Society prided themselves on being able to contribute to the nation's stock of knowledge because they were disinterested, in the sense of being superior to economic concerns, Sprat seems to have registered that merchants, a group that was interested in a specifically economic sense, also—if inadvertently—contributed more to the nation's well-being than natural philosophers currently did. Merchants made their contribution not just by paying excise taxes and not just by circulating the "universal" knowledge the Society was trying to produce, but by modeling a form of sociability that *equated* collective interests with individual interests. This, of course, is precisely what the Royal Society aspired to do, and Sprat's backhanded tributes to merchants may signal his recognition that disinterest might come from this kind of interest as well as from owning land and making science.

I will take up this strand of the genealogy of interests in the next chapter, for it was not until the eighteenth century, and in relation to the emergence of civil society and the elaboration of theories about subjectivity, that a form of self-interest that was specifically economic began to be elaborated as a constructive—indeed, the paradigmatic—contribution to national well-being. For much of the rest of the seventeenth century, English merchants continued to be reviled and the language of interests continued to be generally, though not exclusively, political. It was partly for this reason, of course, that Thomas Sprat characterized the members of the Royal Society as "uninterested"; this was the pledge of their indifference to politics and thus their talisman against those conflicts that continued to simmer around the newly emboldened Parliament and the newly restored king. We can see just how controversial economic self-interest continued to be throughout the rest of the century—and in spite of the continuing efforts of merchant apologists—by turning to William Petty. Petty drew upon the example of natural philosophical matters of fact to authorize another kind of knowledge, because the knowledge he wanted to produce had everything to do with money and with his own monetary interests in particular. Indeed, William Petty's economic theory can be read as an elaborate attempt to offset the charge of interestedness, which was levied because Petty made a fortune drawing a map of Ireland.

WILLIAM PETTY, IRELAND, AND ECONOMIC MATTERS OF FACT

William Petty arrived in Ireland in 1652, in the capacity of physician to the lieutenant-general of Cromwell's conquering army. Within three years this son of a Hampshire clothier had acquired nearly nineteen thousand acres of Irish land, some of which was given to him in lieu of salary, and some of which he was able to purchase from the soldiers to whom the land had been granted because, by the definitions Petty had written into law, it was "unprofitable."[50] Although the land might have been "unprofitable" under the law, it was the primary source of Petty's considerable fortune. In 1652 his total assets had been less than £500; by 1685 his annual income, which came primarily from his now fifty thousand acres of Irish land, was about £6,700.[51]

The path by which Petty achieved such riches in the Irish land market reflects the opportunities opened by the social turmoil of mid–seventeenth-century Britain.[52] Disdaining his father's profession, Petty had signed on as a cabin boy at age fourteen in hopes of finding adventure and advancement. Put ashore in France ten months later with a broken leg, he was taken up by Jesuits in Caen and enrolled in the university there, where he studied languages and mathematics. Petty eventually returned to England and joined the Royal Navy, but, along with many of his countrymen, he fled in 1643 when the English Civil War disrupted the customary routes to professional success. On the Continent, where he studied medicine, anatomy, and mathematics, he was befriended by the mathematician John Pell, who introduced him to Hobbes; Hobbes in turn brought Petty into the Mersenne circle in Paris, which included Descartes and Gassendi. On returning to England in 1646, Petty was introduced to Samuel Hartlib and, through Hartlib, to Robert Boyle. He became a member of the London Philosophical Society then, in rapid succession, a fellow at Brasenose, deputy to the university professor of anatomy, vice principal of Brasenose, professor of anatomy at Oxford, and professor of music at Gresham College.

Petty was apparently not content with an academic career, for in 1651 he obtained two years' leave from his university positions to take up the post of physician in Cromwell's army. Within two years he had shifted directions again, having petitioned to take over the project of mapping the nearly eight million acres of Irish land that Cromwell claimed under the Act of Settlement. This task had already been commissioned to Benjamin Worsley, but Worsley's "gross" survey—which consisted of a written description, probably drawn up largely by guesswork, of the "grosse surroundes" of the confiscated lands— soon proved unsatisfactory; among other reasons, it did not distinguish usable land from wasteland.[53] Making this distinction was necessary because much of the confiscated land was to be assigned to the soldiers as compensation for their service and to the so-called adventurers who had helped finance Cromwell's retaliation against the Irish rebellion of 1642.[54]

The survey of Ireland that Petty took over was one part of the contentious process that gradually forged what was to become the "united kingdom" of Great Britain. Some historians have emphasized the extent to which this undertaking helped eradicate Ireland's indigenous culture, while others have presented the seventeenth-century appropriation as "a gigantic experiment in primitive accumulation."[55] However one interprets it, the details of Petty's contribution are clear: within thirteen months, and with the help of modern surveying instruments and purpose-trained soldiers, he completed a survey and drew up maps of twenty-nine counties, which included about five million acres. Although the Down Survey was not a general map of the country, Petty did include topographical as well as cadastral information, in hopes of earning permission to make a more comprehensive map in the future.[56]

Because the primary purpose of the Down Survey was to settle debts, Petty was required not only to measure and record the cadastral and topographical features of the landscape, but also to create some basis for assigning monetary value to the confiscated property. It was in order to do so that he distinguished between "profitable" and "unprofitable" lands. By his definitions, the former included "arable, meadow, and pasture" lands, and the latter consisted of "wood, bog, and mountaines."[57] While this distinction enabled him to assign monetary values to the Irish acres, however, it immediately introduced problems, first for his surveyors, then for Petty himself. The problems arose because qualitative description did not depend on—and thus could not be referred to—instruments; as a consequence, it seemed to resist method, or "rule," and as an extension of this resistance, it was susceptible to the charge that the person who made the judgment was interested in its results. Quantities, Petty's surveyors complained, could easily be justified by reference to instruments and rules, but quality knew no instrumental measure. "As for the quality of land," moaned the surveyors Smith and Humphreys, "wee had noe rule to walke by, only as aforesaid, but did according to the best of our judgements, and the best information wee could get."[58] This early lesson—that quantification was different from qualitative descriptions *in being less subject to controversy or dispute*—was to remain with him throughout the rest of his varied career.

By 1659, when Petty probably composed his *History of the Down Survey,* the pitfalls associated with assigning interpretive descriptors (like "profitable" and "unprofitable") to quantifiable physical entities (like land) were driven home to him, for the difficulties his surveyors had experienced had mushroomed into legal accusations against him. In 1658 a series of anonymous letters caused the lord lieutenant in Dublin to appoint an investigative committee to inquire into the charge that Petty had profited unfairly from his work. The deliberations of this committee were interrupted in April 1659, however, when he was charged again, this time in London by Sir Jerome Sankey. Petty was ordered to appear before Parliament, but this proceeding was also interrupted when Parliament was dissolved in the confusion surrounding the fall of Richard Cromwell. When the Long Parliament was assembled in May, Sankey filed articles of impeachment against Petty, charging that he had taken bribes, had profited unfairly from his official position, and had represented profitable lands as unprofitable in order to buy them more cheaply. Parliament took these charges seriously enough to refer them to the commissioners who were managing Ireland, but because Parliament was dissolved again in October, no action was taken. In December 1659 Parliament was reconvened, but with the country in a state of virtual anarchy, with various factions vying for power and a population refusing to pay taxes, his case was pushed aside for more pressing concerns. In March 1660 Parliament dissolved itself, after issuing writs for the

Convention Parliament, and in April the monarchy was restored. After the Restoration, we hear no more about Petty's threatened impeachment.[59]

Having completed the Down Survey, Petty turned to the two undertakings that were to occupy the rest of his professional life: his contributions to the scientific projects of the newly chartered Royal Society and his largely theoretical writings about (what we call) economics.[60] In what follows I focus on the latter, for in his economic writing we begin to see how he enlisted aspects of both the experimental method and Hobbes's deduction to create a variant of the modern fact that government officials (although not in Petty's lifetime) would increasingly find attractive for policy formation. The kind of fact Petty promoted differed from the Baconian fact, however, in being neither an observed particular nor a "deviating instance." Petty's facts were conjectural rather than observed, and they described abstractions rather than historical events. Despite these striking differences, he claimed for his facts the same degree of epistemological authority that members of the Royal Society claimed for experimental facts, but he did so based not on collective witnessing but on a peculiar mixture of claims about the precision of numerical representation and the impartiality of expert interpretation. By representing expert interpretation as superior to personal interests, Petty helped forge the relationship between numbers and impartiality that has made the modern fact such a crucial instrument for policy-making. In the complex amalgam he created from experimental philosophy and Hobbesian deduction, expertise linked particulars that seemed to be (but were not) observed to theories that seemed not to be (but were) interested, for his representation of expertise made interpretation (and interest) seem incidental to method and instruments.

Petty's experiences in Ireland—both his firsthand observations about acreage and profitability and the accusations of malfeasance that accompanied his success there—played a complex role in the constitution of this variant of the modern fact. On the one hand, his personal experience in Ireland—and particularly his experience of measuring so much of this country—enabled him to refer to Irish incomes and expenditures, population size and wealth, with an authority that seemed to derive from eyewitnessing and the precision of instrumental measure. As a result of his firsthand experience, he could represent Ireland as a kind of laboratory, where (he claimed) economic "experiments" could be made to yield usable results.[61] On the other hand, however, because Petty's adventures in Ireland had generated both a considerable personal fortune and interminable disputes over taxation, which only the king could settle, it was inevitable that his critics would malign as self-interested any policy recommendations he advanced. It was to negotiate these complex waters that he forged the link between personal experience, mathematics, and impartiality that made his experience in Ireland seem both essential and incidental to the kind of knowledge he produced for the king. Numerical representation was critical to this link, because the credibility of numbers that purported simply to reflect what had been counted was enhanced by firsthand experience, while the precision of "computing" seemed to efface the personal interests of the person who made knowledge from numbers.

Petty's attempts to yoke numerical representation to impartiality gradually altered the form in which modern facts were produced and consumed, especially by governments, which now routinely seek knowledge that seems rigorous, uniform, communicable, and immune from the need for both intimate knowledge and personal trust.[62] Precisely because the relation between numerical representation and impartiality has been naturalized in the decades following Petty's work, we need to recover the stages by which this connection was established. We have to remember that when Petty took up the subject of taxation in 1662, it was not self-evident that numerical representation constituted the sign of impartiality, nor was it obvious that governments should prefer numerical facts about population and domestic production to private communications about religious or political intrigues. Indeed, despite Petty's efforts to convince Charles II and James II that quantifying the losses incurred in Ireland would ground policies capable of enhancing England's greatness, the kings remained more interested in plots brewing on the Continent than in all the numbers Petty could devise. Because his numbers were not actually descriptive, moreover—because they were not typically derived from counting— and because his computations were not consistently mathematical, it was not obvious that he was yoking numerical representation to the rigor of mathematics, much less that his numbers offset the interests he undeniably had in the policies he recommended. To understand how numerical representation gradually acquired the connotations of impartiality and rigor, so that governments (and individuals) came to invoke numbers to settle disputes, we need to see how Petty tried to turn the numerical fact into a neutral instrument by which taxpayers could be persuaded to pay and kings could be assured dominion.

Despite his considerable experience in Ireland, Petty did not substantiate the policies he advanced early in his campaign to win the king's attention by first-person references to his personal experience. In this sense, as in others, he departed from the authenticating practices of Boyle and the Royal Society. Nor, in his early economic writing, did he rely heavily on numbers. Instead of firsthand, first-person accounts of historical events or the numbers that would characterize his later writing, Petty's *A Treatise of Taxes and Contributions* (probably composed in 1662) features lists of theoretical assertions, which range from the causes of Ireland's resistance to taxation, to proposals for increasing revenue collection, to discussions of the penalties the king might impose for tax evasion. His use of these conventions makes *A Treatise of Taxes* seem more like a political theoretical treatise than a report of experimental natural philosophy, and because it makes no reference either to his personal experience in Ireland or to how the policies he recommends would affect his property there, *A Treatise of Taxes* seems more interested in the fiscal well-being of the king and his subjects than in the financial health of William Petty.

Modern readers should not be taken in by Petty's decorum, however, for many of the recommendations he advanced in his *Treatise* would have strengthened his personal claims in Ireland. In 1662 Petty might well have wanted to strengthen these claims, because even though Charles II had pardoned him for his service to Cromwell, awarded him a knighthood, and secured to him by royal letter all of his Irish holdings, his investment in this property was still imperiled both by the insecurity of Irish land registration and by the system of land assessment and tax collection known as tax farming. Petty's proposals that the government establish a land registry capable of securing private titles, that the king levy a regular and equitable tax based on knowledge of the land and its value (so assessment would not be left to the tax farmers), that a survey of the land be conducted to determine its exact value, and that an excise tax be instituted capable of both generating revenue and making record keeping easier can thus be seen as remedies directly applicable to his own situation in Ireland.[63] It was partly to ward off the accusation of self-interestedness (which, remember, had already surfaced in relation to his map) that Petty did not deploy the first-person narration characteristic of natural philosophical writing. By casting his policy suggestions in the form of reason-of-state political theory, which highlighted the nation's (England's) interest and did not allude to his own, he sought to convince the king that, because he seemed to have no personal interests, he could recognize the nation's interests better than other advisers could.

By emphasizing Petty's interest in the recommendations he advanced, I do not mean to diminish the theoretical importance of his ideas. Nor do I want simply to shift our critical gaze from effects to intentions, as if knowing that Petty's policies would have secured his lands in Ireland could tell us something that his contemporaries did not see. I *am* arguing that the formal strategies by which he tried to efface his own interests are important, but I not suggesting either that these strategies fooled his contemporaries or that recovering his motives (if that could be done) would somehow discredit his considerable impact on theories of government and economy. Indeed, from the perspective of effects, Petty's motives are irrelevant. From this perspective, which I want to adopt for just a moment, we can see that his recommendations contributed simultaneously to the reconceptualization of government that had been under way since the early part of the century and to the appearance of a domestic economy, which had not been visible before the Civil War. In recommending that the king establish some method for keeping records—about landowner-ship, domestic consumption and production, taxation, and national productiv-ity—Petty thus advanced Bacon's adage that a strong government needed good information and reinforced the idea that information about domestic produc-tion was essential to national strength.

Because these ideas have become commonplace, let me once more register the novelty that characterized them in 1662. Petty's insistence that the govern-ment keep records about domestic consumption, production, trade, and popu-lation constituted an *interpretation* of both government and wealth; his proposals did not simply conform to a logic of government centralization that was some-how unfolding in 1662, nor did they reflect some natural or necessary link be-tween the growth of the state and information collection. Government record keeping was not altogether unheard of in 1662, of course. Some kinds of gov-ernment records had been routinely kept for some time. Preeminent among these were customs records, which were both relatively easy to collect and use-ful, because customs duties had long proved a more reliable source of Crown revenues than either excise taxes or Parliament. The British government seems not to have assumed that such information could guarantee revenue, however; customs records were not centralized until 1671, and the first official depart-ment charged with keeping track of trade was not established until 1695.[64] By contrast with the customs, the English government devoted relatively little en-ergy to documenting domestic production, and what efforts it did make were often thwarted by small-scale manufacturers, who located cottage industries at the borders of counties or townships to escape the tax farmers.[65] The nature and extent of local economies began to become visible only during the English Civil War, when commissioned militiamen traveling through the countryside brought back reports of these industries. Domestic production at a *national* (English) level began to seem more consequential than local production for consumption, moreover, only when the commonwealth, then the monarchy, was able to mobilize the idea of national economic potential as a rallying cry to counter the localist tendency of the 1640s. These factors help explain why mer-chant apologists like Misselden and Mun would not have been able to relate their accounts of international trade to accounts of domestic production, even had they wanted to. It also explains why the modern concept of a "national economy" was not available in England before the kind of records Petty rec-ommended had been kept for some time.

By the same token, Petty's emphasis on domestic production did not sim-ply reflect the increasing importance of agriculture in Britain's overall wealth.

Just as his idea that good government depended on accurate records was part of a relatively new theory of government in 1662, so his idea that national wealth derived from domestic production, and thus from labor, constituted an intervention in traditional ways of thinking about wealth. It is tempting simply to assume that Petty's emphasis on labor reflected the increasing importance of British agriculture, because, as economic historians have told us, England's agricultural superiority was enhanced during the seventeenth century by the introduction of various forms of convertible husbandry.[66] Despite considerable agricultural improvements, however, in 1662 as in 1620, the most spectacular sector of England's economic activity was still overseas trade, not farming. Economic historians also tell us that England began to overtake Holland in exports, largely because of the considerable advantage England enjoyed in shipping.[67] Thus, for different reasons, contemporary analysts in 1620 or 1662 might have stressed either agricultural production or commerce, although the latter was most obviously linked to England's "greatness" through the dual role that ships played in commerce and in war. That Petty elevated agriculture did not reflect the decreased prominence of commerce or war, then, but was of a piece with his desire to extend the purview of official records.

In 1662 the case Petty made for keeping official records was directed primarily toward taxpayers, whose resistance to the Crown's collection of information seemed the first obstacle to the improvement he envisioned. To make this case, he offered two kinds of arguments: that if the sovereign could be assured that he would eventually be able to collect assessed taxes, it would be to his advantage to keep this money in circulation as long as possible by letting his subjects trade and produce; and that even money given to the king in taxes constituted *national* riches, which would come back to the king's subjects eventually, though possibly in some other form. With the last argument, Petty urged the king's subjects to consider the nation an economic entity that represented individuals *collectively,* even if at any given moment particular individuals might experience a loss. "The Money leavied not going out of the Nation, the same also would remain as rich in comparison of any other Nation; onely the Riches of the Prince and People would differ for a little while, namely, until the money leavied from some, were again refunded upon the same, or other persons that paid it: In which case every man also should have his chance and opportunity to be made the better or worse by the new distribution; or if he lost by one, yet to gain by another" (*Economic Writings,* 32).

In Petty's account, what some saw as a contest between the Crown and the people becomes a common effort directed against other nations, and what could look like a game of chance becomes a circulation of wealth that seemed equitable—as long as the money did not leave the country and if one simply waited for one's turn to come around. Calling into imaginative existence a national alliance that included the king and a confederacy of taxpayers, Petty could then emphasize the importance of making the tax burden fair. This is the point at which the need for exact information enters the equation. In his argument, equitable taxation depends upon "certain knowledge of [every man's] Wealth or true Estate," which obviously requires information about numbers of acres (*Economic Writings,* 53). Establishing the importance of gathering such information was only one stage of Petty's argument, however, for the information he desired was of a very particular kind. Certainly, if exact information was readily available—"if every mans Estate could be alwayes read in his forehead," as he humorously suggests—then trade would advance, individuals would prosper, and the nation would continue to grow in riches and strength (*Economic Writings,* 53). But in order to write the value of a man's estate on his forehead, something more than the numbers of his acres would have to be known. Because Petty included the price of the labor necessary to make the land profitable in his valuation of "a man's estate," he could insist that "computations [about a man's estate] are very hard if not impossible to make" (*Economic Writings,* 52). Because he asserted that these computations are so difficult, in turn, he could also insist that the king employ experts who would be able not just to gather numerical information but to interpret it, so as to compute "value" from it. Petty's emphasis on expert computation demonstrates why the economic facts he endorses could never have been the deracinated, theory-free particulars natural philosophers claimed to collect. It also explains why, no matter how we interpret his motives for making these particular recommendations, he stood to profit from making economic expertise part of the production of economic matters of fact.

Even though we cannot know Petty's motives, then, and even though it might be best to measure the importance of his recommendations by their effects, not their origins, it is critical to recognize that the way he formulated his policies also served a personal agenda: his desire to extend the purview of official records, and numerical records in particular, was a bid to elevate a form of expertise that could be demonstrated by numerate men such as himself over the experience of merchants, who were the best spokesmen for Britain's spectacularly successful overseas trade.[68] Petty linked impartiality to numbers to enhance the authority of one kind of experience over another, but we must remember that the impartiality he associated with numbers both implied and entailed interpretation—of how to use and understand the numbers themselves. In place of the eyewitness testimony that Boyle and his colleagues endorsed or the variant of hands-on knowledge that Misselden and Mun had celebrated, Petty recommended "computations" as the necessary, if difficult, correctives to experience that could be said to be self-serving.

Even in the *Treatise of Taxes and Contributions,* which, by contrast with his subsequent economic writings, does not rely heavily on numerical representation, we can see that Petty installs interpretation as the critical component of making knowledge from numbers. His primary object in the *Treatise* was to find some way to assess both land and labor, so that both could be equitably taxed. Typically, land and labor would have been assigned a monetary value for purposes of taxation, but Petty did not want to use money as a universal equivalent, because he thought that its usefulness as a measure of value was compromised by its other role, as a repository of wealth; if silver fluctuated in value according to its relative scarcity and fineness, then how could it be used to measure something else?[69] Instead of money, Petty wanted to make land and labor themselves the measures of value, because they seemed like "natural Denominations."[70] To do so, he had to find some way to make these two measures comparable, so that values expressed in one would make sense in terms of the other as well. To render land and labor comparable, Petty established a method for determining what he called a "natural Par" between the two.[71] This method includes a number of stages, some of which seem to involve consulting records (about the rent collected on particular lands, for example), some of which involve guesses, but all of which yield numbers that required both interpretation and calculation.

> Having found the Rent or value of the *usus fructus per annum,* the question is, how many years purchase (as we usually say) is the Fee simple naturally worth? If we say an infinite number, then an Acre of Land would be equal in value to a thousand Acres of the same Land; which is absurd, an infinity of unites being equal to an infinity of thousands. Wherefore we must pitch upon some limited number, and that I apprehend to be the number of years, which I conceive one man of fifty years old, another of twenty eight, and another of seven years old, all being alive together may be thought to live. (*Economic Writings,* 45)

This complex statement requires close attention, for even though Petty seems to derive the number of "years purchase" that a fee simple is worth from "nature," actually he simply provides an interpretive figure: "Wherefore I pitch the number of years purchase, that any Land is naturally worth, to be the ordinary extent of three such persons their lives. Now in *England* we esteem three lives equal to one and twenty years, and consequently, the value of the Land, to be about the same number of years purchase" (*Economic Writings,* 45).

None of the numbers Petty uses here could have been accurate, of course; in 1662 no one could have known the average life span of agricultural workers, because nothing remotely resembling modern techniques for gathering such information had been applied outside urban centers.[72] Nevertheless, the numbers serve his end, which was to create a mathematical *formula* for computing the value of land. This formula, and *not* Petty's experience, was to be the basis for his authority, for the formula could draw on the epistemological connotations of certainty associated with mathematics without directly raising the issue of either experience, which was considerably more troubling for a man who wanted to efface the extent to which his experience had yielded monetary rewards, or interpretation, which could always be called self-interested.

In Petty's second set of economic treatises, written a decade later, we see more clearly how he used numerical representation, arithmetical calculation, and mathematical formulas to generate knowledge about fiscal matters that, not incidentally, offset the charge that self-interest motivated his policy recommendations. In these texts we can also better see what kind of numbers he used. On the one hand, as we have already seen, these numbers were not deracinated particulars, they were not derived from measurement or counting, and they could not have been accurate, because the conditions necessary to easy measurement and counting of the entities Petty wanted to describe did not exist. On the other hand, however, these numbers do not belong to what we would recognize as mathematical models; they did not simply model the world but claimed to be derived from it. While some of Petty's numbers do describe entities that could have been counted, the chimneys in Ireland, for example, some of his other numbers cannot even be said to describe entities that existed as things in the world—"the state of the people" or "the value of people" (*Economic Writings,* 270, 454). These numbers, which constitute the characteristic form of Petty's economic matters of fact, describe abstractions that have been brought into being by a method; this method in turn has been designed to create knowledge about something that exists only as an effect of the method—that is, as part of a theory. The numbers he offered to describe the "state of the people" or "the value of people," in other words, constitute evidence for a theory about what "the people" was and how it should be represented and understood—whether or not these numbers accurately reflected actual people in a real world.[73]

The best way to grasp the epistemological nature of these theory-dependent, theory-producing facts is to look at one of Petty's numerous efforts to establish the "value of people."[74] In the instance that appears in his *Political Arithmetick,* his computations take him effortlessly from the aggregate to the individual. From this we see that he does not value particulars for their singularity, nor does he simply create the aggregate by adding the particular individuals together. Instead, what looks like an aggregate is actually an abstraction, which he constructs not by arithmetic but by generalization; what looks like an individual is also a theoretical construct, but it is generated by a combination of arithmetic—in this case, division—and theory (which leads him to value male and female adults equally but at twice the rate of children). Notice also that the values of both "the people" and "the individual" are based on assertions, which are conjectures or estimates, and whose accuracy matters less than the precision of the computation. Finally, note the centrality—although not the rhetorical prominence—of Petty's foundational theoretical proposition: that the "value" of human beings should be figured in monetary, not religious or ethical, terms.

> Suppose the People of *England* be Six Millions in number, that their expence at 7 *l. per* Head by forty two Millions: suppose also that the Rent of the Lands be eight Millions, and the profit of all the Personal Estates be Eight Millions more; it must needs follow, that the Labour of the People must have supplyed the remaining Twenty Six Millions, the which multiplied by Twenty (the Mass of Mankind being worth Twenty Years purchase as well as Land) makes Five Hundred and Twenty Millions, as the value of the whole People: which number divided by Six Millions, makes above *80 l.* Sterling, to be valued of each Head of Man, Woman, and Child, and of adult Persons twice as much; from which we may learn to compute the loss we have sustained by the Plague, by the Slaughter of Men in War, and by the sending them abroad into the Service of Foreign Princes. (*Economic Writings*, 267)

The last part of this computation reveals why this method might be useful to a king: it would enable the king to devise policy according to what we would call a cost-benefit analysis, by weighing the expense of disease prevention, for example, against the cost of an unresisted plague.

Such numbers were useful, moreover, because the policies they could be used to justify seemed impartial; these policies seemed impartial, in turn, because the method by which the final numbers were apparently generated was the rule-bound method of mathematics, which had nothing to do with politics or religion. Even if Petty did not actually compute the "value of people" strictly with mathematical tools, his language, which implies that he did, constituted a bid for the impartiality of numbers. Just as Hobbes used the trope of mathematics to hold the place of certainty in *Leviathan,* so Petty used the trope of mathematical method to promise impartiality in most of his economic writing.[75]

Although the mathematical method constituted a visible basis for claiming impartiality, both for numbers and for Petty's interpretations, by the 1670s Petty had begun to make it clear that (apparently) impartial computation still required something else: before the expert analyst could compute, he needed numbers that were—or seemed to be—accurate. Without such credible numbers, after all, the expert computer might still be subject to the charge that his conclusions simply confirmed theoretical (and possibly self-interested) positions. To generate credible numbers, he began in the 1670s to lobby the king to finance "experiments," which, like Boyle's exploits with the air pump, could be trusted to produce reliable data.[76] Asking the king to sponsor experiments, however, instead of turning to a society of like-minded gentlemen, placed Petty in the position of the suppliant philosopher, which Hobbes had also occupied; and this position rendered Petty's promise to generate impartial knowledge—like Hobbes's promise to utter political truths—dependent on the king to whom he offered what he claimed were impartial—apolitical—facts. Just as Hobbes needed Charles I to enforce the peace that would restore the social ground of certainty, so Petty needed Charles II to make his numbers credible by guaranteeing that experiments could be conducted. Needing Charles's patronage meant that Petty could not be free from political interests—no matter how immune from politics was the method he offered the king.

In his *Political Arithmetick,* which was probably composed in the early 1670s, Petty invoked the authority of eyewitnessing that he had rejected in *A Treatise of Taxes and Contributions,* but he did so in the context of a deductive method that resembles Hobbes's reckoning more than Boyle's experimentalism. This method, as he describes it in the preface to *Political Arithmetick,* was new, impartial, exact, and therefore useful to a king anxious to place his greatness beyond political intrigue. This method, Petty explains,

> is not yet very usual; for instead of using only comparative and superlative Words, and intellectual Arguments, I have taken the course (as a Specimen of the Political Arithmetick I have long aimed at) to express my self in Terms of *Number, Weight,* or *Measure;* to use only Arguments of Sense, and to consider only such Causes, as have visible Foundations in Nature; leaving those that depend upon the mutable Minds, Opinions, Appetites, and Passions of particular Men, to the Consideration of others: Really professing my self as unable to speak satisfactorily upon those Grounds (if they may be call'd Grounds), as to foretel the cast of a Dye; to play well at Tennis, Billiards, or Bowles, (without long practice,) by virtue of the most elaborate Conceptions that ever have been written *De Projectilibus & Missilibus,* or of the Angles of Incidence and Reflection.[77]

Like Bacon and Sprat, Petty describes style where one might expect him to describe method. But his argument for style is not an evasion of method. Instead, by distinguishing his style—which features numbers—from rhetorical ornament, he is making a claim about method: like that of Bacon and Boyle, Petty's style embodies a method that appeals to the "visible Foundations in Nature," and *not* to "mutable Minds, Opinions, Appetites, and Passions." Petty's style, in other words, may place numbers where Bacon and Boyle placed elaborate narrative descriptions, but it does so to achieve the same ends: to elevate what can be seen over what can simply be spoken, to privilege knowledge that is exact over appeals to interest.

Although this paragraph seems to align Petty's method with Baconian induction, however, the next paragraph, which is rarely quoted with the first, aligns it with Hobbes's deduction. In this paragraph he claims more than the experimentalists were willing to do; he claims that his method can produce *certain* knowledge—as mathematics does—*if* the king accords him the authority (and money) necessary to conduct the experiments necessary to produce the numbers.

> Now the Observations or Positions expressed by *Number, Weight,* and *Measure,* upon which I bottom the ensuing Discourses, are either true, or not apparently false, and which if they are not already true, certain, and evident, yet may be made so by the Sovereign Power . . . , and if they are false, not so false as to destroy the Argument they are brought for; but at worst are sufficient as Suppositions to shew the way to that Knowledge I aim at. (*Economic Writings*, 244–45)

In this passage Petty is positioning his new method in relation to *both* of the epistemological instruments that he considered superior to rhetoric: experimental natural philosophy and Hobbes's deduction. Like Boyle's natural philosophy, Petty's method will require experiments. The "Argument they are brought for" is the argument that experiments are necessary. Like Hobbes's demonstrations, however, Petty's method could produce knowledge that is "true, certain, and evident"—if the sovereign power was willing to allow him to conduct the experiments. Unlike Boyle's experimentalism, then, Petty's method aspired to certainty; and unlike Hobbes's deductions, Petty's method aspired to describe what actually existed—what could be measured and counted and experimented on. His claim to combine certainty and accuracy—which constitutes another incarnation of double-entry bookkeeping's claim to derive accuracy from precision—rested on the dual authority of mathematics (which could generate epistemological certainty) and royal power (which had the political power to declare what would count as true, or at least legal). If the first was above or outside politics, the second was decidedly not; thus Petty's promise to produce impartial knowledge remained in tension with his need to convince a king that the knowledge he supplied would be useful—more so by seeming impartial.

The numbers that Petty used so copiously in *Political Arithmetick* thus held open the possibility of both actual experiments, which would yield numbers through measurement and counting, and mathematical certainty, which would result from the calculations conducted on those numbers. In the 1670s Petty

had personal as well as theoretical reasons for lobbying the king to adopt political arithmetic. During the 1660s he had suffered serious financial setbacks on
his holdings in Kerry, and even more disastrously, a clerical error made by the
tax farmers had set the quitrents he owed on these estates too high. Being taxed
not only on what he had designated in the Down Survey as "profitable" lands
but on his total acreage, Petty owed £20,000 for the period 1660 to 1668 alone.
Given the complex system of jurisdiction over Ireland, rectifying this situation
in the early 1670s would have depended on royal action.[78] Instead of simply pe-
titioning the monarch for relief, however, which would have raised the specter
of self-interest, he offered him a method that seemed to serve the king's interests and not those of Petty himself. Because the specific terms in which he promoted political arithmetic promised to remove policymaking from the
contentious domains of politics and religion, Petty seemed to offer the king a
way to place his decisions beyond controversy and resistance.

If the methodological claims Petty made on behalf of political arithmetic
promised to free him from the charge of self-interest and the king from political intrigue, the specific policies he recommended in the 1670s reveal that the
method was never immune from interests—even when it was used specifically
to erase those political and religious affiliations that had proved so divisive for
the past half century. These policies, which illustrate how the modern fact's dependence on theory can always admit politics into apparently impartial numbers, all focused on "transmuting" the recalcitrant Irish into tractable English
subjects. In the 1670s such a scheme might well have appealed to an English
king, both because Irish Catholics continued to be a source of potential political instability and because Ireland's considerable natural resources promised additional revenue to a monarch still engaged with domestic and foreign rivals.
Both to address Ireland's threat and to capitalize on its potential, William Petty
sought to convince the English king that the Irish wanted—or could be made
to want—simply to become English. To do so, he translated the Irish problem
into a set of numbers.

Petty initially applied the method of political arithmetic directly to Ireland
in *The Political Anatomy of Ireland* (1672), although he continued to elaborate the
schemes advanced there throughout the rest of his writing, which culminated
in *A Treatise of Ireland* (1687). Like his *Political Arithmetick* and unlike *A Treatise of
Taxes,* these later texts are filled with numbers—describing acres of Irish land,
both profitable and unprofitable; ratios of Catholics to Protestants; extent of
revenue, and so on. By relying so heavily on numbers, Petty seems simply to
bolster his argument that numbers are by nature impartial and accurate. When
he describes in *The Political Anatomy of Ireland* how his old enemies, the tax
farmers, have typically valued the land, however, he admits that numbers are not

necessarily either impartial or correct. "Only take note," Petty warns the king, "that these Valuations were made as Parties interested could prevail upon and against one another by their Attendance, Friends, Eloquence, and Vehemence; for what other Foundation of Truth it had in Nature, I know not" (*Economic Writings,* 178–79). To counteract such interested valuations (from which he continued to suffer), Petty had to find a way to distinguish his numbers from these "interested" numbers. Thus he devised a "Rule in nature, whereby to value and proportionate the Lands of *Ireland*." "The first [rule] I propose to be; that how many Men, Women and Children live in any Countrey Parish, that the Rent of that Land is near about so many times 15s. be the quantity and quality of the Land what it will. 2. That in the meanest of the 160 M. Cabbins, one with another are five Souls, in the 24,000 six Souls. In all the other Houses Ten a piece, one with another" (*Economic Writings,* 180).

It is difficult to know how these numbers could have been used or even exactly what Petty intended them to mean. In his *History of Ireland* (1689–90), Sir Richard Cox noted that he considered this computation "very strange" and remarked that it "can have noe certainty nor pbability [*sic*]"; calculations that Petty or a copyist performed in the margin of one of the manuscripts, moreover, do not verify or even follow the formula that Petty offers (*Economic Writings,* 180, nn. 2, 3). Because they are even more conjectural than the numbers by which he devised the "value of people," these numbers seem primarily to function as a voucher for Petty's own willingness to use numbers and arithmetic, instead of resorting to the "Attendance, Friends, Eloquence, and Vehemence" to which "Parties interested" appealed. By the time he composed *The Political Anatomy of Ireland,* in other words, he was willing to use numbers as a stand-in and (theoretically, at least) a guarantee for impartiality, instead of trying to demonstrate *why* they should be considered proof of impartiality.

In fact, by the time he composed *A Treatise of Ireland,* Petty had begun shoring up his claim that numbers were a sign of impartiality with his implicitly political claim that numbers constituted the best way to efface politics. In other words, by using numbers to expunge the affiliations that most of his contemporaries considered signs of partiality—religion and politics—Petty tried to argue both that numbers were impartial and that they were impartial because they could erase politics. Not acknowledging that this erasure was itself a political gesture, he displaced religion and politics with a new set of categories, which allowed for computation and which elevated relatively neutral terms over those that had fueled centuries of bloodshed.

> We shall consider the Present Inhabitants of Ireland, not as old Irish, or such as lived there about 516 Years ago, when the English first medled in that Matter; Nor as

those that have been added since, and who went into Ireland between the first Invasion and the Change of Religion; Nor as the English who went thither between
the said Change, and the Year 1641, or between 1641 and 1660; Much less, into
Protestants and Papists, and such who speak English, and such who despise it.

But rather consider them

1. Such as live upon the King's Pay.
2. As owners of Lands and Freeholds.
3. As Tenants and Lessees to the Lands of others.
4. As Workmen and Labourers. (*Economic Writings,* 561–62)

These categories, which emphasize source of income, sought to transform the
Irish into economic beings *instead of* religious, political, or even national subjects. In theory, substituting economic categories for religion, political affiliation, and language would encourage the Irish to see themselves as Petty saw
them; and this perspective in turn would encourage them to realize that their
true interest lay not in clinging to their Irishness but in embracing habits that
would make them like the English. In 1672 Petty assumed that this conversion
could be voluntary, although he devised a plan for forcibly transporting Irish
subjects to England if persuasion did not work. Using "interests" to refer both
to the self-interest of individual Irish men and to a reason-of-state argument
about England's (now Britain's) national interests, he represents the desired
conversion as following from the Irish's recognition of their true—that is, economic—interests. "As for the Interest of these poorer *Irish,*" he explains, "it is
manifestly to be transmuted into *England,* so to reform and qualify their housing, as that *English* Women may be content to be their Wives, to decline their
Language, which continues to be a sensible distinction, being not now necessary. . . .It is their Interest to deal with the *English,* for Leases. . . .'tis their Interest to joyn with them, and follow their Example, who have brought Arts,
Civility, and Freedom into their Country" (*Economic Writings,* 203).

By 1687 Petty realized that "transmuting" the Irish into English subjects
would require more than an appeal to Irish reason. In his *Treatise of Ireland* he
elaborated the scheme for forcibly transporting the Irish that he had first advanced in *Political Anatomy,* but by 1687 the number of Irish he planned to
transport had swelled to a million, and the able-bodied, fertile few he allowed to
remain in Ireland had become veritable slaves of the English government. Consisting exclusively of cowherds and dairymaids "aged between 16 and 60 Years,"
Ireland's Irish were to be "Servants to those who live in England, having no
Property of their own, in Land or Stock" (*Economic Writings,* 569, 568). In Petty's
final grandiose vision, the new Ireland would have no domestic governing
body, no judiciary, and no educational system. The residents would be assigned

colonialism/imperialism at core

English names and taught the English language; they would have scant money ("and that Local"); they would wear uniforms and be tended by English-born priests; they would keep a military force only large enough to protect their borders from foreign invasion. Most important, they would be forbidden to speak of or to assess the injustices of the past. Indeed, in the brave new world Petty envisioned, it would be a crime for the natives to keep the kind of numerical accounts on which he had staked his claim to impartiality and expertise: "It may be offensive to make Estimates of the Number of Men slain in Ireland for the last 516 Years; and of the Value of the Money and Provisions, sent out of England thither; Of the Charge of the last Warr begun Anno 1641; the Value of the Wasting and Dispeopling the Countrey, Charges at Law for the last 30 Years &c." (*Economic Writings*, 569; see also 568–69).

I cite Petty's notorious proposal about eradicating Irish culture for two reasons. First, we must see the link between the method he developed and the production of abstractions like the "value of people," but also like "national wealth" or "the population." Although there is no necessary connection between them and the kind of brutality expressed in Petty's solution to the Irish problem, such abstractions do permit—if they do not encourage—the formulation of policies that overlook the well-being of the individuals they theoretically represent. This is true because such abstractions—especially when they are yoked to economic matters of fact—tend to emphasize the well-being of an aggregate rather than of individuals. It is also true because in order to create abstractions like the "value of people" one has to define "well-being" exclusively in economic terms even if, as in the case of the Irish, other terms seem more meaningful to the individuals so represented. As we will see in the concluding chapter of this book, these issues eventually became visible in another science developed to explain and enhance wealth—political economy. Although no political economists actually solved the problems raised by this tendency to subordinate all considerations to the single criterion of wealth, some nineteenth-century theorists, such as John Stuart Mill, did recognize that such subordination constituted an interpretive gesture, which could generate one kind of knowledge but not others.

The second reason I cite Petty's transportation scheme follows from this point. For all the reasons I have discussed, it served the interests of William Petty to represent the method of political arithmetic as impartial; yet to make this method appeal to a monarch seeking to shore up his claims to absolute authority, he represented this method both as capable of serving the king's interests and as immune from—even above—politics. The paradox of Petty's economic matters of fact stems from the context in which he developed political arithmetic: the cabin boy turned knight needed to persuade two kings that method guaranteed impartiality and certainty—that a method that simulated mathematics could mask *and advance* the interests of monarchs whose absolute claims rested on their ability to foresee or forestall intrigue. Given the historical specificity of the situation in which the connection between numbers and impartiality was forged (after James II, no other British monarch could pretend to absolute authority) and given the elaborate positioning necessary to efface interest from numbers, it seems paradoxical that Petty's bid to make numbers useful to governments should have triumphed so completely. To understand both how this happened—and why it did not happen right away—it will be necessary first to see that in the seventeenth century the cultural prestige of numbers was also being enhanced in other ways. After a brief excursion through the role mathematical instruments played in lending numbers the authority Petty tried to claim for them, we can return to the concept of experiment, to see how the modern fact was gradually transformed again, this time not just into a number but into the kind of abstraction that numbers would eventually be used to describe.

THE AUTHORITY OF MATHEMATICAL INSTRUMENTS

During his own lifetime, William Petty was not consistently successful in persuading either Charles II or James II that the monarch's interests would be best served by implementing the policies he recommended for Ireland. Despite the royal audiences he was intermittently given, Petty never achieved the political offices he sought, and though the Down Survey is still considered a triumph of early cartography, his schemes to establish an Irish land registry and to transport large numbers of Irish to England never met with much support.[79] The one sense in which Petty eventually succeeded was in making quantification an acceptable instrument of knowledge production for a government increasingly entrusted (in this respect at least) to experts who could interpret numbers.[80] Remember, however, that Petty was not solely responsible for making numbers an instrument of rule. To the contrary: his ability to present his self-interested ideas about Ireland as impartial analyses derived from the authority with which numerical representation had begun to be invested by 1660. To see why Petty's claim that numbers were useful to governments eventually triumphed, it is necessary to understand how numerical representation was invested with cultural prestige during the course of the seventeenth century.

To do so, we have to turn for a moment to that set of practices called the mathematicals, which, along with bookkeeping, constituted the practices in relation to which numeracy was typically acquired in the early modern period. The mathematicals, which included astronomy, navigation, surveying, gunnery, horology, architecture, and mensuration, initially attained importance in England in the mid-sixteenth century with the expansion of foreign exploration, international trade, and military engagements. Beyond their obvious contributions to England's growing international power, however, the mathematicals also constituted the site of several critical transitions. First, as practices that were necessary to the state but could also be adopted for private initiatives, they provided skills that enabled some men to move between the patronage-governed court and what we would call the private sector, which required and increasingly rewarded individual enterprise. As a bridge between the court and private enterprise, the mathematicals were transitional in a second sense: they served as one medium in relation to which the traditional ground of civic virtue—a gentleman's willingness to defend the Crown through military service—was reconstituted in new terms, as the willingness to participate in the government, both by service in Parliament and, increasingly after 1660, by participation in some relatively minor protobureaucratic capacity. Finally, the instruments essential to the mathematicals helped make mathematics accessible and acceptable to English gentlemen, and the resulting "gentrification" of mathematics helped confer legitimacy on the kind of knowledge these instruments could produce.

Instruments were critical to all these transitions. As a means of securing royal patronage, mathematical instruments were particularly significant in absolutist courts like those of the sixteenth-century and early seventeenth-century Italian states.[81] English monarchs were also enraptured by mathematical instruments, and makers of ingenious instruments, such as Nicholas Kratzer, John Rotz, John Dee, and Thomas Bedwell, found positions at court through the attention their inventions received.[82] Even though instruments could undoubtedly win royal patronage in many European courts, however, in England royal patronage did not constitute the only avenue for personal advancement as it did elsewhere in northern Europe. As early as 1630, William Oughtred (1575–1660) had established himself as a private instructor in mathematical skills; by that time, in fact, numerous mathematics teachers were selling both instruments and knowledge from their homes in London.[83] Almost every private mathematical practitioner either worked closely with an instrument maker or was himself one, for in London as at court, instruments constituted a critical aid to professional advancement.

In early seventeenth-century England, private teachers of mathematics were riding the wave of public attention that had turned to the mathematical arts in 1588. In that year the threat posed by the infamous Spanish Armada had led the Privy Council to approve the first public lectureship in mathematics, in hopes of providing the captains of the hastily assembled London militia with training in fortification, gunnery, and (to a lesser extent) navigation.[84] Thomas Hood initially delivered these lectures (from 4 November 1588), but since Hood did not teach fortification, gunnery, or martial affairs, the Privy Council soon withdrew its support.[85] At that point Hood's mathematical instruction became a private enterprise, and because he had to tailor his teaching to the specific interests of his clients, his initial emphasis on astronomy was soon replaced by a focus on navigation, which was as useful for the expansion of commerce as for military defense. The defeat of the Armada fueled the interest of English merchants and investors in navigation, and Hood's self-advertisement as a fit instructor for gentlemen "in this our travelling age" yoked adventure to profit and drew pupils to his house in Abchurch Lane.[86]

During his four-year career as a mathematics lecturer, Hood published four books, three of them dealing with mathematical instruments. As Stephen Johnston has argued, Hood used instruments regularly in his lectures, both because they helped make his lessons clear and accessible and because they literally embodied the instrumental reason that mathematical practitioners hoped to promote as essential to national and personal security.[87] As a private teacher, Hood also designed mathematical instruments, advertised them in his books and lessons, and sold them from his home. By making such instruments more widely known and also by presenting them as beautiful objects that were desirable additions to gentlemen's collections, Hood helped create an elite market for mathematical instruments, which in turn helped legitimize mathematical knowledge and familiarize even the innumerate with applications of numerical representation.

As early as the late sixteenth century, then, a gentleman's traditional responsibility to defend the Crown made acquiring some mathematical skills at least theoretically important.[88] At the same time, the renewed enthusiasm for foreign exploration and wealth that followed the defeat of the Armada made gentlemen adventurers more eager to acquire mathematical skills and to employ and honor mathematical practitioners. Sir Walter Ralegh hired the mathematician Thomas Hariot, for example, to help him plan his expeditions; Thomas Cavendish welcomed the Oxford scholar Robert Hues on board his ship; and the earl of Cumberland regularly employed Edward Wright, author of the country's most influential book on navigation.[89] These highly visible testimonials to the increasing prestige of mathematical skills helped enhance the opportunities for much less conspicuous kinds of private enterprise, as practiced, for example, by the numerous teachers of mathematics in London, the instrument makers who supplied the necessities of their craft, and the authors of manuals designed for self-instruction in everything from reading the time off a pocket dial to double-entry bookkeeping.

The representation of mathematical instruments as aesthetic objects also played a central role in enhancing the prestige of mathematical skills and the value of numeracy, for even the gentlemen who found the analemma difficult to operate could take pride in possessing and displaying it.[90] The sumptuous sets of celestial and terrestrial globes created by Emery Molyneux were particularly coveted, but so numerous were the available mathematical instruments that any well-to-do collector could expect to command a range of aesthetically pleasing objects.[91] The pleasure a gentleman took in his collection could be carried over to the kind of knowledge these instruments produced, both because instruments like the mathematical jewel were designed to provide shortcuts to complex calculations and because using such instrumental aids was more entertaining than writing numbers or memorizing tables. The "love" for mathematics that these beautiful instruments inspired could theoretically motivate gentlemen even to attempt calculations that did require writing and memorization, such as deriving square roots, "wch were they to learne in the first place they would never endure the harshness of it. It would be like eating a chopd-hay or taking a bitter medicine," John Aubrey explained in 1683. "They must be enticed only by pleasure and delight."[92]

In addition to the royal and private patronage they attracted, their utility in aiding national and private voyages of adventure, and the aesthetic connotations with which they were invested, sophisticated instruments were also accorded cultural prestige by the Royal Society's claims that instruments like the microscope and telescope placed knowledge never previously available before everyone's admiring eye. According to Joseph Glanvill, instruments were critical in elevating the moderns over the ancients, "so that much greater things may well be expected from *our Philosophy,* than could ever have been performed by *theirs.*"[93] The instruments preferred by members of the Royal Society were not, by contemporary definition, simply mathematical instruments, for they were devoted to producing natural philosophical knowledge instead of simply working out mathematical problems. Nevertheless, contemporaries often spoke of mathematical and natural philosophical instruments as if they were related, and the latter often included measuring devices alongside their optical lenses or tubes. Indeed, the proximity of mathematical and natural philosophical instruments underscores once more the equivocal nature of the facts they were used to generate: in many cases these facts both referred to natural phenomena, as if they were detached from theory, and described them in (numerical) terms that presupposed that measurement was the best index to meaning.

As early as 1686, William Petty was being credited with contributing to the transvaluation of numerical representation that was inextricably bound up with the prestige accorded mathematical instruments. Petty made his first significant contribution to statecraft, we remember, by his skillful use of modern surveying instruments in Ireland, but it was his transfer of the mode of knowledge production associated with mathematical instruments to "all concerns of human Life" that made his intervention so noteworthy. This, at least, was the sentiment of a writer in the *Philosophical Transactions,* who remarked that Petty had "made it appear that Mathematical Reasoning, is not only applicable to Lines and Numbers, but affords the best means of Judging in all concerns of human Life."[94] John Arbuthnot elaborated this point in 1701, when he equated "true political knowledge" with Petty's political arithmetic. Numerical calculation, Arbuthnot wrote, was "not only the great instrument of private commerce, but by it are (or ought to be) kept the public accounts of a nation. . . . Those that would judge or reason about the state of any nation must go that way to work, subjecting all . . . particulars to calculation. This is the true political knowledge."[95]

The rapidity with which numerical calculation became a critical component both of gentlemanly virtue and of effective statecraft can be grasped by two sets of comments, separated by less than a century. In 1622 Henry Peacham had had to argue that geometry was an admissible part of a gentleman's education.[96] In 1698, by contrast, Charles Davenant referred to the newly valued capacity to "compute" as an eternal verity of good government: "The abilities of any minister have always consisted chiefly in this computing faculty," he wrote; "nor can the affairs of war and peace be well managed without reasoning by figures upon things."[97] By 1719 John Jackson could take it for granted that every gentleman would need—and want—mathematics. "The necessity that Gentlemen are under, that would be Considerable in the Art of War or any great Employment (either in Church or State) which cannot well subsist without a considerable knowledge in the *Mathematics;* makes them to throw aside several trifling Amusements, and apply themselves to the *Mathematical Sciences.*"[98]

The increased authority attributed to numerical representation by the end of the seventeenth century can also be conveyed by the institutionalization of some of the kinds of record keeping Petty espoused. One measure of the growing prestige of numbers is the sheer increase in state officials, many of whom were involved in collecting and processing numerical information. Whereas the eleven years of the Interregnum (1649–60) saw the employment of only about 1,200 officials of state, by 1688 there were 2,500 officials busy in the area of tax collection alone.[99] Another measure of the increased power attributed to numerical representation is the burgeoning of demand for records of accountability, some of which were configured as numerical tables. Given its dual roles as agent of government and public watchdog, the Restoration House of Commons issued a steady demand for documentation from its own committees and from particular trades; especially after 1688, individual members of Parliament lobbied their colleagues with reams of information on such social issues as poor law and penal reform; and lobbyists for special interests both demanded and (after the 1740s) received public records of parliamentary transactions.[100]

As we will see in the next chapter, however, William Petty's desire to conduct statecraft by "the Terms of *Number, Weight,* or *Measure*" was not immediately realized in Britain; notably, British citizens and some politicians remained hostile to any form of official census throughout the eighteenth century.[101] *Buck* Nor were some of the more grandiose numerical schemes conceived in the spirit of political arithmetic implemented. For example, an early eighteenth-century plan to collect detailed data on trade, navigation, tax collection, the availability of war supplies, and the state of the market in public funds foundered before it could generate much information.[102] Nevertheless, it is undeniable that by the early eighteenth century one could make a case—as John Arbuthnot did—for considering numerical representation the quintessential form of "useful knowledge." So prestigious were numbers considered, in fact, that when moral philosophers sought tropes to figure their own method, mathematics proved especially attractive. Even though Petty's numerical variant of the modern fact was not immediately adopted by the British government, then, by the early eighteenth century numbers had acquired a set of connotations that would soon make them central to what counted as knowledge in numerous domains.

Experimental Moral Philosophy and the
Problems of Liberal Governmentality

In what may be taken as a paradigmatic moment in the campaign to consolidate the prestige that numbers had lacked a century before, on 19 September 1711 Richard Steele staged a debate in the pages of the *Spectator* between his representative gentleman, Sir Roger de Coverley, and his emblematic merchant, Sir Andrew Freeport. Noting that "the landed and trading Interest[s] of *Great Britain*" constitute "Parties" whose "Unanimity is necessary for their common Safety" but whose "Interests are ever jarring," Steele makes the current disagreement between the two turn on merchants' use of numbers, and accounting in particular. Sir Roger insists that merchants' reliance on numbers approaches the unscrupulous use of an unfair technology and that, even if numbers do give merchants an advantage in trade, their preoccupation with them bespeaks the ignoble nature of their profession. "Indeed what is the whole Business of the Trader's Accompt, but to over-reach him who trusts to his Memory?" Sir Roger asks. "But were that not so, what can there great and noble be expected from him whose Attention is for ever fixed upon ballancing his Books, and watching over his Expences? And at best, let Frugality and Parsimony be the Virtues of the Merchant, how much is his punctual Dealing below a Gentleman's Charity to the Poor, or Hospitality among his Neighbours?"[1] In his rejoinder, Sir Andrew not only yokes numeracy to mercantile virtue but also insists that merchants' facility with numbers underwrites "prudence," which he presents as essential to the gentleman's ability to administer—and thereby retain—his estates. "The Gentleman no more than the Merchant is able without the Help of Numbers to account for the Success of any Action, or the Prudence of any Adventure," Sir Andrew explains. "'Tis the Misfortune of many . . . Gentlemen to turn out of the Seats of their Ancestors, to make Way for such new Masters as have been more exact in their Accompts than themselves."[2]

At stake in this mock debate are several issues that more or less directly informed British attitudes toward knowledge, facts, and numerical facts in particular in the early eighteenth century. First, Sir Roger's implicit defense of aristocratic nonchalance explicitly links a relative indifference to the precision of bookkeeping to the willingness to dispense charity and hospitality. In this account, the aristocrat's tendency to trust to his memory instead of keeping books is of a piece with his spontaneous generosity, for both signal superiority to a rigid calculus of income and expenditure. When Sir Andrew links numbers to prudence, by contrast, he repudiates this defense of aristocratic nonchalance by pointing out that, in a society where money rivals birth, gentlemen may lose the estates that make charity possible if they do not keep good accounts. His response does not attack aristocratic generosity but illuminates the conditions that make such generosity possible—or impossible—in a society undergoing the dramatic changes associated with what historians have called the financial revolution.

The second issue raised by the brief exchange between Sir Roger de Coverley and Sir Andrew Freeport concerns the role that "parties" played in this newly reformed society. As Steele demonstrates, by 1711 Petty's various attempts to efface the role that interest played in the formulation of government policies had been essentially reversed; instead of cultivating an impartiality modeled on the production of natural matters of fact, partisans celebrated shared interests under the banner of political party. Inevitably, although in some unpredictable ways, the rise of modern political parties affected the relation between facts and interests—especially when the facts at issue were those considered useful for knowing, or governing, civil society. Indeed, the emergence of civil society itself, that domain of extragovernmental organizations through which English (then, after the union with Scotland in 1707, British) men governed many of their activities and sought to enhance their nation's strength, was partly a function of new instruments of knowledge production that were used, among other things, to make the case that party interests coincided with national interests.[3]

These new instruments of knowledge production, which included popular periodicals like the *Spectator*, ephemeral publications such as newspapers and pamphlets of all kinds, and public lectures on a variety of subjects, also helped package the modern fact for the market. Typically produced in the private sector and sold on the open market, these new instruments of knowledge production sometimes featured diagrams, pictures, and staged experiments, which helped make knowledge desirable by giving it visual form and by presenting it as entertainment. As Barbara Maria Stafford has argued, the effort to make knowledge entertaining had originated in the "ingenious pastimes" and "mathematical recreations" first marketed across Europe in the seventeenth century. By the third decade of the eighteenth century such devices had generally been replaced by "rational recreations." Unlike ingenious pastimes, which tended to emphasize Baconian singularities—artificial curiosities as clever as nature's own—rational recreations typically directed the consumer to the general principles that particular mathematical exercises or puzzles illuminated.[4]

In this chapter I explore some of the ways modern facts were adapted to a market society governed by the institutions associated with civil society and party politics. I do not explore all the incarnations of this variant of the modern fact, if for no other reason than that the proliferation of books and other kinds of writing after the expiration of the Licensing Act (1695) would make anything resembling coverage of this subject impossible. Instead of trying to capture the variety of rational recreations and modes of writing that helped popularize the modern fact and justify party politics, I limit myself to what may seem like the least entertaining and least political of all eighteenth-century genres of knowledge production: experimental moral philosophy. I do so because I am interested in the epistemological developments that link—and separate—political arithmetic and political economy as disciplines; at the level of epistemology, moral philosophy was the science that connected these disciplines, although, as we will see in this chapter and the next, experimental moralists' interest in mapping human subjectivity instead of national productivity has made it difficult for historians to recognize the disciplinary affiliation between moral philosophy and political arithmetic. I also focus on experimental moral philosophy rather than efforts to popularize science or defend party politics because other scholars have already explored those areas.[5] What follows is intended as a supplement to this important work, not a challenge to it, although I do argue that in experimental moral philosophy we can see clearly two developments not so obvious in rational recreations or overtly partisan defenses of party: the privileging of the universal over the Baconian singularity and the effort to theorize the interiority or subjectivity of the universal subject of science.

In the first section of the chapter, I show why the kind of large-scale, government-sponsored projects intended to generate knowledge that could be conveyed in numerical form did not immediately prosper in eighteenth-century Britain, as Petty clearly hoped they would. This will also help explain why the modern fact has not invariably been yoked to numerical representation. Taking Defoe's *Essays upon Several Projects* (written in 1697, published in 1702) as my primary example of why political arithmetic seemed inadequate to problems that were new after the Glorious Revolution, I show how Defoe used the kind of distinctions that could be made only in—and about—language to ground theories about critical components of market behavior like emulation and custom. Like Hobbes, Sprat, and Petty, Defoe sought to use a difference in style to shore up a distinction that seemed critical to both knowledge and social harmony. In the context of the new market economy, however, his incarnation of stylistic difference did not so much endorse government-sponsored knowledge schemes as suggest the kind of *subjective* motivations that could support—or even substitute for—the rule from above associated with seventeenth-century sovereignty.

I devote special attention to governmentality in this section, for to understand the demise of political arithmetic, one must also understand the mode of governmentality—the technologies and theoretical accounts by which individuals were rendered thinkable as governable subjects—that developed in the vacuum left by the flight of James II. Indeed, one way to explain the demise of political arithmetic would simply be to say that the collapse of absolutism in Britain meant that the king no longer had the power Petty both assumed and supported. Without the kind of absolute monarchy that still existed in France and that had been instituted in some of the German territories after the Thirty Years' War, the "science of police" (*Polizeiwissenschaft*) could not be pursued by royal decree, nor was government by information self-evidently desirable in a society where gentlemen looked after—and jealously guarded information about—so many of their own interests.[7]

The new mode of rule that emerged after the collapse of absolutism in Britain has been called liberal governmentality.[8] Operating in civil society and through the market, liberal governmentality depended (depends) on self-rule rather than rule by coercion; eliciting voluntary compliance through the mechanisms of discrimination and emulation essential to rule by fashion, it did not rely on numbers in the same way that sovereignty did. Thus it is not sufficient simply to say that, with the demise of absolutism in Britain, the monarch no longer had the power to implement political arithmetic schemes. Instead we need to see that, as the monarch ceased to be the primary guarantor or steward of knowledge and as the old machinery of fashion took up some of the burdens of both (self-) rule and knowledge production, theorists reconsidered the kind of knowledge useful for government. Administering self-rule in a market society involved understanding human motivations, including the desire to consume, rather than simply measuring productivity or overseeing obedience. As a consequence, the knowledge that increasingly seemed essential to liberal governmentality was the kind cultivated by moral philosophers: an account of subjectivity that helped explain desire, propensities, and aversions as being universal to humans as a group.

Since the mid-nineteenth century, the domain of subjectivity has been mapped with increasing subtlety by the sciences of psychology and psychoanalysis, and it is tempting simply to apply terms from these discourses to the theories about human motivation developed in the eighteenth century. In the discussion that follows, I take another tack. By showing the extent to which eighteenth-century moral philosophy provided an account of human motivations, I do want to demonstrate that this discipline was one of the antecedents of modern psychology and psychoanalysis. By emphasizing the contrasts between eighteenth-century moral philosophy and its disciplinary descendants, however, I also want to highlight the difference between the eighteenth-century science of mind and its late twentieth-century counterparts, so that we can see what the eighteenth-century philosophers thought they were doing instead of simply applauding (or chastising) them for anticipating Freud.

One of the features that links eighteenth-century accounts of subjectivity to eighteenth-century science more generally was the assumption that subjective "events" are as particularized and observable as phenomenal events. This view meant that the moral philosopher assumed he could conduct "experiments" on subjectivity and that the results would simultaneously describe particular events and contribute to systematic knowledge. While this feature of eighteenth-century moral philosophy resembles assumptions inherent in at least some branches of late twentieth-century psychology, however, at least two subtle differences differentiate the two practices. First, eighteenth-century moral philosophers assumed that science should explore human motivation primarily for its *social* implications, not its implications for individual happiness or misery. Second, and by extension, they assumed that one sought knowledge about the particulars of subjectivity in order to understand the *regularities* of the moral universe, including the principles that underwrote (most) human beings' willingness to submit to government. Stressing regularities (not individual idiosyncrasies) and attentive to the social implications of these regularities, they developed a science of subjectivity that focused on *universal human nature,* not atomistic individuals or even psychological types. Indeed, the very universality of the subject of moral philosophical science explains why theories about it could also be considered contributions to a philosophy of government: because they assumed that individuals were instances of universal human nature, eighteenth-century moral philosophers also assumed that knowledge about human nature contributed to—indeed, was the basis of—a theory about government that emphasized both self-rule and the idea that some individuals could stand for (or represent) others.

In the second section of the chapter, I begin to describe how eighteenth-century attempts to elaborate the interiority or subjectivity of the universal subject of science contributed both to the theory of liberal governmentality and to the reworking of the modern fact. In texts by Francis Hutcheson and George Turnbull, we see that the eighteenth-century variant of the modern fact helped make liberal governmentality seem as reliable as law-governed nature was assumed to be. The variant of the fact produced by these philosophers differed from the Baconian rarity and resembled Petty's numerical fact in several ways: unlike the Baconian rarity, what mattered about both the eighteenth-century universal and Petty's numerical fact was not that they were singular or able to float free of theory, but that they described abstractions grounded on theory. Unlike quantitative abstractions like Petty's "value of a people," however, eighteenth-century abstractions were universals, generated not by arithmetic but by deduction—by generalizing from a set of a priori assumptions. These universals—man, mankind, human nature—theoretically represented everyone, of course, so that the knowledge philosophers produced about them theoretically applied to all individuals everywhere and could be used not only to describe but also to predict human behavior.

This kind of fact seems suspect to late twentieth-century readers, because the claim that such facts are universal now seems implicitly, if not explicitly, coercive and exclusive rather than descriptive and inclusive. It now seems self-serving for a male philosopher to use "man" to describe all human beings, for example, not only because some incarnations of "man" have always been female but also because, in eighteenth-century Britain, "man" did not even include all males, especially if the males were black or brown (or Irish). In the eighteenth century, by contrast—and at least for the philosophers who generated knowledge about these universals—such universal facts generally did not seem troubling or self-serving. This was true for two reasons: first, because the individual the philosopher could most directly observe—the philosopher himself—was taken to be *representative;* and second, because the trope of mathematics converted what were represented as universals into what looked like *aggregates.* Thus British—or more frequently, Scottish—philosophers moved from observed particulars to general claims about universals like "man" by claiming that their universals were somehow derived from an additive process that identified the "greatest good of the greatest number" by looking at the philosopher's (representative) self. By means of such claims, which were shored up by a providentialism that also looks out of place to late twentieth-century readers, Scottish philosophers like Hutcheson and Turnbull were able to argue that the similarities individuals shared were more important than the differences that distinguished them, even though these differences were also essential to government by fashion and to self-rule.

By making this argument, Hutcheson and Turnbull were also able to avoid what modern philosophers call the problem of induction: the challenge that one cannot know in advance of observation that all instances of a single phenomenon will be more alike than different. I do not take up the problem of induction at length until the next chapter, but I do include here two brief discussions of the philosopher who made this question an explicit component of modern philosophy: David Hume. By examining Hume's repudiation of experimental moral philosophy and his turn to the essay, I show that eighteenth-century attempts to produce knowledge about a universal subject through experiment coexisted with another kind of knowledge project, which sought not so much to generate facts about a universal subjectivity as to engage readers' subjective responses in the service of producing something else, which eighteenth-century writers variously called conversation, moral emulation, and self-improvement. What distinguished the knowledge projects that claimed to produce universal facts from attempts to inaugurate such subjective engagements was the method of experiment, which, as we will see, was derived from Baconian induction and also capitalized on an ambiguity inherent in Newton's adaptation of Bacon's method. By following Hume's repudiation of experiment, I illuminate the proximity between universal facts and the kind of conversation entailed by reading an essay, but I also want to point out that government by information, the goal pursued by Bacon and Petty alike, had to compete throughout the middle part of the century with the more subtle, less numbers-based form of government that Hume associated with conversation, sociability, and even wisdom.

Before I turn to Defoe, let me recall some of the political and economic reasons why theories that linked accounts of subjectivity to a defense of liberal governmentality also typically addressed the practices associated with fashion and taste. The first point I want to make about this period is that when the parliamentary faction known as the Court Whigs successfully lobbied for the ouster of James II and the peaceful invasion of William of Orange in 1688, they engineered a political compromise that dramatically reconfigured the nature of political power in Britain. This was true both because this compromise gave birth to political parties and civil society and because the power conferred on the group that Swift called the "Monied Men" decisively shifted power away from the landed gentry and toward the new representatives of finance whom William patronized in order to wage war against France. When London goldsmiths and Tory landowners continued to oppose William's war against France in the first years of his reign, and when the existing tax-gathering apparatus proved inadequate to make up the shortfall in necessary revenue, a group of finance capitalists who were politically affiliated with the Whigs stepped in and offered the king a loan. From the £1,200,000 loan that William accepted, the Bank of England was born, and with the founding of the bank in 1694, the ground was laid for a national debt, the stock exchange, and the association of moneyed men with state power.

In general terms, then, the Whigs associated with the court and the City supported William and his new instruments of finance, while the Tories, the Country Party, and the True Whigs opposed the system of public finance by which William conducted war. Although the opposition to "Dutch finance" did not abate during the eighteenth century, the bank and its attendant instruments of credit continued to gain institutional strength as William continued to borrow. From its beginning, the bank was an instrument of credit, and since it was recognized by parliamentary charter, it always had the potential to tie national security and politics to the system of credit it both promoted and required. By the end of the seventeenth century, the Bank of England had been granted the further rights to receive money directly from the public and to lend it at interest, as well as the right to issue its own credit instruments in the form of paper. When shareholders in these public funds began to trade their shares for profit, it became not only possible but tempting to imagine that the state itself was a marketable commodity.[9]

In this context, the question of interest inevitably surfaced once more, especially since the financier Whigs were rumored to profit most from the new credit economy. In the first half of the eighteenth century, however, as was not always the case in the seventeenth century, "interest" typically connoted economic self-interest in addition to—or even instead of—a political variant of interest that could be justified by allusions to reason of state. Whether or not it dovetailed with party affiliation, however, economic interest could *also* be equated with national interests, for national security increasingly *did* depend on the money lent to the government through shares. Thus in 1711 Defoe acknowledged that "it is a *disputed point,* whether this Levying Money by Loans, upon Funds of Interest, be a Service to the Nation or a Prejudice," but by explaining to his readers how credit worked, he hoped to demonstrate once and for all that good credit supported the nation, because the nation's interests and the creditors' interests were inevitably the same: "I have, I think, plainly lay'd down the great Foundation of Credit among us. I have shewn how the Funds *however* they are call'd Publick, and *however* having Money in them, is call'd having money in the Government, *are our own,* and that the Government are, in this Case, no more than our *Rent gatherers,* Stewarts and Tellers of the Money."[10]

As J. G. A. Pocock has argued, the equation of government with a system of credit did not inspire everyone to assume that profit-driven investment was virtuous, as it did Defoe. Indeed, Defoe's allusion to the *"disputed point"* about "Service" *or* "Prejudice" reminds us of the civic humanist argument that the new credit economy had undermined the very possibility of public virtue, because individuals who were motivated by their own self-interest could not have the autonomy or disinterestedness necessary to practice civic virtue.[11] As we will see, this dispute continued throughout the first half of the eighteenth cen-

tury, and once Mandeville entered the fray it was not even possible simply to oppose disinterested, public-spirited virtue to self-interested and publicly ruinous
vice. For now, let me simply note that one early response to the charge that the
new system of credit had undermined civic humanism was that of Joseph Addison and Richard Steele. In the *Tatler* (1709–January 1710–11), then the *Spectator* (March–December 1712, June–December 1714), Addison and Steele
suggested that the practices associated with taste—sociability, sympathy, and
honesty, among others—could form the basis for a new kind of virtue, which
served national interests by promoting civility and, not incidentally, by
strengthening Britain's commerce with the rest of the trading world.[12] Thus
fashion, which was indirectly related to both taste and credit through its more
direct relationships to consumption and investment, was one link that tied discussions about subjectivity, where taste was presumably cultivated, to discussions about a mode of government that needed both emulation and credit.

Indeed, the *Spectator* coupled taste and refinement not only to personal
virtue but, more specifically, to economic investment understood as civic
virtue. In so doing, it also linked beneficial *national* government to effective *self-*
government, through manners and civility. This in turn meant that, in spite of
the explosion in the number of government administrators under William,
effective national government was not necessarily equated with an increase in
the amount of information the government was able to collect about its citizens. On the one hand, if self-government was critical to social stability, then
understanding how manners helped discipline innate desires was critical to theorizing government. On the other hand, with London coffeehouses and newspapers producing and distributing a flood of information after the repeal of the
Licensing Act, the state was not even able to monopolize the task of knowledge
production as Petty had wanted it to do.

The second point I need to make about this period is that the discourse of
politeness popularized by the *Spectator* also took up and significantly revised the
epistemological claims about disinterestedness that Sprat had made for natural
philosophy and the claims about impartiality that Petty had made for political
arithmetic. As Ronald Paulson has argued, in fact, Addison's ideas about politeness constituted just one of a cluster of attempts to rework civic virtue by elaborating an ideal of disinterestedness; whereas the Addisonian concept of
disinterestedness supported the new credit economy, its more austere counterpart insisted that disinterestedness removes from the idea of virtue *all* considerations of profit—even the prospect of heavenly rewards.[13] This revision of civic
humanism is most frequently associated with the earl of Shaftesbury and the
emergent discourse of aesthetics.

Like Bacon and the members of the Royal Society, and following in the philosophical tradition epitomized by John Locke, Shaftesbury was essentially an empiricist; he maintained that all experience, and hence all knowledge, ultimately derived from observation and the senses. While Shaftesbury can be called an empiricist, however, the twist that his emphasis on appreciation gave to empiricism significantly altered Lockean ethics and Baconian epistemology. Because he insisted that the only actions that express civic virtue are those that are free of the desire for both religious *and* material rewards, he was inclined to privilege aesthetic appreciation over other kinds of observation; this means that he considered the apprehension of beauty equivalent to the apprehension of truth. Emphasizing that both virtue and aesthetic appreciation were subjective responses to perceived objects, moreover, tended to privilege the observer rather than the thing observed and thus to elaborate a protopsychology of the passions that could explain both aesthetic judgment and consumption. Like Defoe and the moral philosophers who would take Shaftesbury as their inspiration, Shaftesbury sought to devise a map of interior actions, like motivations and preferences; in his case, this map of subjectivity resembled the map of attributes that coud be seen, like proportion, harmony, and beauty. For Shaftesbury, as Paulson explains, "virtue and beauty are as homologous as the moral and aesthetic senses. The one shifts the philosopher's attention back to motives as the other does to responses, and in both cases to the most disinterested, the furthest removed from personal gain."[14] Aligning aesthetic appreciation with moral virtue, in turn, was apt to equate—or even to replace—God with beauty, which to Shaftesbury could be defined by the mathematical criteria of proportion and harmony. Thus we arrive at the quintessential Shaftesburian equation: aesthetic contemplation is equated with a disinterested and therefore virtuous reverence for "order and beauty," whether in a work of art or society itself. "This . . . is certain," he wrote in 1699, "that the admiration of love and order, harmony, and proportion, in whatever kind, is naturally improving to the temper, advantageous to social affection, and highly assistant to virtue, which is itself no other than the love of order and beauty in society."[15]

This deist formula had profound implications for theories about the production of knowledge and for how what counted as a fact was understood. In brief, it tended to demote knowledge that was narrowly "useful" and to elevate in its place the attributes that could be assessed only by the quasi-psychological, quasi-social faculty of appreciation. I call this faculty quasi-psychological because, even though Shaftesbury was primarily interested in subjective responses, he did not develop a comprehensive account of the dynamics of subjectivity. I call it quasi-social because, in Shaftesbury's account, the attributes that individuals naturally appreciated were Platonic abstractions like "beauty" or "virtue."

Such abstractions can be understood as something like a bridge between the kind of abstraction Petty had tried to quantify and the universals that experimental moral philosophers claimed to observe: while beauty and virtue could not be derived by measurement or counting, they could be identified by reference to the mathematical properties of proportion and "fit"; and while they were not universals like "man," because only men of a "liberal education" could embody the "perfection and grace and comeliness" essential to Shaftesburian virtue, abstractions like beauty and virtue were theoretically universal in the sense that they held true for all time and in all places.[16]

As this last quotation implies, Shaftesbury's abstractions also had political implications that were more explicit than the political implications of either Petty's numerical abstractions or eighteenth-century universals: Shaftesbury assumed that only men of a "liberal education" could embody virtue because only such gentlemen were superior to the need to work. In elaborating the connection between disinterestedness and this kind of abstraction—and in linking both to a gentleman's ownership of land—he reinforced the traditional distinction between gentlemanly pursuits and other kinds of labor. According to this distinction, gentlemen, who did not pursue a single occupation or work with their hands, were able to achieve greater distance from the things of the world, while workers—whether professional, artisanal, or manual—were necessarily immersed in the specific details of their labor. Thus, and partly because of the influence of Shaftesbury's theories, abstract or theoretical knowledge was considered superior to concrete or detailed knowledge; the latter was held to be "interested" not simply because it involved politics or even the desire for gain, but because it emphasized the immediacy—the materiality—of possession, consumption, or use.[17] This emphasis on abstraction and landed wealth, of course, ran counter to the enterprise that Petty pursued in Ireland, both because it undermined his claims to base expertise on experience and because it tended to discredit the pretensions of a clothier's son, no matter how many acres he owned in Ireland.

The third point I need to make about this period is suggested by Shaftesbury's allusion to "love and order, harmony, and proportion," but the importance of this formula becomes clear only when we turn to its more orthodox variant, that peculiar blend of theology and natural philosophy called "physicotheology."[18] As I will argue in the second section of this chapter, physicotheology reflects the influence of Newton's contributions to natural philosophy. Specifically, Newton's methodological reliance on mathematics along with instrumental experiment tended to elevate mathematically derived regularities (as well as mathematical properties like order, harmony, and proportion) to a status at least equal to that of the knowledge that could be produced simply

by observation and induction. Although a crucial discrepancy between Newton's statements about method and his reliance on mathematics allowed self-
described Newtonians to interpret his work in a variety of ways, it became
possible to argue that the natural world, which visibly displayed order, harmony,
and proportion, embodied God's plan. This was the position articulated in the
Boyle Lectures, which were delivered annually beginning in 1691. The authors
of these lectures, almost all of whom were Church of England divines, argued
that the order discernible in the smallest miracle of nature (say, a fly's eye)
demonstrated God's presence and proved not only his benevolence but also that
all natural phenomena—even those we have yet to understand—are part of a
great design superintended by God.

As it was set out in the Boyle Lectures, especially by Richard Bentley, John
Harris, Samuel Clarke, and William Derham, physicotheology gave new content (and meaning) to those "laws" that Petty had claimed to see in trade.[19]
Whereas Petty had insisted that one could produce the knowledge that made
these laws visible only with adequate instrumental methods and some nationwide scheme of state-sponsored knowledge production, the Boyle lecturers
claimed that at least some incarnations of the laws of nature were visible to
everyone and required only the enhancement of interpretation. This claim
might have been defensible in the realm of natural knowledge production,
where a community of like-minded men agreed that observations and experiments elicited those singularities that revealed nature's hidden laws, but when
moral philosophers like Hutcheson and Turnbull tried to apply physicotheology
in the domain of sociality, not to mention subjectivity, they found they could not
simply appeal to each other or to observation. Because they could neither conduct experiments on interior actions like fear or belief nor assume that the
regularities they claimed to find through introspection would transcend controversy, moral philosophers had to find a new basis for their authority.

As we will see, the authority these philosophers claimed derived from the
same assumption about mathematical—especially geometrical—figures that
Newton held. Indeed, this was the same assumption that had inspired Luca
Pacioli in 1494: the assumption that the harmony, measure, and proportion evident in these figures embodied God's order as truly as did those intricate natural phenomena so dear to the physicotheologist's heart.[20] Unlike natural
phenomena, however, which were always in some sense unique and whose singularity had to be either discounted or interpreted, mathematical figures were
never singular; by definition, one triangle was equivalent to another, even if
their corresponding sides were different lengths. Using mathematical figures as
the image of God's order therefore encouraged moral philosophers to privilege
not singularity but universality. Instead of moving from the observed particu-

lar, whose singularity enhanced its epistemological value for the natural philosopher, to implicit regularities or laws, moral philosophers tended to discount what was singular about observed particulars and to see through them, as it were, to the universals they supposedly incarnated. As self-professed experimentalists, these philosophers did use observation, but their observation was typically focused on the object closest to hand—the philosopher's self. This means that beyond discounting singularity, moral philosophers also elevated the domain of subjectivity over the natural world. By constructing the knowledge they claimed to discover through introspection in the image of mathematics, experimental moral philosophers were able to claim that the moral facts this knowledge was composed of were simultaneously derived from (a kind of) observation, aligned with the (visible) harmony of God's universe, and universally true. Thus the figure of mathematical harmony—not the instrument of numerical representation—was used to ground the kind of knowledge considered useful to ensure self-government in a society ruled by taste, fashion, and the new tyrant of public opinion.

As it was worked out by experimental moral philosophers, the science of man was constructed by analogy to natural philosophy, and by assuming that mathematical harmony characterized the interior domain of human motivations just as it seemed to characterize the natural world. The universals embodied in mathematical principles thus constituted the implied analogue to the image of a universal human nature, although, as I have already noted, this figure also seemed referential because philosophers' use of mathematical language made "human nature" seem like the product of a mathematical operation (addition). This image in turn—universal human nature—constituted the conceptual backdrop against which taste and fashion were supposed to function as instruments of (self-) government, for it was only because moralists assumed that all individuals were alike in wanting the same thing(s) that they could assert that fashion would generate competition, virtue, and social harmony, not a proliferation of atomized consumers seeking to satisfy idiosyncratic wishes at the expense of national prosperity. This unusual partnership—of assumptions about universal human nature, which were grounded on the figure of mathematics, and a discourse of taste that emphasized discrimination and difference—constituted the backbone of that mode of liberal governmentality we associated with consumer society. Unlike the sovereign government of seventeenth-century Britain, this mode of governmentality did not obviously benefit from the collection of numerical information. In order to revive the argument on behalf of numerical facts that Petty had launched, later eighteenth-century theorists would have to marry the numerical fact to both the philosophical universal and the model of human motivation that experimentalists constructed in the image of mathematics. Before we consider how this was done, we need to examine in some detail how political arithmetic clashed with government by taste.

GOVERNMENT BY TASTE IN THE WORK OF DEFOE AND HUME

When Richard Steele defended numeracy in 1711 as an instrument essential to preserving the "common Safety" of Great Britain, he rested his case on the moral and social advantages that good accounts would confer: by promoting "prudence," which was both a personal and a social virtue, good accounting would enhance personal and collective well-being. Steele did not recommend that the government should keep accounts, nor did he assume that it should oversee the collection and distribution of resources. In privileging the ancient prerogatives of charity and hospitality as both signs of virtue and instruments of rule, he may seem to have been trying to resurrect feudal sociality; but he can also be read simply as voicing the commonplace suspicion of central government that was strengthened by the last intrigues of James II and the fallout of the Glorious Revolution. By 1711 this suspicion was such that it had become difficult for the Crown to carry out any nationwide information gathering (especially since the "nation" now included the geographically remote territory of Scotland). As Patricia Cline Cohen has pointed out, the British government's two primary attempts to gather information on a large scale both collapsed by the 1720s.[21] After the Board of Trade was reorganized in 1696, for example, the newly appointed inspector general of imports and exports began to collect raw data about trade from the reformed customs records, but apart from one table drawn up in 1701, no compilations were ever assembled that could make this numerical material useful. In 1694 a similar project was begun that would have advanced another of Petty's ambitions: in that year Parliament required each locality in England and Wales to list all its inhabitants so that taxes could be levied on births, deaths, and marriages and so that bachelors over age twenty-five, as well as childless widowers, could be fined.[22] Although this law stayed in force until 1706, it met with considerable popular opposition and seems not to have yielded either substantial or useful information.

As Cohen also notes, the failure of political arithmetic in Britain was overdetermined. The suspicion of central government, which focused on the unfair extraction of taxes for politically interested projects, was compounded by the even older fears that "numbering the people" would provoke God's wrath and that government-sponsored attempts to standardize weights and measures would subvert local customs for administering justice.[23] Partly because of the resistance to central government in Britain, in the late seventeenth and early eighteenth centuries no apparatus existed for collecting numerical information on a large scale, and no theories had achieved consensus about how to bridge the gap between numbers collected from actual counting and the large-scale numbers that could be used for administrative purposes. As I have already begun to explore, moreover, these obstacles persisted partly because the mode of governmentality associated with the financial revolution solicited behaviors—as well as theories about those behaviors—that seemed to have little to do with counting. To see why political arithmetic clashed with this new mode of governmentality, it is helpful to take up a text in which the ambitions of political arithmetic coexisted with an embryonic recognition that the efficacy of liberal governmentality depended on reconceptualizing—or even retraining—the subjective responses of individuals: Daniel Defoe's *Essays upon Several Projects, or Effectual Ways for Advancing the Interests of the Nation.*

Composed in 1697 and published in 1702, Defoe's *Essays* begins to illuminate what happened when one tried to implement a political arithmetic scheme in the context of an emergent, but not yet fully operational, market system of government by taste. At the same time, the work enables us to see some of the epistemological problems posed by the emergent mode of liberal governmentality. The affinities between Defoe's projects and Petty's political arithmetic are clear. Beyond his direct allusions to Petty's calculations, which I will discuss shortly, Defoe also recommended that the monarch establish a commission of assessment, which would be charged with assessing personal estates for the purposes of taxation, and a land registry, "so that mortgages might be very well kept, to avoid frauds."[24] He supported a centralized state bank (although he had reservations about the Bank of England itself), parliamentary acts devoted to the improvement of highways, a government-sponsored system of maintenance for the mentally disabled, a royal academy for military exercises, and a national registry for seamen (*Essays,* 12, 16, 29, 39, 45). All these projects would have centralized power and enforced compliance through government regulation; most important, each of them would have aided the collection of information and promoted what Defoe calls "universal correspondence," that system of information and product exchange epitomized by the system of highways that he also wanted to extend and improve (*Essays,* 16).

Despite the obvious affinities between these projects and the schemes Petty outlined, however, Defoe's proposals differ from Petty's in some significant ways. Most striking, perhaps, is the very concept of "project," to which he devotes the first pages of his essay. Like Petty, Defoe was interested in large-scale enterprises: "The true definition of a project," he explains early in the *Essays,* "is . . . a vast undertaking, too big to be managed." Like Petty again, Defoe assumed that these projects would ultimately benefit the nation as a whole: "Projects of the nature I treat about, are doubtless in general of public advantage, as they tend to improvement of trade, and employment of the poor, and the circulation and increase of the public stock of the kingdom" (*Essays,* 9, 7). Unlike Petty, however, Defoe associated projects with private enterprise, not with the monarch or state government; specifically, he associated projects with merchants and, even more specifically, with the motivations and moral capacities that trade encouraged merchants to develop. Emphasizing subjective motivation led him to reverse Petty's position on interest. Arguing that modern projects would align "public good, and private advantage," he insisted that the motive of private gain was the primary incentive to that individual ingenuity that would ultimately benefit the nation (*Essays,* 10).

To Defoe, merchants were the first citizens of what he calls the "Projecting Age," both because "every new voyage the merchant contrives, is a project" and because French triumphs at sea early in the current war provoked extraordinary efforts from English merchants: "These, prompted by necessity, racked their wits for new contrivances, new inventions, new trades, stocks, projects, and anything to retrieve the desperate credit of their fortunes" (*Essays,* 7, 6). Since the very nature of trade tended to make the "true-bred merchant, the most intelligent man in the world, and consequently the most capable," English merchants rose to the challenge posed by the French: "To this sort of men it is easy to trace the original of banks, stocks, stock-jobbing, assurances, friendly societies, lotteries, and the like" (*Essays,* 7). As the rest of Defoe's pamphlet makes clear, English merchants generally prospered from these projects, although, as he repeatedly claims, government intervention—especially with regard to banks and highways—was still necessary to consolidate the interests of the merchants.

This, of course, was the problem that Defoe faced (along with subsequent theorists of liberal governmentality): How could one arrive at the proper ratio between mercantile freedom and government regulation so as to enhance both the opportunity for private profit and the likelihood of national prosperity? In his *Essays* he tries to address this problem in relation to two schemes in particular: his plan for a national pension office, and his scheme to set up a society of polite learning. Both of these projects reveal that the problem of adjudicating between too much regulation and too little was not limited to merchants or trade. They also demonstrate that, once one embraced the kind of society that rested on trade—that is, a market society—it became both difficult and imperative to locate reliable mechanisms for ensuring compliance from citizens who wanted simultaneously to be like and different from everyone else.

Defoe's projected pension office was intended to provide a social safety net for honest members of the working class. Essentially a social security scheme,

the pension office was the cornerstone of his grandiose plan to suppress beg-
ging. It is worth noting from the outset that Defoe is typical in applying the
method of political arithmetic to the poor rather than to members of the middling or landed classes; unlike Petty, who would have counted everyone, so as to
manage and tax them, eighteenth-century practitioners of political economy
tended to devote the vast majority of their attention to what contemporaries
called "the great body of the people."[25] Like Petty's projects for assessing taxes,
Defoe's scheme for a pension office aspired to coverage (of the designated
group): he wanted everyone concerned to register by name, trade, and place of
abode in a central office, which would be staffed by a secretary, a clerk, and a
searcher. Unlike Petty's tax plans, however, Defoe's pension scheme resembled
a nineteenth-century friendly society more than a government-sponsored welfare system. All registrants would be required to contribute to what would
eventually be the pool that funded their own maintenance. As one might expect, the problem this scheme presented was how to marry what was essentially
a free enterprise system to the subjective attitude necessary to make it work.
Defoe had to figure out how to ensure that "all sorts of people, who are labour-
ing people, and of honest repute" would actually register, and he also had to devise a way to check whether they had registered.

In the best of all worlds, Defoe believes, the problem of compliance would
be solved by a simple appeal to "sense" and "interest." "Want of consideration is
the great reason why people do not provide in their youth and strength for old
age and sickness," Defoe announces; "all men should have sense enough to see
the usefulness of such a design, and be persuaded by their interest to engage in
it" (*Essays,* 25, 27). Unfortunately, however, not everyone was motivated primarily by sense and interest. Because he recognized that "some men have less
prudence than brutes, and will make no provision against age till it comes," Defoe bolstered appeals to self-interest with a series of measures designed to enforce compliance. As he describes them, however, it becomes clear that even
these measures could not ensure obedience; indeed, because they were not
backed by the law, these measures could derive authority only from the informal and mysteriously subjective system of public opinion. Thus Defoe recommends sending officers to dispense information about the office, and even
though he threatens to have the clergy, justices of the peace, and beadles of the
parish refuse parish relief if the indigent refuse to register, he knows that he has
no authority to make good his threat. "I know that by law no parish can refuse
to relieve any person or family fallen into distress," Defoe admits, "and therefore
to send them word they must expect no relief, would seem a vain threatening;
but thus far the parish may do; they shall be *esteemed* as persons who deserve no

relief, and shall be *used* accordingly; for who indeed would ever *pity* that man in his distress, who, at the expense of two pots of beer a month, might have prevented it, and would not spare it?" (*Essays,* 29, my emphasis; see also 28).

As this passage illustrates, Defoe wanted to manipulate both individual feelings and public opinion more generally, because he recognized that such subjective props were necessary to support—or even substitute for—government coercion. Making people believe that giving their pennies to someone else would provide greater benefits *in the long run* than drinking in the pub today, of course, required more than one man's assurances, especially in the absence of laws that could enforce the rectitude of the pension society and when the entire scheme was based on the combined vicissitudes of the market (inflation and deflation) and the imponderables of mortality rates. In an attempt to anchor the belief he wanted to create in something other than his own assertion that this scheme would work, Defoe initially invoked political arithmetic. Initially, that is, he tried to encourage a new attitude toward saving by using incontrovertible numbers. Thus he resurrected Petty's estimate that one person out of every forty dies every year in order to figure out how much each participant would have to pay to ensure adequate returns without bankrupting the business.[26] Although this section of the *Essays* contains impressive columns of numbers that purport to reach unimpeachable conclusions about the charges the proposed office will incur, however, Defoe finally acknowledges that even he does not trust this method. Repeatedly he admits that his figures are only guesses, and in conclusion he confesses that he is not certain he has performed the calculations correctly: "As to my calculations, on which I do not depend neither, I say this, if they are probable, and that in 5 years' time a subscription of 100,000 persons would have 87,537*l.* 19*s.* 6*d.* in cash, all charges pay, I desire any one but to reflect what will not such a sum do" (*Essays,* 28, 29). Such a qualified statement, in which Defoe's uncertainty focuses both on the method of political arithmetic and on its ability to inspire the proper affective response (confidence), illustrates just how inadequate political arithmetic seemed to be in a context in which attitude mattered as much as information.[27]

Despite the resemblance between many of Defoe's projects and Petty's political arithmetic schemes, then, Defoe was too committed to private enterprise, a market society, and the role that subjective dynamics played in the market to endorse wholeheartedly either government-sponsored schemes for collecting information or the specific calculations that Petty had offered. Although he had begun to imagine that the collective subjectivity that manifested itself as public opinion might supplement law in a market society, however, he was still not certain how to consolidate or control personal attitudes, nor was he sure that, even if such attitudes could be consolidated, public opinion would govern wisely on its own. As his repeated use of the term "custom" implies, he seems to have realized that the desire to imitate was one of the chief motors of public opinion. Because he lacked terms adequate to understand desire, imitation, or custom, however, and because he was uncertain that imitation would really benefit the nation, Defoe repeatedly sought to bolster the subjective dynamic of emulation and opinion with holdovers from the mode of sovereign government—with laws, threats, and even an official instrument of censorship.

We can see the transitional nature of Defoe's *Essays upon Several Projects* in his proposal to found an academy to encourage polite learning. On the one hand, this academy was to be staffed by "private gentlemen—and a class of twelve to be open for mere merit" (*Essays,* 36). On the other hand, it was to be sponsored by the king. Indeed, Defoe pitched his proposal to William in terms reminiscent of the terms Petty used to interest James II, by conflating princely ambition with national reputation: "The present King of England," begins Defoe's panegyric, "as in the war he has given surprising instances of a greatness of spirit more than common, so in peace, I dare say, with submission, he shall never have an opportunity to illustrate his memory more than by such a foundation, by which he shall have opportunity to darken the glory of the French king in peace, as he has by his daring attempts in the war" (*Essays,* 35). One reason Defoe sought state sponsorship was that he considered the English language, whose improvement was one of his primary targets, a national property. Another reason was that he could not imagine any way to enforce compliance with the standards the academy set other than institutionalizing something approaching censorship.

With the Licensing Act so recently expired (in 1695), it may seem odd that in 1697 or 1702 a Dissenter so full of his own projects would call for anything that even resembled state censorship. That he did is a measure of Defoe's uncertainty about how a society governed by the subjective mechanisms of emulation and opinion could preserve its citizens' virtue. Defoe clearly considered the national academy the cornerstone of his ambitious plan for national improvement, for he calls it "the most noble and most useful proposal in this book" (*Essays,* 38). When it came time to describe how the academy would work, however, he fell back on a confused combination of appeals to "encouragement" and allusions to official approval (and by extension disapproval). The confusion he voiced about how the national academy would govern, of course, repeats the problem that the emergent mode of liberal government posed in general; and the strategy by which he sought to resolve this problem reflects the limit of the epistemological resources available to the would-be theorist of human motivation at the turn of the eighteenth century.

In Defoe's initial description of the academy, he seems content with encouraging a change in attitudes, even though his allusion to purgation suggests that the academy's work may also involve more draconian tactics.

> The work of this society should be to encourage polite learning, to polish and refine the English tongue, and advance the so much neglected faculty of correct language—to establish purity and propriety of style, and to purge it from all the irregular additions that ignorance and affectation have introduced; and all those innovations in speech, if I may call them such, which some dogmatic writers have the confidence to foster upon their native language, as if their authority were sufficient to make their own fancy legitimate. (*Essays,* 36)

Defoe's desire for an official authority to displace the upstart claims advanced by "dogmatic writers" implies that, despite his invocation of encouragement and custom, he did not actually want the English language to be the product of either merely subjective motives or market-driven competition. As he continues, it becomes clear that he did not want the language to be shaped by the affective desire behind emulation, because without some official authority to control what writers chose to imitate, there was no guarantee that they would select socially constructive models. Thus Defoe confidently proclaims that "custom, which is now our best authority for words, would always have its original here [in the academy], and not be allowed without it"; but he immediately acknowledges that "custom" was currently ruining the English language by sanctioning an "inundation" of swearing. In a digression remarkable for its vehemence and length, he suddenly launches into a diatribe against "swearing, that lewdness of the tongue, that scum and excrement of the mouth" that highlights the problem inherent in government by emulation.

> 'Tis a senseless, foolish, ridiculous practice: 'tis a means to no manner of end; 'tis folly acted for the sake of folly, which is a thing even the devil himself don't practise. . . .This, of all vicious practices, seems the most nonsensical and ridiculous; there is neither pleasure nor profit; no design pursued, no lust gratified, but is a mere phrensy of the tongue, a vomit of the brain, which works by putting a contrary upon the course of nature. (*Essays,* 37)

Considered in the abstract, it is difficult to understand Defoe's vehement outburst against swearing. Understood in relation to his evolving model of government by emulation, however, his fury makes sense, for swearing represents both the limit of legal sanctions—"to suppress this, laws, acts of parliament, and proclamations, are baubles and banters"—and the limit of the writer's ability to control the subjective dynamics he had identified at the heart of liberal governmentality. In fact, as Defoe acknowledges, the current rage for swearing had been spawned by the very subjective mechanism he wanted to use to control it: custom. "Custom has so far prevailed in this foolish vice that a man's discourse is hardly agreeable without it" (*Essays*, 38, 36).

If custom can no more police people's attraction to swearing than can laws, then how is this "phrensy of the tongue" to be stopped? Once more, Defoe imagines a combination of subjective motivation and sovereignlike coercion. He recommends that gentlemen simply set a good example: "If the gentlemen of England would once drop it as a mode, the vice is so foolish and ridiculous in itself 'twould soon grow odious and out of fashion" (*Essays*, 38). But he immediately points out that someone must set an example for the gentlemen, and he confers on the academy to which he entrusts this work a power that considerably transcends simply modeling good behavior. Although Defoe does not explicitly name censorship here, his reference to the vetting of plays implies that writing that did not pass muster might well be barred from the court of public appeal.[28]

> I believe nothing would so soon explode the practice [of swearing] as the public discouragement of it by such a society; where all our customs and habits, both in speech and behaviour, should receive an authority. All the disputes about precedency of wit, with the manners, customs, and usages of the theatre, would be decided here. Plays should pass here before they were acted, and the critics might give their censures, and damn at their pleasure: nothing would ever die which once received life at this original. The two theatres might end their jangle and dispute for priority no more; wit and real worth should decide the controversy, and here should be the infallible judge. (*Essays*, 38)

Defoe's climactic desire for an "infallible judge" suggests that he was finally unwilling to entrust something as important as the English language to the subjective dynamics intrinsic to government by emulation. In 1702, in fact, he wanted to remove everything that was really valuable from the market where custom and taste ruled, even though to do so he could only fall back on a series of claims about the absolute power of reason, meaning, and sense that his intermittent enthusiasm for market society seemed to challenge. The basis for these claims was a distinction between what he presents as two kinds of language: "words" and "noise" or "words" and "sense." This distinction bears a functional correspondence to Petty's distinction between "superlative Words" and "Arguments of Sense," but whereas Petty used numerical representation to figure his preferred alternative, Defoe resorts to a series of linguistic distinctions to explain how "words" differ from "noise":

> 'Tis true custom is allowed to be our best authority for words, and 'tis fit it should be so; but reason must be the judge of sense in language, and custom can never prevail over it. Words, indeed, like the ceremonies in religion, may be submitted to the magistrate; but sense, like the essentials, is positive, unalterable, and cannot be submitted to any jurisdiction: 'tis a law to itself, 'tis ever the same, even an act of parliament cannot alter it. . . . [T]here is a direct signification of words, or a cadence in expression, which we call speaking sense: this, like truth, is sullen, and the same ever was and ever will be so, in what manner and in what language soever 'tis expressed. Words without it are only noise, which any brute can make as well as we, and birds, much better; for words without sense make but dull music. Thus, a man may speak in words, but perfectly unintelligible as to meaning; he may talk a great deal, but say nothing. But 'tis the proper position of words, adapted to their significations, which makes them intelligible, and conveys the meaning of the speaker to the understanding of the hearer. (*Essays*, 37)

In Defoe's struggle to define the properties of language that were exempt from the subjective dynamics at play in custom, we see an effort to identify a strictly communicative function for language—an effort, that is, to exclude the property of persuasion with which ornamental language was traditionally associated. He struggled with this problem because the opprobrium increasingly heaped on eloquence or rhetoric made it seem critical to distinguish between the communicative and persuasive properties (or uses) of language.[29] Defoe could not solve this problem, however, and the absence of an obvious marker of linguistic difference makes it impossible to see where "words" leave off and "sense" begins. His proposed academy could have institutionalized this distinction, of course, just as the Royal Society ruled on what counted as a natural matter of fact. In the absence of such an academy, Defoe could only gesture toward an absolute standard of judgment that lay beyond both subjective dynamics and the market. In 1702, despite his enthusiasm for the emergent market society, he simply could not understand how subjective factors like emulation and custom could be allowed to govern something as vital as the English tongue.

In the *Essays upon Several Projects,* then, we see the problems that materialized when one attempted to adapt political arithmetic projects to a market society. On the one hand, the emergent market society clearly needed projects conceptualized on the scale implicit in Petty's schemes and with the potential to cover (and unite) the nation; highways constituted the paradigm of the kind of centralized improvement Britain's fledgling market society needed. On the

other hand, however, as Defoe's oscillation between bows to emulation and in-
vocations of the kind of governing instruments more typically associated with
sovereignty demonstrates, it was not clear how—or whether—the subjective
dynamics intrinsic to private enterprise would submit to, much less generate,
governing mechanisms adequate to stimulate both commerce and virtue. For
the rest of his writing life, Defoe intermittently returned to these problems. Although he conjured numerous formulas for the proper ratio between freedom
and government regulation and between desire and censorship, his *Complete
English Tradesman* provides a particularly revealing preview of how he eventually decided to address this characteristic problem of liberal governmentality.

Defoe's *Complete English Tradesman* was published in 1726, just five years
before the prolific author's death. The purpose of this self-described "collection of useful instructions" was to teach the young tradesman how to survive
and prosper in a society permeated by the market values that had been incipi-
ent, but not yet dominant, at the turn of the century. "Tradesmen cannot live as
tradesmen in the same class used to live," he warns; "custom, and the manner of
all the tradesmen round them, command a difference; and he that will not do as
others do, is esteemed as nobody among them, and the tradesman is doomed to
ruin by the fate of the times."[30] Even in this brief sentence, we can take the
measure of the changes Defoe had witnessed. By 1726 custom had come to
"command," emulation was a necessity, and the price of failure to conform was
not only the loss of esteem but "ruin" both social and financial. "There is a fate
upon a tradesman; either he must yield to the snare of the times, or be the jest of
the times; the young tradesman cannot resist it; he must live as others do, or lose
the credit of living, and be run down as if he were bankrupt. In a word, he must
spend more than he can afford to spend, and so be undone; or not spend it, and
so be undone" (*Tradesman*, 4).

Faced with the tyranny of custom, Defoe recommended that the young
tradesman school his personal desires: express but limit your passion for fash-
ionable dress, he cautions; cultivate a deferential mode of addressing your customers; and most of all, develop a style of writing that signifies virtue and
establishes your creditworthiness among your peers. In his discussion of mercantile writing, we can begin to see how Defoe negotiated the subjective component of the mode of government associated with market society. Essentially,
he did so by displacing the problem of subjective dynamics—the problem of in-
citing the desire for virtue—by a solution that did not require changing (or even
knowing) what the tradesman *really* felt. His solution was to shift the emphasis
from the internal domain of subjectivity to the external arena of writing style.
"As plainness, and a free unconstrained way of speaking, is the beauty and excellence of speech," he explained, "so an easy free concise way of writing is the best

EXPERIMENTAL MORAL PHILOSOPHY

style for a tradesman. He that affects a rumbling and bombast style, and fills his
letters with long harangues, compliments, and flourishes, should turn poet instead of tradesman, and set up for a wit, not a shopkeeper" (*Tradesman,* 17).

The distinction between "plainness" and "bombast" should remind us of
the distinction championed by advocates of both double-entry bookkeeping
and mercantile expertise: the distinction between the plain style epitomized by
the merchant's accounts and the *copia* associated with rhetoric. Like the six-
teenth- and seventeenth-century apologists for mercantile writing, moreover,
Defoe claimed that the merchant's plain style would be read as a sign of his
virtue. Unlike John Mellis, however, he insists that style would be read as a sign
of virtue because plain writing communicates ideas clearly, not primarily be
cause the precision of the accounts embodies the divine property of balance.

> The tradesmen need not be offended at my condemning them, as it were, to a plain
> and homely style—easy, plain, and familiar language is the beauty of speech in gen-
> eral, and is the excellency of all writing, on whatever subject, or to whatever per-
> sons they or we write or speak. The end of speech is that men might understand
> one another's meaning; certainly that speech, or that way of speaking, which is
> most easily understood, is the best way of speaking. If any man were to ask me,
> which would be supposed to be a perfect style, or language, I would answer, that in
> which a man speaking to five hundred people, of all common and various capaci-
> ties, idiots or lunatics excepted, should be understood by them all in the same man-
> ner with one another, and in the same sense which the speaker intended to be
> understood—this would certainly be a most perfect style. (*Tradesman,* 23)

According to Defoe, the tradesman's style was universally comprehensible
because it was a form of "exact writing," which captured the details of the commercial transaction and, in so doing, theoretically conveyed the tradesman's
honesty.[31] To a large extent, Defoe seems to have thought that a tradesman's
plain style of writing had a performative, rather than a mimetic, relation to his
subjective feelings. In other words, he believed that a tradesman's plain style encouraged him to be honest, because plain writing inspired *in others* the confidence that underwrote both business and credit; to merit this confidence, the
tradesman had to imitate the honesty his writing expressed. The result is what
Defoe calls "a harmony of business":

> Here is a harmony of business, and every thing exact; the order is given plain and
> express; the clothier answers directly to the point; here can be no defect in the cor-
> respondence; the diligent clothier applies immediately to the work, sorts and dyes
> his wool, mixes his colours to the patterns, puts the wool to the spinners, sends his
> yarn to the weavers, has the pieces brought home, then has them to the thicking or

fulling-mill, dresses them in his own workhouse, and sends them punctually by the time; perhaps by the middle of the month. Having sent up twenty pieces five weeks before, the warehouse-keeper, to oblige him, pays his bill of £50, and a month after the rest are sent in, he draws for the rest of the money, and his bills are punctually paid. The consequence of this exact writing and answering is this—

The warehouse-keeper having the order from his merchant, is furnished in time, and obliges his customer; then says he to his servant, "Well, this H. G. of Devizes is a clever workman, understands his business, and may be depended on: I see if I have an order to give that requires any exactness and honest usage, he is my man; he understands orders when they are sent, goes to work immediately, and answers them punctually." (*Tradesman,* 21)

In this vignette, we see that by 1726 mercantile writing was explicitly being promoted as an instrument of (self-) government, which had implications for, but did not directly address, the question of how attitudes (as opposed to behaviors) could be shaped. The potential of rule-governed writing to encourage rule-governed behavior was implicit in the system of double-entry bookkeeping, of course, but Defoe argues that this potential could be fully realized in the credit economy of eighteenth-century Britain because, once sovereignty was replaced by the market, a publicly recognizable form of self-government was necessary to encourage trust. By emphasizing the behavior exemplified by style—a style of writing, a style of doing business—he essentially identified the feature of the market where public signs mattered more than actual attitudes or beliefs. Style could be trusted to govern, in other words, because, as the basis of credit, it constituted an incentive that no one could afford to ignore.[32] Ideally, individuals would be willing to govern themselves, to refrain from behaviors that might impugn their creditworthiness, because they wanted both to ape their social superiors and to differentiate themselves from the upstarts nipping at their heels. Ideally, and because the good things of life (even land) could more readily be purchased than ever before, individuals would govern themselves because they wanted the credit necessary to buy, because they wanted the goods essential to make their credit visible.

Of course, as Defoe knew so well, government by emulation did not guarantee virtue. Indeed, even though highlighting style instead of subjectivity seemed to set aside the problem of human motivation, it did not overcome the obstacles generated by the absence of a compelling account of subjectivity. If anything, in fact, Defoe's elevation of style made this lack more damaging because it explicitly raised the possibility that what looked like mercantile honesty might be *merely* a matter of style. When Defoe broaches this sticky subject in *Complete English Tradesman,* he initially explains that the duplicity the trades-

man sometimes had to practice was simply a concession to the unsociable behavior of his customers, especially his "gossiping, tea-drinking" *lady* customers.
"A tradesman behind his counter must have no flesh and blood about him, no
passions, no resentment," Defoe explains. Because of such ladies, he continues,
a man "must be a perfect complete hypocrite, if he will be a complete tradesman" (*Tradesman*, 64, 70). By the end of the book, however, Defoe presents the
tradesman's duplicity not simply as a martyrdom to the ladies, but more generally as an inescapable facet of modern society itself: "All the ordinary communication of life is now full of lying; and what with table-lies, salutation-lies, and
trading-lies, there is no such thing as every man speaking truth with his neigh-
bour" (*Tradesman*, 165). He even implies that people had come to *like* being deceived, because pleasure as well as commercial gain accrued to those lies by
which things were made to appear what they were not.

For Defoe, who was struggling to define the mode of (self-) government
appropriate to a market society, the plain style of mercantile writing seemed the
best defense against the excesses he associated with "rumbling" rhetoric. As
these last passages illustrate, however, the excess of linguistic "rumbling" could
not finally be defended against in a society that delighted in novelty and ornament. If the tradesman *had* to lie about his wares (or at least embellish them) in
order to be stylish—in order, that is, to converse with a society that valued plea
sure as much as virtue—then how could the plain style succeed either in guaranteeing virtue or in enhancing trade? Despite his attempts to distinguish plain
style from mere lies, Defoe was finally forced to recommend as a compromise
what he calls a "grave middle way of discoursing to a customer." Even this compromise seems fragile, however, given the predilection of lady customers (in
particular) for "false and foolish words" and the incentive the market offered the
tradesman to profit from cheap and shoddy goods.[33]

Defoe finally attributed the tradesman's almost inevitable descent to deceit to
two features of the market economy, both of which could be exacerbated (or
counteracted) by the tradesman's facility with language. The first feature was
(what we would term) psychological: customers wanted to experience consumption as an affective, even aesthetic exchange, and this desire could be enhanced by the tradesman's "fawning and flattering language." The second
feature was material: "sorry, unfashionable, and ordinary goods" dominated the
market, and these required the tradesman to make up in bombast what his wares
lacked in quality (*Tradesman*, 179). In theory at least, both the psychological and
the material incentives to deceit could have been avoided had the early market
economy had standards: standards for aesthetic appreciation, which could have
distinguished between the "high" affect of aesthetic taste and the "low" affect of

commodity lust, and standards for consumer goods, which would have protected the consumer (and the tradesman) from both shoddy goods and the motive to deceive. Establishing standards for commodities, of course, would have raised the same problem I have already associated with liberal governmentality: how to discriminate between too much government oversight and too little. Establishing standards for aesthetic appreciation, by contrast, did not need to involve state government at all. Indeed, as David Hume contended in "Of the Standard of Taste," many of the problems associated with government by emulation could be addressed if one simply mapped emulation onto taste—if, that is, one could subject individuals' complex desires to both imitate and outshine their neighbors to a trustworthy faculty of discrimination.

"Of the Standard of Taste" was published in 1757, after Hume had abandoned experimental philosophy. To take up Hume's work at this point may thus seem to anticipate the argument I develop in the last section of this chapter. I do so despite this risk, however, because both the problem Hume addresses here and the solution he devised show why eighteenth-century efforts to theorize the human motivations that underwrote government by emulation should be read as *proto*psychological accounts. Because these accounts were contributions to a theory of government and not primarily attempts to decipher individual psychopathology, they focused on *social* interactions, as the arena where personal morality and collective goods (political, ethical, and material) were produced. Like Addison and Steele, who were primarily interested in *conduct,* and Defoe, who focused on *custom,* Hume was interested in social behaviors. Unlike these writers, however, he wanted to move beyond behaviors that could be said to be public to a more comprehensive exploration of subjective motivation. Thus he took as his subject *judgment,* a topic that moved the intersection of social and psychological concerns further in the direction of the psychological but that, in preserving the category of *taste,* retained the social (collective) dimension of what later writers would construe as an idiosyncratic (and fully psychological) process.

The epistemological problem introduced by Hume's attempt to theorize the relation between subjective feelings and collective categories like taste was how to move from idiosyncratic, individual instances of judgment to the kind of generalization that could function as a standard, and thus as a basis for making general claims about human motivation, sociality, and government. This problem, he explains, was visible in the great variety of tastes with which everyone was familiar: "The great variety of Taste, as well as of opinion, which prevails in the world, is too obvious not to have fallen under every one's observation."[34] Because Hume was committed to (a variant of) Baconian induction, he considered it important to begin with such observations. Because he was also committed to producing philosophical (systematic) knowledge, however, he also wanted to move from such commonplace observations, which tended to emphasize the idiosyncrasy of individual instances of taste, to some other mode of analysis that could reveal the "general principles of taste" that presumably underwrote—and somehow canceled out—idiosyncratic instances. To understand how Hume generated universal principles out of individual observations, we need to recover a peculiar use of mathematics that characterized the eighteenth-century practice of experimental moral philosophy in general.

We can begin to see how Hume negotiated the gap between the individual (observation or instance) and philosophical generalization in his references to experience. Like Bacon, he claimed to consider experience the foundation of all rules—in this example, "rules of composition": the foundation for "the rules of composition," he explains, "is the same with that of all the practical sciences, experience." Immediately, however, he defines experience in such a way as to set aside *individual* experience: "Nor are they [these rules] any thing but general observations, concerning what has been universally found to please in all countries and in all ages" ("Standard," 231). In this quotation we can see that Hume assumed that some relation existed between "experience" and "general observations," but it is not yet clear how he gets from "experience," which might belong to an individual, to "general observations," which obviously do not

To follow this move, we need to understand why Hume considered Baconian induction inadequate to the philosophical project. He did so for two reasons. First, he assumed that those qualities Locke called secondary existed in various degrees and mixtures in every object; as a consequence, he claimed that the "disorder" resulting from the mixture of minute quantities of qualities can obliterate one's ability to distinguish individual qualities. In one sense taste is "general" because no individual exercise of taste can make the fine distinctions that scientific instruments can record. "As these qualities may be found in a small degree, or may be mixed and confounded with each other, it often happens, that the taste is not affected with such minute qualities, or is not able to distinguish all the particular flavours, amidst the disorder" ("Standard," 235). Second, Hume believed that human beings are susceptible to prejudice and limited by a combination of their affective response to beautiful objects and the narrowness of their experience; thus he insisted that "the general principles of taste" could be obscured by the very act of observation. "There is a flutter or hurry of thought which attends the first perusal of any piece, and which confounds the genuine sentiment of beauty. The relation of the parts is not discerned: The true characters of style are little distinguished: The several perfections and defects seem wrapped up in a species of confusion, and present themselves indistinctly to the imagination" ("Standard," 238).

To address the problems introduced by the mixture of qualities in the object and the inadequacy of perception in the subject, Hume invokes something resembling a mathematical corrective. That is, he proposes that by *repeatedly* surveying the object so as to "form *comparisons* between the several species and degrees of excellence, and estimat[e] their proportion to each other," the observer can hope to compensate for the limitations imposed by peculiarity and singularity ("Standard," 238). If Hume used something like an additive process to compensate for the limitations of experience, then he invoked something resembling subtraction to correct those prejudices that also confounded judgment. Faced with the imperative to judge, he argues, the critic must forget his individuality: "Considering myself as a man in general, [I must] forget, if possible, my individual being and my peculiar circumstances" ("Standard," 239).

I will call this invocation of mathematical processes and language "gestural mathematics" because it summons mathematical procedures and connotations by gesturing toward mathematics, by invoking a tropical relation between philosophical analysis and mathematics. That it is critical to Hume's method is clear from his explanation of how "models and principles" of taste are generated and recognized. "Wherever you can ascertain a delicacy of taste, it is sure to meet with approbation," he asserts; "and the best way of ascertaining it is to appeal to those models and principles, which have been established by the uniform consent and experience of nations and ages" ("Standard," 237). Presumably, to achieve "uniform consent," one must discount individuals whose judgment departs from that of people in general; presumably, to identify the "experience of nations and ages," one must somehow add up the judgments produced across space and over time. Of course Hume did not really subtract idiosyncratic opinions or add up judgments he had actually counted; nor does he claim, as Petty did, that *if* we counted, then numbers would convey the truth. Instead, he depicts evaluation as being *like* counting in involving an estimation of quantity but *unlike* counting in reaching conclusions through some method other than arithmetic. This method is like that of mathematics, then, but not because it involves counting or real mathematical operations; instead, both enable one to prove—to demonstrate—what one assumes in the first place. What Hume assumed in the first place was that all individuals are somehow the same, that "human nature" is universal.

In theory, then, and if one accepts Hume's foundational assumption (that all individuals are the same), some combination of additive experiences and "a proper violence [imposed] on [the] imagination" can transform a mere individual into the "true judge" who adjudicates the "true standard of taste and beauty": "Strong sense, united to delicate sentiment, improved by practice, perfected by comparison, and cleared of all prejudice, can alone entitle critics to this valuable character; and the joint verdict of such, wherever they are to be found, is the true standard of taste and beauty" ("Standard," 241). This true judge, like Defoe's "infallible judge," would presumably have solved the problem of government written into market society, because by establishing the standard of taste—or by encouraging all individuals to recognize it in themselves—the true judge would have assessed the differences that both fueled competition and imperiled virtue. Of course, as Hume immediately acknowledges, once one has discounted individual judgment, admitted that qualities are mixed and confounded in objects, and cited the limitations of experience, it becomes impossible to specify the criteria by which ordinary individuals can distinguish this true judge from mere pretenders. Hume's gestural mathematics, in other words, could not really solve the epistemological problem *for—or of—the individual,* because, in the absence of some aggregate concept like the statistical population, which might explain how he actually moved from individuals and discrete observations to a philosophical generalization about "uniform consent," he could not describe how individuals could transcend the limitations of their particular experiences.

This acknowledgment provokes something like a spasm in the progress of the essay, which Hume associates with "embarrassment." As soon as he admits to this embarrassment, however, he immediately abandons his effort to describe how one moves from observations of individuals to the general idea of uniform consent in favor of what he considers more important than philosophical rigor: the *social outcome* of consensus.

> But where are such critics to be found? By what marks are they to be known? How distinguish them from pretenders? These questions are embarrassing; and seem to throw us back into the same uncertainty, from which, during the course of this essay, we have endeavoured to extricate ourselves.
>
> But if we consider the matter aright, these are questions of fact, not sentiment. Whether any particular person be endowed with good sense and a delicate imagination, free from prejudice, may often be the subject of dispute, and be liable to great discussion and enquiry: But that such a character is valuable and estimable will be agreed on by all mankind. Where these doubts occur, men can do no more than in other disputable questions, which are submitted to the understanding: They must produce the best arguments, that their invention suggests to them; they must acknowledge a true and decisive standard to exist somewhere, to wit, real existence and matter of fact; and they must have indulgence to such as differ from them in their appeals to this standard. ("Standard," 241–42)

In this crucial passage, we see how a particular philosophical embarrassment, which subsequent analysts were to call the problem of induction, propelled Hume from the kind of gestural mathematics that seemed designed to generate universals from individual experience to another mode of discourse altogether: "Where these doubts occur, men can do no more than in other disputable questions . . . : They must produce the best arguments, that their invention suggests . . . ; and they must have indulgence to such as differ from them." People had to turn from philosophy to something resembling conversation—which seems, in this account, to be derived or modeled on rhetoric—because it was more important *to society* for the philosopher to tolerate disagreement than it was to demonstrate conclusively how he could discount positions that departed from his own.

In "Of the Standard of Taste," Hume does not directly address the problem of induction. As a consequence, it is not quite clear how he could claim to generate even a protopsychology for the universal subject out of flawed observations of flawed individuals like himself. What is clear is that he believed that, to ensure both collective prosperity and individual virtue, the individuals who composed society *had to assume* that "a true and decisive standard [existed] somewhere." It was this necessity, which was finally social and not either psychological or philosophical, that grounded his assertions about "general principles" and "universal consent." The problems of liberal governmentality could be solved, he insisted, only if individuals recognized this necessity and acted accordingly—whether or not they understood how the differences so essential to market society were transcended by the common nature they (presumably) shared.

I will return to Hume and the problem of induction in the last section of this chapter. Before I do so, however, we need to explore in greater detail the philosophical method he was invoking and abandoning here: experimental moral philosophy. Like Hume, experimental moral philosophers deployed what I have called a gestural mathematics in order both to generate a universal subject of philosophy and to begin to sketch the subjective motivations that animated this subject. Doing so tended to confer on moral philosophy the prestige increasingly associated with Newton's mathematical method and to make the universal subject seem less like an a priori assumption than like an aggregate. Because most experimental moral philosophers were not troubled by Hume's skepticism, moreover, they also bolstered the universal subject by a set of assumptions about God that Hume could not endorse. The result was a concept that seemed to be generated from a combination of mathematics, experiment, and faith and that reinforced the assumptions with which the experimental philosophers embarked. In the form of human nature, mankind, and *homo economicus,* this concept not only dominated eighteenth-century British philosophy and historiography; it also anchored the science that was new in this century: political economy.

EXPERIMENTAL MORAL PHILOSOPHY

To appreciate the epistemological contributions of eighteenth-century experimental moral philosophy, we need to understand that it was a practice devised
in the image of natural philosophy but designed as an account of human motivation and thus, indirectly at least, as an instrument of liberal governmentality.
That is, like natural philosophers, experimental moralists claimed to describe
observed particulars and to extract from them the general laws or regularities
that informed them, especially those laws that explained the virtuous behaviors
that made individuals social (governable). As with natural philosophers, then,
one of the primary epistemological problems moral experimentalists faced was
how to assimilate the variations represented in particulars into kinds or types
sufficiently uniform to allow for claims about general laws. By and large, natural
philosophers used taxonomies to address this problem, because taxonomies
distinguished as salient those features that could be used to assimilate a group of
discrete but similar entities into a single commensurate kind. At the same time,
taxonomies allowed for variation because, by definition, many kinds could be
seen to coexist in the natural world, and one could describe regularities either
at the level of kind or at some higher level of abstraction. As we will see, experimental moralists departed from this model when they anchored their laws in
claims about the universality of human nature. These claims did not assemble a
variety of commensurate kinds but rendered entities that were radically heterogeneous instances of a single type.

Partly because the idea of universality essentially disallowed meaningful
departure from the type, the moral experimentalists' claims about regularities
were different in at least one respect from the claims that natural philosophers
made about natural laws. Whereas natural philosophers could claim simply to
describe observed phenomena and to derive regularities from their observations, experimental moralists had to account for the discrepancy between the
varieties of human behavior anyone could observe and the single type the
philosophers claimed was universal. This challenge was compounded by another consideration, which stemmed from moral philosophy's participation in
what was traditionally an ethical, or even a religious, arena. Because they
wanted to preserve the Christian notion of free will without relinquishing the
goal of generating universal laws, eighteenth-century philosophers had to devise a kind of law that could do more than simply describe. With the advantage
of hindsight, we can say that moral experimentalists met this challenge by formulating general laws that were *prescriptive* as well as *descriptive*—although it is
critical to remember that for them this would not have been a meaningful distinction. We would say that, while claiming simply to describe what moral ex-

perimentalists claimed to observe, these laws also functioned implicitly to police behavior by identifying moral and social imperatives and by locating these
imperatives in "nature" itself. Eighteenth-century philosophers, by contrast,
would have said they were discovering God's laws and making those laws visible
to their fallible but educable readers.

Recognizing the differences as well as the similarities between natural and
moral philosophies, and seeing that the latter's agenda can be described in these
two ways, helps explain a rift that has opened in scholarship about these writers.
Whereas some modern analysts, like Nicholas Phillipson, have been almost exclusively interested in the ethical claims of the moral experimentalists (thus taking the philosophers more or less at their word), others, like P. B. Wood, have
tended to emphasize *only* their scientism, as if this was incompatible with their
claims to be performing ethical work.[35] Rather than choosing between the two
approaches, I want simply to note that the second view—that a descriptive
("scientistic") agenda is incompatible with an ethical agenda—is an effect of
the disciplinary division that had barely begun to separate physical science from
ethics in eighteenth-century Britain. Before that division was complete, and
thus by contrast with the modern disciplines, the map of eighteenth-century
philosophy was relatively undifferentiated. As Richard B. Sher has demonstrated, moreover, Scottish universities constituted a privileged site at which
natural and moral philosophies not only coexisted but overlapped. In the
church-sponsored Scottish universities of the eighteenth century, as Sher reminds us, where professors of moral philosophy were entrusted with the spiritual as well as the intellectual development of boys in their middle teens, all
knowledge was conceptualized within a framework that was both implicitly
theological and explicitly moral.[36] In the light of this conceptualization of
knowledge and this population of students, it does not seem so strange that
early in the century the same individual—George Turnbull, for example—often taught both moral and natural philosophy or moved, as Adam Ferguson did,
from the chair of one subject to the chair of the other.[37] Nor should it seem
strange that the eighteenth-century moral experimentalists who were also university professors—Ferguson, Turnbull, Francis Hutcheson, and Thomas Reid
in particular—were concerned with generating knowledge that was *equally*
ethically efficacious (prescriptive) *and* true to observation (descriptive), *equally*
morally improving *and* as systematic as natural philosophical knowledge aspired
to be. That these philosophers could claim to generate both kinds of knowledge, however, should not completely obscure the epistemological difficulties
that arose when they did so.

Chief among these difficulties is what is now known as the problem of induction—that in an experimental philosophy (whether natural or moral), one

cannot know in advance of observation that instances yet to be observed will resemble those the philosopher has seen. For most eighteenth-century moral philosophers, this problem was effectively neutralized by the combination of a methodological ambiguity bequeathed by Newton and a variant of the providentialism also endorsed by natural philosophers. In the late 1740s, as we will see, David Hume finally formulated the problem of induction as a specific question—although he did not consider it ruinous to the production of moral truths. He did not consider it so, as we have already begun to see, because he elevated the social agenda of virtuous (self-) government over the scientific agenda of producing systematic knowledge. Because the stages by which the problem of induction gradually became visible can be glimpsed by comparing the natural philosophical method of Newton with Shaftesbury's theories about the link between sensory apprehension and virtue, I begin my discussion of experimental moral philosophy with the two epistemological projects most often identified as the immediate antecedents of experimental moral philosophy: Shaftesbury's aesthetics and Newtonian physics.

Anthony Ashley Cooper, third earl of Shaftesbury, was neither an experimentalist nor a university professor. Nevertheless, his work was critical to the experimental moralists, because he provided one model for a mode of governmentality that also purported to describe both human motivation and the moral universe more generally. For my purposes, the cornerstones of Shaftesbury's epistemological contribution were his emphasis on self-observation as an instrument of knowledge production (and moral improvement) and his reliance on mathematics as the manifestation of a virtue that was both pleasing and disinterested, both personal and universal.

In that generic mixture published as *Characteristics of Men, Manners, Opinions, Times* (1711), and in the context of the disrepute into which Scholastic philosophy had fallen in the moribund English university system, Shaftesbury sought to resuscitate philosophy so that it could serve as an instrument of both knowledge production and moral government. To do so, he transformed both moral philosophy and its ancient precursor, rhetoric, into an instrument that resembled the "art or science" of surgery. As he describes it, however, this surgery was to be performed not on another's body but on the self, through a kind of introspection that "multiplies" the self by dividing it into segments that can act independently.

> If it be objected against the above-mentioned practice and art of surgery, "that we can nowhere find such a meek patient, with whom we can in reality make bold, and for whom nevertheless we are sure to preserve the greatest tenderness and regard," I assert the contrary; and say, for instance, that we have each of us ourselves

to practise on. "Mere quibble!" you will say; "for who can thus multiply himself into persons and be his own subject? Who can properly laugh at himself, or find in his heart to be either merry or severe on such an occasion?" Go to the poets, and they will present you with many instances. Nothing is more common with them, than this sort of soliloquy. A person of profound parts, or perhaps of ordinary capacity, happens on some occasion to commit a fault. He is concerned for it. He comes alone upon the stage; looks about him to see if anybody be near; then takes himself to task, without sparing himself in the least. You would wonder to hear how close he pushes matters, and how thoroughly he carries on the business of self-dissection. By virtue of this soliloquy he becomes two distinct persons. He is pupil and preceptor. He teaches, and he learns.[38]

In Shaftesbury's account, introspection takes the form of an interior dialogue, which makes self-government resemble social interaction, but which also makes self-government the source of that virtue that moralizes social interaction. Introspection can be a source of virtue, in turn, because the dialogue that introspection provokes gradually yields understanding of the proper relation between the self and the whole of which the individual is but a part. This is the principal lesson Shaftesbury drives home in the long dialogue titled "The Moralists." The "main subject" of this dialogue, announces the sage Theocles, is

that neither man nor any other animal, though ever so complete a system of parts as to all within, can be allowed in the same manner complete as to all without, but must be considered as having a further relation abroad to the system of his kind. So even this system of his kind to the animal system, this to the world (our earth), and this again to the bigger world and to the universe. . . .See there the mutual dependency of things! the relation of one to another; of the sun to this inhabited earth, and of the earth and other planets to the sun! the order, union, and coherence of the whole! and know, my ingenious friend, that by this survey you will be obliged to own the universal system and coherent scheme of things to be established on abundant proof. (*Characteristics,* 2:65)

Moving outward from the observable self, this "survey" resembles natural philosophical observation, although Shaftesbury's insistence that what one sees when one "looks" is order and coherence suggests that observation is a trope here, just as surgery functions as a trope for self-dissection. Indeed, Shaftesbury *assumed* that order and coherence inform the interior universe, just as he assumed that an internal dialogue would eventually produce morality, even though he claimed that his "survey" established this order "on abundant proof." For the "proof" Shaftesbury offered was not experimental, nor did he reach his conclusions through induction. Instead, his proof, like his method, derived from the resemblance he posited between human nature, external nature, and the essential properties of mathematics: proportion, order, symmetry, and—as a consequence of these—truth. The heart of his embrace of mathematics, moreover, is what links his claims about human and external natures: his assumption that human beings naturally love mathematical properties, both because they give us pleasure and because they appear, in the form of beauty, in the external world. For Shaftesbury, the Platonic abstractions of proportion, order, symmetry, and beauty were more significant—because of their higher order of generality—than the particulars revealed by the senses.

> There is no one who, by the least progress in science or learning, has come to know barely the principles of mathematics, but has found, that in the exercise of his mind on the discoveries he there makes, though merely of speculative truths, he receives a pleasure and a delight superior to that of sense. When we have thoroughly searched into the nature of this contemplative delight, we shall find it of a kind which relates not in the least to any private interest of the creature, nor has for its object any self-good or advantage of the private system. The admiration, joy, or love turns wholly upon what is exterior and foreign to ourselves. . . .Having no object within the compass of the private system, it must either be esteemed superfluous and unnatural (as having no tendency towards the advantage or good of anything in Nature) or it must be judged to be what it truly is, "A natural joy in the contemplation of those numbers, that harmony, proportion, and concord which supports the universal nature, and is essential in the constitution and form of every particular species or order of being." (*Characteristics,* 1:296)

As the second half of this quotation reveals, the thesis for which Shaftesbury is (and always has been) best known—the claim that aesthetic contemplation is disinterested—was intimately connected to his understanding of the virtues of and induced by mathematics: contemplation of the proportion epitomized by mathematics does not serve private interests, he maintains; loving proportion in numbers, as we naturally do, simply enhances virtue, which is "no other than the love of order and beauty in society" (*Characteristics,* 1:279). For Shaftesbury, then, human subjectivity—the ground of liberal governmentality—was formed in the image of mathematical order, and thus was naturally attracted to it. Society could be orderly not, as Defoe had asserted, because individuals would dissemble to obtain what they wanted but because they wanted nothing more than to actualize the order they perceived in themselves.

As numerous modern commentators have noted, Shaftesbury may have claimed that the love of mathematical proportion was universal, but he also assumed that the disinterestedness necessary to perceive and appreciate proportion could be attained only by an elite few.[39] In this sense he did not provide a

universal account of human motivation—or more precisely, he did not produce a universal subject—even though he assumed that the mathematical qualities he admired were universally true and even though he advanced his claims
about proportion and beauty as claims about universals.[40] We need to recognize
that he did not produce a universal subject because we need to see that, to preserve a special role for elite, well-educated men, he kept separate the two functions that moral philosophers would soon fuse. To preserve a special role for
elite philosophers, Shaftesbury insisted that disinterested virtue had to be modeled or taught, so that what initially seem like descriptions of a universal love of
proportion turn out to be prescriptions for what (most) men could feel if they
learned from philosophers like him.

This point was not lost on Shaftesbury's contemporaries, especially since
the elite men to whom he attributed the capacity of disinterestedness tended to
belong to his political party, the Whigs. Thus contemporaneous critics, like
Mandeville and Hogarth, targeted Shaftesbury's politicizing of disinterestedness, along with the sensuality and economic motives they thought it masked,
in some of the most pointed satires of the early eighteenth century.[41] To
counter Shaftesbury's attempt to transform "self-interest" into "disinterestedness" by the alembic of aesthetics, Mandeville invoked the old reason-of-state
argument, which, we recall, was traditionally linked to the kind of information
collection about economic matters of fact associated with political arithmetic:
"Every Government ought to be thoroughly acquainted with, and steadfastly
pursue the Interest of the Country," Mandeville insisted in 1723.[42] By the
1720s, of course, would-be political arithmeticians' ambition to acquaint the
government with the interest of the country had itself become susceptible to
charges associated with the emphasis on subjectivity linked to liberal governmentality—that is, to charges of self-interest. When Mandeville argued that
"private vice" could lead to "public benefit," he played directly into this suspicion, for "private vice" was also a subjective motivation; however paradoxical it
seems, Mandeville also inadvertently endorsed the argument that Whigs like
Shaftesbury had already used to undermine reason-of-state apologies for strong
government. If "private vice," which apologists for the credit economy depicted as virtuous self-interestedness, *could* advance "public benefit," as even
Mandeville acknowledged it could, then *self*-government, not central government, was the best guarantor of both virtue and prosperity.

Superintending self-government, of course, was the task Shaftesbury had
assigned to moral philosophers. If he successfully resuscitated philosophy as an
instrument for self-government, however, he also knew that the kind of rhetorical disputation conducted in the interior dialogue or soliloquy no longer con-

ferred authority on conclusions once reached through public debate. To reinforce the validity of his thesis that introspection would lead to virtue (for some men), Shaftesbury invoked mathematics again, this time as a trope for his method. In so doing, however, he introduced another twist that would continue to haunt experimental moral philosophy for most of the eighteenth century. In his summary of method, which he calls "moral arithmetic," he comes close to saying that human beings can know and appreciate the virtuous truths of proportion and order *even if* no external objects exist—even if human subjectivity constitutes the only domain we can know.

> We have cast up all those particulars from whence (as by way of addition and subtraction) the main sum or general account of happiness is either augmented or diminished. And if there be no article exceptionable in this scheme of moral arithmetic, the subject treated may be said to have an evidence as great as that which is found in numbers or mathematics. For let us carry scepticism ever so far, let us doubt, if we can, of everything about us, we cannot doubt of what passes within ourselves. Our passions and affections are known to us. They are certain, whatever the objects may be on which they are employed. Nor is it of any concern to our argument how these exterior objects stand: whether they are realities or mere illusions; whether we wake or dream. For ill dreams will be equally disturbing; and a good dream (if life be nothing else) will be easily and happily passed. In this dream of life, therefore, our demonstrations have the same force; our balance and economy hold good, and our obligation to virtue is in every respect the same.[43]

Shaftesbury's moral arithmetic was emphatically not counting; indeed, he explicitly ridiculed "counting noses" as a mode of determining truth.[44] Instead, moral arithmetic was modeled on mathematics, because mathematics was assumed to produce certain knowledge. Using mathematical demonstration as a model for moral philosophy, Shaftesbury "demonstrates" not that the harmony we experience actually exists in the external world, but that the knowledge produced in the subjective landscape governed by "passions and affections" is trustworthy, *because* it is based on introspection, which cannot err. In making this argument, he appropriated the certainty associated with mathematical demonstration for introspection, and thus for the moral philosopher's claims about human subjectivity. In making certainty an effect of introspection, of course, he simply set aside the question of accuracy—the question whether the proportion we intuit in ourselves exists anywhere other than in the disciplined (elite) mind. Thus Shaftesbury used mathematics to ground an aesthetic model of virtue, which depended *in some sense* on observation (of the self), but he suspended the issue that would preoccupy many of his followers: the question whether the order "we" naturally love exists in the nature we perceive.

Shaftesbury's tendency to ignore the question whether order existed, as well as his not so subtle elitism, limited the appeal of his idea that (self-) government was underwritten by an aesthetic discrimination practiced through introspection and modeled on mathematics. This was true for both philosophical and social reasons. Philosophically, if one allowed, as Shaftesbury did, for a variety of degrees of disinterestedness among individuals, then it became impossible to explain how disinterestedness in general could exist or to distinguish absolutely between disinterestedness and its self-interested twin. Socially, Shaftesbury's aesthetic moralism had limited appeal because once birth and land, the traditional bases for the differential distribution of discrimination, lost their hegemony, it was no longer clear how introspection and aesthetic judgment could be trusted to govern. In a society where wealth, not birth, increasingly dictated who could obtain political power, the old guarantors of disinterestedness no longer exercised the same authority; indeed, in a philosophical context in which philosophers sought the ground of virtue in subjective dynamics like aesthetic appreciation, it was even possible to argue (as Mandeville did) that disinterestedness was not the heart of virtue at all. In the absence of an authority external to the market that could perform the function once performed by land, and in the midst of theories that equated private interest with public good, neither aesthetic appreciation nor mathematical principles seemed reliable enough to inspire trust.

To a certain extent Francis Hutcheson stabilized Shaftesbury's model of (self-) government by anchoring the subjective dynamic of aesthetic discrimination not just in mathematical principles but in providential design. In doing so, he enhanced both the philosophical and the social appeal of a mathematized variant of aesthetics. Yet to a degree Hutcheson repeated Shaftesbury's ambivalence about the ontological status of the anchor he provided for aesthetic discrimination. Whereas Shaftesbury had cheerfully set aside the question whether order existed outside the perceiving subject and had wholeheartedly embraced a notion of differential (class-specific) judgment, Hutcheson insisted that properties like order and proportion did exist, *but only* in the relationship between the observer and the natural (or mathematical) object.[45] As we will see in a moment, moreover, Hutcheson was at least intermittently uncertain about whether every human being experienced things in the same way, much less whether they could be counted on to make the same judgments—although he did not theorize the kind of land- (or class-) based determinant for these differences that Shaftesbury celebrated. Partly because he was able to stabilize Shaftesbury's model of self-government, Hutcheson exercised enormous influence over theorists eager to explain why individuals should be trusted to govern themselves in the emergent market society. And because he was unwilling either to disregard or to reason away the problem of difference, he preserved elements of what Hume would soon theorize as the problem of induction.

As many recent commentators have noted, Francis Hutcheson's intellectual relationship with Shaftesbury was complex.[46] On the one hand, Hutcheson's reliance on empiricism was as qualified (albeit in a different way) as that of his acknowledged master; while he claimed that his primary contribution to moral philosophy, the concept of a moral sense, was modeled on the five external senses, Hutcheson, like Shaftesbury, tended to privilege mathematical figures when he sought examples of beauty or harmony—not the kind of natural objects that empiricists generally emphasized. On the other hand, Hutcheson differed from Shaftesbury on the all-important question of the ontological status of that order and proportion on which both modeled virtue: unlike the deist Shaftesbury, Hutcheson specifically set out to prove that belief in God underwrote the *"kind generous Affections"* with which we respond to nature. Some of the intellectual differences that separate Hutcheson from Shaftesbury may reflect their institutional situations: unlike Shaftesbury, the aristocrat who initially circulated his ideas to an elite audience of like-minded men, Hutcheson was a university professor for sixteen years (1730–46) and was therefore responsible for the moral training of the young Scottish boys Sher has discussed.

Whatever the source of their philosophical differences, the nature of those differences was clear from the very beginning of Hutcheson's career, even before he assumed the chair at Glasgow. Stated succinctly, Hutcheson wanted to correct Shaftesbury's deism by securing aesthetic discrimination to a foundation of orthodox religious belief. He wanted to do so, however, in the context of the epistemological practice that was rapidly gaining authority both within the Anglican Church and in educated society more generally: natural philosophy. The prestige of natural philosophy had been enhanced late in the seventeenth century both by the discoveries of Sir Isaac Newton and by the Boyle Lectures, which called on Newtonian theories to justify the argument from design and which used the pulpit to popularize (and explain) Newton's taxing theorems. Hutcheson's debt to Newton is explicit: in his *Essay on the Nature and Conduct of the Passions and Affections* (1729), Hutcheson specifically drew an analogy between the laws he devised to describe the "nature and conduct of the passions" and those *"Laws of Motion"* that Newton had described (*Essay*, 35). Hutcheson capitalized on an ambiguity inherent in Newton's descriptions of method to align orthodox belief in principles that could not be seen with an experimental practice that privileged observation. By so doing, he transported the kind of claims that the Boyle lecturers routinely made about experiment to the domain of moral philosophy—and to the mapping of subjectivity more specifically—thereby enhancing moral philosophy's credibility as both an epistemological practice and a theory of government.

Hutcheson's eagerness to capitalize on the growing prestige of natural philosophy is clear in his decision to model the human faculty of discrimination on the external senses, especially sight. In using the language of ocular inspection to describe this process of "observation," Hutcheson moved Shaftesbury's tropical allusions to an introspective "survey" closer to literal description, for he posited an actual (presumably physiological) "sense" that apprehended order. Postulating a "moral sense" provided a way to bring subjective, aesthetic responses into the domain of science and to fuse the science of nature with religion, for in Hutcheson's account the moral sense was governed by its responsiveness to God's design: "Our *moral Sense* shews this calm extensive Affection to be the highest Perfection of our Nature; what we may see to be the *End* or *Design* of such a Structure, and consequently what is required of us by the Author of our Nature: and therefore if any one like these Descriptions better, he may call Virtue . . . 'acting according to what we may see from the Constitution of our Nature, we were intended for by our Creator'" (*Essay,* xvi–xvii). By this description, what we "see" when we look into ourselves or out at the (social or natural) world is an order that embodies God's design. *Not* to perceive design when we "look" would be "contrary to Experience," Hutcheson asserts; even worse, it would "lead to a Denial of PROVIDENCE": "To suppose 'no Order at all in the *Constitution* of our Nature, or no *prevalent Evidences* of good Order,' is . . . contrary to Experience, and would lead to a Denial of PROVIDENCE in the most important Affair which can occur to our Observation" (*Essay,* 202).

Claiming that one "sees" design when one observes the self and society and that not to do so risks blasphemy simultaneously privileges the idea of design over Baconian particulars and renders "seeing" an expression of religious belief. "Seeing" for Hutcheson may be modeled on ocular vision (and thus be directed toward the physical world), but it also functions as an expression of belief (because it is directed toward God). For this reason, "seeing" sets aside the welter of details that distinguishes one thing (person) from another in favor of the design that makes all things (persons) alike. Because he conceptualized "seeing" as a moral sense attuned to God's design, Hutcheson could explain why human beings naturally seek order in their social as well as their mathematical relations.

Even if Hutcheson maintained that "seeing is believing" in this very special sense, however, it is still not clear how he demoted those idiosyncratic details

that made the Baconian singularity seem valuable to seventeenth-century natural philosophers. Given that he wanted to capitalize on the prestige natural
philosophers had assigned to observation, how did Hutcheson diminish the
importance of the specific particulars that one could actually observe? The answer, as he set it out in his *Inquiry into the Original of Our Ideas of Beauty and Virtue*
(1728), involved using a variant of the gestural mathematics that I have associated with Shaftesbury to distinguish between "uniformity" and "variety."[47]
Whereas Shaftesbury had simply modeled human virtue on the order, harmony, and proportion evident in mathematics, however, Hutcheson used
mathematical principles to "demonstrate" or derive "laws" of virtue. For him
mathematics provided the analytic method by which one could move beyond
observed particulars to the design that informed them, because mathematics
demonstrated general principles or regularities. By so doing, mathematical
principles linked the kind of observation that was (simply) analogous to ocular
vision to the belief in providential design; in this way mathematics rendered design more consequential than singular details because, like mathematics, design
could be represented as more fixed—more certain—than particulars. Thus, to
explain the qualities in objects that "excite" the idea of beauty, Hutcheson
offers something resembling a mathematical formula:

> The Figures which excite in us the Ideas of Beauty, seem to be those in which there
> is Uniformity amidst Variety. There are many Conceptions of Objects which are
> agreeable upon other accounts, such as Grandeur, Novelty, Sanctity, and some oth-
> ers, which shall be mention'd hereafter. But what we call Beautiful in Objects, to
> speak in the Mathematical Style, seems to be in a compound Ratio of Uniformity
> and Variety: so that where the Uniformity of Bodys is equal, the Beauty is as the Va-
> riety; and where the Variety is equal, the Beauty is as the Uniformity. (*Inquiry*,
> 11–12)

To further illustrate his point, he offers a counterexample:

> let us compare our Satisfaction in such Discoveries [of uniformity amid variety],
> with the uneasy State of Mind when we can only measure Lines, or Surface, by a
> Scale, or are making Experiments which we can reduce to no general Canon, but
> are only heaping up a Multitude of particular incoherent Observations. Now each
> of these Trials discovers a new Truth, but with no Pleasure or Beauty, notwith-
> standing the Variety, till we can discover some sort of Unity, or reduce them to
> some general Canon. (*Inquiry*, 20–21)

In such passages Hutcheson equates the discovery of order—by which he
means the "observation" or belief in both design and laws—with a subjective
state: satisfaction, pleasure, or (the sense of) beauty. His proposal that pleasure is

a barometer of order repeats Shaftesbury's aesthetic, of course; but whereas Shaftesbury considered pleasure the final arbiter of order (and by extension virtue), Hutcheson *explains* pleasure by reference to his belief that the order that occasions pleasure reflects God's providential design; he "proves" that design exists, in turn, by means of mathematical ratios. Thus he reduced the particular details that seventeenth-century natural philosophers valued to "a Multitude of particular incoherent Observations," which he considered essentially meaningless, and he elevated uniformity over variety, because it could be said to manifest the providential order that he claimed to discover by (and in) mathematical formulas. With the same gesture, Hutcheson stabilized subjectivity by making human responses as uniform (hence predictable) as the ratio of angles in a triangle.

Hutcheson bolstered his claims that subjective responses were uniform by representing what he claimed were the laws of the human passions in the language of mathematics. In these representative passages we see how closely his laws resembled Newton's laws of motion:

> In computing the *Quantities* of Good and Evil, which we pursue or shun, either for ourselves or others, when the *Durations* are equal, the Moment is as the *Intenseness,* or Dignity of the Enjoyment: and when the *Intenseness* of Pleasure is the same, or equal, the Moment is as the *Duration.* (*Essay,* 40)

And most telling, because most familiar:

> In comparing the moral Qualitys of Actions, in order to regulate our Election among various Actions propos'd, or to find which of them has the greatest moral Excellency, we are led by our moral Sense of Virtue to judge thus; that in equal Degrees of Happiness, expected to proceed from the Action, the Virtue is in proportion to the Number of Persons to whom the Happiness shall extend; (and here the Dignity, or moral Importance of Persons, may compensate Numbers) and in equal Numbers, the Virtue is as the Quantity of the Happiness, or natural Good; or that the Virtue is in a compound Ratio of the Quantity of Good, and Number of Enjoyers. In the same manner, the moral Evil, or Vice, is as the Degree of Misery, and Number of Sufferers; so that That Action is best, which procures the greatest Happiness for the greatest Numbers; and that worst, which, in like manner, occasions Misery. (*Inquiry,* 117)

It is difficult to imagine how this kind of gestural mathematics might have been applied. What is clear is that by combining mathematical language with descriptions of moral apprehension that equated subjective events like appreciation with ocular vision, Hutcheson made a map of human motivations that seemed reliable because it was derived from *both* experiment and mathematics.

His map is important for two reasons: in producing a scientific account of human subjectivity, it seemed to explain why individuals could be trusted to govern their self-interested passions; and in combining experiment and mathematics, it revealed that the science considered most authoritative for such demonstrations was Newtonian.

Although scholars have frequently noted Newton's influence on eighteenth-century British philosophy, the dramatically different ways that contemporaries interpreted his works have not always featured in this literature.[48] While some eighteenth-century writers associated Newton with experiment, observation, and the repudiation of hypotheses, as modern historians have tended to assume, others, including Shaftesbury, rejected his work because they considered it *too* speculative.[49] Instead of simply calling Hutcheson "Newtonian," then, it is helpful to acknowledge, as James G. Buickerood has done, that eighteenth-century philosophers tended to fall into one of two Newtonian camps. On the one hand, Buickerood notes, some contemporaries, following the principles Newton set out in *Principia Mathematica* (1687), tended to adopt a "mathematicodeductive" method that used mathematical demonstration to describe the character of bodies, motions, and forces in mathematical language. On the other hand, following the lead of the *Opticks* (1704), others adopted a "speculative-experimental" method that emphasized observation but also, at least in the hands of some practitioners, deployed hypotheses—despite Newton's repeated disclaimers about hypothesis and despite the vehement hostility expressed by some self-professed Newtonians.[50]

The distinction Buickerood makes is particularly useful in helping us understand why the second Newtonian method—the speculative experimental method associated with the *Opticks*—was appealing in the early eighteenth century. To moral philosophers, this method was attractive because it did not seem to rely on hypothesis or deduction, which orthodox Britons tended to revile because they associated hypothesis with the atheism of Hobbes and Descartes. Thus when William Emerson, the eighteenth-century author of *Principles of Mechanics,* declared his methodological affiliation with Newton in 1773, he did so based on Newton's repudiation of "hypotheses, conceits, fictions, conjectures, and romances"—all of which, this author declared, were "invented at pleasure and without any foundation in the nature of things."[51] If Buickerood's distinction helps us understand the popularity of the method associated with Newton's *Opticks,* however, it cannot explain why the other Newtonian method—the "mathematicodeductive" method associated with the *Principia*—was also popular in the early eighteenth century, as the success of the Boyle Lectures so amply demonstrates.

To understand the appeal of Newton's mathematicodeductive method, as well as the variant of Newtonianism that appears in Hutcheson's work, we need to appreciate what separating these methods will not let us see: that in his own descriptions of method, Newton *combined* analytic procedures that most philosophers tended to represent as opposites: induction, which Bacon associated with experiment, and deduction, which was associated with Scholasticism, with rhetoric, and—in the form of demonstration—with mathematics.[52] Thus, even though Buickerood may be correct in saying that eighteenth-century moral philosophers tended to *emphasize*—often for polemical purposes—one Newtonian method rather than the other, the very fact that what look to us like theoretically discrepant procedures could both be *called* Newtonian means they had something in common. Indeed, most contemporaries who explicitly adopted Newton's method tended to combine references to "experimentalism" with some use of the language of mathematics, as Hutcheson did. To understand why they did this, and why this seemed to them like Newton's method, we need to turn to Peter Dear's recent discussion of Newton.

To explain how Newton recast the experimentalism practiced by Boyle, Dear describes Newton's famous emphasis on mathematics as a variant of the "Physico-Mathematicall-Experimentall Learning" advocated by John Wilkins and John Barrow in the 1660s.[53] As Dear demonstrates, when Newton defended experiment in the third edition of the *Opticks,* he equated experiment with the first stage of a mathematical demonstration: experiment was like analysis, the subdivision of a problem into its component parts. Confusingly, however, Newton equated the next stage of mathematical demonstration—composition or synthesis—with what he calls induction; this identification is confusing because previous theorists had distinguished between deduction and induction in order to differentiate between the method associated with Scholasticism and that inaugurated by Bacon. Newton avoids this confusion while capitalizing on Bacon's prestige by simply using "induction" without precisely defining it. We see this in what may well be Newton's most frequently cited comment on method:

> As in Mathematicks, so in Natural Philosophy, the Investigation of difficult Things by the Method of Analysis, ought ever to precede the Method of Composition. This Analysis consists in making Experiments and Observations, and in drawing general Conclusions from them by Induction, and admitting of no Objections against the Conclusions, but such as are taken from Experiments, or other certain Truths. For Hypotheses are not to be regarded in experimental Philosophy. And although the arguing from Experiments and Observations by Induction be no Demonstration of general Conclusions; yet it is the best way of arguing which the

Nature of Things admits of, and may be looked upon as so much the stronger, by how much the Induction is more general. And if no Exception occur from Phaenomena, the Conclusion may be pronounced generally.[54]

This description shows that Newton's induction resembled Baconian induction, which moved by steps to ever higher levels of generalization. As we saw in chapter 2, however, Bacon and the experimentalists who followed him tended not to elaborate this component of induction—not to specify how one moved from stage to stage—because they wanted to elevate singular facts in order to avoid the theoretical disagreements associated with Scholasticism and civil strife. By contrast, Newton, who lacked Boyle's investment in rarities because he was not so concerned with theoretical disputes, did specify the stages by which one produced generalizations: one used the method and language of mathematics. As Dear points out, Newton introduced mathematical method— without either abandoning the language of induction or reconciling his practice with that of Bacon—in order to address the problem that Boyle's emphasis on singular facts introduced and could not solve: if facts were valuable insofar as they were idiosyncratic or unique, then how could the philosopher produce general knowledge? When Newton supplemented the Royal Society's emphasis on "experiment" and "observation" with a form of induction supposedly based on the mathematical method of analysis and synthesis, he "gave event experiments a philosophical respectability that they had formerly lacked," because he seemed to combine experimentalism with demonstration and thus to generate philosophical generalizations about a nature that everyone simply assumed to be regular and constant.[55]

By 1728, when Hutcheson composed the first of his two treatises, mathematical demonstration had thus come to seem a more reliable basis for natural philosophical matters of facts than mere observation, because Newton had successfully elevated (invisible) mathematical laws over deracinated particulars. When Hutcheson declared, late in the *Inquiry,* that "from the former Reasonings we may form almost a demonstrative Conclusion, 'That we have a Sense of Goodness and moral Beauty in Actions distinct from Advantage,'" he mobilized this mathematical model so as to borrow its cultural prestige. By so doing, he sought to make the knowledge he produced by means of gestural mathematics as incontrovertible—as certain—as was the knowledge Newton had produced about the natural world.[56]

By the same token, and because Newton's method combined mathematical demonstration with experiment, Hutcheson also needed to position his variant of moral philosophy within experimentalism most generously understood. This not only was important because he wanted to borrow natural phi-

losophy's prestige, although prestige was not to be taken lightly in a market society; it was also necessary that he align his work with experimentalism because
he wanted to anchor Shaftesbury's aesthetic model of self-government in
something beyond (class-specific or idiosyncratic) pleasure. Specifically, and
even though he radically qualified what it meant to "see," Hutcheson wanted to
anchor Shaftesburian aestheticism in empiricism because he wanted to map
human motivations onto something that was reliable because it was as available
through vision as it was propped on belief: the basis for Hutcheson's map of human desire, as we have already seen, was God's design, which is (supposedly)
everywhere manifest in the natural and moral worlds.

Hutcheson's effort to combine mathematics and empiricism helps explain
both the nature of his "experimentalism" and why it seemed to generate reliable knowledge. On the one hand, this experimentalism resembled the practice
of Bacon and Boyle in that it did rely on a variant of observation. On the other
hand, however, like mathematical demonstration, Hutcheson's experimentalism *assumed* or *posited* what it also claimed to prove. One had to assume that all
human beings possessed "an *Instinct toward Happiness,*" he explained, before one
could demonstrate the ratios that defined happiness by showing what human
beings instinctively sought—or could be made to desire. "Without such *Affections* this Truth, 'that an hundred Felicities is a greater Sum than one Felicity,'
will no more excite to study the Happiness of the *Hundred,* than this Truth, 'an
hundred Stones are greater than one,' will excite a Man, who has no *desire of
Heaps,* to cast them together" (*Essay,* 225–26).

In relying on such foundational assumptions—that all human beings possess these instincts, that all human beings are alike—Hutcheson revealed that
both his gestural mathematics and his claims about observation rested on the
very bedrock of belief they were being used to prove. To late twentieth-century
readers, using belief to authorize knowledge seems dubious at best, because a
consensus about the authority (or basis) of belief no longer exists. To most eighteenth-century Britons, by contrast, belief constituted the most authoritative
ground for knowledge because most (though not all) Britons assumed some
variant of Anglican orthodoxy—including the conviction that the subjective
experience called belief was a response to God. Anchoring knowledge about
human motivations in belief in God's order thus paradoxically helped legitimate even experiment, for in Hutcheson's account experiments were judged
reliable when they proved what the philosopher believed.

There are at least two passages in the treatises where Hutcheson reveals
some anxiety about what looks to modern readers like circular logic. Twice he
confesses some uncertainty about whether what he both assumed and sought to
prove—that all human beings are alike in what and how they see—is true. In the

preface to the fourth edition of the *Inquiry*, for example, he admitted that "in the first Treatise, the Author perhaps in some instances has gone too far, in supposing a greater Agreement of Mankind in their Sense of Beauty, than Experience will confirm" (*Inquiry*, xvi); in his *Essay*, he acknowledged that he could claim only that "it is highly probable that the *Senses* of all Men are pretty *uniform*" (*Essay*, 285). With such statements, Hutcheson anticipated the distress later readers would suffer. Although he did not give in to this distress, or even elaborate its implications, his simply acknowledging that his claims to describe a moral faculty as reliable as mathematics admitted doubt opened the philosophical rift into which Hume would soon drive the wedge of skepticism.

The writings of Shaftesbury and Hutcheson demonstrate that early eighteenth-century attempts to anchor liberal (self-) government in a theory about moral discrimination and the faculty that produces taste generated problems at the very site where confidence was most essential: the question whether conclusions produced from observing the philosopher's self held good for all "mankind." In Shaftesbury's work, this question emerges in relation to his suggestion that the harmony we appreciate might be no more real than a dream; if a moralized version of appreciation is that solipsistic, after all, one could not possibly know whether everyone appreciates the same things, and it would not matter whether they did or not. In Hutcheson's work this question emerges in relation to his cautions that he can only *assume* everyone has the same sense of beauty; if he needs to assume this in order to prove anything else, yet he can assume it only because otherwise his proofs have no purchase, then the foundation for his moralism seems shaky at best. As we have seen, variants of what I have called gestural mathematics stopped the erosion that doubt might have caused for both of these philosophers: for Shaftesbury, allusions to the proportion and harmony of mathematics anchored claims about "our" natural attraction to virtue; for Hutcheson, mathematical ratios seemed to describe how one identified the uniformity that gives meaning to variety, and thus how one derived the laws of virtue that theoretically inform everything in God's world.

George Turnbull, who served as regent of Marishal College, Aberdeen, from 1721 to 1727, approached the problem of how one might demonstrate the universality of moral philosophical claims in a slightly different manner than did Shaftesbury and Hutcheson, even though he drew on the work of both philosophers. He also provided the most explicitly self-conscious formulation of the method of experimental moral philosophy before Hume, so his work is particularly valuable for any attempt to understand that distinctively eighteenth-century genre. Indeed, Turnbull described his particular branch of moral philosophy—pneumatology—as "the Natural Philosophy of Spirits," a practice that, according to one of his colleagues, was "founded solely on experiments and observations."[57] As with both Shaftesbury and Hutcheson, the experiments and observations Turnbull conducted focused almost exclusively on himself. Even more adamantly than Shaftesbury or Hutcheson, moreover, he insisted that one could generalize from such observations because one could—indeed, had to—assume that what one observed in the self enacted God's laws.

In emphasizing the lawful nature of God's world, Turnbull made explicit what Newton and Hutcheson stated but did not emphasize: that what all philosophers seek to describe is *invisible*. What all philosophers seek to describe, in fact—what makes them philosophers and not (mere) historians—are laws or regularities that, by their very nature, cannot be seen as such.[58] This claim is not the same as Shaftesbury's idea that the harmony we appreciate might belong to a dream. Like Hutcheson and Newton, Turnbull insisted that invisible laws actually operate in the world of nature as well as the self. Just because we can "see" these laws only with the help of experimental moral philosophy does not mean they are not real; it just means that we need moral philosophy to make sense of the harmony and proportion we perceive with both our bodily and our moral senses. To make sense of what we perceive—to make what we see with our bodily eye more than just a "Multitude of particular incoherent Observations," as Hutcheson put it—we have to *believe* that the laws we "see" through moral philosophy embody God's order, which (we must also believe) is essentially good. The combination of confidence and optimism that emanates from Turnbull's assertions about providential design constitutes his special contribution to eighteenth-century experimental moralism and to the map of human motivation that philosophers were trying to produce. Recognizing the place of his confidence in experimental moralism will help us understand its persistence in both of the disciplines I take up in the next chapter: Scottish conjectural history and political economy.

From the opening pages of Turnbull's major philosophical treatise, *The Principles of Moral Philosophy,* it is clear that by 1740 natural and moral philosophies had come to seem sufficiently different that what had once been considered their commonalities needed to be reasserted. Indeed, the newly perceived difference between natural and moral philosophies is the backdrop against which we must understand all eighteenth-century projects to devise an *experimental* moral philosophy. In the wake of the seventeenth-century scientific revolution, as we have already seen, the natural world was increasingly viewed as the site of a particularly valuable kind of truth, and experiment and observation were held to be privileged modes of gathering nontheoretical facts about this world. In this context it was tempting to demote moral philosophy to a merely theoretical practice, which lacked both an object that could be observed and a method that transcended the problems Newton associated with hypothesis. It was to stave off such accusations of "mere" conjecture that moral philosophers began to emphasize the "reality" of their object and the "experimentalism" of their method. Thus Turnbull asserted in 1740 that the object of moral philosophical analysis was "no less real and exact than . . . the body" and that its method was "the fair impartial way of experiment."[59] Because he and the other moral experimentalists were also formulating their practice in the context of an emergent market society, however, Turnbull did not stop at claiming that moral philosophy was *like* natural philosophy in object and method. Instead, because he wanted to make moral philosophy contribute to the theory of liberal governmentality that was being developed to ground the virtue of consumer society, he represented moral philosophy as superior to natural philosophy. More precisely, by representing the sole aim of natural philosophy as the explication of the "order, beauty, and perfection of the material world," he positioned natural philosophy as *a branch of moral philosophy.*

> 'Tho natural philosophy be commonly distinguished from moral; all the conclusions in natural philosophy, concerning the order, beauty, and perfection of the material world, belong properly to moral philosophy; being inferences that respect the contriver, maker, and governor of the world, and other moral beings capable of understanding its wise, good and beautiful administration, and of being variously affected by its laws and connexions.
>
> In reality, when natural philosophy is carried so far as to reduce phenomena to good general laws, it becomes moral philosophy; and when it stops short of this chief end of all enquiries into the sensible or material world, which is, to be satisfied with regard to the wisdom of its structure and oeconomy; it hardly deserves the name of philosophy in the sense of *Socrates, Plato,* Lord *Verulam, Boyle, Newton,* and the other best moral or natural philosophers. (*Principles,* 8–9)

In this passage we see the heart of Turnbull's philosophical project. By defining the "chief end of all enquiries into the sensible or material world" as satisfaction about "the wisdom of its structure and oeconomy," he made religious conviction the basis and the aim of natural philosophy. This in turn made natural philosophy a branch of moral philosophy, because the latter, which had no phenomenal objects to consider, was so much more obviously suited to demonstrating religious belief than was natural philosophy, which had to relegate different kinds of natural phenomena to different taxonomic positions in order to elevate uniformity over variety. In a sense, of course, Turnbull's argument was simply converting a disciplinary necessity into an epistemological virtue: because of the elusiveness of their object, moral philosophers—even would-be experimentalists—*had to* resort to "reasoning from principles known." To represent this kind of reasoning as superior to observation was thus to claim, implicitly at least, that observed particulars were inferior clues to what constituted genuine knowledge, even for the natural philosopher, because genuine knowledge consisted of "inferences that respect the contriver, maker, and governor of the world"—that is, God.

If Turnbull can be accused of converting a disciplinary necessity into an epistemological virtue, then it must also be acknowledged that the end product of this conversion was identical to the claims made by the Boyle lecturers; both Turnbull and the Anglican lecturers maintained that because God wrote the book of nature, natural philosophy constituted a religious exercise. In stressing that what counts about nature is invisible, however, Turnbull was shifting the emphasis of philosophy away from observed particulars so emphatically that the only phenomena that retained any validity were those that proved what the philosopher assumed. This shift in emphasis foreclosed the possibility of discovery as Robert Boyle had understood that concept; it converted philosophy (both natural and moral) into an instrument devoted to demonstrating God's plan; and it implied that knowledge was progressive, because the more we know about natural and moral laws, the more we know about God. In other words, instead of seeking to generate probable knowledge about the natural world or human motivation from observed particulars, Turnbull wanted philosophy to reveal certain knowledge about God, which could be confirmed (or known in advance) by reading God's other book, the Bible.

To reveal knowledge about God, Turnbull insisted that experimental moral philosophy, like its natural philosophical counterpart, should use "the double manner of analysis and synthesis" associated with geometrical proofs and with mathematics more generally. On the one hand, he advanced this methodological program because, like Hutcheson, he wanted to borrow the prestige of Newtonian natural philosophy; on the other hand, however, he did so because he saw what the natural philosophers who followed Newton practiced but did not preach: that even natural philosophers inevitably relied on both experiment *and* hypothesis. Once more, while Turnbull seems to show deference to the method of natural philosophy, he manages to make moral philosophy, which seems simply to borrow from its counterpart, the disciplinary master. Thus he represented natural philosophy as a variant of the seventeenth-century practice of mixed mathematics; and he constructed moral philosophy as the twin that revealed the pitfalls natural philosophers barely escaped.

> That as in natural philosophy, though it would be but building a fine visionary Theory or Fable, to draw out a system of consequences the most accurately connected from mere hypotheses, or upon supposition of the existence and operation

of properties, and their laws, which experience does not shew to be really existent; yet the whole of true natural philosophy is not, for that reason, no more than a system of facts discovered by experiment and observation; but it is a mixture of experiments, with reasonings from experiments: so in the same manner, in moral philosophy, though it would be but to conceive a beautiful, elegant romance, to deduce the best coupled system of conclusions concerning human nature from imaginary suppositions, that have no foundation in nature; yet the whole of true moral philosophy, will not, for that reason, be no more than a collection of facts discovered by experience; but it likewise will be a mixed science of observations, and reasonings from principles known by experience to take place in, or belong to human nature. (*Principles,* 19–20)

This elephantine sentence, with its convoluted syntax and inordinate dependence on negative constructions, demonstrates the difficulty of what Turnbull was trying to do. Seeking to claim both that the only knowledge that counts is systematic knowledge and that the systematic nature of knowledge does not discredit its claims to accuracy, he wants to hold up moral philosophy as what critics may claim systematic natural philosophy is—*merely* a "beautiful, elegant romance"—and also as proof of what natural philosophy should and could be: a set of principles that are both systematic and true to nature. Turnbull rejects the first charge and claims to prove the second by asserting that moral philosophy enables us to see in nature what we believe in our hearts: that, having ordered nature and made man in his image, God now orders knowledge, so that we know that what we know is true by its systematic accord with what we believe about God.

This two-part belief, Turnbull repeatedly insists—that our faith in a God-given order grounds all philosophical knowledge and that all philosophical practices worthy of the name demonstrate God's order—is what enables human beings to recognize virtue as they generate knowledge. In insisting that belief in a God-given order constitutes the basis of both knowledge and virtue, moreover, he also began to naturalize the link between order and virtue that philosophers like Shaftesbury and Hutcheson had associated with mathematics. Whereas they had claimed that mathematical proportion exemplifies virtue because it embodies something like a Platonic ideal, Turnbull argued that the signal incarnation of order is not mathematical proportion but the natural regularity of cause and effect. Because identical effects invariably follow the same cause, he insisted, human beings can predict future consequences by past experience. And because knowing these consequences gives a moral meaning to human actions, virtue is an effect of the combination of knowledge about and belief in the laws of cause and effect. Thus Turnbull's map of subjectivity opened a space for free will because, in making natural laws and knowledge of those laws the basis for morality, he entrusted human beings with the responsibility of choice.

> The author of nature, with regard to us, may be justly said to be teaching, or forewarning us by experience in consequence, of having endued us with the capacity of observing the connexions of things, that if we act so and so, we shall obtain such enjoyments, and if so and so, we shall have such and such sufferings. That is, the author of our nature gives us such and such enjoyments; or makes us feel such and such pains in consequence of our actions. . . .And, in general, all the external objects of our various, natural appetites and affections, can neither be obtained, nor enjoyed without our exerting ourselves in the ways appointed to have them; but, by thus exerting ourselves, we obtain and enjoy those objects in which our natural good consists. In like manner, our progress in knowledge, in any art, or in any virtue, all moral improvements depend upon ourselves: they, with the goods resulting from them, can only be acquired by our own application, or by setting ourselves to acquire them according to the natural methods of acquiring them. This is really our state; such really is the general law of our natures. . . .
>
> But a capacity and a way of attaining to; and a capacity and way of escaping certain ends and consequences, suppose general fixed uniform connexions in nature between certain manners of acting and certain consequences: that is, they suppose fixed, uniform and general laws with regard to the exercises of powers or actions. . . .The same Author of nature, who hath conferred certain faculties upon us, must have established certain laws and connexions with regard to the exercises of them, and their effects and consequences; otherwise we could not know how to turn them to any account, how to employ them, or make any use of them.
>
> The result of all this is in general, "That we can have no liberty, no dominion, no sphere of activity and power, natural or moral, unless the natural and moral world are governed by general laws: or so far only as they are so governed can any created beings have power or efficiency: so far only can effects be dependent on their will as to their existence or non-existence." (*Principles,* 26, 28)

With such statements, Turnbull detached the account of subjectivity implicit in theories of liberal governmentality from Platonic idealism—and even, to a certain extent, from the claim that human beings have a moral sense—and cast it in terms of natural law. Once one assumes that nature is governed by general laws, one can understand the human capacity for virtuous self-government as an expression of voluntary compliance with what we have learned from the laws of cause and effect. Thus knowing about nature becomes instrumental to moral self-government not only because natural philosophy constitutes a religious practice, but also because the laws of nature constitute the backdrop for virtue. Experience therefore becomes not primarily an instrument for knowledge production, but the critical stage in learning those laws that make human virtue possible.

As David Fate Norton has pointed out, Turnbull claimed that his method rendered morality as "fixed and as real as Newtonian mechanics."[60] For this reason, perhaps we should credit Turnbull with initiating the "Newtonian turn of British methodological thought," as P. B. Wood wants to do.[61] For my purposes, it is less important to evaluate Turnbull's relative importance in the Newtonian cast that British moral philosophy took for much of the rest of the century than to register the critical role that his belief in providential design played in his and subsequent conceptualizations of human subjectivity, virtue, government, and natural law in general. As we will see in the case of Hume, it was not necessary for an eighteenth-century philosopher to believe in God in order to anchor morality in natural laws that could be experimentally derived. In part because of the epistemological problem introduced by Hume's skepticism, however, Turnbull's providentialism continued to exercise enormous appeal to philosophers who wanted to explain why (and how) liberal subjects should be trusted to govern themselves in Britain's emergent market economy. Indeed, even after the account of universal subjectivity posited by eighteenth-century philosophers became the target of anthropological and historical interrogation, Turnbull's providentialism proved attractive to philosophers because it so usefully linked natural laws, which could be conceptualized as universal even after humans began to seem different, to Anglican (British) orthodoxy and to the confident assumption that, as natural knowledge increased, the capacity for human virtue would automatically improve.

DAVID HUME: FROM EXPERIMENTAL MORAL PHILOSOPHY TO THE ESSAY

As its title page declares, Hume's *Treatise of Human Nature* constituted an exercise in experimental moral philosophy; it was, he announced, "An Attempt to Introduce the Experimental Method of Reasoning into Moral Subjects."[62] Like Hutcheson and Turnbull, Hume vowed to ground his "science of Man" on "experience and observation." Unlike them, however, he insisted that observation, experience, and experiment could take the moral philosopher just so far; unlike Hutcheson and Turnbull, in other words, and despite occasional protestations of religious faith, he refused the consolation that came from assuming that the philosopher can deduce the "ultimate original qualities of human nature" from any hypothesis, including the hypothesis of providential design. "And tho' we must endeavour to render all our principles as universal as possible," Hume cautioned, "by tracing up our experiments to the utmost, and explaining all effects from the simplest and fewest causes, 'tis still certain we cannot go beyond experience; and any hypothesis, that pretends to discover the ultimate original qualities of human nature, ought at first to be rejected as presumptuous and chimerical" (*Treatise*, 44, 43, 44).

The scholarship on Hume's skepticism is vast, and in what follows I have little to add to our understanding of the philosophical import of his decisive intervention. I need to register the epistemological implications of Hume's recognition that induction was inherently problematic; but after discussing the problem of induction, I devote my attention to the generic shift that characterized his writing after 1740. By focusing on his turn from experimental moral philosophy to the Addisonian essay, I want to propose that an alternative mode of producing knowledge—a mode that was not Newtonian and that did not aspire to base facts about universals on anything resembling mathematics—was also available to Britons in the 1740s.[63]

Like experimental moral philosophy, the essay was a genre in which eighteenth-century writers explored both human motivation and its relation to liberal governmentality. Unlike philosophers, however, essayists did not produce systematic knowledge so much as they engaged readers in the exercise of that discrimination by which (self-) government was assumed to proceed. Insofar as it solicited the reader's engagement and instructed by example, the eighteenth-century essay resembles the novel as much as it does experimental moral philosophy. Indeed, because it both sought to generate knowledge—in the form of a conversation—and elicited identification with a more or less particularized speaker, the essay constituted the generic bridge between experimental moral philosophy and the novel, where yet another mode of knowledge production was being codified.[64] Although essays cannot properly be said to have contributed to the constitution or vicissitudes of the modern fact, then, they are of at least passing interest to the argument of this book, because the wisdom with which essayists like Samuel Johnson were credited constituted the primary rival to the numbers-based "useful" knowledge championed by those who wanted to revive—or revise—political arithmetic.

Before turning to Hume's essays, we need to explore some of the components of his philosophical skepticism, both because this skepticism formulated the problem of induction for moral and natural philosophers and because it encouraged Hume to assume "diffidence and modesty" in the face of an unresolvable metaphysical dilemma (*Treatise*, 675). To take the measure of Hume's skepticism, it is important to remember that it was part and parcel of a rigorous experimentalism, which followed his assumption that the mind posed an epis-

temological problem as perplexing as the problems posed by phenomenal nature: "The essence of the mind being equally unknown to us with that of external bodies, it must be equally impossible to form any notion of its powers and
qualities other than from careful and exact experiments, and the observation of
those particular effects, which result from its different circumstances and situations" (*Treatise*, 44). Equally to the point, however, is Hume's insistence that
some of the epistemological tools of experimentalism were *un*available to
moral philosophy. Specifically, because moral philosophers could not assume a
position *outside* the subject of their experiments, their observations were always
in danger of being influenced by the self-consciousness that inevitably accompanied introspection. Thus one facet of Hume's skepticism followed his recognition that the map of human motivation produced by the moral philosopher
would inevitably be colored by his own attitude toward himself.[65]

To a certain extent, Hume devised a way to neutralize the threat that self-
involvement posed to knowledge; by consulting behaviors in "the common
course of the world" instead of simply looking inward, he sought to test
whether the philosopher's introspection was reliable. Even if he assured his
readers that this system of cross-checking could help factor out the distortions
caused by self-consciousness, however, Hume almost immediately acknowledged another problem: Once we admit that the subjective ideas of objects are
separable from the objects themselves, as Locke had insisted we do, how can we
know that our ideas correspond to matters of fact? Equally to the point, once
we admit that experience is the only basis of knowledge, how do we know that
the object we have yet to experience will resemble the object we know? For
Hume the answers to these questions boiled down to two concepts, which were
inextricably connected: belief and custom. To demonstrate that ideas correspond to matters of fact and that we can know even what we have yet to experience, he tried to define belief as a special kind of knowledge. Then, having
asserted that this special kind of knowledge assures us that our idea corresponds
to something that is real, he invoked custom to explain how we move from the
experience of one object to a belief that objects we have not experienced will
be of the same kind. Thus he propped belief on custom *and vice versa.* Eventually, as one might already begin to imagine, this abysmal structure opened onto
the problem of induction—a problem that, in theory if not in practice, undermined the credibility of any attempt to map human subjectivity, much less to
explain why liberal governmentality could work.

In book 1 of the *Treatise,* Hume holds the problem of induction at bay by
invoking the language of gestural mathematics that we have already seen in his
and other philosophers' work. Initially, that is, he seeks to distinguish (true) be-

lief from a "simple idea" by a series of references to *degree*. We can distinguish a belief from a fiction, he asserts, because the former has greater "force and vivacity."

> Belief is somewhat more than a simple idea. 'Tis a particular manner of forming an idea: And as the same idea can only be vary'd by a variation of its degrees of force and vivacity; it follows upon the whole, that belief is a lively idea produc'd by a relation to a present impression, according to the foregoing definition. (*Treatise*, 145)

Even if he invokes the language of gestural mathematics to distinguish a belief from a fiction, Hume immediately acknowledges that neither this nor any other language can specify what characterizes belief. To get on with his argument, he finally falls back on references to common sense or experience ("I . . . am oblig'd to have recourse to every one's feeling, in order to give him a perfect notion of this operation of the mind"), although he admits that even with this aid "'tis impossible to explain perfectly this feeling or manner of conception" (*Treatise*, 146). The following section of the *Treatise* (book 1, section 8) contains the densest proliferation of what Hume calls "experiments" of any chapter of the entire work; and it culminates in his claim that, just as Newton acknowledged the adequacy of a single experiment (the *experimentum crucis*) in proving certain natural philosophical problems, so the moral philosopher must grant that in some moral areas a single experiment will have to suffice, especially since custom can step in where experience fails.

> 'Tis certain, that not only in philosophy, but even in common life, we may attain the knowledge of a particular cause merely by one experiment, provided it be made with judgment, and after a careful removal of all foreign and superfluous circumstances. Now as after one experiment of this kind, the mind, upon the appearance either of the cause or the effect, can draw an inference concerning the existence of the correlative; and as a habit can never be acquir'd merely by one instance; it may be thought, that belief cannot in this case be esteem'd the effect of custom. But this difficulty will vanish, if we consider, that tho' we are here suppos'd to have had only one experiment of a particular effect, yet we have many millions to convince us of this principle; *that like objects, plac'd in like circumstances, will always produce like effects;* and as this principle has establish'd itself by a sufficient custom, it bestows an evidence and firmness on any opinion, to which it can be apply'd. The connexion of the ideas is not habitual after one experiment; but this connexion is comprehended under another principle, that is habitual; which brings us back to our hypothesis. In all cases we transfer our experience to instances, of which we have no experience, either *expressly* or *tacitly,* either *directly* or *indirectly.* (*Treatise*, 154–55)

The problem with this solution—Hume's claim that belief grounds custom and custom grounds belief—does not become visible until the "Appendix," which he added to the *Treatise* in 1740. By that time Hume knew that many readers remained unconvinced by the radical conclusions that followed from his skepticism, because the responses to the *Treatise*'s first two volumes had been so disappointing.[66] To a certain extent he tried to attribute the failure of his philosophical project to his infelicitous style ("some of my expressions have not been so well chosen," *Treatise*, 671); but in the final pages of the "Appendix" he identifies what he had come to see as the unresolvable contradiction at the heart of the *Treatise*. "In short there are two principles, which I cannot render consistent," he admitted; "nor is it possible to renounce either of them, viz. *that all our distinct perceptions are distinct existences*, and *that the mind never perceives any real connexion among distinct existences*" (*Treatise*, 678). Even though Hume links these two principles to the specific problem of identity in the "Appendix," this is precisely the dilemma he had earlier sought to solve by contending that belief and custom supported each other.

The reason the issue of identity exposes the contradiction that had seemed manageable in Hume's discussion of experience is that by introducing once more the possibility that the self-consciousness the philosopher must exercise to explore the puzzle of identity can distort his perception of himself, this formulation of identity threatens to distort or problematize itself. If investigating identity distorts one's knowledge about identity, in turn, and possibly even the philosopher's identity itself, then this compromises the basis for producing *any* knowledge (even the knowledge that the observing self is continuous). This is exactly what happens in the last paragraphs of the *Treatise*. Because Hume now represents the epistemological problem posed by identity as analogous to the epistemological issues raised by what had once seemed straightforward, moreover, the problem epitomized by identity becomes the problem of philosophy *tout court*. What he reluctantly forecasts in these final paragraphs is the emergence of the problem of induction.

> Philosophers begin to be reconcil'd to the principle, *that we have no idea of external substance, distinct from the ideas of particular qualities.* This must pave the way for a like principle with regard to the mind, *that we have no notion of it, distinct from the particular perceptions.*
>
> So far I seem to be attended with sufficient evidence. But having thus loosen'd all our particular perceptions, when I proceed to explain the principle of connexion, which binds them together, and makes us attribute to them a real simplicity and identity; I am sensible, that my account is very defective, and that nothing but the seeming evidence of the precedent reasonings cou'd have induc'd me to receive

it. If perceptions are distinct existences, they form a whole only by being connected together. But no connexions among distinct existences are ever discover-
able by human understanding. We only *feel* a connexion or determination of the
thought, to pass from one object to another. It follows, therefore, that the thought
alone finds personal identity, when[,] reflecting on the train of past perceptions,
that compose a mind, the ideas of them are felt to be connected together, and naturally introduce each other. However extraordinary this conclusion may seem, it
need not surprize us. Most philosophers seem inclin'd to think, that personal identity *arises* from consciousness; and consciousness is nothing but a reflected thought
or perception. The present philosophy, therefore, had so far a promising aspect. But
all my hopes vanish, when I come to explain the principles, that unite our successive perceptions in our thoughts or consciousness. I cannot discover any theory,
which gives me satisfaction on this head. (*Treatise,* 677–78)

As he describes it here, the problem of induction turns on the gap between
one's ability to observe discrete qualities by discrete perceptions and the impossibility of explaining why we believe that qualities belong to the same object or
that perceptions belong to the same person. The impasse, in other words, can be
described either as a failure of philosophical analysis to move from particulars to
generalizations or as a failure of moral philosophy to map the subjective process
by which we have come to believe in objects or identities. This impasse concludes Hume's *Treatise,* for it stalemates the moral philosophical agenda of producing general and demonstrable knowledge about the human mind from
observed particulars.

The problem of induction would not have derailed Hume's philosophical
project had he not assumed that observed particulars, whether the secondary
qualities of objects or the subjective perceptions of the observer, carried unique
epistemological significance; if, like Shaftesbury and Hutcheson, Hume had
subsumed particulars into the universals exemplified by mathematical principles, he would not have worried that he could not explain how we believe that
discrete qualities or perceptions are related, much less why we think that things
as yet unseen will resemble what we know. Nor would this problem have mattered if he had simply referred what he could not understand through observation and introspection to a providential plan, as Hutcheson and Turnbull did.
The method of induction generated this problem for Hume, and not for his
immediate predecessors, for two reasons: first because, having taken the full
force of the epistemological revolution inaugurated by Bacon and Boyle,
Hume insisted that the observed particular was vital; and second because, refusing to assume anything, he would not back up observation and experience with
claims about any kind of invisible or inherent order, whether that order was Pla-

tonic or Christian. As an experimental philosopher, he would not supplement experience with belief because he could not explain belief by reference to experience, even though he claimed to *feel* what made belief true and not a fiction.[67]

It is critical to realize that even though the problem of induction was to become a major stumbling block to subsequent attempts to produce philosophical knowledge, Hume was not completely undone by uncovering it. Although he rendered the problem of induction visible within the domain of moral philosophy, in fact, his response to it led him out of the philosophical enterprise altogether; as a consequence, the problem of induction did not constitute a difficulty for Hume as it did for the generations that followed. In the "labyrinth" of experimental moral philosophy, he admits, the problem of induction constitutes a dead end: "I neither know how to correct my former opinions, nor how to render them consistent." Instead of trying to do either, however, he takes another tack: he presents the reader with something that resembles a debate, in that it contains "the arguments on both sides" (*Treatise,* 675). Although this debate is never resolved in the *Treatise,* the method of providing first one side, then the other begins to engage the reader in a shared project of discrimination that generates not what Hutcheson would have recognized as systematic knowledge *about* human subjectivity but something else—call it conversation or even pleasure. In the form of the essay, to which he turned immediately after relinquishing the unfinished *Treatise,* Hume brought this mode of knowledge production to a new level of refinement, which he used to justify refinement itself as a mode of governmentality more efficacious than philosophy's attempt to explain human beings to themselves.

Nicholas Phillipson has argued that Hume took up the essay form when a plan he had devised with Henry Home, Lord Kames, fell through shortly after Hume's return from France in 1740. This plan was to publish a series of weekly essays modeled on Addison and Steele's *Tatler* and *Spectator,* which had been reprinted in Scotland as soon as they appeared in London. The essays of Addison and Steele were valued in Scotland, Phillipson argues, because the mode of governmentality they articulated was as applicable to collective as to individual improvement. As the flourishing of numerous Scots clubs in this period makes clear, many lowland Scots were committed to finding a way of using voluntary associations to capitalize on the union with England without succumbing to the moral excesses they associated with London and the south.[68]

Whatever his specific motive for turning to the essay—whether he was disillusioned with experimental moral philosophy or simply seeking to expand his literary options—it is clear that Hume knew *before* he published the first volume of the *Treatise* that he was engaged in a dialogue with a reading public, which materially affected what and how he could write. Realizing that it might give offense, Hume excised the section of the *Treatise* titled "Of Miracles" before giving the manuscript to the printer; and he included on the first volume's title page an epigraph from Tacitus that can be read either as an expression of gratitude to his public or as a lament for the constraints imposed by public taste: translated, the Latin phrase reads, "Seldom are men blessed with times in which they may think what they like, and say what they think." However one interprets this epigraph, it clearly signals Hume's recognition that he was submitting his *Treatise* to the marketplace of ideas, where writers competed for readers and respect. Without a university position and acutely aware of the rewards and punishments meted out to writers in the burgeoning age of print, Hume increasingly sought to turn what might have seemed like an unfortunate necessity—the imperative to please his audience—into a stylistic practice infused with philosophical and moral import.

In the discussion that follows, I focus on an argument that Hume developed obliquely in the numerous volumes of essays that appeared between 1740 and his death in 1776. Essentially this is an argument about style and, more specifically, about the ability of a dialogic, self-refining style to supplement, or even compensate for, the limitations of experimental moral philosophy.[69] Because Hume's development of this argument was oblique—indeed, because his choice of the essay form enhanced its indirection—the argument is not presented systematically in any single essay. Thus I begin with "Of the Delicacy of Taste and Passion" in order to illuminate in miniature the mode of argument by which Hume advanced his thesis. I then turn to what may be considered an interrogation of the limits of this mode of analysis—the essays on politics also published in his first volume of essays (1741). In raising, then setting aside, the idea "That Politics May be Reduced to a Science," Hume hints that systematic knowledge—or science—may not be adequate to understand or improve human behavior. Instead, as he proposes in "On Civil Liberty," metaphysics may have to be supplemented with something else—with a certain style, which, while it may not produce incontrovertible truth, can cultivate in an engaged readership a mode of self-education and self-adjustment that resembles the mode of Hume's essayistic reasoning as a whole. Finally, I take up the subject of style as Hume most self-consciously discussed it in "Of Essay-Writing," which appeared in his second volume of essays, published in 1742.[70] My own method in the pages that follow is to focus closely on the turns in Hume's oblique argument. To follow Hume, I believe, requires close attention, and following him—or more to the point, engaging with him—inaugurates the conversation that constitutes the essay's characteristic mode of producing knowledge about the nature of human subjectivity and sociality.

In the brief "Of the Delicacy of Taste and Passion," Hume uses the essay's first two paragraphs to establish what his title suggests: a *parallel* between two subjective functions or capacities, delicacy of passion and delicacy of taste. Hume sums up the substance of this parallel in a sentence whose syntactic complexities highlight another series of parallels, which branch off from only one part of the foundational parallel but do so in such a way as to make the effects of a delicate taste imply identical effects for delicate passion. "In short, delicacy of taste has the same effect as delicacy of passion: It enlarges the sphere both of our happiness and misery, and makes us sensible to pains as well as pleasures, which escape the rest of mankind."[71]

Having established this parallel, however, Hume immediately complicates it. In the essay's next two paragraphs, he first asserts, with reference to common experience, that delicacy of taste is *unlike* delicacy of passion in being more desirable; then he announces that the former can be used to "cure" the latter, because delicacy of taste enables us to develop judgment, which can mitigate excesses of temper, which constitute one articulation of delicacy of passion. Judgment, Hume explains, is both requisite to and cultivated by "regard to the sciences and fine arts," but when he describes our pursuit of these subjects, he notes that "fine taste," which seems to be the basis for judgment, becomes "in some measure" identical to, or even dependent on "strong sense," which presumably belongs to the passions that were "cured" by taste. As perception gives way to judgment, which then becomes discrimination, the faculties Hume initially considered simply parallel become inseparable:

> A greater or less relish for those obvious beauties, which strike the senses, depends entirely upon the greater or less sensibility of the temper: But with regard to the sciences and liberal arts, a fine taste is, in some measure, the same with strong sense, or at least depends so much upon it, that they are inseparable. In order to judge aright of a composition of genius, there are so many views to be taken in, so many circumstances to be compared, and such a knowledge of human nature requisite, that no man, who is not possessed of the soundest judgment, will ever make a tolerable critic in such performances. ("Delicacy," 6)

Given what Hume depicts as an inseparable union of taste and passion, the conclusion of this paragraph comes as a surprise, for suddenly he returns to the idea of "cure," which had seemed to be negated by the image of fusion just introduced. "Our judgment will strengthen by this exercise," he announces of studying the liberal arts. "We shall form juster notions of life: Many things, which please or afflict others, will appear to us too frivolous to engage our attention: And we shall lose by degrees that sensibility and delicacy of passion, which is so incommodious" ("Delicacy," 6).

What looks like a summary statement is immediately cast into doubt once more, this time as a result of what Hume refers to as his "farther reflection." "Perhaps I have gone too far," he admits, "in saying, that a cultivated taste for the polite arts extinguishes the passions.""On farther reflection, I find, that it rather improves our sensibility for all the tender and agreeable passions; at the same time that it renders the mind incapable of the rougher and more boisterous emotions" ("Delicacy," 6). Here Hume *refines* the conclusion he reached in the previous paragraph, just as he now associates discrimination with the refinement of the sensibility. Discrimination as refinement yields *improvement of the observer,* not *judgment of the object;* and the subject's improvement both creates and requires the elaboration of that instrument by which one discriminates in the first place—the delicacy of taste that depends on, but then chastens, "strong sense." The human being improved by—but also in—delicacy of taste, finally, is fit for the arena Hume represents as superior to both studious retirement and "the hurry of business and interest": the arena of "love and friendship." In this arena, he concludes, the man of refined taste finds a company few in number but select in kind grow more precious, as years convert appetite into appreciation and youth into an aesthetic variant of passion. "The gaiety and frolic of a bottle companion improves with him into a solid friendship: And the ardours of a youthful appetite become an elegant passion" ("Delicacy," 7, 8).

Hume's concluding invocation of aesthetics implies that he has turned the facets of subjectivity that Shaftesbury associated with aesthetics—contemplation, discrimination, and appreciation—into aspects of an analysis that proceeds by self-refinement and qualification toward a conclusion that is greater than the sum of its parts. "Of the Delicacy of Taste" is an enactment of contemplation and discrimination, and it culminates in an appreciation for a refined version of the morbid sensitivity to others that it initially castigated. In the course of the essay, Hume has reconsidered—and recast—human sociality, not by providing a philosophical map of human motivation but by leading the reader through the stages by which passion is refined to friendship. By slowly turning the parallel with which he began into a more complex figure, whose foundational units are no longer separable or reducible to what they initially were, he has exemplified the *process* by which the essayist produces the kind of knowledge that contemporaries called wisdom, which cannot be summarized in a single proposition or proved by mathematical demonstration.

Turning and retouching an argument, like a jeweler polishing a stone, may have seemed to eighteenth-century readers an approach appropriate to a treatment of the moral faculties because, as even the moral experimentalists admitted, these faculties were imbricated in the elusive, secondary nature of the qualities they apprehended. For the moral experimentalists, however, exploring moral faculties—or mapping subjectivity—was always conceptualized as part of a larger project: explaining how and why individuals in a market society should be trusted to govern themselves. Hume also kept this larger project in view even after he gave up moral philosophy. Thus most of the essays in his first volume dealt with subjects he considered political; these include "Of the Liberty of the Press," "That Politics May be Reduc'd to a Science," "Of the First Principles of Government," "Of the Independency of Parliament," "Whether the British Government Inclines More to Absolute Monarchy, or to a Republic," "Of Parties in General," "Of the Parties in Great Britain," and "Of Liberty and Despotism."[72] In these political essays, Hume repeatedly raised the problem that lay at the heart of any attempt to produce general knowledge about the self or society: the problem of induction. Given that our experience and observation can yield only *particular* details, can we produce *general knowledge*—especially about human institutions that we experience through a distorting cloud of interest and passion?

Whereas Hume had sought (and failed to find) a definitive answer to this question in the *Treatise,* in his essays he offered two contradictory answers. The first was yes. Particularly in the essay titled "That Politics May be Reduc'd to a Science," he argues that, even though what we see when we look at governments past and present is the influence of "accidental" factors, we can identify the general principles that underwrite the accidents, because the force of institutionalized law is so strong. Indeed, Hume takes this epistemological model, which conflates a principle that he applies to institutions with assumptions about the laws of nature, as a criterion for evaluating various kinds of governments: in a government that is ruled by (human) laws, he says, the force of those laws will be so great that the "accidental" tempers of individual administrators will make no difference in the operation of the general law. This view of things initially gives rise to considerable interpretive confidence. In both the epistemological domain and the domain of government, Hume asserts, "so great is the force of laws . . . that consequences almost as general and certain may sometimes be deduced from them, as any which the mathematical sciences afford us" ("That Politics," 16).

As Hume's telltale "sometimes" discloses, however, he cannot long maintain the existence of "eternal political truths, which no time nor accidents can vary" ("That Politics," 21). For as we discover in this and the adjacent essays on politics, the general laws he claims to have discovered reveal that the best governments are those that allow liberty; but because liberty "naturally" gives rise to factions, it creates the conditions for civil war and thus brings about its own destruction.[73] Because both liberty and the inclination to form factions (or political parties) are theoretically emanations of universal human nature, moreover, it seems that one of the general laws of human nature has a tendency to contradict or thwart another.

The heart of this problem, as Hume exposes it in these essays, is that individual men are drawn, by what he assumes to be human nature, to causes particular to their interests, not to the general principles that might make those interests seem short-sighted. Thus the heart of the problem that undermines epistemological certainty about political generalizations is the problem of the particular, now embodied in the individual (or party) who is engaged by the immediacy of particular interests. This immediacy, moreover, which is ruinous to the certainty of generalizations, is coterminous with the experience (the particular perceptions) by which one knows. Thus we are back to the problem that terminated the *Treatise:* How can particular perceptions be joined into a single whole *except by assuming some theoretical principle of order, which originates not in experience but somewhere else?*

Instead of trying to solve this problem, Hume offered a second answer to whether one can produce general knowledge about politics. This answer, of course, was no. In "Of Civil Liberty" he attributes this failure to time; the implication is that, once we can see more historical experiments, we will have more data, and we will be able to produce more certain knowledge. "I am apt . . . to entertain a suspicion, that the world is still too young to fix many general truths in politics, which will remain true to the latest posterity. We have not as yet had experience of three thousand years; so that not only the art of reasoning is still imperfect in this science, as in all others, but we even want sufficient materials upon which we can reason" ("Of Civil Liberty," 87).

In the revised version of an essay titled "Of the Parties of Great Britain," however, Hume offers another explanation for the philosopher's failure to produce general knowledge about politics from the particulars he has observed about men. In the one-volume edition of the *Essays* published in 1758, Hume added a note to the end of a much worked-over discussion of the difference between the contemporary political parties. In this note, and writing of himself in the third person, he admits that some of the "opinions" included in the earlier versions of the *Essays* were wrong, that they were in fact biased by "his own preconceived opinions and principles."

> Some of the opinions, delivered in these Essays, with regard to the public transactions in the last century, the Author, on more accurate examination, found reason to retract in his *History of* GREAT BRITAIN. And as he would not enslave himself to the systems of either party, neither would he fetter his judgment by his own preconceived opinions and principles; nor is he ashamed to acknowledge his mistakes.[74]

This retraction amounts to an admission that even the philosopher who claims to be superior to preconception and interest may be blinded by a bias he cannot see. Even though Hume's reference to "more accurate examination" seems to hold open the prospect that the philosopher can refine, perhaps even perfect, self-knowledge, it also opens the door for the reader to ask how, if the philosopher's judgment was once "enslave[d] . . . by his own preconceived opinions and principles," we can ever be certain that he is not still enslaved.[75]

Even before Hume added this note, the volumes of the *Essays* published before 1758 simply juxtaposed his two answers to whether political science was possible, for all of these volumes included both "That Politics May be Reduc'd to a Science" and "Of Civil Liberty." Instead of reconciling these two positions, in fact, Hume seems to have been inviting the reader to turn on him the same process of discrimination that he detailed in "Of the Delicacy of Taste and Passion." Instead of spoon-feeding the universal principles of politics to a passive readership, that is, Hume seems to have been initiating a process by which *collective* knowledge could be *improved*. This process, as he describes it in "Whether the British Government" and "Of the Rise and Progress of the Arts and Sciences," resembled a *conversation* or, as Jerome Christensen has noted, a "correspondence" like those in which men of letters participated in the mid-eighteenth century.[76] Such correspondences, as Hume pointed out throughout the rest of his career, militated against the overweening attachment to parties that interest naturally inspired, for they cultivated in men the subjective qualities that led to compromise and negotiation. They also helped enlarge the fund of experiences men could draw on to produce knowledge about self and society; and most important of all, they encouraged men not to assume that their experiences were sufficient to claim that the knowledge they articulated could stand without qualification. While men of letters might have been the principal beneficiaries of such correspondences, however, the "sovereigns" who oversaw them were not men, but women. If the epistemology associated with the essay and with correspondence could generate an alternative to that knowledge produced by moral and natural philosophies, Hume proposes, this kind of knowledge might aptly be designated feminine.

Hume's most extensive treatment of women's dominion over conversation, correspondence, and manners appears in an essay printed only once during his lifetime, in the second volume of *Essays, Moral and Political* (1742). Significantly, the subject of this essay is essay writing itself, thus making it his most self-conscious reflection on the mode and dynamics of this genre. "Of Essay-Writing" opens as if it will simply be an exercise in discrimination. In the essay's opening sentence, Hume distinguishes between those "immers'd in the animal Life," whom he dismisses from further consideration, and "the elegant

Part of Mankind," whom he immediately divides again, into "the *learned* and *conversible.*" He moves from this initial distinction to a series of definitions. "The Learned," he explains, "are such as have chosen for their Portion the higher and more difficult Operations of the Mind, which require Leisure and Solitude, and cannot be brought to Perfection, without long Preparation and severe Labour. The conversible World join to a sociable Disposition, and a Taste of Pleasure, an Inclination to the easier and more gentle Exercises of the Understanding, to obvious Reflections on human Affairs, and the Duties of common Life, and to the Observation of the Blemishes or Perfections of the particular Objects, that surround them" ("Of Essay-Writing," 533–34). After lamenting that "the last Age" tended to separate these parties, to the detriment of both, Hume then congratulates his contemporaries for forming "a League" between philosophy and conversation. In a passage dense with metaphors that map learning onto politics and commerce, he assigns himself a prominent position in this new "League" and gives the essay a critical office.

> 'Tis to be hop'd, that this League betwixt the learned and conversible Worlds, which is so happily begun, will be still farther improv'd to their mutual Advantage; and to that End, I know nothing more advantageous than such *Essays* as these with which I endeavour to entertain the Public. In this View, I cannot but consider myself as a Kind of Resident or Ambassador from the Dominions of Learning to those of Conversation; and shall think it my constant Duty to promote a good Correspondence betwixt these two States, which have so great a Dependence on each other. I shall give Intelligence to the Learned of whatever passes in Company, and shall endeavour to import into Company whatever Commodities I find in my native Country proper for their Use and Entertainment. The Balance of Trade we need not be jealous of, nor will there be any Difficulty to preserve it on both Sides. The Materials of this Commerce must chiefly be furnish'd by Conversation and common Life: The manufacturing of them alone belongs to Learning. ("Of Essay-Writing," 535)

In this passage, Hume has already assumed his office: by representing what happens in the world of tea tables and coffeehouses as "Intelligence" and the learned world's knowledge as "Commodities," Hume enjoins each party to see the other's characteristic product in familiar terms. By translating philosophy and conversation into a common commercial trope (the balance of trade), he also solicits them to view their activities as both exchange and the form of cooperation that distinguishes domestic production—the division of labor.

If the ambassador is charged with devising such translations, which reconcile opposite positions by giving them a common language, then he must also embody the deference one ruler displays to another. Such deference is the cur-

rency of diplomacy; it is the outward sign of respect for another's sovereignty
and of one's willingness to urge one's own sovereign to forgo pride for the sake
of cooperation. However contrived, deference is the ambassador's badge of sin-
cerity; however self-effacing, it is his instrument of rule. When Hume pursues
the trope whose mantle he has just assumed, then, he acts out this deference in
an obsequious bow to his female readers:

> As 'twou'd be an unpardonable Negligence in an Ambassador not to pay his Re-
> spects to the Sovereign of the State where he is commission'd to reside; so it wou'd
> be altogether inexcusable in me not to address myself, with a particular Respect, to
> the Fair Sex, who are the Sovereigns of the Empire of Conversation. I approach
> them with Reverence; and were not my Countrymen, the Learned, a stubborn in-
> dependent Race of Mortals, extremely jealous of their Liberty, and unaccustom'd
> to Subjection, I shou'd resign into their fair Hands the sovereign Authority over
> the Republic of Letters. As the Case stands, my Commission extends no farther,
> than to desire a League, offensive and defensive, against our common Enemies,
> against the Enemies of Reason and Beauty, People of dull Heads and cold Hearts.
> ("Of Essay-Writing," 535–36)

Although this last sentence carries on the work of discrimination with which
the essay began, the dominant mode has shifted from analysis to something like
play, or improvisation, or posturing. Because the donning of his own trope renders his tone playful, because subordinating himself to the figure of ambassador
encourages him to improvise along the lines it suggests, and because the
metaphor itself represents one who only represents another, Hume can pursue
the project of discrimination in another way, not by laying down rules but by
soliciting his female readers' participation through a deft mixture of flattery and
correction. "My fair Readers may be assur'd," Hume purrs, "that all Men of
Sense, who know the World, have a great Deference for their Judgment of such
Books as ly within the Compass of their Knowledge, and repose more Confidence in the Delicacy of their Taste, tho' unguided by Rules, than in all the dull
Labours of Pedants and Commentators" ("Of Essay-Writing," 536).

While Hume's reference to the limited range of women's reading considerably qualifies the authority he allows them, other passages reinforce the impression created here of respect for the office that lady conversationalists
perform. In "The Rise of Arts and Sciences," for example, he offers the man-
nerly exchanges that ladies oversee as both a paradigm of polite conversation
and actual instruments for improving the "polish" of men's minds. "What better school for manners," he asks rhetorically, "than the company of virtuous
women; where the mutual endeavour to please must insensibly polish the mind,
where the example of the female softness and modesty must communicate it-

self to their admirers, and where the delicacy of that sex puts every one on his guard, lest he give offence by any breach of decency?" ("Rise of Arts and Sciences," 134).[77] Because society has enjoined decorum upon women, Hume submits, feminine delicacy has become a weapon in the campaign against those "Enemies of Reason and Beauty, People of dull Heads and cold Hearts."

Hume does not actually grant women dominion, of course, even over the domain of polite conversation. His deference established by his fidelity to the ambassador trope, Hume seizes the opportunity to school those readers who, he claims, school men: enlarge your range of reading, he advises; don't fall for stylistic embellishments, especially in books of "Gallantry and Devotion"; and above all else, promote the project that Hume has identified with the improvement of the modern age: "Concur heartily in that Union I have projected betwixt the learned and conversible Worlds" ("Of Essay-Writing," 537).

The complex role that Hume assigned to women—he considered them both the school where refinement is acquired and in need of schooling themselves—reflects a pervasive cultural ambivalence repeatedly expressed toward women by would-be arbiters of culture and morality in this period. As recent feminist treatments of the work of Swift, Pope, and Richardson have made clear, men who wanted to produce knowledge about or rules for the new institutions of British civil society expressed a mixture of loathing and admiration for the women whose consumption and production so indelibly marked the emergent consumer society.[78] Such ambivalence was not new, of course. Participants in its seventeenth-century incarnation, in fact, the notorious *querelles des femmes,* can be said to have reviled "the sex" for its immunity from the kind of rules that natural philosophers were busy applying to nature; and proponents of the emergent market economy used the feminine Fortuna to stand for the hopes and fears sparked by the unpredictable turns of commerce.[79] In both of these examples, men tended to associate women with what was unruly or ungovernable—sexuality, risk, fortune—and their ambivalence may have been a carryover of the ambivalence many men also felt toward those objects and subjects that natural and moral philosophy sought to subdue. Whatever the source of those ambivalences—and they were no doubt overdetermined—Hume's ambivalence toward women seems slightly different. For however begrudging and qualified, his admiration for the civilizing function women performed constituted another version of the prestige he wanted to give the essay as a form of knowledge production. If readers appreciated how the essay worked to the degree that Hume appreciated women's ability to refine, they would accord the essay the authority the genre deserved; and the kind of knowledge it could produce—a knowledge that replicates experience instead of demonstrating propositions—would rival the philosophical knowledge Hume had cast aside.

Hume's attempt to accord the essay prestige equivalent—or even superior—to that of moral and natural philosophy was not successful, of course. Indeed, as the rest of this book makes clear, the systematic and general knowledge projects associated with philosophy continued to gain authority in the second half of the eighteenth century, and when reunited with numerical representation, as projects designed to produce general knowledge were in the early nineteenth century, these philosophical enterprises even managed to submit subjective states like happiness to actual (not gestural) quantification. Nevertheless, the proliferation of some new genres of imaginative writing in the eighteenth century (the novel) and the persistence of others (poetry) demonstrates that Britons still cultivated modes of knowledge production that departed from the systematic ideal of philosophy. Indeed, during the eighteenth century, then increasingly at century's end, these imaginative modes briefly came to seem at least as appropriate as moral philosophy to the crucial task of exploring the human motivations that underwrote liberal governmentality. The reign of the modes of writing that we call literature over the domain of subjectivity was relatively short-lived, however—or perhaps more accurately, literary writing soon had to share this project with new sciences developed specifically to map subjectivity without regard to the concerns of liberal governmentality. Most of the variants of psychology developed during the nineteenth century combined the systematic aspirations of eighteenth-century moral philosophy with an increasingly medicalized vocabulary modeled on biology, although one could argue that psychoanalysis—the science that eventually challenged both literature and nineteenth-century psychologies for jurisdiction over the subjective domain—was more like poetry, or even the essay, than the natural sciences in whose image the sciences of mind were formed.[80]

From Conjectural History
to Political Economy

Except in the form of rhetorical figures—gestural mathematics or the trope of mathematics itself—numerical representation all but disappeared from my discussion in the previous chapter. There is a good reason for this, beyond my desire to illuminate some of the obstacles that proponents of numbers-based knowledge faced in the first half of the eighteenth century. One might catalog the kinds of numbers that were collected in eighteenth-century Britain or the uses to which they were put, as some historians have done,[1] but to do so without understanding either what *counted as* useful knowledge at the time or what numbers *meant* as an epistemological instrument—as an instrument of knowledge production—would be to overlook the conditions that both promoted and limited the British government's commitment to collecting numerical data.[2] Thus we must recognize that, even though the British government collected considerable numerical information in the first three-quarters of the eighteenth century, these data were not gathered or used in the context of a coherent theory about the relation between numbers and rule that could rival the relatively well-theorized relation between accounts of subjectivity and liberal governmentality.

As I noted in chapter 4, there were at least two reasons for the relative neglect in Britain of the more theoretically coherent German "science of police," to which Petty's political arithmetic can retrospectively be assimilated.[3] First, the emergence of a market society in Britain and the concomitant dilution of the monarch's power meant that theorists devoted more attention to liberal governmentality and the subjective dynamics of self-rule than to devising plans for the use of numbers-based information, which might have been collected by and used to strengthen a central government. Second, the priority Newtonian philosophers assigned to universals and to the (invisible) laws of nature went hand in hand with a devaluation of the observed particular. Since one could only *count* observed particulars, this meant that counting was devalued too. Indeed, instead of promoting the elaborate infrastructure necessary to collect numerical information, British theorists of wealth and society developed a mode of analysis that could be used in the *absence* of numerical data. In general terms, as we have seen, what counted as knowledge about these subjects in Britain—at least before about 1776—was a form of theoretical generalization that devalued observed particulars in favor of something that could not be seen and, in so doing, made collecting numerical data all but redundant. In this chapter I argue that the model by which this kind of knowledge was constructed without numerical data not only preceded the collection of such data but also informed their treatment after this information had begun to be gathered on a wider scale.

In what follows I initially take up the two eighteenth-century disciplines that drew on the method of experimental moral philosophy to create new sciences of society and wealth. The first of these disciplines—Scottish conjectural history—did not contribute directly to the curious use of numbers that epitomizes the paradoxical nature of the modern fact, but it did explicitly raise—and find a way to address—the epistemological problem that lies at the heart of this paradox. As we have already begun to see, this difficulty—the problem of induction—challenged the assumption that particulars one had yet to observe would resemble the particulars one has already seen; to address the problem of induction, the philosopher had to explain how one could assume that systematic knowledge could be generated from what was inevitably an incomplete survey. For the so-called conjectural historians this challenge was especially great, because what they wanted to describe—the origins of modern society, and especially how "rude" societies became "civilized"—had not been reported by witnesses in a position to record what they had presumably observed.

To supplement the lack of eyewitness evidence, the Scottish historians used the experimentalists' assumptions that some system organizes the phenomenal world and that human nature is universal to "conjecture" what they could not document. Such conjectures constitute *assertions* that what one has not seen resembles what one can observe, but to understand the basis for these assertions, we need to see two things: the place that belief in providential design played in the conjectural histories, and how some of these historians began to convert the universalized assumption embraced by the experimental moralists (that human nature is everywhere the same) into another kind of theoretical entity—a form of abstraction that seemed capable of acting in the world. This new kind of abstraction—William Robertson's figure of "the human mind," for example—was conceptualized as the agent of history; as a historical agent, "the human mind" could be inferred from its effects, many (though not all) of which had been documented by eyewitnesses who recorded particulars whose larger significance they did not understand. Thus abstractions like "the human mind" helped meet the challenge that the problem of induction posed to systematic knowledge because they enabled philosophical historians to know what no one could actually see: the invisible (but consistent) agent whose agenda was realized in phenomena both observed and yet to be seen.

Because they were conceptualized as historical agents, the kind of abstractions that evolved (and carried over assumptions) from the experimental moralists' universals also produced effects that, at least in theory, could be measured or counted. This is especially evident in the abstraction that Adam Smith placed at the heart of political economy: the "market system." In Smith's account, abstractions like the market system created a new role for numerical representation, for as descriptors of the products (actually or theoretically) created by the institutions associated with the (idea of a) market system, numbers seemed to refer to entities that had been (or could be) counted. At the same time—and this is critical both to the paradigmatic use of numbers in the modern fact and to that epistemological entity's paradoxical nature—Smith's numbers also embodied his a priori assumptions about what the market system *should be*. The numbers Smith called for in *Wealth of Nations,* in other words, were not descriptive as we typically use that term; he did not want exclusively numbers that actually reflected the state of commerce in 1776, even if these could have been collected, because the market system had yet to be freed from interfering legislation. Instead of descriptively accurate numbers, the numbers he wanted were those that would help legislators see the system that was not yet visible, the market system operating as it should.

Having said that Smith's numbers were not descriptive in the sense of being accurate, let me note that Smith *would* have called these numbers descriptive, because his criterion for descriptive adequacy involved the ability to create a certain feeling in the observer—the feeling of satisfaction—which was a response to the systematic nature of the description itself. By setting "descriptive" in opposition to "rhetorical" and by aligning "satisfaction" with system, he created an epistemological space for an apparently nonsuasive mode of representation (whose form could be numerical but did not have to be), whose credibility came from its internal coherence as much as from its truth to nature. Thus Smith did not so much elaborate the old idea of "precise, and *therefore* accurate," as elevate precision (systematic coherence) over accuracy. At the same time, he made precision (systematic coherence) the bridge that linked observed particulars to the still unrealized potential that only the philosopher could see. Thus the idea of a market system, which was generated by the systematic science of political economy, also helped address the problem of induction *in protopsychological terms,* for it brought what had not yet been observed (or counted) into relation with what the theorist could imagine, and it signaled that it had done so by conferring satisfaction where there had once been doubt.

In addressing the problem of induction in protopsychological terms, those abstractions that Smith considered satisfying, like the market system, provided a new basis for linking the theories about subjectivity that I discussed in chapter 4 with apologies for liberal governmentality. Although I do not deal extensively with the much-discussed relation between the particular account of subjectivity that Smith developed in the *Theory of Moral Sentiments* and liberal governmentality, I do show why his assumptions about subjectivity are intrinsic to the theory of government implicit in *Wealth of Nations*.[4] In passing, I also comment on another subject that has been extensively reviewed in the scholarly literature: the way Smith's political economy constitutes the logical end of those reason-of-state theories of government developed in the sixteenth and seventeenth centuries.[5] Instead of taking up the relation between Smith's various accounts of subjectivity and governmentality per se, however, I focus on the epistemological issues he raised in his attempts to link these two topics. I argue that, in insisting that society constitutes a *system* visible only to the moral philosopher cum political economist, and that we know that system by the satisfaction that assumptions about it confer, Smith helped explain both how "human nature" manifests itself in the market system that (ideally) governs itself and how legislators can come to know (and actualize) the system they cannot see.

Even though Smith's science of wealth explicitly supported liberal governmentality—even though his most adamant demand was that legislators leave the market alone—both the kind of abstraction whose effects could be quantified and the political economic variant of the modern fact by which these abstractions were produced soon proved attractive to the British government. These epistemological entities appealed to government officials seeking to consolidate (and theorize) the government's relationship with its subjects at home and abroad for at least two reasons. First, as Smith described them, political economic facts embodied the qualities of impartiality, transparency, and methodological rigor that have made numerical information so attractive to modern governments. And second, while abstractions like the market system set limits to some kinds of legislative interference, they mandated the implementation and enforcement of other kinds of laws and policies. Even while we register the arguments by which Smith forged a science of liberal governmentality out of the moral experimentalists' universals, then, we should also be alert to the susceptibility that political economic facts would have to an argument Smith would never have endorsed: the argument that the central government should grow strong precisely by collecting and using the kind of fact he helped naturalize.

If Smith's political economic facts epitomize the late eighteenth-century variant of the modern fact, then the text I consider in the final section of this chapter—Samuel Johnson's *Journey to the Western Islands of Scotland*—holds these facts, as well as the epistemological assumptions that underwrote them, up to the scrutiny of an emergent cultural relativism. Paradoxically, this epistemological position was adumbrated in the work of some of the conjectural historians, who, for all their interest in the universal subject of science, were nevertheless fascinated by the different "stages" of human development visible in the various races spread across the globe. A full-blown version of what we call cultural relativism did not materialize either in the writings of the Scottish historians or in Johnson's travelogue, of course, for just as the historians were primarily interested in what made humans alike (or at most incarnations of different stages in a single line of development), so Johnson was still committed, late in his life, to a variant of the wisdom literature he had written throughout his career. Nevertheless, Johnson's last work, like Hume's essays, provides an example of the forms of knowledge production that rivaled numerical knowledge and in relation to which proponents of numerical knowledge constructed their defenses of numbers; and it does so, moreover, in terms that specifically questioned the epistemological assumptions that were increasingly being used to defend political economic facts. Before turning to the institutionalizing of political economy, as I do in the next chapter, and to Malthus's resuscitation of the numerical fact, let us pause for a moment over one dissenting opinion in what increasingly became, after 1776, a general (if gradual) acceptance of the validity and utility of numerical information.

SCOTTISH CONJECTURAL HISTORY

The word "conjecture" was introduced into the English language in the late fourteenth century. From the Latin *conjectûra*—a throwing or casting together—it initially referred to a mode of producing knowledge about the future from signs or omens believed to be portentous. Beginning in the mid-sixteenth century, perhaps as one indication that probabilism was beginning to rival such magical thinking, "conjecture" began to carry a second meaning. As the *OED* tells us, "conjecturing" also began to mean "offering an opinion on grounds *in*sufficient to furnish proof, . . . guessing" (my emphasis). Thus, from the mid-sixteenth century "conjecture" has carried two antithetical meanings: on the one hand, the word can refer to a mode of generating knowledge considered legitimate because it respects current epistemological conventions; on the other hand, it can mean irresponsible speculation—mere guesswork.

During the second half of the eighteenth century, the ambiguity written into "conjecture" intersected with an evolving debate about how moral philosophers should generate knowledge about society. British philosophers, particularly before the 1790s, generally held the method of conjecture in low esteem; among other things, conjecture, especially when construed as mere hypothesis, was the method associated with Scholasticism, Hobbes, and Descartes. The scorn heaped on conjecture by philosophers like Thomas Reid gained additional weight from the beloved Newton's famous (but famously enigmatic) repudiation of hypothesis, which appeared in the second edition of the *Principia* (1713): "Hypotheses non fingo." Yet in 1790 Dugald Stewart, the widely respected professor of moral philosophy at Edinburgh University, coined the phrase "conjectural history" to describe a kind of historiography practiced by Adam Smith, David Hume, and Henry Home, Lord Kames.[6] Stewart clearly intended to praise this groundbreaking work, and his use of "conjectural" implies that by praising Smith, Hume, and Kames he also wanted to rescue both the method and the term from the disrepute they had fallen into.

In the contest over conjecture, we can glimpse both one of the directions taken by the sciences of wealth and society in late eighteenth-century and early nineteenth-century Britain and the role played in these sciences by the conflation of description and theory that I have associated with the modern fact. Indeed, for late twentieth-century readers, the eighteenth-century philosophical controversy over conjecture serves to expose the problematic of the modern fact, although, as we will see in the next chapter, conjecture was also offered as a solution to the complication that had driven Hume from philosophy altogether—the problem of induction.

In the next chapter I will examine the role that Stewart's attempt to redeem conjecture played in the early nineteenth-century revision of political economy, but first we need to understand both why conjecture seemed so dubious to the eighteenth-century historians to whom Stewart assigned this term and why he could call these historians' work "conjectural" even though they would not have done so. In his 1777 meditation on historical method, William Robertson helps us identify one reason conjecture was held in such low esteem for most of the eighteenth century: because it indulged the philosopher's propensity for speculation, conjecture opened the door to partisanship and opinion. "Without indulging conjecture, or betraying a propensity to either system, we must study with equal care to avoid the extremes of extravagant admiration, or of supercilious contempt for those manners which we describe," Robertson warned in *The History of the Discovery and Settlement of America*.[7]

Robertson implies that the repudiation of conjecture belonged to a longing for an *impartial* philosophical position, a position that would be inoculated against charges of self-interest. Later in the century Thomas Reid, who was professor of moral philosophy at Aberdeen, then Glasgow, cast further aspersions on conjecture when he called this method "contraband and illicit." L. L. Laudan has argued that Reid's hostility to what he alternately called theory, hypothesis, and conjecture was probably so extreme because in the science he sought to legitimate (pneumatology, or the science of mind), a hypothesis was very difficult to disprove because the phenomena at issue could not be observed.[8] Although this may well be true, Reid's language suggests that, like Robertson, he was also trying to forestall complaints about philosophical self-interest or even grandiosity. Beyond violating the strictures bequeathed by "the great Newton," Reid charged, the philosophical use of conjecture drew what ought to be a sober practice based on reason perilously close to a grandiose indulgence of the philosopher's overweening ambition. Reid cultivated the image of the "mad" philosopher by invoking the specter of "Indian philosophy," although one must suspect that "Indian philosophy" stands in here for "French philosophy" (Cartesianism).

> In the operations of nature, I hold the theories of a philosopher, which are unsupported by fact, in the same estimation with the dreams of a man asleep, or the ravings of a madman. We laugh at the Indian philosopher, who to account for the support of the earth, contrived the hypothesis of a huge elephant, a huge tortoise. If we will candidly confess the truth, we know as little of the operation of the nerves, as he did of the manner in which the earth is supported; and our hypothesis about animal spirits, or about the tension and vibrations of the nerves, are as like to be true, as his about the support of the earth.—His elephant was a hypothesis, and our hypotheses are elephants. Every theory in philosophy, which is built on pure conjecture, is an elephant; and every theory that is supported partly by fact, and partly by conjecture, is like Nebuchadnezzar's image, whose feet were partly of iron and partly of clay.
>
> The great Newton first gave an example to philosophers, which always ought to be, but rarely hath been followed, by distinguishing his conjectures from his conclusions, and putting the former by themselves, in the modest form of queries. This is fair and legal; but all other philosophical traffic in conjecture, ought to be held contraband and illicit.[9]

If eighteenth-century historians and philosophers tended to associate conjecture (especially "pure" conjecture) with self-interest, irresponsible speculation, and partisanship, then why did Dugald Stewart—who was, after all, a student and professed admirer of Reid's—call the historical practice exemplified by Robertson "conjectural"? Assuming (as is almost certainly the case) that Stewart did not intend to disparage these histories (or the philosophy that resembled them), and allowing for his own desire to redeem conjecture (a desire to which I will return), we can see that the term "conjectural" was warranted by the nature of the project in which many mid- to late eighteenth-century historians and philosophers were engaged. The most obvious reason Stewart used "conjectural," in other words, was that the kind of history Robertson, Smith, Hume, Ferguson, and Lord Kames wrote—like experimental moral philosophy—could not rely on written records, eyewitness testimony, or any kind of evidence that met the strictest definition of "experience." Just as one could not literally see the nerves or the moral sense, so one could not see, or read accounts of anyone who had seen, the transition from hunter-gatherer to agricultural society.

Whereas the early eighteenth-century experimental moralists like Hutcheson and Turnbull were (relatively) content with metaphorical vision, because they were ready to demote the observed particular, Reid and the conjectural historians wanted to shore up the Baconian components of their practices. But if one wanted to generate knowledge about phenomena one could not literally see, as both Reid and the conjectural historians did, then how was one to move from the particulars one could observe to phenomena one could not see to systematic knowledge about both? Stewart's answer, which was published in his *Account of the Life and Writings of Adam Smith,* was that one used conjecture.

> When, in such a period of society as that in which we live, we compare our intellectual acquirements, our opinions, manners, and institutions, with those which prevail among rude tribes, it cannot fail to occur to us as an interesting question, by what gradual steps the transition has been made from the first simple efforts of uncultivated nature, to a state of things so wonderfully artificial and complicated. . . . Whence the origin of the different sciences and of the different arts, and by what chain has the mind been led from their first rudiments to their last and most refined improvements? Whence the astonishing fabric of the political union, the fundamental principles which are common to all governments, and the different forms which civilized society has assumed in different ages of the world? On most of these subjects very little information is to be expected from history, for long before that stage of society when men began to think of recording their transactions, many of the most important steps of their progress have been made. . . . In this want of direct evidence, we are under a necessity of supplying the place of fact by conjecture; and when we are unable to ascertain how men have actually conducted themselves upon particular occasions, of considering in what manner they are likely to have proceeded, from the principles of their nature, and the circumstances of their external situation.[10]

That Stewart is explicit about the role of conjecture in systematic knowledge demonstrates that by 1790 it had come to seem necessary to name (if not defend) what almost all British eighteenth-century moral experimentalists routinely did: supplementing accounts of observed particulars with conjectures. Stewart's explicitness reflects the delayed but increasingly powerful effect of Hume's skepticism, which Reid was also trying to refute. That is, conjecture came to seem like a method that had to be either reviled *or* defended only after the problem of induction posed by Hume—and, in a different key, by Adam Smith—came to *seem like a problem*. The criticism of conjecture launched by Robertson and Reid, then, is one sign of the effect of Hume's articulation of the problem of induction, for that difficulty lay behind the imperative to conjecture: if one could not observe every instance of a phenomenon, then one had to make inferences about what one had not seen and to devise accounts of why one's speculations were neither self-interested, irresponsible, nor grandiose.

Before the last quarter of the eighteenth century, British philosophers generally did not admit the role conjecture played in their method, for it simply did not seem unreasonable either to postulate an origin for society or to deduce from this postulate how things that could not be observed were "likely to have proceeded, from the principles of their nature." Nor did such deductions seem antithetical to the induction that these historians, like Reid, tended to embrace because it was Baconian, Newtonian, and British. Even though they wrote after the publication of Hume's *Treatise,* in other words, and even though the furor about conjecture signals to us a sensitivity to philosophical method articulated most clearly by Hume, most of the eighteenth-century historians considered their method perfectly defensible. Before we turn (again) to Hume and then to Smith, we need to identify the assumptions that informed the versions of conjectural history produced by writers like Adam Ferguson, John Millar, William Robertson, and Lord Kames to see how they held the problem of induction at bay in a discipline to which it now seems so obviously related.

The few twentieth-century historians who have explicitly addressed Scottish conjectural history have struggled to identify a set of essential features that might be said to unify works as different as Kames's *Sketches of the History of Man* (1774) and Ferguson's *Essay on the History of Civil Society* (1767).[11] Although it seems to me unnecessary—and potentially misleading—to try to name *the* features that make these texts a single body of work, it is necessary to understand what kind of project these historians considered themselves to be pursuing. In retrospect, we might emphasize that identifying the origins and tracing the development of modern institutions like manners or commerce was part of the ancient/modern debate, which had raged in England since at least the last third of the seventeenth century; but to the mid-eighteenth-century historians themselves, the more relevant project was the characteristically Scottish effort to develop an experimental moral philosophy, so that the study of society and subjectivity could be made as systematic as that of the natural world.

The methodological affiliation between experimental moral philosophy and conjectural history is clear in John Millar's *Observations concerning the Distinction of Ranks in Society* (1771). "By real experiments," Millar announces in his prefatory remarks, "not by abstracted metaphysical theories, human nature is unfolded; the general laws of our constitution are laid open; and history is rendered subservient to moral philosophy and jurisprudence."[12] As we saw in chapter 4, "real experiments," especially when undertaken by eighteenth-century moral philosophers, tended to combine elements of what now seem like different methods (but did not seem so to contemporary Newtonians): induction, or reasoning from observed phenomena; and deduction or, by analogy to mathematics, reasoning from postulated axioms. That the coexistence of these methods was considered as noncontroversial to most of the conjectural historians as to the experimental moralists can be seen from the former's tendency to rely both on eyewitness accounts, generally of so-called rude societies, and on a priori principles, which were derived from a combination of introspection and assumptions about providential design and the laws of human nature. Indeed, so evident did most mid-eighteenth-century Scottish historians consider their method that they tended to identify their chief problem as one of *sources*, not method. The difficulty, Millar and Robertson explained, was determining which eyewitnesses of rude societies were reliable—that is, which sources provided evidence that proved the principles the historian assumed—not whether induction was compatible with deduction.[13]

In one sense, the project of the conjectural historians was to extend the moral experimentalists' map of subjectivity so as to explain more precisely why human beings formed societies and why these societies changed over time. Thus we can see the continuity between experimental philosophy and conjectural history not only in their method, but also in the object whose "laws" Miller wanted to illuminate: "human nature" or "our constitution." In another sense, however, and especially as William Robertson and Adam Smith elaborated the project of conjectural history, this enterprise sought to *discover* or even to *create* abstractions that could explain *how* "human nature" realized itself in those social arrangements both recorded and still unknown. This created abstraction—"the human mind," for example—could function as a historical agent in a way that the universal "human nature" could not, and because it was an additive rather than an essentialist concept, it allowed the historian to explain that, even if particular individuals did not contribute to the progress of "the human mind," human beings taken as a whole did. Thus to narrate the history of "the human mind" required the historian both to assemble as much information as he possibly could and to dismiss any evidence that failed to show "the human mind" at work. "In order to complete the history of the human mind, and attain to a perfect knowledge of its nature and operations," Robertson explained, "we must contemplate man in all those various situations wherein he has been placed" (*History of America,* 91).

To a certain extent, of course, "human nature" and "the human mind" are the same kinds of philosophical concepts. Both, for example, entailed Aristotelian analyses: both "human nature" and "the human mind" assumed that the operations of the general type were governed by its nature, just as the *telos* of an organism was assumed to be given by its characteristic essence.[14] Thus, if one discovered (or postulated) the nature or essence of the type, one could deduce its behavior even in the absence of written records, in large part because one knew (or assumed) its *telos* or function. Despite the Aristotelian analytic that was applied to both objects of analysis, however, "human nature" and "the human mind" constitute slightly different kinds of philosophical concepts. As I have noted, "human nature" was an essentialist concept, whereas "the human mind" was (in theory at least) derived from adding up numerous examples; this suggests that the Aristotelian analytic I have just described worked better for "human nature" than it did for "the human mind" because, even if the latter obeyed the law of type *when taken as a whole,* one could not assume that each member of the aggregate would do so. By the same token, "human nature" could be represented as a product of at least metaphorical observation because one could claim that a combination of introspection and extrapolation could confirm one's belief about this essence, whereas "the human mind" as an aggregate was a philosophical construct that no one could claim to see. At most one could maintain one saw the *effects* of an abstraction like "the human mind," but even the claim to identify certain events *as effects* of an aggregate reflects the way the philosopher had derived such abstractions in the first place: unlike universals ("human nature"), which derived from assumptions that preceded analysis (and informed metaphorical observation), abstractions like "the human mind" were *produced by the method of conjectural history itself* in order to make something that exceeded any individual incarnation available to intellectual contemplation.[15] In this sense "the human mind" resembles those other great abstractions generated out of the theoretical systems that support the modern fact: society, poverty, commerce, the market system, the economy.

When I take up Adam Smith later in this chapter I will examine the effects produced by abstractions like the human mind, which substantially exceeded the effects of universals like human nature. For now, however, let me note that when conjectural historians focused on either philosophical entity, they privileged uniformity over variety or difference. Having said that, I should also point out that in distinguishing "rude" from "civilized" societies, most of the conjectural historians also began to open a space for noticing—though not yet for valuing—cultural and racial differences. Although none of the other conjectural historians went as far as Lord Monboddo, who speculated that dark-skinned races (and even apes) might be the species ancestors of white Europeans, Lord Kames and, to a lesser extent, Robertson and Ferguson did offer theories about racial difference, which began to propose that environmental factors and not some essential quality might account for differences in skin color.[16] While one should not overstate the cultural relativism articulated in the work of these historians, we need to see that simply by examining racial difference they laid the groundwork for what would eventually become an interrogation of the eighteenth-century assumption that the human subject of science was by nature everywhere the same.

By and large, the conjectural historians were less interested in cataloging marks of difference than in extrapolating from the uniformity they assumed to exist in the general laws of human nature, as these laws were revealed by (unfolded in) human history. That this agenda dictated the method of conjectural history is clear in Robertson's insistence that "in a general history of America, it would be highly improper to describe the conditions of each petty community, or to investigate every minute circumstance which contributes the form to the character of its members. Such an inquiry would lead to details of immeasurable and tiresome extent. The qualities belonging to the people of all the different tribes have such a near resemblance, that they may be painted with the same features" (*History of America,* 91). As we saw in our investigation of Hobbes, "details of immeasurable . . . extent" had traditionally been the province of the natural historian; in repudiating these details as "tiresome," Robertson was aligning the history of "the human mind" with moral philosophy instead of natural history. In so doing, and despite his professed Baconianism, he was also demoting observed particulars. This philosophical gesture was perfectly consistent with taking "the mind" as the object of historical analysis, for if "the mind" or "the commercial spirit" was the agent of history, then individual actors and particular events were less important than what (only) the philosophical historian could describe by adding all these actors together and subtracting those that did not conform to the emerging pattern.

Perhaps incidentally, one effect of privileging abstractions like "the mind" over observed particulars was to enhance the status of the conjectural historian. If the proper object of historical analysis was an abstraction that only the philosophical historian could describe, then only he could identify the effects that counted. Unlike Roman historians, that is, who had sought to identify the intentions of politicians and statesmen as well as to chronicle their acts, the conjectural historians focused on the unintended effects of past events. This emphasis privileged the historian over the historical actor, for it elevated interpretive hindsight over intention. It also privileged philosophical analysis over reason of state, for the best-laid plans could not be trusted to bring about the desired effects when the perspective that mattered was the retrospective gaze cast back from the future.

Focusing on unintended consequences makes history less a teleological narrative of some essence reappearing across time than a sequence of events whose significance only the historian can infer. Thus the Crusades, which contemporaries considered "wild expeditions," had the unexpected and unintended effect of awakening in Europeans new tastes for learning and commodities: "To these wild expeditions, the effect of superstition or folly, we owe the first gleams of light which tended to dispel barbarism and ignorance."[17] Robertson considered the "spirit of chivalry" another "wild" institution that had unanticipated but monumental consequences: "Perhaps the humanity which accompanies all the operations of war, the refinements of gallantry, and the point of honour, the three chief circumstances which distinguish modern from ancient manners, may be ascribed in a great measure to this institution, which has appeared whimsical to superficial observers, but by its effects has proved of great benefit to mankind" (*View of the Progress,* 4:85). For Adam Ferguson, every ruler's attempt to imagine outcomes or to devise profitable schemes is subject to this rule of unintended consequences and thus to the superior judgment of the historian: "Mankind, in following the present sense of their minds, in striving to remove inconveniences, or to gain apparent and contiguous advantages, arrive at ends which even their imagination could not anticipate, and pass on, like other animals, in the track of their nature, without perceiving its end."[18]

As we will see more clearly when we turn to Adam Smith, the conjectural historians' tendency to privilege abstractions over either recorded intentions or the natural phenomena that they (or a trusted eyewitness) could (or might have been able to) observe required them to hold two quite different understandings of "nature": one included phenomena one could actually see, and from which the historian (presumably) constructed the aggregate; the other consisted of the implicit significance of these phenomena, which one could know only retrospectively and only by discounting whatever particulars diverged from type. Smith used his theory of description to reconcile (or at least explain the relation between) these two understandings of nature. Robertson and Ferguson, by contrast, who were less troubled by possible discrepancies between what they (or others) saw and what they inferred from their theoretical assumptions, did not have recourse to an explicit theory of description. Instead, they simply claimed that observation and theory revealed the same thing: a four-stage history of human progress in which "the human mind" moved from a state of savagery to hunter-gatherer communities to agricultural associations to commercial society.[19]

In addition to its epistemological implications, this historical schema is interesting for at least two other reasons: first, because it focused on the intersection of subjectivity and sociality, this account tended to assign new importance to domesticity, manners, and women;[20] and second, because it considered commercial society the most sophisticated incarnation of human sociality, it suggested that the dynamics of commerce revealed a (collectively realized) tendency inherent in "the human mind" that was more significant than the old question whether individuals were self-interested or self-denying "by nature."[21] The combined effects of manners, women, and commerce could thus be used to explain how liberal governmentality worked, and the argument that commercial society embodied "the human mind" could be used to ban government "interference" in a domain that was essentially self-governing because it realized something greater than the will of any parties or individuals.

As my reference to the conjectural historians' two views of nature illustrates, privileging abstractions like "the human mind" required support beyond mere observation. That is, if constructing the history of "the human mind" involved not only collecting as much evidence as possible but also discounting instances that did not fit one's assumptions about such abstractions, then something had to ground the assumptions that guided the historian's interpretations. Simply put, discrepancies inevitably occurred between what the historian observed and what he believed must be true. Thus, for example, even though the conjectural historians thought that the history of "the human mind" unfolded in four stages, when they looked to Europe for evidence of those stages, they saw not a four-part progression but the overlap of one stage by another—the persistence, for example, of an agricultural society within the full-fledged commercial society of Great Britain. Looking at America, by contrast, they observed what they expected to find: a society still in its least civilized stage and apparently preparing to make the great transition. With the exceptions of David Hume and Adam Smith, to whom I will return, they typically closed the gap between what they observed and what they assumed with a single belief: the conviction that providential design was working itself out in history, through those great abstractions that the historian brought into sharp relief. This strong providentialism, which constituted the bedrock of most conjectural histories, is most explicit in the work of Lord Kames. Indeed, in Kames's *Sketches* it becomes clear that belief in a beneficent providential order was not only what closed the gap between what one saw and what one knew but also what allowed the conjectural historian to write history even though he lacked the kind of data a Baconian might have desired.

> As men ripen in the knowledge of causes and effects, the benevolence as well as wisdom of a superintending Being become more and more apparent. . . .Beautiful final causes without number have been found in the material as well as moral world, with respect to many particulars that once appeared dark and gloomy. Many continue to have that appearance: but with respect to such, is it too bold to maintain, that an argument from ignorance, a slender argument at any rate, is altogether insufficient in judging of divine government? How salutary is it for man, and how comfortable, to rest on the faith, that whatever is, is the best! (*Sketches*, 2:219)

Like Turnbull's, Kames's providentialism implied two things: that a systematic order underwrote the particulars individuals could see, and that this order was both beautiful and good. Thus his belief in providential design enabled him to direct his readers' attention away from their immediate woes, as well as away from the gloomy vision of cyclical refinement and decay that he observed in the historical record.[22] Indeed, in Kames's account, individuals must learn to discount their immediate fortunes, to look beyond what they experience, and with the help of sciences like experimental philosophy and conjectural history, to recognize the pattern of which individual experience is merely an insignificant part. "The system of Providence differs widely from our wishes," Lord Kames admits; but "from what is known of that system, we have reason to believe, that were the whole visible, it would appear beautiful" (*Sketches*, 2:204).

Kames's reliance on Providence to explain what he could not see makes explicit how thoroughly the kind of fact produced by the conjectural historians depended on a priori assumptions, belief, or what Stewart was to call conjecture. All of them both constructed and interpreted such abstractions as "the human mind" not through induction or additive observations but simply by acting on their Christian belief in a benevolent and lawful God. As we will see in the next chapter, the epistemological model that resulted—in which general knowledge was constituted *in the absence of* collected data—continued to govern the treatment of data even after such information had begun to be gathered. As we see from Hume's contribution to conjectural history, moreover, as soon as one set aside the foundational assumption that God's Providence was the ground of what one failed to understand, then those abstractions that the midcentury historians treated as real entities ("the human mind," "the commercial spirit," "the market system") could also be understood as mere fictions—indeed, as effects of the mania for systematic analysis (and belief) that was the backbone of moral philosophy and conjectural history alike. By examining one of Hume's conjectural histories in some detail, then, we can see both how foundational belief was to the conjectural variant of the modern fact and what happened when one called into question the foundation of belief.

Although David Hume wrote several texts that have been classified as conjectural histories, the earliest and most revealing is that section of book 3 of *A Treatise of Human Nature* titled "Of the Origin of Justice and Property."[23] Jerome Christensen has brilliantly analyzed the intricacies of this history, and in what follows I will not have much to add to his analysis. Placing Christensen's observations in the context of conjectural history and the emergence of the idea of a market system more generally, however, does enable us to see two things that he does not stress: first, that Hume constructed the idea of a commercial (or market) system *as an alternative to* providential design; and second, that replacing Providence with the metaphor of system laid the ground both for subsequent vivifications of the market (including those that made its components quantifiable) *and* for the kind of methodological interrogation that would challenge the entire enterprise of political economy.

History was important to Hume partly because it constituted one laboratory (along with introspection) for experimental philosophy.[24] In *A Treatise of Human Nature,* Hume offers two historiographical models. In the first, the historian has as evidence "the unanimous testimony of historians," which can be traced back through an unbroken "chain" to the testimony of "those who were eye-witnesses and spectators of the event" in question.[25] Such histories do not seem like conjectural histories, for obvious reasons: the unbroken chain of testimony obviates the need for conjecture. In the second model, no such unbroken chain of evidence exists. This is true partly because the events in question—in this case the origins of justice and property occurred before written records of eyewitness spectators were kept. Because no chain of evidence is available, Hume does resort to conjecture; and like the other historians I have examined, he fills in the lacunae with postulates about human nature derived from introspection, observation, and speculation.

In Hume's conjectural history, however, it becomes clear that the lack of an evidentiary chain does not constitute the exceptional difficulty, which the conjectural historian uniquely faces, but typifies the dilemma intrinsic to the collection of any kind of evidence whatever. In other words, the lack of an evidentiary chain exemplifies the problem of induction, for the breaks that sever the chain require a version of the extrapolation by which one assumes that unobserved phenomena will resemble what one has already seen. Without this extrapolation, which Hume identifies as the *belief in system,* general knowledge would be impossible. Thus he contends that induction works not by moving stepwise from observed particulars to ever greater levels of generalization but *by immediately effacing the specificity of the particular in favor of what one believes about the system.* In so doing, induction both acts like metaphor (it insists on likeness, not difference) and imposes metaphors (the tropes of system, law, design). In his recognition of the place that metaphor—and fiction more generally—occupies in induction, Hume departs from the conjectural historians I have examined, for whereas they referred "design" to Providence, he grounds it in the human capacity to invent fictions.

In his analysis of Hume's "Of the Origin of Justice and Property," Christensen points out how Hume repeatedly installs contradictions at the site of origin.[26] As we will see in the next section of this chapter, these contradictions resemble the two views of nature that Adam Smith held side by side. Thus, for example, Hume explains that society was initially formed because, *by nature,* human beings are *unnatural:* only in human beings do we see the "unnatural conjunction" of expansive need and bodily infirmity (*Treatise,* 536–37). He immediately complicates this picture by pointing out that, even though that collectivity we call society was necessary to satisfy the individual's needs, no individual could have known this before society existed because, before "cultivation," no human being could imagine society's advantages: "In order to form society, 'tis requisite not only that it be advantageous, but also that men be sensible of these advantages; and 'tis impossible, in their wild uncultivated state, that by study and reflection alone, they should ever be able to attain this knowledge" (*Treatise,* 537). At first Hume proposes that human beings resolved such contradictions naturally, by satisfying another appetite: the "natural appetite betwixt the sexes" led one to mate with another, and the resulting family made the children "sensible of the advantages, which they may reap from society" (*Treatise,* 538). As soon as he offers this possibility, however, Hume retracts it, because the family, in his account, is too closely tied to the individual to promote recognition of those "remote and obscure" connections that constitute the real advantages of society.[27]

Having rejected nature (the family) as the basis for the initial recognition by "man" that he needs society to fulfill his needs, Hume then turns to "artifice" and "convention." "The remedy" for the partiality of family affections, he explains,

> is not deriv'd from nature, but from *artifice;* or, more properly speaking, nature provides a remedy in the judgment and understanding, for what is irregular and incommodious in the affections. For when men, from their early education in society, have become sensible of the infinite advantages that result from it, and have besides acquir'd a new affection to company and conversation; and when they have observ'd, that the principal disturbance in society arises from those goods, which

we call external [possessions], and from their looseness and easy transition from one person to another; they must seek for a remedy by putting these goods, as far as possible, on the same footing with the fix'd and constant advantages of the mind and body. This can be done after no other manner, than by a convention enter'd into by all the members of the society to bestow stability on the possession of those external goods, and leave everyone in the peaceable enjoyment of what he may acquire by his fortune and industry. . . .After this convention, concerning the abstinence from the possessions of others, is enter'd into, and everyone has acquir'd a stability in his possessions, there immediately arise the ideas of justice and injustice. . . .A man's property is some object related to him. This relation is not natural, but moral, and founded on justice. . . .The origin of justice explains that of property. (*Treatise*, 540–42)

In this passage Hume simply assumes what he initially set out to explain: instead of telling us how individuals realized they needed society, here he shows us individuals already in society and seeking, through the "convention" of private property, to preserve it. Property preserves society, he continues, because it gives us the idea of justice (don't take my property and I won't take yours); and the idea of justice gives us the idea of that "moral" relation that is property.

In such statements, Hume repeats the operation we examined in the previous chapter. Just as he props belief on custom and vice versa at the end of the *Treatise*, here he props the origin of property on the origin of justice and vice versa. By doing so, Hume also displaces the question of how society originated. When he returns to his first question, which now must be viewed in the light of our understanding that property and justice not only stabilize society but ground each other, Hume announces that how society originated is moot. It is moot because of something essential to human nature: human beings *automatically* recognize that, in order to satisfy the "passion" for property, they have to embrace justice; and embracing justice means they are already social. "'Tis evident," Hume asserts, "that the passion [to acquire property] is much better satisfy'd by its restraint, than by its liberty, and that in preserving society, we make much greater advances in the acquiring possessions, than in the solitary and forlorn condition. . . .[Thus] 'tis utterly impossible for men to remain any considerable time in that savage condition, which precedes society; but that his very first state and situation may justly be esteem'd social" (*Treatise*, 544).

As Christensen realizes, the pertinent question in this dizzying deferral of origins is this: *To whom* is it "evident" that the passion for property is best satisfied by restraint? The answer cannot be the savage, for Hume has already told us that primitive man cannot see beyond his own interests and the gratification of immediate needs. The answer must be Hume, as Christensen argues. Hume, the

historian, sees what primitive man cannot: that the individual's needs can be satisfied only by seeing that his own needs have some relation to the needs of others, and thus that immediate needs will be satisfied eventually, and possibly in
some other form, through the artifice or convention of society.[28] Society, then,
is merely the *idea* of relationship and likeness (I need you because I am like you);
society is a trope of connectedness available only—and only in retrospect—to
the historian.

We can see that Hume is *implying* that society is the historian's trope; what
he actually *says* is that the "state of nature," which supposedly preceded society,
is "a mere philosophical fiction, which never had, and never cou'd have any reality." Even though it is a fiction, however, the idea of a state of nature is necessary, for "nothing can more evidently shew the origin of those virtues, which
are the subjects of our present enquiry" (*Treatise,* 544, 545). In fact, Hume admits that poets, who are the chief chroniclers of a state of nature, "have been
guided more infallibly, by a certain taste or common instinct, which in most
kinds of reasoning goes farther than any of that art and philosophy, with which
we have yet been acquainted." By postulating a state of nature, *even though it is a
fiction,* poets have been able to deduce both that nature is insufficient to human
needs and that human beings are not by nature disinterested (*Treatise,* 546). Of
course these are precisely the postulates Hume began with; thus the idea of a
state of nature with which he began this conjectural history—and that gave him
the question he eventually declared moot (What is the origin of society?)—is
like the poet's fiction. Indeed, Hume's initial representation of man "only in
himself" (that is, before society; that is, as an incarnation of "human nature") is
also a fiction, although it too is necessary to his demonstration about the origins
of property and justice. Thus, by an intricate series of displacements, deferrals,
and self-corrections, Hume has installed fiction making, which Christensen
calls composition, at the heart of both the idea of society and theorizing about
society (and, by extension, human nature). This fiction making affects deduction because, in Hume's account, deduction reasons from a fictive premise; and
it affects induction because, by this account, induction proceeds by subsuming
the individual (human being, detail) into the (social, theoretical) whole. The result, as Christensen argues, is a peculiarly "inconsequential" theory of society.

> When theory proclaims itself as not merely conjectural but otiose, a curious rela-
> tion between theory and practice is established. Theory, which in the best of cases
> is for the empiricist an inference that follows evidence, is here cast as entirely su-
> perfluous to a practice (the formation of society) which is going on now and will
> go on anyway regardless of the way it is formulated or how we theorize about it.
> An antihistorical history, Hume's theory of the origin of justice is also an antithe-

oretical theory, which applies its force to the displacement of theories that do grant some privilege to the theoretical, such as contract theories of the origin of society. Hume's "theory" of the origin of society, like his passages of uninhibited skepticism, has no consequences. It is incapacitated by its conjectural character, by its inconsistencies, by its admission that the concept of nature which it presupposes is a fiction, and by its conviction that none of that really matters since things will go on pretty much as they always have anyway.[29]

The point at which I part company from this impressive deconstruction is signaled by Christensen's passing reference to the inconsequential nature of Hume's "passages of uninhibited skepticism." Although I agree that all the maneuvers by which Hume demonstrates that both the idea of society and theories about society are fictions belong to his skepticism, I believe the only way to understand why his account is so intricate is to recognize that this skepticism was *not* "uninhibited." Had Hume been willing simply to declare that design was not providential, as Voltaire did, he could have begun his conjectural history at the point where it concludes: that the state of nature is a fiction. Had he been willing to assert that providential design led the first humans to embrace society or that Providence would eventually supply the links historians could not see, as Lord Kames did, he could have dispensed with the problems inherent in conjectural history much more expeditiously. Hume's conjectural history is so interesting because he takes neither of these positions and because his refusal to do so tells us something about the state of experimental philosophy in the 1740s: practitioners had to assume that order existed in the objects of analysis in order to claim for their practice the status of "science" (or philosophy); yet in the middle of the eighteenth century, no socially viable ground for order existed other than providential design.[30] We need to understand the complexities of Hume's argument as signaling the difficulties inherent in the attempt to locate an alternative to providential design; and we need to understand his determination to find some alternative as an expression of his adamant refusal either to accept the methodological dependence on postulates and deduction historically associated with Scholasticism (as well as Christianity) or to give up the idea of producing systematic knowledge.

Just because Hume's theory about the origin of society was both intricate and skeptical does *not* mean it was inconsequential.[31] While the conclusion Hume reached about how society originated may have just been one among many theories generated by midcentury philosophers, both the method by which he reached this conclusion and the nature of the trope in which he embedded it had enormous implications for the production of general knowledge and for political economy more specifically. As I have already pointed out, in

Hume's account of the (non-) origins of society, he figures society as a system (of human needs and conventions), and he represents it as becoming visible through another system (the historian's theory). Thus the ability to see—or impose—likeness, the ability to make a system, retrospectively recognizes the advantages *of* system for individuals who cannot see either the advantages or the system themselves. For the individual, the advantages that system offers are considerable. In practical terms, it augments individual incapacity with sufficient power to acquire food, harvest crops, and defend against enemies. In subjective terms, it enables the individual to discount immediate experiences of loss or injustice because he knows that something greater than the individual is at stake: because he belongs to society, he can rest assured that his own loss is dwarfed by society's gain.[32]

For the theorist of governmentality, the implications of a model that discounted individual experience for more general gains proved considerable, for this model submits that policies should support—and could be defended as supporting—the general and long-range effects that experts recognized rather than what individuals experienced as their immediate interests. In the next chapter we will see that Thomas Malthus applied this model to sexuality in order to dissuade individuals from early marriage and that J. R. McCulloch invoked it to elevate "national prosperity" over individual well-being. Valuing the general effects that only the expert could see diminished the significance of all but the most abstract version of those subjective motivations that early eighteenth-century philosophers had sought to describe. In so doing, it also enabled the philosopher to differentiate his work from mere observation or simple natural history, thereby justifying both the occasional failure of theory to account for particulars and the occasional (or chronic) absence of particular data. In Hume's apology for the intricacy (or "fineness") of "general reasoning," he alludes specifically to the connection between the philosopher's discounting "superfluous circumstances" and his enhancing his own status and authority.

> When we reason upon *general* subjects, one may justly affirm, that our speculations can scarcely ever be too fine, provided they be just; and that the difference between a common man and a man of genius is chiefly seen in the shallowness or depth of the principles upon which they proceed. General reasonings seem intricate, merely because they are general; nor is it easy for the bulk of mankind to distinguish, in a great number of particulars, that common circumstance in which they all agree, or to extract it, pure and unmixed, from the other superfluous circumstances. Every judgment or conclusion, with them, is particular. They cannot enlarge their view to those universal propositions, which comprehend under them an infinite number of individuals, and include a whole science in a single theorem.

Their eye is confounded with such an extensive prospect; and the conclusions, derived from it, even though clearly expressed, seem intricate and obscure. But however intricate they may seem, it is certain, that general principles, if just and sound, must always prevail in the general course of things, though they may fail in particular cases; and it is the chief business of philosophers to regard the general course of things.[33]

If privileging system at the level of knowledge in general meant privileging "the business of philosophers," a modern reader might ask if self-interest could possibly have influenced this representation of knowledge. For Hume and most of his contemporaries, this question had no purchase, in large part because philosophy had always been both devoted to producing general and systematic knowledge and a practice limited to an elite few. What did (intermittently) concern Hume, if not most of his contemporaries, was whether the foundational metaphor of system was itself grounded on anything other than the philosopher's ability to desire and *imagine* a system. One might very well say that postulating systematicity in the object of analysis is necessary to systematic knowledge (or science), and that systematic knowledge is necessary if one wants to distinguish "a man of genius" from "a common man." But how can one know that the system the philosopher postulates really exists in the world, rather than simply being an effect—an effect of the desire for system (and prestige) institutionalized in philosophy itself?

For Hume, of course, who refused to refer the metaphor of system to Providence, the only answer comes from experience over time—that is, from history. "General principles, if just and sound, must always prevail in the general course of things." As we already know from his willingness to embrace the fiction of the "state of nature," moreover, he considered the *effect* of the system that philosophical history generates to be just as efficacious as system itself, for he maintained that imagining the effects of any given system was the first stage in bringing that system into existence. That this is true is evident not just in the history of society, but in the history of commerce, for as Hume points out in the essay I have just quoted, once individuals and rulers imagine interests beyond their own, they can create (although they do not always do so) the institutions and laws by which both individual and state will benefit.[34]

For Hume, the impossibility of determining whether the metaphor of system was grounded on anything beyond desire and a certain ideal of knowledge was not sufficiently troubling to drive him away from writing altogether. This is true because he considered the metaphor so useful, whether one wanted to address history or politics, and whether one wanted to pose as the authority or invite readers into a conversation. Also, Hume was not paralyzed by "the

impossibility of explaining ultimate principles" because, as he notes in the introduction to the *Treatise* and as we saw in chapter 4, he thought that arriving "at
the utmost extent of human reason" confers personal and social benefits that
surpass the false consolations of Scholastic philosophy. According to Hume, the
contentment peculiar to skepticism comes from its humility, its willingness to
acquiesce and share—all characteristics that he considered essential to society,
whose maintenance was ultimately more important than philosophical certainty. "As this impossibility of making any further progress is enough to satisfy
the reader, so the writer may derive a more delicate satisfaction from the free
confession of his ignorance, and from his prudence in avoiding that error, into
which so many have fallen, of imposing their conjectures and hypotheses on
the world for the most certain principles. When this mutual contentment and
satisfaction can be obtained betwixt the master and scholar, I know not what
more we can require of our philosophy" (*Treatise,* 45).

By the second decade of the nineteenth century, as we will see in the last
chapter of this book, the blow that Hume's skepticism dealt to the philosophical method of induction had come to seem problematic. Paradoxically, one
place the effects of skepticism were most visible was the discipline that Hume's
metaphor of system helped found—political economy. Paradoxically again,
one of the complaints directed against political economy in the nineteenth
century was not that it was insufficiently systematic, or even that its system had
no grounds, but that once the science of wealth abandoned its providential
foundation it had no basis for authority beyond the self-interest of its practitioners (or those the political economist served). Before we consider why political economy came under fire, of course, we need to understand the terms in
which it was initially articulated. For that, we turn to Adam Smith.

DESCRIPTION AND SYSTEM: THE CONSTITUTION
OF POLITICAL ECONOMY

David Hume's reliance on the metaphor of system constructed a bridge over
the gap between written records of the past and the unrecorded origins of society, which otherwise could be crossed only by belief in Providence. As we
have seen, Hume suspended the question whether this metaphor was grounded
in anything beyond the human wish for order, but as long as one believed that
the effects of fiction included actualizing the order one wished to find, then it
did not matter if the metaphor of the system originated simply in philosophical speculation. Adam Smith, whose work dealt more extensively with the relation between wealth and governmentality, felt compelled to probe the effects
of the metaphor of system more thoroughly. More focused on policy than

Hume, Smith was interested in how the theoretical entities philosophy generated actually worked. One way they worked, he thought, was by producing real effects whose products could presumably be measured or quantified. In the progress Smith charts—from systematic philosophy to claims about universals (human nature) to descriptions of abstractions (the market system) to the quantification of the effects or products of these abstractions (labor, national prosperity), we see how the trope of system helped produce the entity it claimed simply to describe (the market). In essence this was the work of political economy, the preeminent science of wealth in eighteenth-century Britain.

As Donald Winch has pointed out, Adam Smith did not use the term "political economy" frequently, and from the few accounts he did offer of this new science, it is not even clear whether he intended the emphasis to fall on riches or power—on wealth or governmentality.[35] Smith's most explicit description of political economy appears in the introduction to book 4 of *Wealth of Nations* (1776): "Political oeconomy, considered as a branch of the science of a statesman or legislator, proposes two distinct objects: first, to provide a plentiful revenue or subsistence for the people, or more properly to enable them to provide such a revenue or subsistence for themselves; and secondly, to supply the state or commonwealth with a revenue sufficient for the public services. It proposes to enrich both the people and the sovereign."[36] Because Smith neglected to specify *how* the new "branch" could make knowledge useful to legislators, it has always been necessary to *interpret* his version of political economy, both by consulting his own texts and by working one's way through the numerous editors and commentators who have already interpreted him. In the next chapter I will deal in earnest with some of Smith's earliest and most prominent interpreters. For now we should recognize that *interpretation* constitutes an essential component of Smith's own practice. Indeed, his insistence that interpretation is essential to the production of systematic knowledge may well be his most significant contribution not just to the emergent science of wealth but to the political economic variant of the modern fact more generally. Certainly it is true that the tension inherent in interpretation—the tension between description and system—constitutes the central epistemological issue of Smith's work.

Several recent commentators have argued that Smith's *Wealth of Nations* was the logical end of the reason-of-state argument that only the statesman (or sovereign) could recognize and represent the common good of a nation, which was increasingly defined in terms of wealth.[37] This argument hovers in the background of what follows, but instead of focusing on the implications of Smith's work for governmentality or for the modern state's efforts to accumulate wealth, I highlight the epistemological dimension of his writing. Of course, even placing Smith's work in the tradition of writing on jurisprudence or economics has required commentators to address epistemological issues obliquely, for Smith's assertion that the market is governed by a combination of natural, but abstract, systems—a "system of natural liberty" and a system of artificial plenty (5.11.651; 1.1)—implies that to know what is vital about trade and exchange one must supplement observation with something else—a combination of introspection and (systematic) assumptions. One can know, in other words, that the self-love that creates the division of labor (which in turn creates plenty) belongs to a system only by looking into the self and by assuming that the sympathy one (theoretically) finds there is common to all human beings. For this reason it is fair to say that Smith's political economy rests on the map of subjectivity he laid out in his *Theory of Moral Sentiments,* and though I do not discuss this much belabored topic directly, everything that follows assumes a constitutive relation between Smith's assumptions about sympathy and his codification of political economy.[38]

Despite the proximity of Smith's jurisprudential and economic conclusions to the epistemological problematic that yokes both assumptions about subjectivity and the privileging of discrete and observed particulars to an assumption about system, Smith's epistemology has received relatively little attention in the literature. This omission is all the more striking because he makes the inadequacy of empiricism and the necessity of postulating a system the subjects of the opening chapters of *Wealth of Nations.* As he explains in the first chapter, empirical observation was inadequate in 1776, because by that time "great manufactures" had begun to involve so many workmen "that it is impossible to collect them all into the same workhouse. We can seldom see more, at one time, than those employed in one single branch [of the operation]" (1.1.4). We should note that seeing seemed inadequate to Smith because what he wanted to see—the division of labor—was an abstraction. Like Robertson's "the human mind," "the division of labor" was an abstraction produced by Smith's philosophical assumption that knowledge should be systematic; indeed, the idea of a division of labor, which enables one to imagine physically separated workers as somehow participating in a single activity, was the philosopher's unique contribution to that epistemological division of labor that enabled manufacturers to reconceptualize (and modernize) their enterprises. Philosophers, according to Smith, are those individuals "whose trade it is not to do any thing, but to observe every thing; and who, upon that account, are often capable of combining together the powers of the most distant and dissimilar objects" (1.1.10).

Even though it is a truism of the critical literature that Smith naturalized concepts like the division of labor, then, it should also be remarked that the most capacious concept he naturalized—"the market system"—was an abstraction generated by political economy *as a philosophical enterprise that subordinated description to systematic analysis.* Insofar as the market system was an effect of a philosophic system, it resembles those abstractions we examined in chapter 2— "trade" and "commerce." But Smith was more self-consciously interested in the interpretive act by which abstractions are generated, as well as in their effects, than were Malynes and Misselden. Like Hume, Smith believed that creating abstractions was essential to the production of general knowledge, not simply because a single eye could not see all the workmen at the same time, but because what one needed to analyze was nowhere visible as such—nowhere, that is, except in the writing of political economists like Smith.

By the same token, however, Smith also had access to more empirical evidence about trade than did Misselden or even Mun, and unlike Hume, he implied that the evidence that did exist, much of it numerical in form, had to be somehow taken into account. Indeed, whereas the conjectural historians and even Hume were content to construct theories from a combination of abstractions, introspection, and experience, Smith also wanted to base political economic knowledge on numbers. We see his predilection for numbers in the tables he reproduces in *Wealth of Nations.* These tables chronicle changes in the prices of wheat (along with a translation into current prices; 1, conclusion, 251–58); the amount of taxes collected on various liquors (5.2.841); amounts and kinds of revenue collected against the national debt (5.3.876); and the tonnage bounty paid on the white herring fishery (appendix, 901–23). As we will see, Smith intermittently expressed grave reservations about the reliability of most available numerical information and was openly skeptical about political arithmetic because its computations could not have been exact.[39] Nevertheless, in these tables and in the countless less graphically marked uses of numbers that punctuate *Wealth of Nations,* Smith implied that this kind of information could be of use—indeed, that it was indispensable to the kind of knowledge he wanted to create.

Partly because of the way Smith positioned numbers within a discussion of governmentality—and despite his explicit reservations about these numbers—numerical representation gradually came to seem like an essential form of evidence about abstractions like the market, especially if this evidence was to be useful to legislators. To the extent that Smith helped make numerical information seem useful to legislators, he can be seen as participating in a variant of the project in which Petty was also engaged (albeit with different models of government in mind); and to the extent that Smith's use of numbers helped confer on them connotations of impartiality, transparency, and methodological rigor, he can also be seen as participating in the prolonged campaign to neutralize the old connotations (of necromancy and sorcery) that had once made numerical representation suspect. Rather than simply assimilating his work to either of these projects, however, it seems more useful to highlight the tensions inherent in the political economic use of numbers, for these tensions emanated from the epistemological dilemma that Smith identified as essential to philosophy in general. On the one hand, like the conjectural historians, he recognized that it was essential to invoke systematic assumptions about human nature (*homo economicus*)—partly, but not exclusively, because he lacked experiential data sufficient to ground his claims about the market. On the other hand, like a good Baconian, he wanted to use the available firsthand evidence, especially when this existed in the form of numbers. In this context of insufficient but essential numerical data, Smith wanted to create a way to *interpret* the data that did exist, so that legislators would know both what to make of the available numbers and what kind of information to collect in the future.

Smith's treatment of numbers, then, provides one way into the epistemological problematic that characterizes his work and this variant of the modern fact more generally. We can see how he typically dealt with numbers in his demonstration that "the liberal reward of labour" (high wages) "is the natural symptom of increasing national wealth" (1.8.73). To reach this conclusion, Smith initially sought to establish some way to determine relative costs of living at different times and in different places so that he could determine what the cost of living had to do with both wages and national wealth. To establish these costs, he notes, it would be useful to have something like a price list of commodities. Immediately, however, he admits that no such lists have ever been kept, and that even if they existed, the numbers they would most likely contain ("the different quantities of silver for which [an item] was sold") would be the wrong numbers. What he really wanted were numbers that showed "the different quantities of labour which those different quantities of silver could have purchased." In 1776 these records did not exist either, so Smith contents himself with the only numerical records that did exist—or at least the only records historians had noticed—records of the prices of corn. "Those of corn, though they have in few places been regularly recorded, are in general better known and have been more frequently taken notice of by historians and other writers. We must generally, therefore, content ourselves with them, not as being always exactly in the same proportion as the current prices of labour, but as being the nearest approximation which can commonly be had to that proportion" (1.5.38).

Even though the numbers Smith longed for were not available, the way he sets them up here implies that, if records had existed of the amount of labor that various quantities of silver could have purchased, he would have been able to calculate comparative "real values" from them. In implying that he would have used them had they been available, Smith's claim resembles Petty's: if we had the numbers, this is the kind of results we could produce. The way Smith uses the numbers that did exist, however, suggests that, unlike Petty, he would never have performed actual calculations. As he repeatedly insists, he considered calculations "tedious or doubtful" and therefore superfluous, because they were both too particular and too speculative to constitute the basis for adequate analysis.

What looks like an endorsement of numerical records and arithmetical calculation, then, turns out to be considerably more complex than it initially seemed. To prove that high wages were a symptom of national prosperity, in fact, Smith simply questioned the available numerical records and dismissed arithmetical calculation in favor of a series of generalizations that seem to be based on extensive travel and careful observation.

> In Great Britain the wages of labour seem, in the present times, to be evidently more than what is precisely necessary to enable the labourer to bring up a family. In order to satisfy ourselves upon this point it will not be necessary to enter into any tedious or doubtful calculation of what may be the lowest sum upon which it is possible to do this. There are many plain symptoms that the wages of labour are nowhere in this country regulated by this lowest rate which is consistent with common humanity. (1.8.74)

The "plain symptoms" Smith goes on to cite are all visible only to the observant spectator *who already has a theory* about what defines or gives content to the abstraction "common humanity." This theory, *not* calculations about "what is precisely necessary to enable the labourer to bring up a family," enables Smith to determine that current wages exceeded necessities and could therefore be called high. At the conclusion of the "symptoms" he cites, he moves from the series of observations that have been informed by his assumptions about the abstraction "common humanity" to the general case: if wages are currently high (as Smith has "proved"), then can we read this as a sign that high wages benefit the nation as a whole? The answer, he asserts, is "abundantly plain," because it accords with the *assumptions about equity* with which he embarked.

> Is this improvement in the circumstances of the lower ranks of the people to be regarded as an advantage or as an inconveniency to the society? The answer seems at first sight abundantly plain. Servants, labourers and workmen of different kinds, make up the far greater part of every great political society. But what improves the circumstances of the greater part can never be regarded as an inconveniency to the whole. No society can surely be flourishing and happy, of which the far greater part of the members are poor and miserable. It is but equity, besides, that they who feed, cloath and lodge the whole body of the people, should have such a share of the pro-

duce of their own labour as to be themselves tolerably well fed, cloathed and lodged. (1.8.78–79)

In the rest of this chapter Smith shores up his endorsement of high wages with other justifications, but they all rest, as this example shows, on his principled support of the assumptions with which he initiated the analysis, not on anything that numbers could reveal. In this demonstration, which is typical of Smith's treatment of numbers in *Wealth of Nations,* numerical records are summoned, then dismissed, just as eyewitnessing is invoked only to be so supplemented by his foundational theoretical suppositions that one wonders whether looking is not as superfluous as calculation. In *Wealth of Nations,* Smith repeatedly voices the distrust of numerical records he implies here; he dismisses the customs records as an inadequate index to the balance of trade (because they are so often inaccurate), and he throws out figures that show "the course of exchange" because they are based on an inadequate understanding of the difference between real and computed exchanges (4.3.442, 443). At the same time, however, he continues to use numbers, not just in the tables I have already cited, but wherever he wants to make the market come alive. If he was so skeptical of the available numbers, as he undeniably was, and if his references to them were so frequently as cosmetic as my example demonstrates, then why did Smith repeatedly insert numbers into *Wealth of Nations?* To answer this question, we need to understand not only what he thought was wrong with the numbers that existed in 1776 but also what he imagined more adequate numerical information could supply.

The reason Smith considered existing numerical records inadequate was that simple enumerations could not make visible what really counted: the self-regulating market system. This was true because any numbers that simply tabulated current market transactions could not display the market in its essential form; and *this* was true because the current form of the market did not correspond to its "natural" form, since interfering legislation kept the market from being what it would have been "in nature." If numerical information was gathered with a systematic understanding of what this system should be, however, and if the resulting numbers were interpreted in the light of this theoretical understanding, then numbers could provide useful evidence about how the market system should operate—if all the transactions that constituted "the market" were left alone, as they ought to be. Indeed, numbers so gathered and interpreted could make the market system become what it ought to be, and they could make the vision of what the market ought to be *seem* to correspond to what it was, for numbers had a peculiarly compelling quality: they appeared to be simply descriptive, as explicitly theoretical accounts obviously were not. Despite his skepticism about the numbers that were available in 1776, then, Smith implied that a new kind of numerical information, informed by theory at every level, could provide what legislators currently lacked—both information about an abstraction that could not be seen and proof that, even though no one could see it, the market system was (or rather could be) lawful— that is, able to govern itself.

To understand Smith's curious use of numbers, it is helpful to grasp two points. First, his treatment of numbers implies that what initially seems like a practical question (How can one use the numerical information that already exists?) was actually part of a theoretical question (What kind of epistemological instrument can produce numerical information that will make what is otherwise invisible appear?). Second, the epistemological instrument Smith championed—the kind of abstraction that draws on but considerably exceeds universals like "human nature"—should be considered "descriptive" in a very special sense. Understanding the link between the numerical information that had already been collected in 1776 (and that presumably aspired to reflect existing conditions) and the kind of numbers that would have been able to show what the market system *could* be requires us to reconstruct Smith's quite specific notion of description.

Smith's account of description, which appears in his *Lectures on Rhetoric and Belles Lettres* (and which does not deal specifically with numbers), sets up an opposition between description and history on the one side and the "Rhetoricall" mode of writing on the other.[40] According to Smith, history "proposes barely to relate some fact," whereas rhetoric proposes "to prove some proposition": "The Rhetoricall endeavours by all means to perswade us; and for this purpose it magnifies all the arguments on the one side and diminishes or conceals those that might be brought on the side conterary to that which it is designed that we should favour" (*Lectures,* 62). As this passage demonstrates, Smith's taxonomy of composition participates in that reduction of rhetoric that I discussed in chapters 3 and 4. (Smith's criticism of ornament per se appears in lectures 7–11, in his analysis of Shaftesbury's style.) In distinguishing between history and rhetoric, Smith echoes the distinction that Locke had promoted, between writing *about* something (an object or a passion) and using language to *do* something (persuade or instruct). For Smith, description, which is the primary instrument of history, is relatively neutral; it aspires to transparency rather than distortion. Given what looks like an endorsement of descriptive transparency, however, it may come as a surprise to modern readers to learn that for Smith, descriptive writing ought to be transparent not to phenomenal objects but to the sentiment, passion, or affection of the writer. According to Smith, the qualities that distinguish effective writing *about* something are conciseness, propriety, and precision *as judged in relation to the writer's emotional state.* "The perfection of stile consists in Express[ing] in the most concise, proper, and precise manner the thought of the author, and that in the manner which best conveys the senti-ment, passion or affection with which it affects or he pretends it does affect him and which he designs to communicate to his reader" (*Lectures,* 55).

The criteria by which Smith judged description, in other words, belong to the domain we call psychology. This is another reason his account of subjectiv-ity is implicit in all his accounts of systematic knowledge, including his expla-nation of the kind of description political economy was created to provide. Just as history writing, which relies most heavily on description, rests its case on its adequacy to the writer's sentiment, so the reader evaluates such writing by what *feels satisfying.* Smith's endorsing such subjective criteria makes his version of description seem more like rhetorical ornament than what we think of as im-partial or "objective" language, for rhetoric was also judged by a set of pro-topsychological criteria (the ability to please and persuade), whereas a modern "objective" language—like numerical representation—is generally evaluated by its ability to *eliminate* subjective factors. To grasp Smith's conception of de-scription, then, we have to recognize that, far from thinking it suspect, he con-sidered language's ability to convey and appeal to psychological considerations its greatest strength. He valued this kind of appeal (but not rhetorical orna-ment) because he assumed that subjectivity was the feature that all individuals shared; to him, subjectivity was "human nature." To mobilize human subjectiv-ity properly was thus to appeal to what was essentially human in writer and reader, and this common humanity was more telling than (what we would think of as) neutral depictions of objects.

If assumptions about the subjective dynamics of the universal subject in-formed Smith's description of description, then another, related assumption informed his account of historical narrative more generally. This assumption focused on the importance of system in the narrative account. Once more, the criteria Smith nominates for judging the adequacy of historical narrative be-long to the domain of the psychological; one can identify a good historical nar-rative by the satisfaction it confers, and a narrative confers satisfaction when it contains no "gaps"—when, that is, it constitutes a closed system. Achieving the effect of systematicity, Smith admits, requires considerable skill, for the chain of causes and effects potentially exceeds the requirements of the subject at hand. In order not to be "reduced to the necessity of tracing the whole back even to the fall of Adam," he suggests, the writer must modulate the reader's responses by "touching" remote events ever "more slightly" (*Lectures*, 92, 93). "The Cause of the Event makes a less impression than the Event itself and so excites less curiosity with regard to its Cause; that cause therefore is to be touched upon more slightly, and by being so it excites but very little Curiosity about its Cause, which therefore may be still more superficially mentioned." By tracing effects back to causes whose presentation arouses no further curiosity, the historian "leaves us fully satisfied that we know all that is necessary of the matter" (*Lectures*, 93).

In Smith's account, the historian renders his historical narrative adequate—subjectively satisfying—by using description to incite, then quiet, the reader's curiosity. But what if, as with those works Stewart christened conjectural, the historian wants to trace events back to causes too remote to identify? For such histories, the historian cannot simply ratchet down the reader's interest by ever slighter descriptive touches, both because remote origins are precisely the point of the enterprise and because records do not chronicle all the links between events. In lecture 18 Smith acknowledges that the "gaps" historians might be tempted to leave in such circumstances would be ruinous to their credibility, precisely because such gaps would undermine the satisfaction historians aspire to create. "We should never leave any chasm or Gap in the thread of the narration even tho there are no remarkable events to fill up that space. The very notion of a gap makes us uneasy for what should have happened in that time." Although he does not tell the historian what to do when these gaps yawn, Smith does propose "an other way of keeping up the connection" between event and cause: "that is, the Poeticall method, which connects the different facts by some slight circumstances which often had nothing in the bringing about the series of the events" (*Lectures*, 100).

Like Hume, Smith installed fictions in the lacunae where one might expect to find data. Like Hume again, Smith believed it was inevitable that what counted as data would be informed by what looks like a fiction. In his account, this was true for two reasons: first, because the human desire for a systematic account of history and nature inevitably informs—and sometimes overwhelms—even the most scrupulous philosopher's ability to observe or describe the kind of phenomenal particulars that Bacon denominated facts; and second, because Smith considered the most important particulars to be psychological—that is, subjective and universal.

If Smith resembled Hume in valuing the human need for satisfying—that is, systematic—narratives over accounts that were either ornamental or simply faithful to the plethora of phenomenal objects, then the two philosophers were also alike in their willingness to suggest that the human desire for system might be its only ground. As his backhanded tribute to Descartes makes clear, Smith believed that a systematic presentation of an internally coherent system of ideas could *seem* credible simply because it was satisfying, even if it was not true.

> It gives us a pleasure to see the phaenomena which we reckoned the most unaccountable all deduced from some principle (commonly a wellknown one) and all united in one chain, far superior to what we feel from the unconnected method where everything is accounted for by itself without any reference to the others. We need [not] be surprised then that the Cartesian Philosophy . . . tho it does not perhaps contain a word of truth . . . should nevertheless have been so universally received by all the Learned in Europe at that time. The Great Superiority of the method over that of Aristotle . . . made them greedily receive a work which we justly esteem one of the most entertaining Romances that has ever been wrote. (*Lectures,* 146)

That Smith's skepticism about the ground of systematic philosophy was not mobilized only by the dubious (because French) Descartes is clear from the report he gives of Newton's laws in his *History of Astronomy.* So persuasive is Newton's systematic account of the operations of the celestial bodies, Smith explains, that the philosopher is

> insensibly . . . drawn in, to make use of language expressing the connecting principles of this [system], *as if* they were the real chains which Nature makes use of to bind together her several operations. Can we wonder then, that it should have gained the general and complete approbation of mankind, and that it should now be considered, *not as an attempt to connect in the imagination* the phaenomena of the Heavens, but as the greatest *discovery* that ever was made by man, the *discovery* of an immense chain of the most important and sublime truths, all closely connected together, by one capital fact, of the reality of which we have daily experience.[41]

In such passages, Smith does not make it clear whether what looks like order in the world is merely an effect of the human desire for system, but he does leave this possibility open. Unlike Hutcheson and Lord Kames, he was only intermittently willing to refer the desire for system to a providential order ("the invisible hand").[42] What he was willing to do—indeed, what he offered as the solution to the epistemological necessity of reconciling accounts of discrete particulars with the human desire for system—was to propose two understandings of "nature," which, when held side by side, helped explain why what one sees so frequently differs from what one believes.

This is the point at which to refer once more to Smith's use of numbers, for the odd application we see in *Wealth of Nations* articulates this double understanding of nature.[43] As I have already explained, the numbers he *wanted* the legislator to use in *Wealth of Nations* are descriptive in the sense that Smith defines description in the *Lectures on Rhetoric:* they described what *should* be, according to Smith's conviction that the market was naturally systematic; or more precisely, they described what the market *could* be if legislators left it alone. The numbers actually available to Smith, however, described an *unnatural* state of affairs, in which the true nature of the market had not been allowed to realize itself, as the analyst could see from its unsystematic or irregular dynamics. Only the political economist could identify the true nature of the market, Smith believed, because only political economic theory created the abstraction (the market system) whose regularities one had to assume in order to realize this system.

Recognizing that Smith tended to discount those numbers that were available is not the whole story, then, for he also longed for numbers that did not yet exist. This helps illuminate what I mean by Smith's double understanding of nature: in one understanding, "nature" was logically deduced from what philosophers believed about that universal central to eighteenth-century philosophy—human nature; in the other, "nature" consisted of the phenomena that Bacon loved to catalog. In Smith's account, neither of these understandings is sufficient, for although adhering to the former might remind the philosopher of what should be, abandoning the latter would not help him devise policies for moving from the current "unnatural" state of things to the nature that should exist. By the same token, while adhering to Baconian phenomena would enable the philosopher to catalog the world, abandoning the nature that was deduced from introspection and logic would deprive him of the ability to remember what policies he should help bring about. Thus Smith wanted both to reveal the current unnatural state of the market and to stress that, if the market's natural liberty was restored, numbers could tell legislators what they needed to know: that the market was capable of governing itself. Smith's double understanding of nature, then, illuminates a feature essential to the political economic variant of the modern fact: the theory or belief with which one sets out to observe always informs what one can see, as well as how one interprets; there is no observation outside of theory or belief.

It should be obvious by now that one of the effects of transforming philosophical universals (human nature) into abstractions that could be conceptualized as historical agents (the market system) was a model that enabled one to name and quantify the effects of the abstractions themselves. That is, once one declared the market system to be an incarnation of human nature, and once one devised epistemological instruments for making the market system visible, then one could name its products and components as (if they were) entities, many of which could be quantified. As we will see in the next chapter, what we might call the second-order abstractions of "labor" and even "happiness" soon came to seem quantifiable for just this reason. Labor and happiness seemed to be quantifiable not just because the philosopher had at his disposal the analogical language of gestural mathematics, but also because Smith's understanding of description created a nonrhetorical (nonsuasive) place for a kind of representation that described what *could be* as if this potential was simply waiting to materialize.

The numbers for which Smith created an epistemological place in *Wealth of Nations*—numbers that could seem simultaneously transparent to phenomenal entities and evidence for what (the philosopher believed) the market system would be if it were free—typify the variant of the modern fact produced by political economy. It must be noted that, although numbers constitute a particularly clear example of the political economic fact, numerical representation is not the distinguishing feature of this epistemological unit. More central than form is the peculiar combination of claims about descriptive accuracy and dependence on a priori belief that I have already identified. As we have also seen, Smith helped make it possible to imagine that representation could be both accurate and informed by a priori belief by elevating the distinction between description and persuasion over the distinction that now seems more salient to us—the difference between claims that are grounded in empirical evidence and claims that are simply asserted or assumed. Instead of operating according to the dichotomies of theory/observed particular, true/false, or even accurate/precise, Smith's political economy aligned the observations one's eye could provide with the assumptions the philosopher held and opposed both to language intended simply to persuade. Thus the opposition between belief and rhetoric, which had gradually been stabilized in the developments I have been charting, made it possible to conceptualize seeing and believing as synonymous even as one forgets that seeing has anything to do with belief.

Many of Smith's specific policy recommendations (especially about how to deal with Britain's colonies) were extremely controversial in the decades following the publication of *Wealth of Nations,* and though the text did find some champions, what subsequent practitioners were to call the science of wealth did not flourish until the beginning of the nineteenth century. Despite the delayed impact of much of the content of *Wealth of Nations,* however, it must be acknowledged that, along with the experimental moralists and the conjectural historians, Smith laid the groundwork for the method of theoretical-cum-descriptive analysis that rapidly became the preferred instrument for producing knowledge about wealth and society. In the next chapter I examine the early nineteenth-century elaboration of political economy and Malthus's reinvigoration of numerical representation. Before doing so, however, it seems instructive to take one last detour so that we can glimpse again one of the alternatives to the numerical version of the modern fact that flourished in late eighteenth-century Britain.

THE DETOUR THROUGH SCOTLAND: JOHNSON'S *JOURNEY TO THE WESTERN ISLANDS*

There is no question that Samuel Johnson's *Journey to the Western Islands of Scotland* (1775) belongs to a different genre than the other texts I have examined in this chapter;[44] and as with my discussion of Hume's essays, I can only warn readers interested exclusively in the relation between the modern fact and numbers that the discussion that follows will seem digressive. Despite Johnson's almost complete indifference to numbers, however, the *Journey* does add something to our understanding of the modern fact, both because Johnson's lifelong commitment to producing general knowledge tested whether the genre of travel literature could be made to accommodate systematic knowledge and—more relevant to this argument—because what he discovered in Scotland made him question the limits of all systematic knowledge projects. Like the conjectural historians, Johnson wanted to understand how modern commercial society had evolved out of its "barbaric" predecessor; like them again, he encountered a gap between what he had expected to find in Scotland—"old traditions and antiquated manners"—and what he actually saw: he and Boswell arrived in Scotland "too late to see what we had expected," Johnson ruefully acknowledged.[45] Instead of invoking Providence, as the conjectural historians characteristically did, however—and despite his devout belief in God—the old man responded to the gap between what he believed and what he saw by questioning whether a priori assumptions should govern how we interpret what we see. In so doing, Johnson did not abolish the problematic of the modern fact—as long as one insists on valuing both observed particulars and systematic knowledge, one cannot know what one is seeing except in the light of some systematic belief. Even if he did not abolish the issue, however, in questioning the limits of systematic knowledge he did open a space for the kind of cultural relativism that would eventually challenge the marriage between theory and observation that characterizes the modern fact—even though the divorce between these two would be delayed until long after the articulation of cultural difference itself.

Although he does not pursue this connection as far as one might like, Pat Rogers has recently suggested that there is much to be learned from placing Johnson's *Journey* in the tradition of Scottish conjectural history. For Rogers, Johnson's *Journey* constitutes a methodological variant on the method of conjectural history; according to him, the work "tests many of the ideas adumbrated in the works of Enlightenment authors, but it does so by means of field-work and first-hand observation."[46] That Johnson could use fieldwork to test Enlightenment ideas—or even that he thought he could—already tells us something about the limitations that the conjectural historians' object of analysis imposed on their analytic method: because they wanted to investigate abstractions ("the human mind," "the market system"), they were not really interested in visiting places where "rude"societies were presumed to exist, even when those places were close by—as the Scottish Highlands undeniably were.[47]

Of course, as Johnson discovered when he visited the Highlands in 1773, this was not exactly the "rude" society he and the conjectural historians assumed it was. After the battle of Culloden in 1746, most of the traces of "savage virtues and barbarous grandeur" were forcibly eradicated from the Highland clans, largely to curtail the threat posed to the English monarchy by the Jacobites. After the Pretender's defeat, English laws were used to undermine the clan system by prohibiting clan tartans, forbidding chieftains to carry weapons, and launching initiatives to replace ancient Erse with English. So successful was the English policy that Johnson was repeatedly struck by the *lack* of difference between the Highlanders and the English, although, having expected "a people of peculiar appearance, and a system of antiquated life," this lack of difference also made him realize the belatedness of his visit.[48] "The clans retain little now of their original character, their ferocity of temper is softened, their military ardour is extinguished, their dignity of independence is depressed, their contempt of government subdued, and their reverence for their chiefs abated. Of what they had before the late conquest of their country, there remain only their language and their poverty" (43).

Reading Johnson's *Journey* alongside the conjectural histories, then, does not tell us what kind of knowledge the latter might have produced had they been more "rigorously"Baconian. What this exercise does begin to illuminate is why experimental philosophy's dependence on a priori analysis eventually began to seem questionable to the inheritors of this tradition. When confronted with phenomena that did not conform to the postulates he set out with, Johnson did not simply look through them to universal principles, subordinate them to some greater abstraction, or dismiss them as anomalies; instead, he modified his original opinions. When faced with the impossibility of discovering the truth about even simple statements, he did not propose conjectures based on the principles he already believed; instead, he meditated on the difficulty of producing systematic knowledge and the incentives that might lead an interlocutor to distort even firsthand accounts.

The features that distinguish Johnson's treatment of such methodological and epistemological issues from that of the conjectural historians might well be an effect of genre, of course. Since Johnson's *Journey* was based on a notebook and a series of letters that he composed while traveling, that he modified his views or reflected on the elusiveness of evidence may simply reveal the dynamic *process* of generating philosophical positions rather than reflecting a different set of *conclusions* about how to produce philosophical knowledge. Because the methodological dilemmas he described were eventually interpreted as challenges to the method of experimental moral philosophy, however—because by the 1840s some of the most self-conscious theorists of philosophical method considered it necessary either to mistrust induction or to clarify the distinction (and redefine the relationship) *between* induction and deduction—we can see Johnson's struggle to understand Scotland as one harbinger of a transformation in the understanding of philosophical analysis itself.[49]

Like the Scottish conjectural historians, Samuel Johnson inevitably examined Scotland from the viewpoint of modern commercial society—that is, from a distinctively *English* perspective. In the opening paragraphs of the *Journey,* which chronicle events that preceded his arrival in the Highlands, Johnson repeatedly places what he sees in the context of what he knows, and more often than not the resulting contrast leads him to interpret Scottish arrangements as inferior—or as he phrases it, "unimproved." When he and Boswell discover the ruins of a small fort on the island of Inchkeith, for example, Johnson comments that the ruin is "not so injured by time but that it might be easily restored to its former state." That he sees the little fort through the eyes of a culture already engaged in collecting (and commodifying) "antiquities" and "picturesque retreats" is apparent from his parting observations: "We left this little island with our thoughts employed awhile on the different appearance that it would have made, if it had been placed at the same distance from London, with the same facility of approach; with what emulation of price a few rocky acres would have been purchased, and with what expensive industry they would have been cultivated and adorned" (4).

Johnson's tendency to view—and judge—Scotland by English lights is also visible in his early complaints about the scarcity of trees in Hibernia. To him the lack of trees signals the residents' "improvidence," their inability (or unwillingness) to think beyond the present. Characteristically, he accounts for this failure with a moral generalization, which in this case concerns "custom."

> Of this improvidence no other account can be given than that it probably began in times of tumult, and continued because it had begun. Established custom is not easily broken, till some great event shakes the whole system of things, and life

> seems to recommence upon new principles. That before the union the Scots had
> little trade and little money, is no valid apology; for plantation is the least expensive
> of all methods of improvement. To drop a seed into the ground can cost nothing,
> and the trouble is not great of protecting the young plant, till it is out of danger;
> though it must be allowed to have some difficulty in places like these, where they
> have neither wood for palisades, nor thorn for hedges. (8)

Apart from the two final clauses, this passage contains no reference to the surroundings Johnson inhabited at the moment of composition. In the signature
technique of his earlier writing, he uses his immediate situation to launch a series of reflections that were ostensibly self-evident ("no other account can be
given") but that climax in a succinct maxim that only the philosopher could
formulate ("established custom is not easily broken"). The moral observations
that follow also generalize from what pass as universal principles ("to drop a seed
into the ground can cost nothing"), and the final lesson cuts two ways: because
the Scots did not plant trees, their failure to improve their country is still visible;
retarded by "custom," they needed the English ("some great event") to start
their country's life anew.

When Johnson again brings up the scarcity of trees, we begin to see that the
generalizations he derives are not simply self-evident and that his responses are
not limited to philosophical reflections on the unimproved Scots. Now in the
Highlands proper, he clearly peers through the lens of classical literature, for he
reads what he perceives as an absence by the perspective provided by Homer's
text: "Of the hills many may be called with Homer's Ida *abundant in springs,* but
few can deserve the epithet which he bestows upon Pelion by *waving their
leaves*" (29). When Johnson synecdochically generalizes both his own "eye" and
its responses, it becomes plain that what he initially presented as a self-evident
inference is a subjective response informed by reading, experience, and tem-
perament. Through a combination of appeals to common experience, indefi-
nite and first-person-plural pronouns, and abstract nouns, Johnson continues
to present his personal responses as if they both derived from and culminated in
general truths. Significantly, he interrupts this moralizing generalization only
to inform the reader that this was the context in which he "first conceived the
thought of this narration."

> An eye accustomed to flowery pastures and waving harvests is astonished and re-
> pelled by this wide extent of hopeless sterility. The appearance is that of matter in-
> capable of form or usefulness, dismissed by nature from her care and disinherited
> of her favours, left in its original elemental state, or quickened only with one sullen
> power of useless vegetation.
>   It will very readily occur, that this uniformity of barrenness can afford very

little amusement to the traveller; . . . and that these journeys are useless labours, which neither impregnate the imagination, nor enlarge the understanding. It is true that of far the greater part of things, we must content ourselves with such knowledge as description may exhibit, or analogy supply; but it is true likewise, that these ideas are always incomplete, and that at least, till we have compared them with realities, we do not know them to be just. As we see more, we become possessed of more certainties, and consequently gain more principles of reasoning, and found a wider basis of analogy. . . .

I sat down on a bank, such as a writer of Romance might have delighted to feign. I had indeed no trees to whisper over my head, but a clear rivulet streamed at my feet. . . .Before me, and on either side, were high hills, which by hindering the eye from ranging, forced the mind to find entertainment for itself. Whether I spent the hour well I know not; for here I first conceived the thought of this narration.

We were in this place at ease and by choice, and had no evils to suffer or to fear; yet the imaginations excited by the view of an unknown and untravelled wilderness are not such as arise in the artificial solitude of parks and gardens, a flattering notion of self-sufficiency, a placid indulgence of voluntary delusions, a secure expansion of the fancy, or a cool concentration of the mental powers. The phantoms which haunt a desert are want, and misery, and danger; the evils of dereliction rush upon the thoughts; man is made unwillingly acquainted with his own weakness, and meditation shews him only how little he can sustain, and how little he can perform. (29–30)

Despite the process of stylistic generalization by which Johnson struggles to convert the "hopeless sterility" of the Scottish landscape into a useful and inspiring moral principle ("as we see more, we become possessed of more certainties"), the closed in "barrenness" of his surroundings drives Johnson's thoughts not outward, to universal principles, but inward, to those idiosyncratic preoccupations familiar to any reader of his work: a dread of "want, and misery, and danger"; the fear that he can sustain little and perform less. To some extent writing is an antidote to these fears; through a form of narration that always aspires to generalizations, Johnson seeks to master both the "sullen power" of landscape that seems "useless" because it is so unfamiliar and the "phantoms" that were all too familiar to this traveler in his sixties.[50]

Johnson's periodic submission to the kind of anxieties he describes here qualifies the authority of his generalizations—not because (what we would call) psychological doubts vitiate the moral lesson he seeks to impart, but because such passages hint that producing generalizations involves the perceiver as well as the object of analysis. The implications had already begun to appear in contemporary novels, where Sterne and MacKenzie were exploring the moral complexities of sentimentalism.[51] With the exceptions of Hume and Smith, by contrast, the Scottish historians typically did not call attention to the role their own hopes and fears might have played in generating knowledge. When they did mention that ambition or ignorance could influence testimony, Robertson, Millar, and Ferguson focused on the problem this influence posed for the historian; in their accounts, historians were insulated from the distorting effects of subjectivity by a combination of distance, training, and their comparative method.[52]

Passages like the one I have quoted, where Johnson briefly overlays (or undermines) moral generalizations with his own habitual preoccupations, belong to a series of observations in the *Journey* about the nature of knowledge itself. While the implications of some of Johnson's epistemological reflections remain implicit, as with his tacit assumption that anxieties qualify moral truisms, others form the subject of extensive meditations on the difficulty of procuring even the most basic information uncolored by habit or prejudice. When Johnson expresses interest in the Highlanders' characteristic footwear, for example, he is treated to "an early specimen of Highland information": one informant tells him that every Highland male makes his own shoes and that "a pair of brogues was the work of an hour"; another informant, by contrast, reports "that a brogue-maker was a trade, and that a pair would cost half a crown" (38). The answer was of some importance to Johnson, for he was trying, among other things, to determine how extensively commercialization had transformed Highland society. He could not reconcile these answers, however, because he "had both the accounts in the same house within two days." Confounded, Johnson produces a theory that attributes the epistemological uncertainty of the would-be historian to the "negligence" or "ignorance" of "an ignorant and savage people" paid scant attention by their interlocutors.

> Many of my subsequent inquiries upon more interesting topicks ended in the like uncertainty. He that travels in the Highlands may easily saturate his soul with intelligence, if he will acquiesce in the first account. The Highlander gives to every question an answer so prompt and peremptory, that skepticism itself is dared into silence, and the mind sinks before the bold reporter in unresisting credulity; but, if a second question be ventured, it breaks the enchantment; for it is immediately discovered, that what was told so confidently was told at hazard, and that such fearlessness of assertion was either the sport of negligence, or the refuge of ignorance.
>
> If individuals are thus at variance with themselves, it can be no wonder that the accounts of different men are contradictory. The traditions of an ignorant and savage people have been for ages negligently heard, and unskilfully related. Distant events must have been mingled together, and the actions of one man given to an-

other. These, however, are deficiencies in story, for which no man is now to be censured. It were enough, if what there is yet opportunity of examining were accurately inspected, and justly represented; but such is the laxity of Highland conversation, that the inquirer is kept in continual suspense, and by a kind of intellectual retrogradation, knows less as he hears more. (38)

In the struggle to generate reliable knowledge, Johnson holds out the possibility that modern accuracy might replace historical negligence ("it were enough . . ."). Immediately, however, this possibility vanishes in the unreliability of the medium ("the laxity of Highland conversation"). This pattern appears repeatedly in Johnson's *Journey,* and the distorting medium is not always Highland perversity. Thus, for example, while he attributes mistakes about the size of Loch Ness to men's tendency to "exaggerate to others, if not to themselves" (22), he must acknowledge that his inability to report how far he and his companions ventured into a cave was his own fault; he had failed to bring adequate instruments. "No man should travel unprovided with instruments for taking heights and distances," Johnson belatedly remarks of his makeshift measure (109). Johnson interprets the absence of instruments not simply as a failure of foresight, but also as a trope for a more endemic impediment to epistemological certainty: the failure of memory.

> There is yet another cause of errour not always easily surmounted, though more dangerous to the veracity of itinerary narratives, than imperfect mensuration. An observer deeply impressed by any remarkable spectacle, does not suppose, that the traces will soon vanish from his mind, and having commonly no great convenience for writing, defers the description to a time of more leisure, and better accommodation.
>
> He who has not made the experiment, or who is not accustomed to require rigorous accuracy from himself, will scarcely believe how much a few hours take from certainty of knowledge, and distinctness of imagery; how the succession of objects will be broken, how separate parts will be confused, and how many particular features and discriminations will be compressed and conglobated into one gross and general idea.
>
> To this dilatory notation must be imputed the false relations of travellers, where there is no imaginable motive to deceive. They trusted to memory, what cannot be trusted safely but to the eye, and told by guess what a few hours before they had known with certainty. (109–10)

If memory depends on writing, as Johnson thinks it does, then there can be no reliable evidence of a preliterate society like that of the ancient Highlanders.[53] It is largely for this reason—because Erse was never a written language—that Johnson was so skeptical of the validity of the Ossian poems, which James Macpherson had published in 1760 to a flurry of international interest.[54] Johnson's dismissal of Ossian may have fueled a contemporary controversy, but his observations about memory had more serious implications for knowledge production. When Johnson notes in passing that Boswell had to purchase paper in the only standing shop in Col or that he "had some difficulty to find ink" in Skye, he begins to suggest that, just as knowledge depends on writing, so writing depends on an entire chain of material conditions, including shops, consumer goods, and the roads necessary to convey the latter to the former.[55]

The reflections about epistemological difficulties sparked by Johnson's journey culminate in what would have been unthinkable for the conjectural historians: in the course of his journey, Johnson changed his mind about some of the assumptions he had set out with. As I have already noted, his willingness to record such changes was partly a function of the genre in which he wrote. As a travel narrative, *Journey to the Western Islands* did not violate generic conventions by chronicling the responses of the traveler as well as universal truths. Interpreted as a record of how philosophical principles were produced from an accumulation of experimental observations, moreover, the work might even have served as a defense of the conjectural historian's method. Nevertheless, and even as an example of travel writing and not philosophical history, *Journey to the Western Islands* so powerfully dramatized the way subjectivity could affect knowledge production—not to mention the effects experience could have on the observing subject—that it implicitly cast doubt on some of the basic premises of eighteenth-century moral philosophy. Central to the premises Johnson questioned was the moral philosopher's assumption that the goal of philosophy was to assimilate cultural otherness.

Eighteenth-century British moral philosophers sought to assimilate cultural otherness because, like the Newtonian natural philosophers whose method they emulated, they assumed that "facts" emerged at the level of universals or abstractions, not individuals. The taxonomies that constitute such a prominent feature of the natural and moral sciences of this period supported the typification essential to the search for universals, for taxonomies subordinated differences by arranging them in relation to some general category (the type); taxonomies provided a spatialized conceptual schema that supported the idea of a universal while noting and ordering differences by assimilating them to types and prototypes. As we have seen, when conjectural histories elaborated the universal subject of science, they drew on but departed from taxonomic writing. Because they assumed that the relatively "advanced" societies of Western Europe constituted types of the abstraction of "society" in general, conjectural historians transformed the taxonomy into a hierarchical ordering system: they tended to see less sophisticated social arrangements, such as those of American Indians, as (inferior) prototypes for this typical society, not as alternatives that might belong to a different type.

In virtually everything he wrote, Samuel Johnson endorsed this preference for universality and typification, and he made a career as a public spokesman for the kind of moral generalization that universalization underwrote. Refusing to count the streaks of the tulip, Johnson sought to make a variant of moral philosophy's universalism accessible to the society of readers nurtured on new periodicals like the *Gentleman's Magazine* or his own *Rambler* and *Idler*. Throughout his life, in published prayers, poems, philosophical fictions, and moral essays, Johnson tried to ensure that the knowledge produced and consumed in the public print market would participate in the larger philosophical project of discovering truths that were universal, fixed, eternal. Despite his lifelong commitment to philosophical universalism, however, his *Journey to the Western Islands* placed the effects of epistemological uncertainty alongside the longing for certain knowledge. In so doing, his text helped expose what subsequent philosophers were to interpret as the inadequacies of eighteenth-century moral philosophy. In his repeated allusions to a form of knowledge production capable of generating another kind of certainty, moreover, Johnson foreshadows one direction that architects of the new social sciences were to take in the nineteenth century.

At the beginning of his Scottish travels, Johnson was so confident the principles he assumed to be universal were self-evident that he did not even spell them out. In his description of the effects of Cromwell's civilizing conquest, for example, Johnson simply embeds his assumptions in phrases that unself-consciously conflate description with an evaluation consistent with the hierarchical taxonomies of the conjectural historians. Before Cromwell, Johnson notes with some surprise, "men thus ingenious and inquisitive [so as to cultivate Latin] were content to live in total ignorance of the trades by which human wants are supplied, and to supply them by the grossest means. Till the Union made them acquainted with English manners, the culture of their lands was unskilful, and their domestick life unformed; their tables were coarse as the feasts of Eskimeaux, and their houses filthy as the cottages of Hottentots" (20). According to this interpretation, which equates "progress" with the form of society epitomized by Western Europe and associates "improvement" with both manners and commerce, the Union of 1707 constituted the beginning of modern Scottish history, for it provided incentives to emulate what Cromwell had tried to impose forcibly. "Since they have known that their condition was capable of improvement, their progress in useful knowledge has been rapid and uniform. What remains to be done they will quickly do, and then wonder, like me, why that which was so necessary and so easy was so long delayed" (20).

Initially, Johnson was certain that any features associated with ancient Highland culture must be inferior to the modern improvements that were so rapidly replacing them. Thus he denigrates the "irregular justice" by which the old clans governed themselves, even though he notes that such justice might have been "necessary in savage times" (33); he applauds the "strict administration" of English law (34); and he scorns the "ostentatious display[s]" in which "a warlike people" repeatedly indulge their appetite for violence: "The Highlanders, before they were disarmed, were so addicted to quarrels, that the boys used to follow any publick procession or ceremony, however festive, or however solemn, in expectation of the battle, which was sure to happen before the company dispersed" (34).

Of course, as Johnson repeatedly acknowledges, such customs had already vanished by the time he arrived. What he actually encountered in the Highlands was not violence but "civility" (21), "hospitality" (36), and most striking of all, a culture that survived without commerce, manufacturing, or in some remote areas, money. According to the assumptions he set out with, civility should have been an effect of civilization, and societies that lacked commerce should have been miserable, improvident, and barbarous. Although he found a way to reconcile civility with the primitive structure of Highland society ("politeness, the natural product of royal government, is diffused from the laird through the whole clan," 21), Johnson could initially find no way to appreciate, or even understand, the Highlanders' apparent complacency. In Ostig, a settlement on Skye, Johnson could only interpret the residents' lack of a paid labor force, reluctance to sink mines, and failure to cultivate "improvements" as a puzzling combination of necessity and constitutional laziness. "Neither philosophical curiosity, nor commercial industry, have yet fixed their abode here, where the importunity of immediate want supplied but for the day, and craving on the morrow, has left little room for excursive knowledge or the pleasing fancies of distant profit. . . .Having little work to do, they are not willing, nor perhaps able to endure a long continuance of manual labour, and are therefore considered as habitually idle" (60, 63).

As Johnson continued to chronicle what he observed, however, his tendency to evaluate the Highlanders' habits by the standards of English improvement began to give way to something approaching appreciation—if not for what lay before him, then for what (presumably) had passed away. What Johnson saw in the Highlands, in other words, was neither the ancient civilization of the clans nor modern commercial society, but the effects of what he increasingly represents as a violent suppression, which had conferred all the liabilities of modern society but few of its benefits. "Where there is no commerce nor manufacture," Johnson comments of the transformation he sees on Skye, "he that is born poor can scarcely become rich; and if none are able to buy estates, he that is born to land cannot annihilate his family by selling it. This was once the state of these countries. . . . Since money has been brought amongst them, they have found, like others, the art of spending more than they receive; and I saw with grief the chief of a very ancient clan, whose Island was condemned by law to be sold for the satisfaction of his creditors" (63–64).

Johnson's sojourn on Skye, which was prolonged by a spell of stormy weather, produced a series of meditations remarkable for their interrogation of the assumptions with which the philosopher had embarked. While on Skye, Johnson reconsidered his initial criticism of the "warlike" habits of the ancient Highlanders: "Every man was a soldier, who partook of national confidence, and interested himself in national honour. To lose this spirit, is to lose what no small advantage will compensate" (68). He worried about the effects of English justice and English rents: "There was paid to the Chiefs by the publick, in exchange for their privileges, perhaps a sum greater than most of them had ever possessed, which excited a thirst for riches. . . . The Chiefs . . . expect more rent. . . . The tenant . . . refuses to pay . . . and is ejected; the ground is then let to a stranger, who . . . treats with the Laird upon equal terms, and considers him not as a Chief, but as a trafficker in land. Thus the estate perhaps is improved, but the clan is broken" (70–71). He even questioned whether every nation naturally seeks—or ought to seek—commercial institutions: "It may likewise deserve to be inquired, whether a great nation ought to be totally commercial? whether amidst the uncertainty of human affairs, too much attention to one mode of happiness may not endanger others? whether the pride of riches must not sometimes have recourse to the protection of courage? and whether, if it be necessary to preserve in some part of the empire the military spirit, it can subsist more commodiously in any place, than in remote and unprofitable provinces, where it can commonly do little harm, and whence it may be called forth at any sudden exigence?" (68–69).

These philosophical reconsiderations seem to have been mobilized by two conditions in particular, both of which confronted Johnson on Skye. The first was depopulation, which was an effect of the second: the emigration everywhere under way on the Western Isles. Unlike many eighteenth-century philosophers, Johnson deplored what he called "this epidemick desire of wandering," for he viewed emigration as a blow to national (British) strength, and thus to national happiness.

> Many have departed both from the main of *Scotland,* and from the Islands; and all that go may be considered as subjects lost to the *British* crown; for a nation scattered in the boundless regions of *America* resembles rays of light diverging from a focus.

> All the rays remain, but the heat is gone. Their power consisted in their concentration: when they are dispersed, they have no effect.
>
> It may be thought that they are happier by the change; but they are not happy as a nation, for they are a nation no longer. As they contribute not to the prosperity of any community, they must want that security, that dignity, that happiness, whatever it be, which a prosperous community throws back upon individuals. (98)

To halt emigration, Johnson was even willing to restore to the Highland chieftains the right to bear arms. To halt emigration, in fact, he was willing to block the "progress" of "civilization" by supplementing rents with government-sponsored pensions: "If the restitution of their arms will reconcile them to their country, let them have again those weapons, which will not be more mischievous at home than in the Colonies. That they may not fly from the increase of rent, I know not whether the general good does not require that the landlords be, for a time, restrained in their demands, and kept quiet by pensions proportionate to their loss" (73).

If the recognition that the encroaching money economy was driving people and culture from the Highlands led Johnson to reassess his initial assumptions about the self-evident superiority of commercial society, then the contrast between what he had assumed and what he saw also generated a new ability to see cultural otherness as substantial difference. By the time he and Boswell landed on the island of Mull, Johnson was even ready to revise his earlier interpretation of the scarcity of Scottish trees. Whereas he had initially attributed their absence to Scottish improvidence, Johnson now acknowledges that the conditions of Highland existence are so arduous that *noticing* the scarcity of trees reveals as much about the observer as the lack of trees tells about daily hardships. It is noteworthy that Johnson obliquely alludes to, so as to qualify, his earlier opinion ("the first thought that occurs") but then casts his reconsidered judgment as a series of generalizations that tend once more to universalize the lesson he draws.

> It is natural, in traversing this gloom of desolation, to inquire, whether something may not be done to give nature a more cheerful face, and whether those hills and moors that afford health cannot with a little care and labour bear something better? The first thought that occurs is to cover them with trees, for that in many of these naked regions trees will grow, is evident, because stumps and roots are yet remaining; and the speculatist hastily proceeds to censure that negligence and laziness that has omitted for so long a time so easy an improvement. . . .
>
> Plantation is naturally the employment of a mind unburdened with care, and vacant to futurity, saturated with present good, and at leisure to derive gratification from the prospect of prosperity. He that pines with hunger, is in little care how oth-

ers shall be fed. The poor man is seldom studious to make his grandson rich. It may soon be discovered, why in a place, which hardly supplies the cravings of necessity, there has been little attention to the delights of fancy, and why distant convenience is unregarded, where the thoughts are turned with incessant solicitude upon every possibility of immediate advantage. (104)

Johnson's reconsidered response to cultural difference—here he acknowledges difference instead of assimilating it to a universal type—belongs to what we might call an anthropological enterprise, which was being pursued by just a handful of philosophers in the 1770s. Lord Kames's attempts to attribute racial and temperamental differences to climate and geography constituted one branch of this inquiry; Lord Monboddo's theory that human beings and apes were descended from the same ancestor constituted another.[56] Although it would be misleading to attribute Johnson's newfound respect for difference to this emergent cultural relativism—he was, after all, still eager to generate universal truisms about "the poor man"—it would be equally mistaken to overlook the extent to which casting himself as an overhasty "speculatist" qualifies the very truths he seeks to universalize. If he was initially misled by his own inexperience, after all, who was to say that one might not always be misled by experience that was always insufficient?

If Johnson had begun to assume, however intermittently, that particulars—including the particulars of cultural difference—constituted an important component of knowledge production, then this was of a piece with his consistent respect for counting and measurement. Although he did not elaborate a theory about the relation between newly valued particulars and counting, he did repeatedly point out that the latter belonged to a more modern mode of knowledge production that was beginning to displace the traditional habits of estimation, guessing, and generalizing from scant information. One cannot say that Johnson anticipated the nineteenth-century practice of statistics any more than one can legitimately argue that he abandoned his lifelong commitment to universals, but it does seem fair to say that, just as he acknowledged that epistemological difficulties surrounded both claims about universal types and observed particulars, so he thought that counting both troubled some of the epistemological claims of the moral experimentalists and offered an alternative to them.

Measurement and counting were constant concerns to Johnson, both because he repeatedly encountered discrepancies between what he had been led to expect and what he saw and because, unlike questions about the ancient Highlanders, these inconsistencies could presumably have been corrected by something resembling political arithmetic. In Johnson's text, then, counting and measurement are sites of the epistemological murkiness by which so many of his inquiries were thwarted *and* instruments by which at least some of that murk might have been dispelled. From the beginning of his journey, Johnson associated "accuracy of narration"—especially accurate measurement and counting—with philosophical rigor; and he expressed hope that the new Scottish rage for "natural philosophy" would dispel existing exaggerations—about the size of Loch Ness, for example (22, 23). At Raasay Island, Johnson bemoans the lack of any reliable way to measure the "populousness of the place" (49);[57] and confronted with the specter of emigration on Skye, he longs for some way to describe accurately the variance between past and present populations. The "true numbers" of Highlanders under Roman occupation are unavailable, Johnson acknowledges, and the estimates we do have say more about Scottish pride than about the size of the population. "Those who were conquered by them [the Romans] are their historians, and shame may have excited them to say, that they were overwhelmed with multitudes. To count is a modern practice, the ancient method was to guess; and when numbers are guessed they are always magnified" (73–74). Johnson believes that if he had had accurate numbers, he might have been able to understand why Scotland was so much less populous in 1773, just as he contends that if he had been able to measure the cave on Icolmkill accurately, subsequent visitors would not have had to venture inside. Although he never draws these comments into a systematic theory about the modern practice of counting, this is the only mode of knowledge production that he does not relegate to the radical uncertainty he associates with both observation and conversation.

Counting and measurement—the basic tools of political arithmetic—might have represented to Johnson one alternative to the epistemological dilemmas that challenged the theory- and experience-based practice of experimental moralism—but the *Journey to the Western Islands* is not simply a plea for mensuration over theory-based experiment. Nor does he reject the moral philosopher's goal of generating universal moral truths from observed particulars, despite his recognition that knowledge may be blocked by all sorts of impediments, and even despite his willingness to revise the universals he had assumed when he embarked. Johnson's gestures toward counting and measurement do suggest that an alternative mode of knowledge production might have put an end to (some kinds of) investigation; but it is not completely clear that ending investigation was his paramount goal. Epistemological certainty, after all—even if only the certainty produced by precise measurement—might have momentarily halted that "rush" of phantoms that drove him to counter the Highlanders' "sterility" with the fecundity of writing, but it is not clear that he considered losing such uncertainty—or the need to write—altogether good things. Like cultural otherness, he explains, uncertainty makes humans struggle with what they cannot understand; if the "speculatist" was satisfied with his initial self-aggrandizing assumptions, he might bask in moral ignorance forever.

At the end of his Scottish journey, when Johnson reentered the familiar urban landscape, he abruptly stopped writing. "To describe a city so much frequented as *Glasgow*, is unnecessary," he flatly states (119). Having long anticipated the scintillating social exchanges he had engineered for Johnson in Glasgow, Boswell was deeply disappointed by the old man's refusal to narrate these verbal adventures; his supplemental *Journal* was at least in part an attempt to rectify Johnson's unaccountable reluctance to chronicle his conversational prowess. The sudden curtailment of Johnson's travel narrative reflects something other than just modesty, however. Whether or not he intended to make this point, his sense that familiarity renders description superfluous implies that the discomfort inflicted by uncertainty and cultural difference produced an unforeseen benefit: it provoked and therefore prolonged narrative. This oblique suggestion is another variant of Adam Smith's meditation on wonder: as long as the mind is discomforted, by curiosity or unassimilable difference, the philosopher will continue to theorize and the reader will continue to read.[58] Whereas novelists had already begun to mobilize this illimitable drive by narrative techniques designed to prolong suspense, moral philosophers continued to seek conclusive answers, or at least to develop a method that might be used to settle disputes that were simultaneously epistemological and political.

Reconfiguring Facts and Theory: Vestiges of Providentialism in the New Science of Wealth

Despite their explicit commitment to the production of general—even universal—knowledge, the three writers who figured most prominently in chapter 5 all introduced epistemological principles that called into question the very possibility of such knowledge. Even though Hume was able to set aside the demoralizing effects of the problem of induction by celebrating sociality, his skepticism typically provoked more consternation than philosophical humility among his readers. By the same token, Adam Smith was able to neutralize the questions raised by the idea that philosophical systems reflected nothing more substantial than the philosopher's desire, but subsequent philosophers worried over the element of subjectivism that he admitted into the discipline. Perhaps only Samuel Johnson began to register the blow that his rudimentary cultural relativism dealt to philosophical generalization when he refused to chronicle what people already knew at the end of his *Journey*. Once more, however, his nineteenth-century legatees, who could not seek refuge in silence, struggled to fit cultural relativism into the ideal of systematic knowledge by creating the new disciplines of ethnography and anthropology.

For most British philosophers for most of the second half of the eighteenth century, some variant of providentialism or the argument for design cushioned the impact of these challenges to the feasibility of general knowledge. In this chapter I argue that vestiges of this providentialism continued to underwrite the new science of wealth, which was refined and institutionalized in England between 1790 and the early 1830s. Whereas the providentialism we saw in Hutcheson's and Turnbull's writing simply expressed what these philosophers took to be self-evident commonplaces, however, the residual providentialism that appears in the work of Dugald Stewart, Thomas Malthus, and J. R. McCulloch should initially be read as a defense. For Stewart and McCulloch this providentialism, which gained credibility from the rise of evangelicalism in Britain, was primarily a defense against the three epistemological challenges I have just described; for McCulloch, it was a defense not only against philosophical skepticism but also against the notoriety that Malthus cast on political economy it-self. Despite the providential strain clearly identifiable in the first edition of his *Essay on the Principle of Population* (1798), Malthus was almost immediately vili-fied by his readers, and the science he stood for seemed in danger of losing the credibility that Smith and Stewart had conferred on it. In his campaign to use Malthus's conclusions but distance himself from the ensuing controversy, J. R. McCulloch developed a strain of providentialism that managed to seem both moral and secular so that political economy became both the paradigm of use-ful knowledge and the incarnation of a science so dismal and self-serving that it seemed like a threat to national morality.

If the providentialism visible in the writing of Stewart, Malthus, and Mc-Culloch should initially be read as a defense against philosophical skepticism and debilitating pessimism, then the science that preserved this residual provi-dentialism should also be read as the harbinger of a new kind of disciplinary knowledge, which drew its authority not primarily from its theological basis but from something else. One of the aims of this chapter is to acknowledge the complex configuration of educational agendas, government infrastructure, and protocols for professional credentialing that ultimately constituted the institu-tional support for claims about knowledge; but because this configuration did not achieve a workable form until later in the century, I will only outline its el-ements here. Instead of examining all of the institutional supports for this new epistemological authority, I primarily address the impact that the changes and controversies engulfing the old discipline of moral philosophy had on the re-working of the science of wealth and on the meaning of numerical representa-tion more specifically.

Alongside the gradual erosion of providentialism, I suggest, and despite the persistence of providentialism in political economy, the decades after 1798 re-veal a rapid revision of the cultural connotations of numbers. Paradoxically, as numerical representation was removed from the field of Christianized Platonic associations that we saw mobilized by Shaftesbury, Hutcheson, and Turnbull and consigned to a more secular field, numbers increasingly came to seem like the mode of representation *least* imbued with providential overtones or theo-retical prejudices. Paradoxically again, this reworking of the meaning of nu-merical representation made numbers appealing to the British government, whose representatives saw themselves as secular and administrative officials, not agents of God. And it also made numbers appealing to champions of liberalism, who saw opportunities to calculate profits, adapt means to ends, and align (self-) government with a form of progress increasingly conceptualized in fiscal rather than spiritual terms.

The overlapping but divergent tendencies I describe in this chapter—

toward preserving providentialism and toward secularizing an increasingly numbers-based science of wealth—should obviously be read within the context of larger historical events. While I do give some attention to two of the most important contemporary events—the French Revolution and the constitution of poverty as a new kind of problem in the 1790s—I generally leave such historical matters to others. Because I am most interested in the epistemological dimension of these tendencies, I concentrate on how the three writers whose work I examine here addressed the challenges bequeathed by Hume and Smith. (The cultural relativism that Johnson adumbrated did not become a pressing epistemological problem until the 1830s, although, in the form of abolitionist discourse, it remained politically salient throughout these decades.) Specifically, I focus on how what we might call misreadings of Hume and Smith simultaneously preserved (or restored) the providentialism that the eighteenth-century philosophers bracketed *and* began to liberate from all but the most vestigial traces of providentialism a form of knowledge that seemed both general and useful. The taxonomy of knowledge that J. R. McCulloch proposed in 1825, which separated the collection of numerical information about particulars (statistics) from the production of general knowledge (political economy), can be read as a solution to the problem of induction; but it could function as a solution only because McCulloch removed the problem of induction from the *philosophical* field of epistemology and transformed it into a problem of *professional organization*.

To call McCulloch's taxonomy a misreading of Hume, as I do, is to posit both disciplinary continuity and disciplinary divergence. Because the method implicit in this gesture is critical to the enterprise of this book as a whole, it might be useful simply to summarize the stakes of this kind of historical analysis before taking up this crucial episode in my history. Certainly, the more typical approach to these materials is to identify a relatively autonomous discipline (political economy, statistics, or even the social sciences), then trace its "prehistory" or genealogy. The shortcoming of such analyses, as Stefan Collini, Donald Winch, and John Burrow have argued, is that they tend to superimpose the map of present disciplines on the past, so that it becomes difficult to recover the contemporary questions that past authors might have been addressing, much less to appreciate what look like disciplinary dead ends as paths that actually led elsewhere.[1]

Instead of identifying some relatively autonomous discipline that makes sense in late twentieth-century terms, I have been proposing that cultural configurations of knowledge should be understood as historically specific ensembles, whose parts are reconfigured and reordered in a dynamic interaction with the discovery or consolidation of new objects of inquiry. Thus the ancient configuration of knowledge, in which rhetoric played a foundational role, was so altered by Bacon's elevation of the observed particular that rhetoric was eventually consigned to a marginal role in what counted as knowledge. Or even worse, and insofar as rhetoric had been equated with ornament, it was relegated to an oppositional position, as what descriptive, neutral (hence authoritative) representation was not.

Throughout this book I have argued that these dynamics within the ensemble of knowledge-producing instruments can best be understood retrospectively, as episodes in a process that gradually renders methods and questions that once made sense to an intellectual community so problematic that a radical reconsideration seems in order. This analytic method focuses on texts that realign the cultural configuration of knowledge practices or of individual disciplines, because such realignments alert us to problems that may well never have been posed in the terms in which solutions are eventually offered. Thus we can best make sense of Newton's recasting of induction in the light of the problem that Baconian particularism bequeathed to the philosophical ideal of general knowledge; and we can understand the painstaking distinction that McCulloch drew between the sciences of politics, political economy, and statistics in the light of the problem that Hume introduced but did not pursue. By focusing on the moment when McCulloch redrew the relation between two existing disciplines and called for importing a third, we can see that something in the available ways of making sense of the world no longer seemed adequate in 1825, even though neither McCulloch nor his contemporaries formulated the situation as a methodological crisis or an epistemological impasse.

Interpreting McCulloch's disciplinary taxonomy as a solution to a question that Hume and Smith had introduced *but had not experienced as a problem* also enables us to identify tendencies inherent but not realized in the eighteenth-century philosophers' work. This mode of historical analysis, in other words, permits us both to respect the intentions of past writers and to see how developments that these writers could not have foreseen prompted interpretations of their work that stressed meanings at odds with the writers' intentions. When McCulloch distinguished between the two "theoretical" sciences—"the sciences of Politics and Political Economy"—and the new practice of statistics, he emphasized the prospect that Hume's skepticism and Smith's subjectivism had introduced but not pursued: that the empirical collection or description of particulars might fundamentally challenge the philosophical imperative to produce theoretical generalizations, which supposedly constituted the very heart of modern "science." At the same time, however, McCulloch also proposed a professional solution: let one group collect particulars while another, more highly qualified group produces general knowledge. And because these men will all be professionals, let them create a working relationship with each other.

The analytic method I have been deploying requires us to read retrospectively and proleptically rather than superimposing modern taxonomies or constructing genealogies; it requires us to interpret one text as a (mis)reading that diverts the effect of an existing text from its author's intention rather than simply as an anticipation of or contribution to some unilinear prehistory or development. Although this kind of analysis might tell us less about ourselves and our own disciplinary division of knowledge than some readers would like, it does reveal in the process of history writing and the history of disciplines the same tension that I am arguing is constitutive of the epistemology of the modern West: the tension between the desire to construct (or discover) some order that supports meaningful generalizations and the kind of sloppiness or disorder that inheres in discrete particulars. In revealing (and to a certain extent repeating) this tension, this method unabashedly preserves the Enlightenment dream that human systems of knowledge somehow correspond to a phenomenal world, while holding that dream up to the interrogation mandated by the postmodern claim that the entire Enlightenment project, like every system of knowledge, has been constituted in a historically specific context of particulars that have left their indelible mark on the theoretical systems by which we try to discipline them.

In organizing the topics of this book (and this chapter) chronologically, I run the risk of endorsing the Enlightenment dream, not to mention the genealogical reading, that I also want to examine; but chronology offers the opportunity to counteract two excesses that characterize much postmodern analysis. The first interprets the past only retrospectively, as a failed approximation (or faint anticipation) of the present. At its best, such anticipatory analyses help us understand how the knowledge we embrace is always marked by its antecedents; at its worst, it devolves into a quest for political correctness, which can lead late twentieth-century readers to lambaste eighteenth-century philosophers for their classist, sexist, and racist assumptions, even though the modern meanings of class, gender, and race were then just beginning to be formulated. The second excess I want to avoid abandons altogether the project of finding the terms by which past actors knew themselves, because such efforts, this argument goes, are so marked by our own biases that what we claim is history is really only self-projection. While I realize that disentangling the present from (what we imagine to be) the past is as difficult as seeing the stark alterity of the past, it seems to me to be an effort worth making, if for no other reason than that an increased ability to recognize (and tolerate) otherness is one of the epis-temological feats that the philosophical history I am describing has bequeathed to late twentieth-century knowledge producers. In this chapter, as in the book as a whole, my analytic method thus remains in what I hope is a productive tension with the order in which I present the material. The readings I offer here of Stewart, Malthus, and McCulloch look both backward and forward, to the texts and contexts these writers reworked and to the effects these reworkings had on the way we now understand them and their eighteenth-century predecessors. Apart from occasional methodological comments, however, my presentation largely remains faithful to chronology, both because I have wanted to demon-strate that issues became problems only as a result of developments extraneous to authorial intention and because I have wanted to foreclose the kind of judg-ment that too frequently leads modern readers to dismiss rather than under-stand writers of another age.

INSTITUTIONALIZING POLITICAL ECONOMY:
DUGALD STEWART AND THE REPUDIATION OF PARTICULARS

Dugald Stewart is critical to any account of the relation between eighteenth-century moral philosophy and the nineteenth-century science of wealth, be-cause his classroom constituted the site where both moral philosophy and political economy were reworked and passed along to the intellectual leaders of the next generation.[2] Stewart's own educational pedigree gave him a unique place in the lineage of Scottish university-trained philosophers. Having at-tended Adam Ferguson's lectures at Edinburgh in 1765, Stewart took up his studies with Thomas Reid at Glasgow. Then in 1785 he succeeded Ferguson as professor of moral philosophy at Edinburgh, a position he held until 1810. At Edinburgh, the extraordinarily popular Stewart taught moral philosophy to large numbers of young men, and he occasionally supplemented his primary subject with classes in natural philosophy, mathematics, rhetoric and belles let-ters, and most important, political economy.[3] When Henry Brougham, Francis Jeffrey, Francis Horner, and James Mackintosh—all of whom had been Stew-art's pupils—established the *Edinburgh Review* in 1802, they created a vehicle capable of disseminating the lessons Stewart had taught throughout literate Britain. When J. R. McCulloch joined the *Review* in 1818 as the journal's chief economic writer, he began to popularize—and interpret—the principles Stewart had set out, even though it is not clear that McCulloch ever attended his lectures.[4]

Dugald Stewart was thus the single most significant influence on the

British understanding of political economy promulgated by the most prestigious economic journal of the first third of the nineteenth century, as well as the primary interpreter of eighteenth-century moral philosophy for nineteenth-century British intellectuals. Despite his critical position, however, Stewart has received surprisingly little attention from modern historians.[5] Those analysts who have accorded him the prominence he deserves have tended to emphasize either his role as interpreter of Adam Smith or his contributions to what Thomas Macaulay called "that noble science of politics." We can move beyond this impasse in the secondary literature if we begin with Stewart's work on epistemology instead of turning directly to his political philosophy, for his epistemology constitutes both the heart of his contribution to the modern fact and the key to his revisions of Adam Smith's philosophical agenda.

In essence, Stewart's contribution to the modern fact consisted of an elaboration of the generative—or performative—potential of the problem of induction. If it is impossible to know in advance of observation that hitherto unseen phenomena resemble what one has already seen, in other words, and if the resulting uncertainty places belief (or even conjecture) at the center of knowledge, then one can interpret the imperative to believe either as a blow to philosophical confidence or as an opening for the philosophical imagination. Whereas Hume adopted the first interpretation, Stewart embraced the second. When he praised the midcentury historians for realizing that they could "suppl[y] the place of fact by conjecture," Stewart opened the potential for a new mode of philosophical knowledge, which generated not certainty but hope. This hope could lead to action, and if the action was diligently pursued, it could actualize the future of which the philosopher was the first to dream.

Making philosophy performative instead of descriptive entailed two related moves, both visible in chapter 4 of Stewart's *Elements of the Philosophy of the Human Mind* (1792). The first move consists of elevating the signs used to designate phenomena over the phenomenal objects that these signs signify. The second involves demoting the distinguishing features of particular objects in favor of features they can be said to share, so that the philosopher can emphasize—and name—the class to which a series of phenomena belong instead of being distracted by particulars. Stewart subsumes these philosophical operations into the method he calls "abstraction." Abstraction, in his account, is related both to Cartesian analysis and to nominalism more generally. According to Stewart, abstraction breaks objects down into their component features so that the philosopher can "attend . . . to some . . . qualities or attributes, without attending to the rest."[6] Having broken objects down into the particular features they are composed of, the philosopher can then classify; that is, he can name and group objects based on the likeness he deems salient. From such classifications, he begins to generate systematic knowledge.

> This power of considering certain qualities or attributes of an object apart from the rest; or, as I would rather choose to define it, the power which the understanding has of separating the combinations which are presented to it, is distinguished by logicians by the name of *abstraction*. . . . As abstraction is the ground-work of classification, without this faculty of the mind we should have been perfectly incapable of general speculation, and all our knowledge must necessarily have been limited to individuals; and . . . some of the most useful branches of science, particularly the different branches of mathematics, in which the very subjects of our reasoning are abstractions of the understanding, could never have possibly had an existence. (*Philosophy,* 99)

As the last clause of this passage suggests, Stewart treated the substantive abstractions I discussed in the previous chapter ("the human mind," "the market system") as products of the philosophical method of abstraction: once one has separated those combinations that look like entities into their component parts, one is able to create new conceptual entities ("abstractions of the understanding"), which then become the objects of "useful" philosophical thought.

To Stewart, privileging the method of abstraction over either observation or pure deduction offered several advantages. The first was that it enabled the philosopher to attain that impartial or disinterested stance that Robertson and Reid had also sought. This impartiality was a function of Stewart's nominalism—his tendency to credit words (or even algebraic symbols) with the power to dictate meaning—for his nominalism turned on the ability of language to rein in the tendency that Smith had associated with the imagination: the tendency to sympathize or even identify with others. For Stewart, "confining" the imagination to signs was critical, because more vital entities—whether people or phenomena—engaged, distracted, and (presumably) interested the philosopher, who should aspire to the status of judge, not advocate.

> As the decision of a judge must necessarily be impartial, when he is only acquainted with the relations in which the parties stand to each other, and when their names are supplied by letters of the alphabet, or by the fictitious names of Titus, Caius, and Sempronius; so, in every process of reasoning, the conclusion we form is most likely to be logically just, when the attention is confined solely to signs; and when imagination does not present to it those individual objects which may warp the judgment by casual associations. (*Philosophy,* 110)

Freed from imaginative engagement, the philosopher basks in a domain of pure language where, Stewart continues, he can enjoy the second advantage that abstraction offers: participation in a project that exceeds any individual's ability. In leaving observed particulars behind, philosophy thus reaches the heights where "comprehensive theorems" reside. "The same faculties which, without the use of signs, must necessarily have been limited to the consideration of individual objects and particular events, are, by means of signs, fitted to embrace, without effort, those comprehensive theorems, to the discovery of which, in detail, the united efforts of the whole human race would have been unequal" (*Philosophy,* 135).

If the capacity of signs to leave "individual objects and particular events" behind distinguished philosophy from more "interested" and practical endeavors, then the mode of knowledge production that privileged signs over observed particulars could also be turned to practical use. This practical application, which Stewart addresses in the section of *Philosophy* titled "The Use and Abuse of General Principles in Politics," fused philosophy with politics in such a way as to render the latter precisely the kind of science whose promise Hume had raised. Stewart believed that politics could become a science if the philosopher passed over "peculiarities" for those "comprehensive views" that illuminated "fixed and certain maxims."

> It is necessary (as well as in mechanics) to pay attention to the peculiarities of the case; but it is by no means necessary to pay the same scrupulous attention to minute circumstances, which is essential in the mechanical arts, or in the management of private business. There is even a danger of dwelling too much on details, and of rendering the mind incapable of those abstract and comprehensive views of human affairs, which can alone furnish the statesman with fixed and certain maxims for the regulation of his conduct. (*Philosophy,* 149)

Stewart's dismissal of particulars obviously belonged to a revaluation of what counted as experience. To him the experience that mattered was "extensive experience," not individual experience, because only "extensive experience" could illuminate "the known principles of human nature."[7] From the discussion he offers in *Philosophy of the Human Mind,* it is not absolutely clear what Stewart meant by "extensive experience"—whether, for example, he thought an individual could acquire such experience over the course of a lifetime or through training, or whether he considered it a collective repository of observations and reflections. What is clear is that, for him, "extensive experience" was related to "extensive induction"; both phrases drew on but (mysteriously) elaborated the Baconian method of collecting deracinated particulars in such a way as to thoroughly imbue what counted as a fact with theoretical presuppositions. The theoretical component of what counted as a fact for Stewart began with—but was not limited to—the theoretical concept (the "abstraction of the understanding") that was both the product and the object of philosophical analysis ("the human mind," "the human constitution"). Thus, in the process Stewart called "extensive induction," one created this abstraction by subdividing and categorizing an entity's properties, then one derived its principles from a kind of speculative observation. "Those principles which we obtain from an examination of the human constitution, and of the general laws which regulate the course of human affairs . . . are certainly the result of a much more extensive induction, than any of the inferences that can be drawn from the history of actual establishments" (*Philosophy,* 149).

In Stewart's reworking of experience and induction, we see just how susceptible the eighteenth-century variant of the modern fact was to the kind of theoretical component from which Bacon had sought to insulate his deracinated particulars. What counted as a fact was already being drawn toward its theoretical pole by Hutcheson's and Turnbull's providentialism, of course, and Smith's privileging of system intensified this tendency. But for all these eighteenth-century philosophers, the drift toward theory was halted—whether by the philosopher's effort to model the moral faculty on the physical senses, by his commitment to history or experiment, or as in the case of Smith, by his determination to open a space for a mode of description that could eventually be mimetic. In Stewart's work, by contrast, nothing halts the drift of fact toward theory, for Stewart was trying to create a way to make what counted as a fact inaugurate a future that the philosopher had already imagined. For him, in other words, what counted as a fact ("fixed and certain maxims") derived from the philosopher's commitment to a theory about the future, rather than following from observations about the present or the past.

We are now in a position to understand why Stewart called the method he attributed to Smith and his contemporaries "conjectural" history. He did not do so primarily to draw out these historians' reliance on hypotheses about human nature—although, with the light cast backward by his term, that tendency stands out sharply. Instead, he used this term because he believed the philosopher should be a visionary who contributed to a new social arrangement rather than simply describing the present or the past. While Stewart resembled the conjectural historians in his commitment to a universal subject of science, in his reliance on abstractions, and in his strong providentialism, he differed from them in his determination to make the professor of moral philosophy an agent of change. His model of change also distinguished him from his teachers, for just as he imagined the philosopher inaugurating a future of which he now only dreamed, so he interpreted the past as consisting of revolutionary about-faces that changed the course of history forever. Unlike the conjectural historians, Stewart did not embrace models of repetition, cycles, or uniformity. Instead, he insisted optimistically that the future could radically depart from the present because such ruptures had already occurred.

Stewart anchored his optimism in a variant of providentialism even more emphatic than what historians like Lord Kames had endorsed. Celebrating the wisdom of a feminized nature in the final pages of the chapter on abstraction, Stewart subsumes individual passions and particular circumstances to a grandiose design governed by a beneficent deity.

> So beautifully, indeed, do these passions and circumstances act in subserviency to her [nature's] designs, and so invariably have they been found, in the history of past ages, to conduct him [mankind] in time to certain beneficial arrangements, that we can hardly bring ourselves to believe, that the end was not foreseen by those who were engaged in the pursuit. Even in those rude periods of society, when, like the lower animals, he follows blindly his instinctive principles of action, he is led by an invisible hand, and contributes his share to the execution of a plan, of the nature and advantages of which he has no conception. (*Philosophy,* 168)

By invoking and recasting the conjectural historians' theory of unintended consequences, Stewart subtly adjusted the philosophical gaze. Not only were the consequences he imagined in the future as well as the past, but like Smith, he also focused on an abstract *system*—the beneficial arrangements form part of a plan—instead of just the fate of the universal subject, "mankind."[8] Unlike Smith, however, Stewart argued that the system of morality actually existed—albeit at a level of abstraction ascertainable only by philosophical thought; this strong objectivism enabled him both to assign the philosopher an active role in human improvement and to refute the subjectivism he associated with Smith. Whereas Smith had cautioned that we cannot know the ontological status of the theoretical systems we generate and crave, Stewart insisted that a beneficent design exists; the very philosophical systems by which we gratify ourselves, he maintained, are the instruments by which philosophers know and realize this design. And whereas Smith had argued that even the philosopher cannot understand individual moral judgments unless he considers the particular society in which individuals live, Stewart insisted that particular circumstances are irrelevant, because morality inheres in the general system that the philosopher comprehends and describes.[9] If the philosopher considers human actions at a sufficiently high level of abstraction, he can identify their relative contributions to the providential design he believes in.

In one respect, Stewart did acknowledge the importance of particular past events—although the only historical component of his philosophy all but rendered historical analysis irrelevant. Like contemporary theologians who embraced the doctrine of particular Providence (or miracles), Stewart believed

that "a variety of events have happened in the history of the world, which render the condition of the human race *essentially* different from what it ever was
among the nations of antiquity, and which, of consequence, render all our reasonings concerning their future fortunes, *in so far as they are founded merely on their
past experience,* unphilosophical and inconclusive" (*Philosophy,* 164; my emphasis). In 1792 Stewart named only one such event, the invention of printing:
"This single event, independently of every other, is sufficient to change the
whole course of human affairs" (*Philosophy,* 164). By the next year, when he delivered his eulogistic *Account of the Life and Writings of Adam Smith* to the Royal
Society of Edinburgh, he had added a second historical rupture to his account
of progress: the development of commerce. What distinguishes modern nations from their predecessors, Stewart explains, is

> the general diffusion of wealth among the lower orders of men, which first gave
> birth to the spirit of independence in modern Europe. . . . Without this diffusion
> of wealth among the lower orders, the important effects resulting from the inven-
> tion of printing would have been extremely limited, for a certain degree of ease
> and independence is necessary to inspire men with the desire of knowl-
> edge. . . . The extensive propagation of light and refinement arising from the in-
> fluence of the press, aided by the spirit of commerce, seems to be the remedy
> provided by nature, against the fatal effects which would otherwise be produced,
> by the subdivision of labour accompanying the progress of the mechanical arts,
> nor is anything wanting to make the remedy effectual, but wise institutions to fa-
> cilitate general instruction, and to adapt the education of individuals to the stations
> they are to occupy.[10]

This passage constitutes Stewart's specific rejoinder to Adam Smith, who
had lamented that the division of labor tended to reduce workers to brute stu-
pidity.[11] Stewart manages to correct Smith without repudiating his work by focusing on the importance of the discipline Smith had helped create: political
economy. Stewart saw political economy as the instrument that philosophers
needed to actualize their own vision of God's design, for political economy enabled scientists of society to know, so as to realize, the "order of things represented by nature." In his account, political economy was a science of wealth, but
as a performative philosophical practice it was a science that dignified the study
of wealth. Political economy dignified the modern obsession with wealth, according to Stewart, by subordinating the desire to accumulate to an understanding of a larger, divine plan that was unfolding in time. "It is this view of
Political Economy [as a science of virtue] that can alone render it interesting to
the moralist, and can dignify calculations of profit and loss in the eye of the
philosopher" (*Account,* 59).

In 1793, with the example of the Terror confirming British suspicions that "French" theory would lead to ruinous political experiments, both Stewart's optimism and his commitment to theory at the expense of Baconian (British) empiricism were extremely controversial. Indeed, given the respect he had professed for the French *économistes* in 1792, he had to find some way to distinguish his conjectural philosophy from that of theorists like Condorcet.[12] In the original version of his *Account of the Life and Writings of Adam Smith,* Stewart doggedly pursued what he calls "the science of Politics" through the method of abstraction to which he was committed; but by 1802, when the lecture was published, Stewart had added two substantial notes, which considerably qualified whatever affiliation with French theory auditors might have identified in 1792 or 1793. In the first, he tried to redirect suspicions that might have focused on the method of philosophical "conjecture" toward the specific doctrine of free trade, which had been interpreted as embodying "a revolutionary tendency."[13] In the second note, he downplayed the originality of the French theorists by citing a long list of British precedents and by reminding his readers that Smith's ideas about free trade formed the subject of lectures delivered as early as 1748 (before Condorcet wrote).[14]

If Stewart tried in 1802 to defend his philosophical commitment to theory by distinguishing the specific doctrine of free trade from legitimate philosophical speculation and by repatriating political economy, then by 1820 he had located another defense for theoretical abstraction. Although we have already seen intimations of this defense in his *Philosophy of the Human Mind,* Stewart drew out its implications only in his triumphal *Dissertation: Exhibiting the Progress of Metaphysical, Ethical, and Political Philosophy.* Initially published in another triumphalist project, the *Encyclopaedia Britannica,* the *Dissertation* emphasized two points that had long remained implicit in his work. The first is that political economy, which Stewart now calls "the boast of the present age," epitomizes modern metaphysics.[15] The second is that believing in the existence of a providential design enables the philosopher to participate in realizing what he seems merely to describe. Stewart's commitment to philosophical activism cemented his lifelong opposition to what he here describes as the "dampening" effects of Humean skepticism.

> A firm conviction . . . that the general laws of the moral, as well as of the material world, are wisely and beneficently ordered for the welfare of our species, inspires the pleasing and animating persuasion, that by studying these laws, and accommodating to them our political institutions, we may not only be led to conclusions which no reach of human sagacity could have attained, unassisted by the steady guidance of this polar light, but may reasonably enjoy the satisfaction of consider-

ing ourselves, (according to the sublime expression of the philosophical emperor,) as *fellow-workers with God* in forwarding the gracious purposes of his government. It represents to us the order of society as much more the result of Divine than of human wisdom; the imperfections of this order as the effects of our own ignorance and blindness; and the dissemination of truth and knowledge among all the ranks of men as the only solid foundation for the certain though slow amelioration of the race. (*Dissertation*, 491–92)

For Dugald Stewart, who was more committed to the moral improvement of his student readers than to the accumulation of national wealth, political economy was a science capable of both expressing and confirming the optimistic providentialism that inspired the only kind of development that mattered: the inexorable unfolding of God's beneficent plan. It was to further this enterprise, finally, that philosophy had to transcend the observation of particular details—for attachment to "*particular facts,*" like skepticism more generally, could "disturb our tranquillity, without bringing any accession of good to compensate the uneasiness which it occasions." In his climactic celebration of theory as the only route to the "*general* laws of nature," Stewart struggled one last time to extricate philosophical generalizations from the aspersions cast on theory by a public leery of things French.

The *general* laws of nature, as far as they have yet been traced, appear all so wisely and beneficently ordered, as to entitle us to reject, on this very principle, every theory which represents either the physical or the moral order of the universe, in a light calculated to damp the hopes, or to slacken the exertions of the friends of humanity. This is a conclusion, not resting on hypothesis, but on an incomparably broader induction from particular instances, than what serves as the foundation of any one of the *data* on which we reason in natural philosophy. (*Dissertation*, 520)

The "broader induction" Stewart refers to here, like the "more extensive induction" he conjured in his *Philosophy of the Human Mind,* was obviously intended to invoke but surpass Baconian method and Baconian facts. To a certain extent, as I have already noted, such gestures did not actually solve the problem of induction, for they did not specify either how "broader induction" differed from Baconian induction or how the maxims that rested comfortably in the self-contained systems of language or philosophy were related to the "particular instances" from which they supposedly were derived. In the passages I have quoted from his *Account of the Life and Writings of Adam Smith* and the *Dissertation,* Stewart does nod toward something that might have begun to bridge the gap between the particulars that currently existed and the theory he claimed the future was designed to realize. This something is education—"wise institutions to facilitate general instruction," "the dissemination of truth and knowledge among all the ranks of men." Although Stewart did not elaborate on either the kind of education he envisioned or how it would differ from the church-sponsored schooling typically available in Britain in the first decades of the nineteenth century, he clearly recognized that, in the wake of the revolution caused by printing and if God's design was to be advanced by philosophical enquiry, some mechanism had to be devised to disseminate the "extensive experience" of humankind to a population literate in ever greater numbers. As we will see in the last section of this chapter, by 1825 it became possible to imagine not only that the philosophical problem of induction and the social problem of education were inextricably linked but also that the former could be solved by making political economy the keystone of everyone's education. Before taking up J. R. McCulloch's attempt to address the problem of induction with a solution that involved professionalizing political economy, however, let us return to the subject of numbers, so we can see how the kind of theoretical fact that Stewart liberated from observed particulars was sutured again to the numerable things of the world.

THOMAS MALTHUS AND THE REVALUATION OF NUMERICAL REPRESENTATION

Given political economy's disciplinary roots in moral philosophy, as well as the strong providentialism that characterized Stewart's influential treatment of this science, it initially seems incomprehensible that to many nineteenth-century Britons political economy came to represent the most dismal tendencies of systematic knowledge. For the most part, modern historians have not helped us understand—or even see—this conundrum, for with a few notable exceptions they have tended to read political economy either as a harbinger of modern economics or simply as the eighteenth-century antithesis of more humane discourses like the novel.[16] Viewed from the perspective of eighteenth-century moral philosophy instead of twentieth-century economics, however, early nineteenth-century political economy looks not like an immature effort to determine the regularities of production and consumption but like an attempt to accommodate the laws by which wealth was assumed to accumulate according to God's providential plan. If not fully devoted to moralizing the accumulation of wealth, in the sense of inculcating Christian virtues in manufacturers and capitalists, many of Stewart's disciples—who were the earliest champions of political economy in Britain—were nevertheless devoted to understanding how the happiness that wealth (supposedly) conferred upon societies contributed to the aggregate virtue of the nation.

Despite Stewart's strong and influential providentialism, however, it is undeniable that contemporary critics of political economy targeted precisely the science's indifference to moral issues in their vicious attacks on what Carlyle dubbed the dismal science. To understand why political economy was vilified in this way, it is necessary to take up the case of Thomas Malthus, whose version of political economy was so influential that "Malthusianism" became the most common invective hurled against the science of wealth. To understand why Malthus provoked such a turn in the popular (mis)understanding of political economy, moreover, it is helpful to examine his treatment of numbers in more detail than most readers have done. Malthus was by no means solely responsible for the revaluation of numerical representation that became visible in Britain after the 1790s. But the way a vision of natural laws more bleak than any but the most embittered philosopher had been willing to voice intersected in his work with a use of numbers apparently purged of their Christian Platonism contributed to the impression that numbers had no moral dimension. It seemed, as Dickens so bitterly complained, that matters of the utmost moral importance could be treated as "a case of simple arithmetic."[17]

To understand Malthus's complex relation to the revaluation of numerical data, we need to pose several questions, most of which do not admit easy answers. One line of inquiry must address the kinds of numerical data available in the last decades of the eighteenth century and the social authority assigned to these numbers. This will lead us to restate and expand the questions about counting I raised in the first chapter of this book: Who counted? What did people count? For what social or institutional purposes did people count? And with what instruments and funding were large-scale counting projects implemented? A second line of inquiry concerns the meanings and relative importance assigned to numerical data. As we will see, language that cast assessments of "value" and national well-being in what looked like quantifiable terms was used throughout the eighteenth century, but advocates of theological utilitarianism did *not* typically call for the collection of numerical data to determine *what* behaviors would contribute most to what William Paley called "the greatest sum of human happiness." Indeed, even when Jeremy Bentham began to advocate an empirical variant of utilitarianism in the last decades of the century, he did not wait for data before deciding what policies to recommend. To do Bentham justice, however, he did have a passion for numerical information and, frustrated by its unavailability, recommended establishing (or reforming) the centralized agencies that would have made collecting such data possible.[18] The important point is that until the turn of the century, most government officials and individual philanthropists did not consider it necessary to base philosophical principles on numerical data, much less to ask about "values" or "happiness" in terms that would admit of numerical answers.

By 1825 this situation had just begun to change. Although I will delay my discussion of this change until the final section of this chapter, let me pose the question this line of inquiry evokes in its starkest terms: How and why did happiness and value come to be understood as concepts that could be quantified (and by extension commodified)? How and in what arenas, in other words, did what I have called gestural mathematics give way to literal counting, so that "value," "worth," and "riches"—not to mention "sum," "calculation," and "calculus"—gradually abandoned their older ethical and theological connotations for a rationality designed to adapt means to ends and to quantify productivity and yield?

The questions about counting may never be answerable to our complete satisfaction. Only a few scholars have tackled subjects related to the history of numerical representation in eighteenth-century Britain, and even those who have done so have found it almost impossible to assemble evidence. Moreover, any answer we give must remain partial because the determinants of this semantic revolution are so numerous and diverse. What follows, then, should be read as a barometer of the current state of scholarship and a prolegomenon to further research. Even if my observations are incomplete, however, approaching Malthus's much-interpreted *Essay* along these lines will help us to see why its many incarnations constitute a crux in the history of the modern fact and to understand how the form of representation that late twentieth-century readers associate with disinterestedness or impartiality acquired those connotations partly through Malthus's revision of the modern science of wealth.

Several kinds of numerical records were routinely kept in Britain during the eighteenth century. In addition to the government's excise and trade records, companies as well as many individuals kept records of business or household costs, purchases, and sales; parishes recorded expenditures on the poor; magistrates kept track of criminal arrests and the debts incurred by prison inmates. Beyond such representations of daily, monthly, or annual activities (which make these records resemble the accountant's memorial and serve the same function), another kind of numerical document was regularly produced in eighteenth-century Britain: bills of mortality, which served as the basis for life tables and thus for pricing insurance. William Petty either wrote or promoted a very early interpretation of the bills of mortality in 1662; in 1694 Edmond Halley constructed a life table from the births and deaths recorded in the German city of Breslau; and in 1772 William Price used records kept by All Saints parish in Northampton to construct a new life table.[19]

As important as these numerical records were, from a modern perspective their limitations are even more striking than their extent. The first feature to note about all these records (with the exception of excise and trade figures) is their localism. In keeping with the structure of parish administration, which characterized Britain in the period, records of expenditures on poor relief were kept at the parish level. Even these were collected erratically, however: returns from parishes were made only for the years 1748–50, 1776, and 1783–85.[20] Not until 1775 was a parliamentary committee appointed to collect figures for the nation as a whole, and in 1796 such information was still so incomplete that when a private philanthropic society wanted to help remedy the condition of the poor, its members had to make collecting such information one of their charter projects.[21] Bills of mortality were also, of necessity, kept at the local level, and even those who consulted the bills to generate life tables for general use were uncertain whether these bills really provided generalizable information; before the kind of localized but large-scale information that Edwin Chadwick collected in the 1840s was available, it was not clear whether life expectancy varied with where one lived, much less whether location should influence the price of insurance. Even more striking from a modern perspective are the kinds of records about commercial enterprises that were not kept in eighteenth-century Britain. Records about bankruptcies were scanty and imprecise before the establishment of a bankruptcy court in 1831, for example, and although sophisticated new methods of accounting had been developed, even as innovative a businessman as Josiah Wedgwood did not consider it worthwhile to collect the kind of information that would have made cost accounting possible.[22] The relative indifference to modern modes of accounting among eighteenth-century tradesmen helps explain why the history of accounting is less relevant to the vicissitudes of the modern fact in eighteenth-century Britain than is the history of moral philosophy.

To a certain extent, the relative scarcity of numerical information in this period points to the lack of an infrastructure sufficiently developed or centralized to collect it. The parish system of administration certainly tended to impede the collection of national figures, and in the absence of a centralized body of law that expressly promoted business, it may have seemed wise to avoid providing tax collectors with the kind of records that a money-starved government might seize. Such an explanation goes only so far, however, for the very parish system that might be seen as having impeded national information collection was successfully used in the 1790s to implement the first such project, which culminated in the publication of Sir John Sinclair's twenty-one-volume *Statistical Account of Scotland*.[23] By the same token, even if national laws did not systematically protect private enterprise from taxation, since no laws specifically required businesses to hand over their books to government representatives, entrepreneurs might well have used regular accounts to increase their profits without worrying about undue oversight or taxation.

Instead of simply attributing the relative lack of numerical records to an inadequate infrastructure, we should recognize that many Britons did not consider counting particularly relevant to knowledge or see costing as essential to value. Beyond overt resistance to counting, standardization, and a national census, which has been well documented in the scholarly literature,[24] eighteenth-century Britons also manifested a pervasive indifference to numerical information, which seemed irrelevant to what many people recognized as "truth" or "value." The natural philosophical developments I have already discussed tended to subordinate observed particulars (which could be counted) to universals (which could not), and Britons also tended to neglect counting because the "truths" that most associated with "value" were ethical or even theological. Insofar as the "value" of some action could be measured, "measurement" had less to do with quantification than with determining the "fit" between the action and God's laws.

We have already seen that Shaftesbury and Hutcheson used a mathematical vocabulary to discuss ethical issues. Examples of a mathematized conception of virtue can certainly be found earlier, especially in Puritan literature (in 1627, for example, Richard Cumberland advocated determining "the greatest Good" by "the fittest natural Measure"),[25] but the tendency to discuss virtue in mathematical terms reached its apogee in the eighteenth-century movement known as theological utilitarianism.[26] Beginning in 1731 with John Gay's *Dissertation concerning the Fundamental Principles of Virtue and Morality* and achieving its most influential articulation in William Paley's *Principles of Moral and Political Philosophy* (1785), theological utilitarianism generally sought to identify the greatest "common good" by a "felicific calculus." An Anglican outgrowth of the tradition of natural jurisprudence associated with Hugo Grotius and Samuel von Pufendorf, theological utilitarianism deployed what I have called gestural mathematics to assess how well phenomena "fitted" the providential design that (theoretically) informed the natural and moral worlds. As William Paley explained:

> The fitness of things means their fitness to produce happiness: the nature of things, means that actual constitution of the world, by which some things . . . produce happiness, and others misery: reason is the principle, by which we discover or judge of this constitution: truth is this judgment expressed or drawn out into propositions. So that it necessarily comes to pass, that what promotes the public happiness, or the happiness of the whole, is agreeable to the fitness of things, to nature, to reason, and to truth.[27]

As numerous modern historians have noted, Thomas Malthus belongs to the tradition of theological utilitarianism.[28] What few have noticed, however, is that when Malthus supplemented the largely deductive assertions that characterize the first edition of his *Essay on the Principle of Population* with the numbers that swelled all subsequent editions, he all but gutted the strain of theological utilitarianism that had underwritten the first edition and that had conferred upon numbers (though not on counting) the overtones of Christian Platonism that we saw in Shaftesbury's work. This was true not simply because Malthus substituted induction, which could (but did not always) promote counting, for deduction (which resembled mathematics, not counting). Instead, the revisions he made to his *Essay* robbed theological utilitarianism of its providential and ethical dimension because the numbers he used supported a thesis that made it all but impossible to argue, as theological utilitarians did, that whatever is, is right. Although this thesis was present in the first edition of the *Essay*, its impact was somewhat cushioned in 1798 by the presence of a residual providentialism that took the form of a theodicy. In the first edition, Malthus blunted the brutal implications of his thesis that population growth always tends to outstrip the production of food with a theological account that supposedly explained how this situation advanced God's plan. It was perhaps coincidental that he omitted his heterodox theodicy from the very edition in which he began to increase his reliance on numerical information (the 1803 edition). Whether coincidental or not, the combination of his effacing what remained of his original providentialism and his adding tables and numbers that theoretically demonstrated what he had previously been able only to assert helped make the *Essay* appear to reveal the essential heartlessness of political economy and to make numbers seem like the indifferent handmaiden to moralism's demise.

The revisions that Malthus made to his *Essay* after 1798 were not solely responsible for reworking the cultural connotations of numerical representation, of course. To fully understand why the Christian Platonism so prominent in the gestural mathematics used by Shaftesbury, Hutcheson, Hume, and Paley gave way to an understanding of numbers that stressed their impartiality and methodological rigor, it would also be necessary to investigate contemporary Continental efforts to standardize weights and measures, as well the impact of British technological innovations like the spinning jenny, which required more precise calibrations but rewarded such precision with greater output.[29] Although some scholars have begun to open these vital subjects, however, the epistemological question I am addressing could never be answered simply by saying that the Scandinavian countries had already begun to adopt standardized measures by the middle of the eighteenth century or that some British entrepreneurs realized by 1780 that precision instruments increased yields and profits. Although it is true that Adam Smith visited a pin factory to observe the division of labor and that Malthus went to Scandinavia to collect numerical information for his initial revisions of the *Essay*, the simple existence of isolated examples of this more secular understanding of numbers does not explain why Smith and Malthus thought they should consult them, nor does it explain what they thought they saw when they did. In offering Malthus's revisions as one site where the meanings of numerical representation were reworked at the end of the eighteenth century, then, I do not mean to attribute undue significance to this text. I do want to propose that the combination in his various revisions of a less obviously providential interpretation of contemporary society with a more copious use of numbers *that claimed actually to refer to particulars that had been counted* helped make numerical representation seem immune from the very theoretical strains with which the Christian Platonism of theological utilitarianism so thoroughly imbued it.

Before turning to Malthus's crucial revision of his *Essay,* it is important to recover the providentialism of the first edition, both because it is so often overlooked and because its peculiar quality helps explain why it could so easily have been excised. To appreciate the peculiarity of Malthus's providentialism, we have only to contrast his depiction of contemporary society with Stewart's optimistic account of human progress unfolding according to God's plan. Unlike Stewart, who was able to interpret even the French Revolution as simply an instructive experiment in politics, Malthus viewed the events of the 1790s as one pole of a "perpetual oscillation between happiness and misery."[30] When he looked around him—whether at the developments in France or at the disastrous effects of the grain scarcity at home—what Malthus saw were not signs of progressive improvement but human suffering and pain. To make the brutality and starvation so visible in the middle years of the decade seem like part of a providential plan, he had to adopt a variant of the interpretive device that Adam Smith had also used. Whereas Smith invoked the argument about unintended consequences to reconcile what he saw with what he believed should be true, however, Malthus explained the discrepancy by calling the misery he witnessed an exemplum of God's pedagogy: in order to inspire human beings to labor virtuously for their bread, Malthus asserted in the last two chapters of the *Essay,* God inflicted the principle of population on them; in order to wrest spirit out of the fallen flesh, he made suffering the daily lot of humankind.

I will return to Malthus's unorthodox theodicy in a moment. For now let me point out that even though Malthus used this explanation to make the misery he witnessed "fit" a providential plan, he did not share the methodological commitment to abstraction or prophetic optimism that went along with providentialism in Stewart's work. Instead, he sharply criticized philosophers who envisioned "illimitable, and hitherto unconceived improvement," both because they were blithely indifferent to the suffering around them and because they deployed an "unphilosophical mode of arguing." Indeed, in Malthus's account these two errors were inevitably linked, for it was the "unphilosophical mode of arguing" epitomized by deduction from a priori principles that led philosophers either to dismiss observed particulars as signs of some greater and more momentous tendency or else to take "the higher classes" as types of humanity as a whole (1798; 67, 70, 78). Although Malthus's immediate targets were William Godwin and the French philosopher Condorcet, many of the criticisms he leveled at these spokesmen for perfectibility could have applied to Stewart as well. Indeed, the term on which Malthus concentrated his wrath is the word we have seen Stewart defend: conjecture. Having "shut [their] eyes to the book of nature," Malthus complained, such philosophers advance "the wildest and most improbable conjectures . . . with as much certainty as the most just and sublime theories, founded on careful and reiterated experiments" (1798; 126).

As this statement illustrates, instead of conjecture, Malthus recommended "experiments," which he insisted would corroborate and be confirmed by "experience, the true source and foundation of all knowledge" (1798; 72). Like a good Baconian, Malthus claimed he wanted to go out and look—especially at the poor, and especially at the most physical aspects of the impoverished Britons' lives, their food consumption and their rate of sexual reproduction. If one went and looked, Malthus asserts, it would be difficult to maintain that "the human mind" progresses, or even that "human nature" is the proper subject of philosophy. Indeed, it would be impossible even to *see* the universal subject that had so concerned eighteenth-century philosophers, for when one looked at the poor, one came face-to-face with the physical difference that inadequate nutrition made in the stunted bodies of those who looked back. "The sons and daughters of peasants will not be found such rosy cherubs in real life as they are described to be in romances," Malthus wrote. "It cannot fail to be remarked by those who live much in the country, that the sons of labourers are very apt to be stunted in their growth, and are a long while arriving at maturity" (1798; 93–94).

Although Malthus has been criticized for the limits of his empiricism (this is one of the very few passages in the first edition of the *Essay* that implies he actually observed the poor),[31] he knew what questions would be relevant to a science that *was* thoroughly grounded in "actual observation and experience" (1798; 82):

> Some of the objects of inquiry would be, in what proportion to the number of adults was the number of marriages, to what extent vicious customs prevailed in consequence of the restraints upon matrimony, what was the comparative mortality among the children of the most distressed part of the community and those

who lived rather more at their ease, what were the variations in the real price of labour, and what were the observable differences in the state of the lower classes of society with respect to ease and happiness, at different times during a certain period. (1798; 78)

This passage is significant for several reasons. First, it suggests that Malthus's interest in the poor deconstructed the universal subject of eighteenth-century philosophy in the name of a cultural relativism written onto (what we would call) class. Second, it hints at a methodological claim that subsequent editions of the *Essay* would elaborate: that if one could obtain such proportions, comparative numbers, and variations in prices—if one could actually measure and count—then this would solve the problem of induction, because measuring, counting, and figuring proportions and variations would bridge the gap between the observed particular and general knowledge. And third, the passage suggests that the kind of information Malthus wanted to collect about the poor would have brought into being another kind of collective subject. Like the universal subject of eighteenth-century philosophy, this subject would be seen to obey regularities—but not because the observer simply *assumed* it was universal; instead, this subject's regularities would be the effect of the very mathematical operations that made counting seem to bridge the gap between particulars and generalizations. The regularities this subject displayed, in other words, would follow not from postulates but from observations and mathematical calculations.

Modern statisticians call this new kind of collective subject a "population." For them the population is a conceptual unit constituted in order to study an inherently various or heterogeneous object. With the help of hindsight we can see that the statistical population is produced by the same method that Stewart called abstraction; by means of some principle of classification, *which suits the philosopher's immediate purpose,* some features of the heterogeneous object are highlighted while others are ignored. Thus the modern statistician might intentionally ignore distinguishing factors like class or race, but *not* because she assumes these differences do not matter—*not because she assumes human nature is everywhere the same.* Instead, and for the purposes of a particular calculation, the analyst wants to control for one factor and not others. Unlike Stewart, in other words, who vacillated between maintaining that the substantive abstractions generated by the method of abstraction highlighted features that were *actually* more essential (according to a God-given teleology) and hinting that these abstractions might simply serve the philosopher's need, modern statisticians insist that the statistical population is simply an analytic instrument. It does not purport either to identify the essential characteristics of its object of analysis or to reveal the universality of the results it produces.

Although Malthus implied such an analytic instrument when he argued that a combination of empirical and mathematical analyses would reveal the regularities of "the poor," the aggregate he called into being was *not* a statistical population. By the same token, of course, remember that when Malthus used "population," he was referring not to a statistical population, but variously to the number of people in Britain (what eighteenth-century commentators called "populousness")[32] and to something else, which lies somewhere between the eighteenth-century universal subject and the modern statistical population. What he seems to have had in mind is an aggregate, many of whose most meaningful symptoms were amenable to numerical representation (possibly even mathematical formulation). But this aggregate was not constituted from some mathematical sense of representativeness ("statistical significance"), nor did all its truths yield to quantification. Indeed, we can see the transitional nature of Malthus's concept in the twin agendas he sought to mobilize in relation to this aggregate: he wanted to count and measure behaviors whose regularities could be calculated; and also he wanted to assess this aggregate's "happiness," an operation that required defining happiness instead of just manipulating numbers.

By claiming that if one *could* count, measure, and calculate one would generate a new kind of knowledge about a new kind of object, Malthus widened the gulf between observation and belief that Smith's two versions of nature had also disclosed. If one could count, in other words, and if what one counted showed a population outstripping the production of food (as Malthus insisted it did), then what one saw could not possibly simply embody the benevolent interpretation of God that had been handed down from the eighteenth-century argument for design. In widening this gulf between one kind of knowledge production (counting) and another (reasoning from a priori principles or beliefs), Malthus helped strip numerical representation of the moral connotations that its eighteenth-century affiliation with Christian Platonism had preserved. And even though Malthus insisted that observing and counting were *not* antithetical to Christian belief, this approach helped make numbers (and the science that used them) seem both amoral and antitheoretical—hence both endlessly susceptible to manipulation by self-serving arithmeticians and infinitely useful to governments that wanted to insulate policy from the kind of a priori interestedness associated with party politics.

In 1798, of course, the kind of comprehensive counting that Malthus imagined had yet to be conducted in Britain. Equally to the point, and because he wanted to stress his own providentialism as much as he wanted to gather new data, Malthus did not wholeheartedly endorse collecting numerical data over reasoning from one's philosophical and religious convictions. While he asserted that *if* one could collect the kind of information he wanted, it would "probably prove the existence of the retrograde and progressive movements that have been mentioned" (1798; 78), Malthus also insisted that one should hold on to one's beliefs. Unfortunately, even this was not an unambiguous position, for as he unfolded his argument it became clear that even though the information one could collect would "probably" support one's philosophical beliefs (that population was outstripping food), it still would not bolster one's religious convictions (that God was not arbitrarily hurting humankind). It was to reconcile these two levels of belief that Malthus introduced the infamous theodicy that occupies the final chapters of the first *Essay*.

At the urging of more orthodox Anglicans, Malthus omitted this theodicy from all subsequent editions, and scholars have long debated its importance even to the one edition where it appears.[33] When considered from the perspective of methodology, however, the theodicy seems essential to Malthus's work, for it not only preserved the providentialism that Stewart had made so central to political economy but did so in such a way as to highlight the transitional nature of the method Malthus advocated in the *Essay*. On the one hand, in chapters 18 and 19 of the first edition, Malthus continued to assert the importance of empiricism; on the other hand, and often in the same sentence, he also insisted that a certain understanding of empiricism would support what we do and want to believe. Thus he rejected both theories that conjure perfection and those that consider life "a state of trial and school of virtue" on behalf of a belief he considered "more consistent with the various phenomena of nature which we observe around us and more consonant to our ideas of the power, goodness, and foreknowledge of the Deity" (1798; 200). This belief is that God is lawful; and it is corroborated, he continues, by what natural philosophy has proved: that nature is lawful too. Thus what one would see if one could gather sufficient data—people procreating faster than they can grow food, and famine, war, and pestilence checking population growth when people refuse to delay marriage—would reveal what the Christian philosopher believes: not that God made nature to serve human desires but simply that God manifests himself in the lawfulness of nature. By this interpretation, the apparent evil of the principle of population actually "produces a great overbalance of good," because the "strong excitements" occasioned by seeing this principle at work create both labor and self-control.

> Strong excitements seem necessary to create exertion, and to direct this exertion, and form the reasoning faculty, it seems absolutely necessary, that the Supreme Being should act always according to general laws. The constancy of the laws of nature, or the certainty with which we expect the same effects from the same causes, is the foundation of the faculty of reason. If in the ordinary course of things, the

finger of God were frequently visible, or to speak more correctly, if God were frequently to change his purpose (for the finger of God is, indeed, visible in every
blade of grass that we see), a general and fatal torpor of the human faculties would
probably ensue; even the bodily wants of mankind would cease to stimulate them
to exertion, could they not reasonably expect that if their efforts were well directed
they would be crowned with success. The constancy of the laws of nature is the
foundation of the industry and foresight of the husbandman, the indefatigable ingenuity of the artificer, the skilful researches of the physician and anatomist, and
the watchful observation and patient investigation of the natural philosopher. To
this constancy we owe all the greatest and noblest efforts of intellect. To this constancy we owe the immortal mind of a Newton. (1798; 205)

In content, this statement resembles innumerable eighteenth-century
claims about "the laws of nature"; indeed, the claim that nature is lawful constituted the basis for all philosophical enterprises after the seventeenth century.[34]
Coming as it does, however, in the last two chapters of the first *Essay* and in the
form of an unorthodox defense of what could also be read as a pessimistic description of the inevitable misery of human existence, to many contemporaries
Malthus's providentialism seemed too little too late. Because he had widened
the gulf between what he claimed observations would reveal and what he said
we must and do believe, many of them interpreted this theodicy as an after-
thought, a belated attempt to undo the damage he had already done. Thus,
whereas modern readers might argue that there is something wrong with his
method—that his empiricism cannot support the providentialism from which it
actually derives[35]—early nineteenth-century readers complained that there
was something wrong with the *content* of the argument itself. They charged that
Malthus's pessimism undermined hope, that his claims that misery served a spiritual use sanctioned the neglect of the poor by those more interested in wealth
than in virtue.

As we will see, some of Malthus's earliest readers welcomed his suggestion
that more empirical data be collected, even if it was possible to read his use of
numbers as dangerously amoral, as many of his contemporaries did. By and
large, by contrast, his contemporaries did *not* welcome his conclusions, no matter how he claimed to have reached them. That he not only combined these
conclusions with calls for collecting and consulting data but formulated his
most fundamental principle *as a mathematical ratio,* moreover, encouraged critics who rejected his conclusions on theological or ethical grounds to target his
method, even though what they had to say about it often simply ignored the
providential underpinnings that seem so odd to modern readers. Many of
Malthus's contemporaries did not agree with his conclusions, in other words,

and because they had no way to evaluate the numerical data he did present, much less to tie such numbers to the infamous ratio, they focused on the method that seemed, in form at least, related to the numbers.

The first two editions of Malthus's *Essay* attracted some critical notice, but the vitriolic controversy that converted political economy into Malthusianism did not erupt until 1806, after the publication of the third edition.[36] As I have noted, Malthus had deleted the theodicy from the second edition, which was published in 1803, so what sparked the firestorm could have been neither the heterodoxy of the last two chapters nor the discrepancy between what he claimed were empirical observations and his attempt to fit these to a providential plan. Instead, some combination of Malthus's heavy-handed dismissal of his critics (he declared them "beneath notice") and what he added to the second and third editions must have provoked the public outcry. What he added to the second and third editions, of course, were numbers: tables of annuities, population growth, and food prices, which were assembled from records as diverse as bills of mortality, census returns, and agricultural records. I suggest it was these numbers, *coupled with* the elimination of his explicit providential scheme, that helped make Malthus seem the demonic spokesman for the end of moral knowledge and numbers the amoral vehicle of indifferent facts.

In the preface to the *Essay*'s second edition, Malthus admitted that the first edition was "written on the spur of the occasion, and from the few materials which were then within my reach in a country situation. The only authors from whose writings I had deduced the principle, which formed the main argument of the *Essay*, were Hume, Wallace, Adam Smith, and Dr Price."[37] Immediately upon completing the first edition, Malthus determined to test what had essentially been a deduction against the kind of numbers he had wanted all along. Thus he began to collect and read every book he could obtain on demography, population, and commerce; and he and three college friends conducted a study tour through all the countries then open to Britons: Norway, Sweden, Finland, and Russia. In 1802, after the Peace of Amiens, Malthus extended his investigations to France and Switzerland. The results of his investigations are immediately apparent in the size of the second edition: whereas the first *Essay* was published as an octavo volume of 396 pages and 55,000 words, the second edition was a quarto of 610 pages and 200,000 words; and whereas Malthus consulted only four texts for his initial volume, the bibliography for subsequent editions contains close to two hundred entries.[38]

The number and kind of sources Malthus cites in the later editions of the *Essay* constitute a fair sampling of the extent and nature of the sources of numerical data available to a private, liberally educated individual at the turn of the century. Among his sources are Alexander Humboldt's *Essai politique sur le royaume de la Nouvelle Espagne,* the *Philosophical Transactions* of the Royal Society, Adam Seybert's *Statistical Annals . . . of the United States,* the United States census returns beginning in 1790 (published as *Population Abstract*), the American *National Calendar,* Joshua Milne's *Treatise on the Valuation of Annuities and Assurances,* John Bristed's *America and Her Resources,* William Tooke's *View of the Russian Empire,* and Suessmilch's *Göttliche Ordnung.* As this list illustrates, most numerical data available in print had been collected outside Britain; and until the first census returns (from the census of 1801) and the first official statistics on pauperism (in 1818, from the survey of 1812–15) were available, the scant data that had been collected about Britain had typically been generated by private organizations or individuals. This bibliography makes it clear that Malthus's revised *Essay* was not the only, or even the first, source of numerical information in or about Britain; not only had the figures gathered by Henry Fielding and Jonas Hanway long been in print, but in the 1790s John Sinclair had compiled his enormous *Statistical Survey of Scotland,* and in 1797 Sir Frederick Morton Eden had published *The State of the Poor,* which contained voluminous numerical data.[39] Despite the availability of these texts, however, and especially in the context of the irregular nature of the data available about Britain, Malthus's revised *Essay* played a particularly significant role in the revaluation of numerical representation to which all these texts contributed, if for no other reason than that the *Essay* aroused such public controversy.

Not surprisingly, Malthus used the numerical data he added to the second edition to prove the principles he asserted in the first *Essay:* that, *if unchecked,* population will increase at a geometrical rate; that, *if unimproved,* agriculture will increase food supplies arithmetically; and that, since population has not yet outrun food supplies, some combination of "checks" must be capable of curtailing population growth. Even though his general conclusions were the same, however, the method by which he reached these conclusions opened the space for an analysis that might lead another investigator to results he did not expect. Thus, for example, when Malthus explains why the information he includes in book 2 of the revised edition ("Of the Checks to Population in the Different States of Modern Europe") reveals both regularities in the course of population considered as a whole and variations among the countries he examined, he proposes that, if one looked at "different places of the same country," not to mention different (and non-European) countries, one might find data one had not expected:

> The habits of most European nations are of course much alike, owing to the similarity of the circumstances in which they are placed; and it is to be expected therefore that their registers [of births, deaths, and marriages] should sometimes give the

same results. Relying however too much upon this occasional coincidence, political calculators have been led into the error of supposing that there is, generally speaking, an invariable order of mortality in all countries; but it appears, on the contrary, that this order is extremely variable; that it is very different in different places of the same country, and within certain limits depends upon circumstances, which it is in the power of man to alter. (1826, 1803; 259/260)

While the numerical data included in book 2 finally support Malthus's original thesis, then, passages like this also encouraged other advocates of numerical data to imagine that numerical information could be used to challenge their opponents' theoretical presuppositions or to defend their own, precisely because numbers seemed not always to support the thesis one set out with—precisely because numbers seemed to be divorced from theory.

Although Malthus's copious numbers, as well as his references to numbers' disproving the errors that "political calculators" had cherished, encouraged readers to imagine that he was divorcing numbers from their theoretical (and moral) underpinnings, this was clearly not so in any of the editions of the *Essay.* Even without the theodicy, and even though he cited numerous calculations to document his controversial claims, Malthus's revised *Essay* used only numbers that constituted *evidence* for his thesis. Although the role he assigned numbers in all the revised editions implied that counting might solve the problem of induction by bridging the gap between observed particulars and general knowledge, moreover, Malthus was never explicit about how one might move from such numerical data to general conclusions (instead of working the other way round, from one's foundational assumptions to choosing the numbers that constituted evidence to support them). Malthus seems to have been uncertain whether *averages* were identical to *arithmetical limits,* for example, and he was not sure whether averages should be treated as *norms.*[40] Without the concept of a statistical population, whose regularities *create* statistical norms, Malthus lacked a way to conceptualize the relative importance of averages in his argument about population laws. He compounded the effects of this methodological ambiguity when he construed his fundamental principle as a set of mathematical formulas, for this cast a spurious aura of certainty over his conclusions. Almost immediately, even sympathetic readers complained that mathematical rigor was inappropriate to the kinds of claims he advanced. In 1840 John Stuart Mill was still trying to undo the damage Malthus inflicted on his thesis by "the unlucky attempt to give numerical precision to things which do not admit of it."[41]

Even if Malthus did not divorce numerical data from the providential postulates that were so prominent in the first edition, however, and even if some readers argued that his mathematical formulas were expendable, the combina-

tion of such densely marshaled numbers and Malthus's elimination of the
theodicy made it possible for unsympathetic critics to interpret the facts he presented as if they had been altogether denuded of theory (and hence morality).
Thus, even though some readers praised Malthus for using numbers—and
some, like John Weyland, even called for more numbers[42]—many targeted the
wedge that he seemed to have driven between numbers and morality as the
heart of the problem. To appreciate the nature of this complaint, we have only
to turn to the romantic campaign against Malthus, which was conducted without intermission from at least 1812 until the 1830s.[43]

Led by Robert Southey and Samuel Taylor Coleridge, who were joined by
William Wordsworth, Thomas De Quincey, and William Hazlitt, the romantic
attack on Malthus targeted many subjects: his initial failure to acknowledge that
human beings were sufficiently virtuous to control their sexual appetites, his inconsistency in adding "moral restraint" to the list of "preventive checks" in the
second edition, his fear of a growing population (which, Southey maintained,
was essential to national strength), his criticism of the poor law. For our purposes, however, the most powerful criticisms were those that focused on his use
of numbers. Significantly, Malthus's romantic critics did not charge simply that
he had chosen the wrong numbers or that the numbers he used did not prove
what he claimed they showed; instead, these critics complained that numbers
were irrelevant to the kind of knowledge he claimed to produce. When
Southey set mathematical demonstration in opposition to "religious argument," for example, we can see just how radically Malthus's use of numbers
seemed to have departed from that of the eighteenth-century philosophers.

> The Malthusians observe, in reply to such objections, that the new discovery is [a]
> matter of science, and that religious argument cannot be permitted to stand in the
> way of demonstration. . . .If the two things were incompatible the consequence
> could not be avoided; the argument of the geometrical and arithmetical series was
> a demonstration, and Divine Providence must go to the wall! But there is a moral
> *reductio ad absurdam* which the man of enlightened piety feels to be demonstrative
> wherever it applies: he knows in his heart that whatever opinion is wholly and fla-
> grantly inconsistent with the goodness of creating and preserving wisdom, must
> necessarily be false; and in this knowledge he cannot be deceived, for it is the voice
> of God within him which tells him so.[44]

In 1728, as we saw in chapter 4, "science" and geometry were not considered
antithetical to religious argument; indeed, Francis Hutcheson claimed that his
gestural mathematics was the science capable of demonstrating those religious
principles the philosopher "knows in his heart." In 1812, by contrast, Southey
could accuse the Malthusians of setting science against religion, both because

the conclusions Malthus had reached did not echo what Southey knew in his heart and because he could interpret Malthus as claiming that his use of numbers was incontrovertible *because* numbers were impartial and thus had nothing to do with what one finds in one's heart.

Coleridge drew out the ethical implications of the claim (or complaint) that numbers are impartial when he accused Malthus of moral relativism. "It is this accursed practice of ever considering *only* what seems *expedient* for the occasion, disjoined from all principle or enlarged systems of action, of never listening to the true and unerring impulses of our better nature, which has led our colder-hearted men to the study of political economy," Coleridge charged.[45] In Coleridge's eyes, Malthus's method seemed dangerously close to casuistry, the old juridical mode of making case decisions in the absence of absolute standards. Even worse, as the word "expedient" implies, this method seemed to him to encourage people to accommodate means to ends. It wasn't just that Malthus's version of political economy "disjoined" actions "from all principle or enlarged systems of action" (although this enraged Coleridge too); beyond this, political economy seemed to *replace* the providentialism that Coleridge thought should guide philosophy with another agenda—one that served not the "better nature" that supposedly united all humans but the most self-serving desires of those who were most powerful.

If numbers were impartial—if they could be made to serve any agenda, no matter how heartless or amoral—then the few who used them could inflict actual damage on the many who were powerless to resist. According to Southey, the political economists who used numbers did just this: by means of and on behalf of the "manufacturing system," the economists (who included but were not limited to Malthus and Smith) defended the capitalists, who piled up wealth in pursuit of a religion as cruel as it was false.

> The manufacturing system . . . has enabled us to raise a revenue which twenty years ago we ourselves should have thought it impossible to support, and it has added even more to the activity of the country than to its ostensible wealth; but in a far greater degree has it diminished its happiness and lessened its security. Adam Smith's book is the code, or confession of faith, of this system; a tedious and hard-hearted book, greatly over-valued even on the score of ability, for fifty pages would have comprised its sum and substance. . . . That book considers man as a manufacturing animal,—a definition which escaped the ancients: it estimates his importance, not by the sum of goodness and of knowledge which he possesses, not by the virtues and charities which should flow toward him and emanate from him, not by the happiness of which he may be the source and centre, not by the duties to which he is called, not by the immortal destinies for which he is created; but by the gain

which can be extracted from him, the *quantum of lucration* of which he can be made the instrument.[46]

In such passages we can see how effectively Malthus's claim that numbers might reveal something other than God's plan could be used not only against him, but against the entire "manufacturing system" that he (and Adam Smith) were charged with defending. Given the providentialism that was explicit in the first edition of Malthus's *Essay* (and residual in all subsequent editions), it seems ironic that he should have been one of the targets of such venomous attacks.[47] When placed alongside the work of David Ricardo, after all, who vaulted into the front rank of political economists in 1817, Malthus's *Essay* looks like an effort to make political economy accommodate, not throw out, virtues, good ness, and happiness. Ricardo could not understand Malthus's desire to retain both a theological vocabulary and a moral component;[48] Southey did not even see that he had done so. That both responses to Malthus were possible in the sec ond decade of the nineteenth century reflects the range of positions about what constituted useful knowledge that had become available by 1817—and just how controversial numerical representation had become.

POPULARIZING POLITICAL ECONOMY:
J. R. MCCULLOCH AND THE TAXONOMY OF
MODERN KNOWLEDGE

During the 1820s, proponents of political economy engaged in a heated debate about the nature of political economy itself. Generally known to historians as the Malthus-Ricardo debate, this dispute turned on two questions: Was na tional well-being primarily a matter of "wealth" or of "happiness"? And by what method should political economists generate knowledge?[49] This debate about the proper object and method of political economy was sparked by the publication in 1817 of David Ricardo's *Principles of Political Economy,* but it was fueled by the same issues that provoked Ricardo: in the wake of the Napoleonic Wars, rising poor rates made poor law reform seem increasingly pressing, the suspension of cash payments by the Bank of England raised new questions about credit, and increases in taxes and the national debt led politicians and philosophers to reassess the relation between national prosperity and security. Ricardo's response to these developments was to recast political economy as a mathematical science, whose exclusive concern was the mathematically deter mined behavior of wealth. By converting policy issues into mathematical for mulas, and by capitalizing on the impartiality and rigor associated with mathematics (more than counting), Ricardo sought to place the conclusions of political economy beyond dispute. Not incidentally, he also wanted to protect political economists from those charges of political interest and self-interest that the romantic critics had leveled against "Malthusians." Despite his critics and in defiance of Ricardo, by contrast, Malthus continued to insist that "the science of political economy bears a nearer resemblance to the science of morals and politics than to that of mathematics."[50] Unlike Ricardo, moreover, who wanted to calculate economic principles exclusively by mathematical deduction, Malthus continued to advocate a combination of induction and reasoning from religious principles as the best means of assessing national well-being.

As we will see in the concluding chapter, the debate about the ends and means of political economy echoed debates about method that erupted among British natural philosophers as well. While these debates are of considerable interest to my investigation of historical epistemology, it must also be admitted that to the vast majority of Britons, the distinctions that loomed so large for political economists and philosophers barely registered. Thanks largely to the hostile reviews by romantic critics like Southey and Coleridge, political economy presented a single face to most British readers: under the epithet "Malthusianism," political economy seemed to many simply an apology for the manufacturing system, a rationale for low wages, and an excuse to abandon the laboring poor to the mercies of an increasingly free but decidedly amoral market.

Beginning in 1818, a young Scotsman began to try to change the public image of political economy, not by taking sides in the Malthus-Ricardo debate, but by representing political economy as the most accessible and most valuable modern science. In order to make political economy over in the wake of Malthus's *Essay,* J. R. McCulloch had to do two things: he had to make political economy seem useful to an audience that included people who were not professional philosophers, and he had to restore the providentialism that Malthus's revisions seemed to have deprived it of. To make political economy seem useful, McCulloch first made the science more teachable by publishing editions of the texts he considered essential, by writing the discipline's history, and by supporting all kinds of educational venues through which the science could be disseminated. To restore political economy's providentialism, McCulloch repeated Stewart's account of the history of commercial society as the unfolding of a divine plan. But because McCulloch also wanted to make the conclusions that liberal political economists had reached seem like discoveries of natural laws, not impositions of some a priori theory, he also departed from Stewart by making this plan seem *natural*—that is, not necessarily divine. McCulloch's defense of political economy, then, both preserved the providentialism that Malthus seemed to have destroyed and liberated the science of wealth from all but the most attenuated vestiges of its original affiliation with moral philosophy. Not incidentally, his reworking of political economy also professionalized the science and, in so doing, offered one kind of solution to the philosophical problem of induction.

As we will see in a moment, McCulloch's position on the issues raised in the Malthus-Ricardo debate was somewhat contradictory: he explicitly wanted to make Ricardo's mathematical formulas "perfectly intelligible to the generality of readers";[51] but he also echoed Malthus in claiming that political economy was a "*science* of values" rather than simply the science of wealth.[52] To a certain extent, the contradictions we see in McCulloch's writing follow from his campaign to popularize political economy: beyond the kind of moral objections raised by the romantic critics, political economy aroused skepticism because Ricardo in particular made it seem so *difficult;* to make the science appear accessible, McCulloch inevitably *misrepresented* Ricardo's work, if only in the sense of dramatizing mathematical abstractions in concrete situations.[53] To a certain extent, however, the inconsistencies that surface in McCulloch's writing reveal how hard it was to adapt what had initially been a branch of moral philosophy to a study of abstractions that seemed amoral by definition. Even more than "commerce" or "the market," which still appeared to refer to transactions among people, who were (theoretically) moral agents, "the economy" looked like a self-contained system that might—and according to McCulloch did—simply run by itself, without the kind of agency that could make moral decisions or be held morally responsible.[54] Thus the contradictions we can identify point to the stresses that accompanied the reification of "the economy," a process that involved not only disciplining its science but also removing the very idea of an economic system from the ethical and theological rationalities to which fiscal matters had traditionally belonged.

McCulloch had been the chief economic writer for the *Edinburgh Review* for two years when he launched his first modest attempt to extend the appeal of political economy; in 1820 he offered a private course on political economy in Edinburgh, which attracted nine students (at a cost of £10 each). By 1824 he had considerably expanded his scope: the first Ricardo Lectures, which he delivered in London at the invitation of James Mill, among others, drew a large and appreciative audience that included members of Parliament, successful businessmen, and titled lords.[55] In 1825 McCulloch published the text of the Ricardo Lectures—which also constituted an advertisement for lectures he planned to give in the future—as *A Discourse on the Rise, Progress, Peculiar Objects, and Importance of Political Economy.* Later that year he published a considerably expanded version of his *Discourse* as *Principles of Political Economy,* a text that, by taking the same title Ricardo had used in 1817, established the convention of successive volumes of (differently authored) *Principles.*[56] In 1827 McCulloch's campaign to confer authority on political economy was considerably advanced by his appointment to the first chair of political economy at the newly formed London University. Not incidentally, and though this was not the first chair of political economy in Britain, McCulloch's assuming the much-desired chair also helped establish him as one of the leading spokesmen for the science.[57] Between 1828 and his death in 1864, he continued to try to make political economy accessible by collecting, editing, and publishing a series of texts that he hoped would encourage other people to teach the science. He published editions of Smith's *Wealth of Nations* in 1828, 1838, and 1863; and in the 1850s he financed the printing of several collections of "scarce and valuable tracts"—on commerce, money, paper currency and banking, the national debt, and other economic subjects.[58]

McCulloch told essentially the same story in all his lectures and publications on political economy. At its heart, this was a story of a discipline misunderstood and underappreciated. McCulloch claimed that political economy had been misunderstood because its practitioners disagreed among themselves, and the "differences . . . among the most eminent of its professors" had made it impossible for Britons to realize that the science had "risen" and "progressed."[59] To remedy this situation, he returned to what he considered the origins of political economy; but he also recommended that political economists get their professional house in order, for no revisionist history could redeem political economy if its practitioners were unwilling to agree about what they were trying to do.

In McCulloch's history, the first intimations of political economy appeared in the seventeenth century, in the writings of the British merchant apologists I examined in chapter 2. According to McCulloch, however, these apologists for international commerce launched the study of wealth on the wrong track, because they advocated "the dark, selfish, and shallow policy of monopoly" instead of "those sound and liberal doctrines, by which it has been shown, that the prosperity of states can never be promoted by restrictive regulations" (*Discourse,* 37). To McCulloch, in other words, what counted as true political economy— "the modern theory of commerce"—consisted of a particular content: the doctrines of free trade most systematically described by Smith in 1776. Before Smith, McCulloch continued, the rudiments of liberalism had appeared only in scattered form in the work of other Britons, including Josiah Child, William Petty, and Dudley North.

McCulloch attributed the failure of the first writers on commerce to discover its true (that is, liberal) principles to the methodological immaturity of the infant science.

> Instead of deducing their general conclusions from a comparison of particular facts, and a careful examination of the phenomena attending the operation of different principles, and of the same principles in different circumstances, the first cultivators of almost every branch of science have begun by framing their theories on a very narrow and insecure basis. Nor is it really in their power to go to work differently. Observations are scarcely ever made or particulars noted for their own sake. It is not until they begin to be sought after, as furnishing the only test by which to ascertain the truth or falsehood of some popular theory, that they are made in sufficient numbers, and with sufficient accuracy. It is, in the peculiar phraseology of this science, the *effectual demand* of the theorists that occasions the production of the facts or raw materials, which he is afterwards to work into a system. (*Discourse*, 20–21)

Considered as a contribution to the history of the modern fact, this passage is extremely revealing. Clearly, McCulloch is struggling here with the ambiguity I have identified in all the variants of the modern fact: on the one hand, he associates facts with both observations and "raw materials," as if facts really were the deracinated particulars that Bacon described; on the other hand, he claims that facts can never be separated from the theoretical agenda that leads one to observe or collect raw materials in the first place. By the end of the *Discourse*, McCulloch was to offer a professional solution to the dilemma this ambiguity occasioned. At the beginning of this account of the history of the discipline (and all his other accounts), by contrast, he treats the ambiguity inherent in the modern fact as the factor that propelled his discipline out of its immature stage—that is, as both the incentive and the explanation for political economy's "progress." To McCulloch, the ambiguity captured in the idea of the modern fact occasioned science, for only some systematic body of knowledge (like political economy) could enable people to make sense of particulars or even lead them to notice them in the first place.

If the ambiguity captured in the modern fact sparked the first human efforts to notice and make sense of what they saw, then the earliest attempts to collect raw data and organize these materials into a system simply followed (and reflected) principles inherent in "the original constitution of man and of the physical world." According to McCulloch, the human desire to know in this way—that is, both to collect observed particulars and to make theoretical systems from them—reproduces "the desire implanted in the breast of every individual of rising in the world and improving his condition" (*Discourse*, 10). In making this claim, he was building on the science of subjectivity that eighteenth-century moralists had forged. We know what human beings naturally want, McCulloch claims, because our knowledge reflects (embodies) these desires. Thus "human nature" becomes both the origin of knowledge and its outcome, and the "desire . . . of rising in the world" becomes both what the philosopher knows (believes) and what he seeks to communicate (describe).

Just as the human desires to *acquire* and *make* knowledge reproduce the human desires to "save and accumulate," so the sciences by which humans understand themselves and their world—the sciences of knowledge, of nature, and of wealth—all resemble each other in being both attentive to observed particulars and systematic in nature. For McCulloch, however, the science of wealth differs in one regard from philosophies both moral and natural: whereas the conclusions of philosophy "apply in *every* case," the conclusions of political economy "apply only in the *majority* of cases. The principles on which the production and accumulation of wealth depend are inherent in our nature, and exert a powerful, but not always the *same* degree of influence over the conduct of every individual; and the theorist must, therefore, satisfy himself with framing his general rules so as to explain their operation in the majority of instances, leaving it to the sagacity of the observer to modify them so as to suit individual cases" (*Discourse,* 10).

Thus the tension captured in both the modern fact and the problem of induction returns to political economy, as the tension between "general rules" that explain "the majority of instances" and "individual cases," which do not always conform to these general rules. The tension that anchored this formulation helped account for the discrepancy between particulars and generalizations, of course. In so doing, it constituted a theoretical advance over the kind of abstraction embraced by the conjectural historians, Smith, and Stewart. Whereas those abstractions ("the human mind," "the market system") invariably conformed to the principles the theorist "discovered," because the abstractions were effects of the theory, McCulloch's abstractions were aggregates. Like Malthus's "population," the aggregates McCulloch examined were composite subjects, whose members were not necessarily identical. "It is not required of the economist, that his theories should quadrate with the peculiar bias of the mind of a particular person," he declaims. "His conclusions are drawn from observing the principles which are found to determine the condition of mankind, as presented on the large scale of nations and empires. He has to deal with man in the aggregate—with states, and not with families—with the passions and propensities which actuate the great bulk of the human race, and not with those which are occasionally found to influence the conduct of a solitary individual" (*Discourse,* 11).

Of all the composite subjects that McCulloch mobilized, the most power-

ful might well be "the public." On the one hand, as "public interests" this concept drew the analyst's attention away from vested interests (including, presumably, his own). "The *public interests* ought always to form the exclusive object of
[the economist's] attention," McCulloch declares. "He is not to frame systems,
and devise schemes, for increasing the wealth and enjoyments of *particular
classes;* but to apply himself to discover the sources of *national wealth,* and *universal prosperity,* and the means by which they may be rendered most productive"
(*Discourse,* 12). On the other hand, as "public opinion," this composite provided
an instrument through which the political economist could influence the legislator, for as McCulloch described it, "public opinion" was the most powerful
agent of government. As an agent capable of giving "an impress to all the acts of
government" (*Discourse,* 84), "public opinion" retained the power that the early
eighteenth-century theorists of liberal governmentality accorded individual
subjects, but because "the public" was a composite, this concept required no
elaborated subjectivity. In fact, McCulloch did not develop a complex account
of the subjective dynamics by which "public interest" was mobilized or motivated. Instead, he described a program of education that, if institutionalized,
would have sidestepped both the question of how subjectivity worked and the
problem of governing self-interest. McCulloch's educational program would
have evaded these questions because it would have trained every citizen to recognize and act on interests that were simultaneously personal and national. To
make personal and national interests coincide, McCulloch represented "expressing [one's] opinion" as a citizen's highest duty, and he depicted this duty as
the effect of learning the "general and fundamental principles" of political
economy. "It is the duty of all who do not voluntarily choose to relinquish the
noblest and most valuable privilege enjoyed by the citizens of a free state—that
of expressing their opinion on the conduct of public affairs—to qualify themselves for its proper exercise" (*Discourse,* 80).

McCulloch's picture of a (nonpsychological) "public" learning political
economy in order to govern themselves assigned enormous importance to the
political economist, of course. Because political economy was a difficult science, because "there is no short road—no *via regia*" to understanding it (*Discourse,* 80), the public required the assistance of trained political economists.
The education that McCulloch outlined combined activities that could have
been undertaken on one's own (such as reading the texts he edited), but to grasp
the full implications of political economy, McCulloch strongly recommended
expert tutelage—preferably in the kind of public lectures he had delivered and
hoped to continue to give (*Discourse,* 96–97). To ensure that such instruction
was available, McCulloch lobbied his readers to take an active role again: he encouraged them to petition the government to fund professorships of political

economy in Britain, as governments in Naples, Milan, and Russia had done. Marshaling a blatantly nationalistic argument, McCulloch declared that "England is the native country of Political Economy; but she has not treated it with a kind and fostering hand" (*Discourse*, 90).

Almost all the facets of McCulloch's campaign to redeem the reputation of political economy can be assimilated to this educational agenda. By collecting and publishing hitherto unavailable tracts, by editing Smith's *Wealth of Nations,* by representing political economy as "the one thing needful" (the phrase is Thomas Carlyle's), and by celebrating a public that both would benefit from political economy and could save this science from neglect, McCulloch sought to use education, as a mechanism of liberal governmentality, to secure the reputation of his discipline. The kind of education he envisioned, moreover, because it relied on a few well-chosen (canonical) texts, would theoretically make the subjects of political economy—the individuals who made up the public— more like each other than different. Thus, by the alchemy of education, McCulloch sought to solve the problem of induction again, this time by making those individuals whose idiosyncratic subjectivities might undermine the general knowledge that political economists produced into instances of the same kind of subject—Smith's *homo economicus* enlightened by self-knowledge, thus empowered by self-control. McCulloch held that making members of the reading public more like each other than different would even counteract Malthus's principle of population, for once that subset of "the public" called "the body of the people" knew that (sexual) self-control served "its" (collective, if not individual) interests, this composite subject would voluntarily curtail (sexual) productivity in favor of work. "Make the body of the people once fully aware of the circumstances which really determine their condition," McCulloch confidently declared, "and you may be assured that an immense majority will endeavour to turn that knowledge to good account. . . .The harvest of sound instruction, though late, will, in the end, be most luxuriant" (*Discourse*, 87).

Even more than Smith and Malthus before him, McCulloch counted on public education to supplement the freedom he attributed to the market.[60] A nation like Britain could free economic transactions from restrictive regulations and limit government interference, that is, because educated individuals would recognize that they could fulfill their common desire ("to accumulate and save") by obeying the laws of commerce, which were (theoretically) written in their nature as well as in the natural world. As McCulloch described it, moreover, the laws that governed the system of commerce (which systematic political economy described and supported) embodied the design that moral philosophers had learned to see and celebrate. In a passage notable for its tendency both to preserve the providentialism whose history we have been tracing and to free this providentialism from God, he summoned "Providence" only to replace it with a decidedly more secular agent: commerce.

> Providence, by giving different soils, climates, and natural productions, to different countries, has evidently provided for their mutual intercourse and civilization. By permitting the people of each to employ their capital and labour in those departments in which their geographical situation, the physical capacities of their soil, their national character and habits fit them to excel, foreign commerce has a wonderful effect in multiplying the productions of art and industry. When it is not subjected to restrictions, each people naturally devote themselves to such employments as are most beneficial to each. This pursuit of individual advantage is admirably connected to the good of the whole. By stimulating industry, by rewarding ingenuity, and by using most efficaciously the particular powers bestowed by nature, commerce distributes labour most effectively and economically; while, by increasing the general mass of necessary and useful products, it diffuses general opulence, and binds together the universal society of nations by the common and powerful ties of mutual interest and reciprocal obligation. . . .Commerce has given us new tastes and new appetites, and it has also given us the means and the desire of gratifying them. (*Discourse,* 104–5)

In this passage agency originally resides, as it did in Stewart's work, with "Providence" ("Providence . . . has evidently provided"). Almost immediately, however, and by means of a shift from a geographical description to a prescription for policy ("when it is not subjected to restrictions"), McCulloch shifts the agency to "commerce" ("commerce distributes labour"). With this shift, commerce becomes not only the agent that distributes the bounty of Providence but also an independent creative force, which "has given us new tastes" and thus directs humans to carry out what now seems only distantly a providential plan ("by using most efficaciously the particular powers bestowed by nature, commerce distributes labour"). By the end of this passage, commerce has become the moral agent that providential design was to Stewart; and what would have been an expression of faith for the eighteenth-century historians and prophetic projection for Stewart now looks very much like a straightforward description of a natural process.

Even though McCulloch represented political economy as the science of such lawful "progress," of course, many of his contemporaries were not convinced by his optimism. Just as Malthus was vilified for drawing the starvation of the poor into a providential plan, so McCulloch was ridiculed for trying to force-feed the principles of political economy (McCulloch was the original of Mr. M'chokumchild in Dickens's *Hard Times*); and he was denounced for abandoning the working poor to their own devices ("the lower classes are in a very great degree the arbiters of their own fortune"; *Discourse,* 62). Despite McCulloch's multifaceted campaign to redeem political economy, in fact, the discipline seemed to attract more criticism the louder its spokesmen proclaimed its promise. It was in hopes of foreclosing this continuing criticism, which had been so intense in 1824 that the sponsors of the Ricardo Lectures had been afraid to advertise for subscriptions, that McCulloch offered the last of his solutions to the problem of induction. As an institutional—and ideally a professional—solution, this measure was clearly intended to remove the internal business of political economists from the (critical) public gaze.

What I am interpreting as an institutional or professional solution to the problem of induction consisted of a taxonomy of knowledge, which separated the collection of data from the production of general—that is, theoretical—knowledge. This is also the point at which McCulloch's writing returns us to the subject of numbers and numerical representation, for as he imagined it, the data that statisticians would collect would sometimes (though not always) take the form of numbers. As McCulloch described it, the modern taxonomy of knowledge was to consist of three branches: political science, which would consider the principles of government; political economy, which would examine the principles of wealth; and statistics, which would "describe the condition of a particular country at a particular period" (*Discourse,* 78, 79). As even this brief summary makes clear, McCulloch wanted to distinguish between the descriptive and theoretical functions of what Smith, Stewart, and Malthus had represented as a single endeavor. Whereas the distinction he drew between the political scientist and the political economist helped free the latter from volatile debates about the best form of government, his distinction between the political economist and the statistician was intended to remove the former from controversy altogether. By distinguishing between the statistician and the political economist, he wanted to curtail the criticism that Malthus had aroused by making the description of observed particulars seem neutral, as Bacon had said it was, and making the production of general knowledge seem like the more "expansive induction" Stewart had referred to.

> The object of the statistician is to describe the condition of a particular country at a particular period; while the object of the political economist is to discover the causes which have brought it into that condition, and the means by which its wealth and riches may be indefinitely increased. He is to the statistician what the physical astronomer is to the mere observer. He takes the facts furnished by the researches of the statistician, and after comparing them with those furnished by historians and travellers, he applies himself to discover their relation. By a patient induction—by carefully observing the circumstances attending the operation of particular principles, he discovers the effects of which they are really productive,

and how far they are liable to be modified by the operations of other principles. It is thus that the relation between rent and profit—between profit and wages, and the various general laws which regulate and connect the apparently conflicting, but really harmonious interests of every different order of society, have been discovered, and established with all the certainty of demonstrative evidence. (*Discourse,* 79–80)

In 1825, as we have seen, political economists were not professionalized, they did not even agree on the objects or the method of their science, and their hold on an institutional position—either within the universities or in relation to government—was anything but secure. To recommend such an institutional solution to both the philosophical problem of induction and the controversies sparked by political economy was thus to engage in a certain amount of wishful thinking. Indeed, one might argue that in dividing the functions that Smith, Stewart, and Malthus had assimilated into "political economy" McCulloch was actually recommending *to political economists* that they discipline themselves, that they form themselves into something resembling the old professions, with their systems of credentialing and self-government. Certainly, an example of avoiding controversy by voluntarily curtailing the production of theory was available to McCulloch in 1825, for early in the century the Geological Society of London had declared that "facts" could—and for the time being should—be collected in the absence of theory.[61] The professional benefits that geologists had reaped from their self-imposed moratorium on theory might well have inspired him to hope that political economists also could divorce data collection from theory and gain professional strength.

Of course McCulloch did not want to suspend the production of general—that is, political economic—knowledge altogether; nor did he want to divorce facts from theory. Instead, he simply wanted to separate the collection of the former from the production of the latter, so that political economists, in conjunction with statisticians, could determine what would count as the principles of political economy and how to evaluate the method by which these principles were reached. I will return in the next chapter to the form that McCulloch imagined statisticians' data would take. For now I will simply make two points: first, by establishing this taxonomy of knowledge, McCulloch made it clear that the production of general knowledge should proceed independently of government; and second, by making political economy depend—no matter how imprecisely—on the collection of data of every kind, he made political economy dependent on government, at least insofar as political economists needed the government to sponsor the enormous volume of data their science required.

Just as Adam Smith's liberal model of governmentality finally depended on support from the state government, then (to ensure national security, mandate education, and supply public utilities), so McCulloch's program for professionalizing and popularizing political economy required government assistance to collect the statistics from which general knowledge could be made. As we have already seen in the list of books that Malthus consulted, private individuals and a few voluntary societies had generated some numerical data by the end of the first decade of the nineteenth century. But in Britain, at least, the only large-scale project that had been launched to collect such data was the census. Between 1812 and 1815 a parliamentary select committee did sponsor a survey on pauperism, and the House select committee that reported in 1817 did solicit information; but as J. R. Poynter has demonstrated, even select committees tended to write their reports before those haphazard empirical investigations that were conducted began to yield (what we would consider) usable results. Only with the 1832–34 royal commission, which was appointed to investigate the poor law, were numerical data collected on a large enough scale to be useful for formulating—or at least defending—government policy.[62]

In order to recover all the stages by which the British government was finally persuaded that numerical data would aid the generation of policy, we would need, at the very least, more scholarly work on John Rickman, who lobbied tirelessly to promote the utility of numbers.[63] The story of the government's embrace of numerical representation lies beyond the scope of this book, however, both because others have already written copiously about administrative reform after 1830 and because, by the time numbers came to seem attractive to the British government—once they came to seem like instruments that could generate impartial, rigorous, value-free knowledge—many of the epistemological questions that interest me had been set aside. Before ceding this story to the historians of administrative and educational reform, however, I want to examine in a bit more detail the characteristic form that modern facts took in nineteenth-century Britain: statistics.

7

Figures of Arithmetic, Figures of Speech: The Problem of Induction in the 1830s

Despite McCulloch's attempt to use the distinction between statistics and political economy to quiet the criticism directed at the science of wealth (and, not incidentally, to solve the problem of induction), both political and philosophical controversy continued to swirl around British political economy for much of the first half of the nineteenth century. More to the point, perhaps, by 1840 statistics too had become the subject of controversy, precisely because even its (theoretically) neutral facts were viewed as sites where theory could— or should—surface. Even after McCulloch tried to divorce fact from theories about what constituted general knowledge, then, the modern fact, at least as it was embodied in statistics, continued to display its provocative double nature.

Statistics was not the only practice that prompted discussions about the relation between observed particulars and general knowledge. Beginning in 1830, the leading practitioners of British natural philosophy (christened "scientists" in 1840) raised the problem of method in the physical sciences. Although the complexities of this debate go beyond the subject of this book, I briefly consider John Herschel's contribution to the debate about scientific method, because Herschel focused on number as the paradigm of the problems raised by the effort to produce general knowledge from observed particulars. I also give some attention to John Stuart Mill's effort to elaborate the method of what he called the "social sciences," both because his meditations on the methodological challenge posed by abstractions like society devised a new relation between induction and deduction and because his embrace of statistics reflects the excitement—and the confusion—this new practice occasioned in the 1830s. Finally, in what must be read as a coda to this book and a provocation for another, more capacious study, I briefly consider how the problem of induction figured in the work of three romantic poets: William Wordsworth, Percy Shelley, and John Keats. Although the poetic statements I consider in the few paragraphs I devote to these writers predate the scientific and philosophical statements on method that I discuss in much greater detail, I have included this brief treatment of poetry because I want to make it clear that, although contemporaries increasingly distinguished the knowledge that poetry could produce from that generated by all kinds of science, the epistemological problem of induction was a concern for poets too. Indeed, even though some champions of figures of speech have credited poetry with providing a unique solution to the problem of induction, the very fact that they have wanted to do so implies that this epistemological challenge continues to unite poetry and the sciences rather than forming the ground of their difference.

STATISTICS IN THE 1830s

When interpreting the 1830s debate about statistics, it is important to remember that what most Britons understood as "statistics" was not the rigorous mathematical practice deployed by late twentieth-century statisticians. Lambert Adolphe Quételet, who is generally credited with applying mathematical principles to what he called "social physics," did have some influence over two of the earliest statistical organizations, but most Britons did not understand the relation Quételet drew between numerical data and mathematical formulas; nor, judging from the confusion many displayed about how the law of large numbers affected free will, did they grasp the implications of his method.[1] In Britain, not even the meaning of "statistics" had been stabilized in 1830. Adapted from the German *Statistik,* which was first used as a noun by Gottfried Achenwall in 1749, the word was initially imported to Britain by John Sinclair, whose *Statistical Account of Scotland* (1791–99) compiled an enormous body of data from questionnaires administered by clergymen in every Scottish parish. Sinclair acknowledged the connection between statistics and state government, which was implied by the German usage, but in keeping with the influence that eighteenth-century moral philosophy exercised over British efforts to understand society, he insisted that he wanted to use this method to determine the "quantum of happiness" revealed by such factors as the availability of education rather than simply to produce accounts useful "for the purposes of taxation and of war."[2]

By 1830 the word "statistics" seems to have carried connotations of both substance (statistics recorded the kind of information about national resources that would be useful to the state) and form (statistical information was sometimes, though not inevitably, conveyed in numbers and tables). As Theodore Porter has noted, it is extremely difficult to determine exactly when "statistics" began to refer to a combination of substance and form—when, that is, most people would have begun to lay the stress equally on content and on numerical representation, as Bisset Hawkins did when he defined medical statistics in

1829 as "the application of numbers to illustrate the natural history of man in health and disease."[3] Whatever the date when this usage became common, the understanding of statistics as a practice that necessarily involved numbers was certainly enhanced by two campaigns launched almost simultaneously in 1833. On the one hand, a group of Cambridge intellectuals wanted to save statistics from being equated with mathematical deduction, as Ricardo, Nassau Senior, and (to a lesser extent) McCulloch were trying to do. On the other hand, the British Association for the Advancement of Science (BAAS) wanted to curtail the political controversy that statistics aroused by limiting its purview to the collection of numerical data. These two attempts to define statistics as a numbers-based but nondeductive practice dovetailed in the creation of two institutions: the statistical section of the BAAS (Section F) and the Statistical Society of London. Although these institutions neither insulated statistics from controversy nor settled the debates about method this practice continued to provoke, they did help make numbers seem essential to statistics' description of the "great variety of objects" with which it was typically associated.[4]

In what follows, I draw heavily on Lawrence Goldman's analysis of the immediate origins of the "statistical movement" and on Jack Morrell and Arnold Thackray's treatment of the BAAS. Goldman's work is particularly valuable because he shows us how the Cambridge intellectuals' repudiation of Ricardo's mathematical deduction drew on both natural scientific procedures and the new interest in numerical data aroused (in part) by Malthus and Quételet. Morrell and Thackray, by contrast, are more interested in debates internal to the BAAS, but by demonstrating what was at stake for this fledgling organization in limiting the purview of statistics, they remind us just how controversial statistics was in the 1830s. By turning from the debates within the BAAS to some of the reactions to statistics published after the founding of the earliest statistical organizations, I argue that beyond telling us something about how the statistical movement was inaugurated, the controversy over statistics illuminates how the problem of induction surfaced even in the practice McCulloch tried to place beyond its reach.

Following the lead of Susan Cannon, Lawrence Goldman attributes the origin of the statistical movement in Britain to five men: Richard Jones, professor of political economy at King's College, London; Charles Babbage, champion of the manufacturing system and inventor of the first calculating machine; Adolphe Quételet, Belgian mathematician and astronomer; William Whewell, fellow of Trinity College and a leading theorist of natural science; and Thomas Malthus, who had held the chair of history and political economy at the East India College, Haileybury, since 1805.[5] According to Goldman, these individuals were united in their conviction that Ricardo's reliance on

mathematical deduction had encouraged practitioners of political economy to
rush too precipitously from observed particulars to oversimplified versions of
general laws. As Whewell explained, political economists like Ricardo

> have begun indeed with some inference of facts; but, instead of working their way
> cautiously and patiently from there to the narrow principles which immediately
> inclose a limited experience, and of advancing to wider generalities of more sci-
> entific simplicity only as they become masters of more such intermediate truths—
> instead of this, the appointed aim of true and permanent science—they have
> begun endeavouring to spring at once from the most limited and broken observa-
> tions to the most general axioms.[6]

Beyond the rush to theory, Goldman continues, the Cambridge intellectuals
charged Ricardo with unduly limiting the reach of political economy as a discipline. By subordinating actual observations to "a few very general propositions," the Cambridge group complained, Ricardo and his supporters had
stripped political economy of the very capacity that Smith and Malthus had
emphasized: the ability to relate questions of wealth to the social issues of the
day.[7]

Instead of mathematical induction, the Cambridge group wanted to use a
variant of empiricism that drew on both Bacon and, more generally, the Continental method of analysis associated with Alexander von Humboldt. When
Richard Jones committed himself to the observation of discrete particulars, by
this interpretation, he was invoking Humboldt's effort to use "the accurate,
measured study of widespread but interconnected real phenomena in order to
find a definite law and a dynamical cause" of events.[8] In theory this method,
which was interpreted as a refinement of naive Baconianism, would yield general knowledge, but it would do so in such a way as to qualify generalizations by
the kind of cultural specificity that Samuel Johnson had begun to register in the
1770s. The variant of induction the Cambridge group endorsed, then, both required international cooperation and promised to yield culturally specific results, which would considerably complicate the effort to find "principles that
are truly comprehensive." At the very least, this variant of induction would locate comprehensive principles at a different level of abstraction from the observed particulars than had eighteenth-century moral philosophy.[9]

According to Goldman, the statistical section of the BAAS and the Statistical Society of London originally embodied this methodological agenda, because Babbage and Jones in particular played leading roles in the founding of
both organizations.[10] To distinguish between the method of statistics and both
mathematical deduction and politically motivated theorizing, Jones and Babbage encouraged the Statistical Society of London to disavow "speculation"

and "opinion." Partly on the advice of these men, the Statistical Society adopted as its emblem a sheaf of wheat ringed with the phrase *aliis exterendum* ("to be threshed out by others"), and the founders announced in their statement of purpose that statistics "does not discuss causes, nor reason upon probable effects; it seeks only to collect, arrange, and compare, that class of facts which alone can form the basis of correct conclusions with respect to social and political government."[11]

This tactic did not foreclose criticism, however. Nor did it guarantee that the members of the two societies would be willing—or able—to live up to their motto. No matter how determined its founders were to avoid the simplifications they associated with Ricardian deduction, after all, Jones and his colleagues also wanted statistics to use the mathematical principles Quételet had devised; and no matter how intent they were to avoid "even the appearance of party bias" ("Introduction," 8), the Cambridge group wanted to preserve the engagement with social issues that had been so central to Smith's political economy. By attempting to position statistics between mathematical deduction and partisan theory without succumbing to the worst excesses of either *or* losing the capacity to use mathematics to generate socially useful knowledge, the founders of the two societies essentially wrote their new science into a corner. They promised what it could not deliver and left it dangerously open to all the pitfalls the Cambridge men wanted to avoid. Before the end of the decade, as Goldman points out, the Statistical Society of London had begun to admit that it had generated hardly any results, and individuals more committed to social reform (like Babbage) had pushed the London Society in the direction of partisan politics already taken by the Manchester Statistical Society.[12]

However contemporaries judged the accomplishments of the Statistical Society of London, its self-professed agenda is relevant to my argument because this agenda did help equate statistics with numerical representation. "The Statist commonly prefers to employ figures and tabular exhibitions," the founders explained in the first issue of the Society's journal, "because facts, particularly when they exist in large numbers, are most briefly and clearly stated in such forms, and because he is not satisfied with giving deductions, which admit of question, but supplies the material which each individual may himself examine and compare" ("Introduction," 3). As Jack Morrell and Arnold Thackray have argued, this equation was also forged by arguments within the BAAS, for when the Cambridge group tried to introduce a statistical section in 1833, Adam Sedgwick agreed to allow the unorthodox maneuver only if the section members voluntarily limited statistics to the collection of numerical information.[13] Initially the leaders of the BAAS, who had embraced mathematics as the heart of an impartial, disinterested scientific method, had not wanted a statistical section, for fear of "open[ing] a door of communication with the dreary wild of politics," as Sedgwick phrased it. Babbage countered by invoking Quételet, who was present at the 1833 meeting, and whose mathematical work was unassailable. Sedgwick's response was to admit statistics to the Association, but to link its legitimacy to its numerical form. In Sedgwick's definitions of "science" and "statistics," we see how difficult it was in 1833 to promote mathematics and "abstractions" without straying into the dangerous areas of deduction and theory.

> By science, then, *I understand the consideration of all objects, whether of pure or mixed nature, capable of being reduced to measurement and calculation.* All things comprehended under the categories of space, time, and number belong to our investigations, and *all phenomena capable of being brought under the semblance of law are legitimate objects of our inquiry.* . . .
>
> Can then statistical inquiries be made compatible with our subjects, and taken into the bosom of our society? I think they unquestionably may, *so far as they have to do with matters of fact, with mere abstractions, and with numerical results.* Considered in this light they give what may be called the raw material for political economy and political philosophy; and by their help the lasting foundations of these sciences may be perhaps ultimately laid.[14]

In his concession to Babbage, Sedgwick was obviously reproducing the taxonomy of knowledge that McCulloch had introduced in 1825. Because Sedgwick essentially supported the Cambridge group on the questions of the method and the aims of political economy, however, it was imperative for the BAAS, as for the London Statistical Society, *not* merely to stress the difference between statistics and political economy, as McCulloch had done. To preserve the salient distinction—between collecting raw material and producing general knowledge—without simply repeating McCulloch, the founders of the two organizations invoked another distinction, which stood in for the disciplinary one that McCulloch had made. The Cambridge group and the BAAS emphasized statistics' numerical form, so that they could oppose statistics not to political economy but to a mode of representation whose reputation had already been impugned: rhetoric. "It is indeed truly said that, the spirit of the present age has an evident tendency to confront the figures of speech with the figures of arithmetic," the spokesmen for the Statistical Society of London proclaimed in its Fourth Annual Report; "it being impossible not to observe a growing distrust of mere hypothetical theory and *a priori* assumptions, and the appearance of a general conviction that, in the business of social science, principles are valid for application only inasmuch as they are legitimate inductions

from facts, accurately observed and methodically classified" (quoted in "Introduction," 8).

In such statements, the founders of the Statistical Society of London and
Section F of the BAAS not only elevated the distinction between numerical
representation and rhetoric over the distinction McCulloch drew between statistics and political economy, they also equated the devalued term of the opposition ("figures of speech") with deduction ("hypothetical theory and *a priori*
assumptions"). By the same token, of course, they also equated the valued term
of the opposition—numerical representation or "figures of arithmetic"—with
both induction *and* mathematics, for statistics' affiliation with mathematics was
what got this practice admitted to the BAAS in the first place. Thus the lability
of "numbers," which allowed the eighteenth century philosophers to link
mathematical deduction to counting, Christian Platonism, and poetry, allowed
the defenders of an anti-Ricardian variant of political economy to link statistics with induction and mathematics and to oppose it to deduction and
rhetoric.

As the debate about statistics spread beyond the contest between Ricardo
and his opponents, the meaning of this critical opposition was further elaborated. In 1835, for example, William Cooke Taylor endorsed statistics in the
pages of the *Foreign Quarterly Review* by citing a frustrated attempt by manufacturers to use statistics to defend themselves against their operatives' complaints.
"The manufacturers answered the charges made against them by an appeal to
incontrovertible facts, the tables of mortality, the records of hospitals and po-
lice-offices, the registers of parishes and courts of justice," Taylor explained;
"but there are still people in the world, who prefer the figures of speech to the
figures of arithmetic, and the rules of Longinus to those of Cocker. Pathetic
tales, more than sufficient to supply a whole generation of novelists, prevailed
over a dull, dry parade of stupid figures, and a Committee of the House of
Commons was appointed to examine the state of our manufacturing population" ("Objects," 109). Here the devalued term of the opposition is linked to
pathos and to fiction; and though Taylor admits that the valued term of the opposition is "dull" and "dry," he associates it with "incontrovertible facts." By
means of such associations, "dull" and "dry" numbers continued to accumulate
cultural value, precisely because their dullness seemed to be a guarantee against
the undue embellishment associated with fiction, hyperbole, and rhetoric.

By emphasizing the opposition between figures of arithmetic and figures
of speech, champions of statistics—*no matter where they stood in the all-important
methodological debate*—were able simply to set aside the problem of induction.
By stressing the incontrovertible nature of statistical "facts," that is, *by way of con-*

trast to the excesses and deceits associated with fiction and rhetoric, apologists for statistics were able to downplay the methodological problem of moving from whatever numbers were collected to general principles. Indeed, this shift of emphasis worked even when the apologist complained that Britain currently lacked sufficiently copious and reliable statistics to provide "incontrovertible facts." This was McCulloch's point in his 1835 *Edinburgh Review* article. In "The State and Defects of British Statistics," he conceded the Cambridge group's argument about the deductive rush to conclusions, but he laid the blame for this error not on the deductive method, but on the scarcity of reliable "facts." If political economists had such facts, McCulloch insisted, they would use them. Making this point so strenuously, of course, set aside the real methodological dilemma: If such numbers were available, how would the political economist produce general principles from them?

> It is frequently objected to our political philosophers, that they are too much disposed to deal in hypothesis, and that they too often leave facts out of view in their reasonings. But what can they do else? When there are either no facts, or few, except such as are false or misleading, if they are to reason at all, they must reason principally on hypothesis. If we had possessed circumstantial, and at the same time really accurate accounts of the various changes, however minute, in the wages, habits, accommodations, and condition of the population since the peace of Paris in 1763, we should now have been able to try principles and doctrines by the test of experience; and to appreciate, with considerable accuracy, the influence of particular systems and measures. But we have no such information. ("State," 176–77)

When apologists for statistics did try to explain how numerical data about observed particulars could ground general principles, typically by reference to Quételet, they became ensnared in a set of problems that were more troubling to most readers of the quarterlies than was the philosophical problem of induction. Thus, for example, when Taylor tried to use Quételet's discussion of probability to describe how numbers generated a more reliable version of "the general laws of human action" than introspection did, he found himself engaged in a discussion of the ethics of description. Even if numbers could reveal the probable incidence of some crime, Taylor worried, was it ethically responsible to do so—especially when even dry and dull figures might incite imitation and especially when the readers might be impressionable young boys?[15] By the same token, when Herman Merivale tried to use the law of large numbers to explain how one might determine "the comparative amount of morality" in various places, he inadvertently raised questions about free will that would have disturbed almost any orthodox Anglican.

By examining the particulars of a great many cases, we arrive at conclusions sufficiently accurate to influence our conduct, and are enabled to subject what is
roughly called accident, or destiny, to general rules of calculation. The life of one
man is liable to a thousand contingencies which mock our powers of divination.
Compare a thousand more lives similarly circumstanced, and the influence of contingencies seems to disappear before that of general laws. The case is precisely the
same with those effects of which the proximate cause is the free will of man. Nothing at first sight seems more arbitrary or uncertain than the course which any one
man will pursue, where circumstances, so far as they are known to us, do not seem
to act with any compulsory force on his judgment. Take ten—one hundred—or
one thousand men, whose choice is made under similar circumstances; and the
greater the number of individuals compared, the more does the slightest pressure
of external influence—the mere balance of motives—seem to amount to an irre-
sistible force, effacing all varieties of human choice or caprice. The results of an individual will seem to disappear, it has well been said, before the mean results of
innumerable wills: in other words, under the weight of the vast machinery of
moral causes; and the differences of temper and disposition sink into mere modifi-
cations of general laws, subject to calculation equally with those laws themselves.[16]

In 1844 Robert Chambers adopted an even more dismissive approach to
free will to defend his protoevolutionary account of human "progress." The
heated controversy raised by his *Vestiges of the Natural History of Creation* illustrates just how scandalous many middle-class Victorians (including members of
the BAAS) considered such arguments to be; and the compatibility of this use
of statistics with an evolutionary narrative foreshadows its reappearance in
Darwin's work.[17] Even before 1840, such statistical devaluations of the issues
that were more typically considered ethical, or even religious, had come under
fire, most famously perhaps by Charles Dickens, who published a hilarious par-
ody of the BAAS's statistical section in 1837, and by Thomas Carlyle, whose
treatment of statistics in "Chartism" appeared in 1839.[18]

For the purposes of my argument, the most telling criticism launched
against statistics in the 1830s appeared in an essay published in 1838 in the
London and Westminster Review. In his review of the *Transactions of the Statistical
Society of London*, G. Robertson, the outspoken subeditor for the radical periodical, specifically targeted the Society's repudiation of opinion. In so doing, he
simultaneously resurrected the problem of induction and pointed to the ambiguity written into the statistical fact. The heart of Robertson's charge was that
"facts" could not be distinguished from "theories" and, equally to the point,

that anyone who tried to make this distinction (as both McCulloch and the Cambridge group had done) deprived statistics of any epistemological power it might wield. "Theories themselves are not only facts but the kind of facts about the truth of which most care is taken, and which naturally therefore are oftener true or have more truth in them than details, and particulars[,] about the accuracy of which less pains are taken," Robertson asserted. Because theory is inseparable from facts and vice versa, it is crucial to use theory—or opinion—to guide the collection of facts, as the Society expressly claimed its members were not to do. "Opinion is what is most wanted where truth is the object, it is the parent and precursor of truth . . . the exclusion of opinions is the exclusion of the only guides which can conduct . . . researches to any useful end." Robertson wrapped up his attack on the claim that statistics could (or should) exclude opinion by identifying the word *facts* as the site of ambiguity: "There is an ambiguity in the word facts which enables the council to pass off a most mischievous fallacy: it either means evidences or it means anything which exists. The fact, the thing as it is without any relation to anything else, is a matter of no importance or concern whatever: its relation to what it evinces, the fact viewed as evidence, is alone important."[19]

To view the fact as evidence, of course, mandated that one consider the relation between particular observations and the theories by whose light these observations were made. Robertson charged that the Statistical Society of London not only failed to theorize this critical methodological step; it also tied its members' hands in the effort to avoid having to face it. As a consequence, he observes, in the four years of its existence, the Society had produced only one slim volume of numbers plus a spate of excuses for its lack of productivity. What the Society had not been willing to admit, Robertson concluded, is that statistics could never attain the epistemological status its proponents claimed for it, because it was not a science or even a kind of knowledge, but simply a mode of representation. "Statistics is not a science, and cannot be one. . . . Statistics is not even a department of human knowledge; it is merely a form of knowledge—a mode of arranging and stating facts which belong to various sciences" ("Exclusion," 37).

Robertson's philosophically devastating assault on statistics did not completely undermine the credibility of this mode of representation, of course, nor did it seriously impede the activities of the statistical societies that were founded in Britain in the 1830s.[20] For the most part, despite the controversy it provoked, statistics proved too useful—both to voluntary organizations dedicated to specific reforms and, increasingly, to the British government—for such criticisms to make much of a dent. Beginning with the Poor Law Commission in 1832, in fact, the British government increasingly used the argument that statistics were necessary to avoid "legislating in the dark" to defend its own growth—thereby displacing the problem of induction again, this time by a controversy about whether—or how—central government should grow.[21] By 1834, when the New Poor Law was passed, the machinery of government in Britain was indissolubly tied to the collection of numerical information, even though the methodological problems that persisted in the statistical variant of the modern fact had yet to be solved.

To all intents and purposes these problems were never solved, because from a philosophical perspective they were unsolvable. As long as one assigned the phenomena of nature—or, even more questionably, an abstraction like the economy or society—the kind of prominence that Bacon had done, it was impossible to devise any method *except a mathematical one for moving* from observed particulars to general principles. As we will see, the kind of mathematical models that Quételet adumbrated in concepts like the "average man" and the law of large numbers heralded the advent of an entirely new epistemological paradigm, which now dominates the late twentieth-century world. Before turning from the modern fact to its postmodern descendant, however, it is useful to pause over two additional Victorian attempts to address the problem of induction: John Herschel's explanation of the limits of scientific method in his *Preliminary Discourse on the Study of Natural Philosophy* (1830) and John Stuart Mill's effort to adapt this method to the social sciences in book 6 of his *System of Logic* (1843).

JOHN HERSCHEL AND JOHN STUART MILL: INDUCTION, DEDUCTION, AND THE LIMITS OF SCIENTIFIC METHOD

In terms of the narrative I have been developing, the most telling aspect of Herschel's influential *Discourse* is his treatment of numbers. To appreciate this treatment, however, we first have to grasp what Hershel was trying to do in the *Discourse* as a whole. Herschel's primary objective was to explain the method of natural philosophical investigation in such a way as to do justice to its complexity. Despite his nods to Bacon (he declared experience "the great, and indeed [the] only ultimate source of our knowledge of nature and its laws"),[22] Herschel insisted that it was no longer sufficient simply to celebrate induction. Instead he wanted to demonstrate that induction was actually dependent on deduction, just as a responsible application of deduction required induction. By representing induction and deduction as distinct yet interdependent, he separated the methodological strands that many eighteenth-century philosophers wove together in the name of Newton. By suggesting that induction and deduction are stages in a single method, he laid the groundwork for specifying the steps by which one moved back and forth between observed particulars and theoretical generalizations to produce ever more inclusive versions of knowledge.[23]

Two steps were critical in Herschel's account of scientific methodology: classification and verification by repetition. Classification was crucial because the infinite differences that distinguished particulars made it necessary for the naturalist to group them in some way. For Herschel, as for Dugald Stewart, classification was a cumulative and discriminating activity, which generated knowledge by moving toward ever higher levels of abstraction. Assigning observed particulars to categories thus created "general facts," and these general facts then became the "objects of another and higher species of classification, and are themselves included in laws which, as they dispose of groups, not individuals, have a far superior degree of generality, till at length, by continuing the process, we arrive at *axioms* of the highest degree of generality of which science is capable" (*Preliminary Discourse*, 102). As essential as he considered classification, however, Herschel also reminded readers that this was basically an act of interpretation, and that it was too often motivated by the "rage for arrangement" that scientists in particular feel; too often we forget, he cautioned, that "in nature, one and the same object makes a part of an infinite number of different systems" (*Preliminary Discourse*, 139).

Verification by repetition was critical to the method Herschel described because repetition enabled the scientist to check his emerging theories against new data. Just as classification was haunted by the "rage for arrangement," however, so the imperative to repeat was troubled by what Herschel considered the near impossibility of exact replication. Beyond the material impediments to replication, Herschel continued, even experiments devised simply to confirm a preliminary finding could generate new data, which then had to be factored into the conclusion the scientist was trying to confirm.

As Herschel's acknowledgment of these dangers illustrates, the *Preliminary Discourse* is as noteworthy for the limitations he placed on scientific knowledge as for the claims he made on science's behalf. Indeed, a large measure of his confidence about the explanatory power of science emanates from the limits he eagerly placed on the claims one can make on behalf of method. Both features of Herschel's representation of science appear in his treatment of numerical data. For him, precise numerical data constituted the optimal form of empirical data and signaled the limit of observation, for even though he considered it imperative to attain numerical precision whenever possible, he knew that the eye could not register exact measures: indeed, he admitted, "none of our senses . . . gives us direct information for the exact comparison of quantity" (*Preliminary*

Discourse, 124). Thus the scientist had to resort to "instrumental aids" to supplement the senses, and because "observations once made should remain as records to all mankind," scientists had to agree to adopt standardized weights and measures (*Preliminary Discourse,* 125). As desirable—and necessary—as he considered this goal, however, Herschel also noted the difficulty it introduced: "The selection and verification of such standards . . . will easily be understood to be a matter of extreme difficulty, if only from the mere circumstance that, to verify the permanence of one standard, we must compare it with others, which it is possible may be themselves inaccurate, or, at least, stand in need of verification" (*Preliminary Discourse,* 125–26).

Instead of evading the problem generated by the dependence of precise numerical data on necessary but unreliable instruments or on necessary but controversial standards, Herschel explicitly formulated the conundrum as a problem, which provoked two responses.

> But, it may be asked, if our measurement of quantity is thus unavoidably liable to error, how is it possible that our observations can possess that quality of numerical veracity which is requisite to render them the foundation of laws, whose distinguishing perfection consists in their strict mathematical expression? To this the reply is twofold. 1st, that though we admit the necessary existence of numerical error in every observation, we can always assign a limit which such errors cannot possibly exceed; and the extent of this *latitude of error of observation* is less in proportion to the perfection of the instrumental means we possess, and the care bestowed on their employment. In the greater part of modern measurements it is, in point of fact, extremely minute, and may be still further diminished, almost to any required extent, by repeating the measurements a great number of times, and under a great variety of circumstances, and taking a mean of the results, when errors of opposite kinds will, at length, compensate each other. But, 2dly, there exists a much more fundamental reply to this objection. In reasoning upon our observations, the existence and possible amount of quantitative error is always to be allowed for; and the extent to which theories may be affected by it is never to be lost sight of. In reasoning upwards, from observations confessedly imperfect to general laws, we must take care always to regard our conclusions as conditional. (*Preliminary Discourses,* 129–30)

Hershel's first recommendation—use precise instruments carefully and deploy mathematical models to limit the damage caused by human error—points to the path that scientists increasingly took in the last quarter of the nineteenth century. As Lorraine Daston and Peter Galison have argued, after midcentury scientists who wanted to control for the fallibility of the human observer—for the effects of subjectivity on particular observations—increasingly turned to precision instruments and mathematical modeling in hopes of achieving a standard of mechanical objectivity.[24] Herschel's second recommendation—that the scientist acknowledge *and create a procedure to accommodate* the limitations of both method and conclusion—illuminates what most of his contemporaries failed to do. Read in the context of Herschel's modesty, for example, the claims that McCulloch made for political economy begin to seem both overstated and epistemologically naive.

The epistemological conundrum of numerical precision plays a central role in Herschel's explication of scientific method because for him, as for the eighteenth-century philosophers, number constituted both "an object of sense, because we can count," and the signifying unit of mathematical formulas (*Preliminary Discourse*, 124). As for almost every philosopher I have examined in this book, moreover, the numerical datum was critical for Herschel because it seemed simply to reflect objects one had observed and counted (that is, number seems to be accurate), and it could also be used to generate mathematical formulas, which, though rule bound or precise, might not accurately reflect observable reality at all. Theorizing the method by which the scientist bridged the gap between numerical data and mathematical formulas, then, became a way of theorizing scientific method *tout court;* at the same time, of course, it also recapitulates Herschel's solution to the problem of induction. As we have just seen, this solution involved factoring the limits of epistemological certainty into the method of science itself.

We see how acknowledging limits constituted a solution to the problem of induction in Herschel's second attempt to address the issue of numerical data, this time with regard to the verification of quantitative laws.

> In their simplest or least general stages . . . they [quantitative laws] usually express some numerical relation between two quantities dependent on each other, either as collateral effects of a common cause, or as the amount of its effect under given numerical circumstances or *data*. . . . To arrive inductively at laws of this kind, where one quantity *depends* on or *varies with* another, all that is required is a series of careful and exact measures in every different state of the *datum* and *quaesitum*. Here, however, the mathematical form of the law being of the highest importance, the greatest attention must be given to the *extreme cases* as well as to all those points where the one quantity changes rapidly with a small change of the other. (*Preliminary Discourse*, 176)

According to this statement, taking account of extreme cases could enable the scientist to generate a mathematical law; but Herschel immediately acknowledges that this mathematical law may not be "true," because the observations it is based on may not have taken account of every possible case. "After all, unless our induction embraces a series of cases which absolutely include the whole scale of variation of which the quantities in question admit, the mathematical expression so obtained cannot be depended upon as the true one, and if the scale actually embraced be small, the extension of laws so derived to extreme cases will in all probability be exceedingly fallacious" (*Preliminary Discourse,* 177).

Herschel could not solve this problem. In fact, he did not even try to solve it; instead he gave it a name, referred it to a common example, and reiterated the limitations it imposed on scientific knowledge production.

> Laws thus derived, by the direct process of including in mathematical formulae the results of a greater or less number of measurements, are called "empirical laws." A good example of such a law is that given by Dr. Young . . . for the decrement of life, or the law of mortality. Empirical laws in this state are evidently *unverified inductions,* and are to be received and reasoned on with the utmost reserve. No confidence can ever be placed in them beyond the limits of the data from which they are derived; and even within those limits they require a special and severe scrutiny to examine *how nearly* they do represent the observed facts. . . . When so carefully examined, they become . . . most valuable. . . . On the other hand, when empirical laws are unduly relied on beyond the limits of the observations from which they were deduced, there is no more fertile source of fatal mistakes. (*Preliminary Discourse,* 178–79)

This admission came as close as it was possible to come to solving the problem of induction, because it mandated both constant scrutiny of the fit between the data one collected and the actual world and a ruthlessly honest acknowledgment of the limits of the data themselves. Of course, even this was not a philosophical solution. Instead, like McCulloch's taxonomy of knowledge, Herschel's demand that scientists use—but interrogate—empirical laws constituted a professional solution; by agreeing among themselves to use empirical laws and by assuring their public that they were honest about the limitations of the knowledge such laws generated, scientists could hope to earn the authority necessary to make their picture of nature seem more plausible than what ordinary observation revealed.

As we have already seen, the collection of data about mortality constituted one of the earliest applications of political arithmetic, and the generation of mortality tables, which supposedly expressed "the law of mortality," was a staple not only of the burgeoning insurance industry but also of claims made, in early nineteenth-century Britain, on behalf of statistics and political economy more generally. McCulloch tried to enhance the prestige of political economy by distinguishing between the collection of data (about mortality, for example) and the production of theories (about the relation between population and national wealth), but it would have been ruinous to admit that the law of mortality inferred from the data was simply a fallible "empirical law." This was true in part because, as the spokesperson for a controversial practice that was inextricably embroiled in volatile political questions about wages, trade restriction, and the poor law, McCulloch could ill afford to admit that his method was potentially a "source of fatal mistakes." He could not allow the limitations of his method to stand alongside its potential because, however it was defined or delimited, political economy (and McCulloch) lacked the institutional support that natural philosophy (and John Herschel) enjoyed. Because he was lobbying his readers for a version of the support that the BAAS consolidated for natural philosophy, McCulloch stressed only the positive contributions this new science could make. Because he could not afford to acknowledge political economy's limitations, he did not devise a theoretical account of them, much less a method capable of allowing for error.

Beginning in the 1830s, John Stuart Mill developed a definition of political economy that helped disentangle this "social science" from the most devastating effects of the problem of induction. Working from Herschel's and Whewell's meditations about natural philosophical method, Mill argued that "the phenomena of Society" constituted a composite analytic object that resembled, but was not identical to, a mathematical model of natural phenomena. Because "the circumstances . . . which influence the condition and progress of society, are innumerable, and perpetually changing," and because it is not possible to experiment on society, Mill thought that such social phenomena as the production and distribution of wealth required a mode of analysis that differed from even an expanded version of Baconian induction.[25] The mode of analysis he recommended was deduction, or reasoning a priori. Knowing the hostility directed to "mere conjecture" or "pure theory," Mill was quick to insist that deduction did *not* neglect "experience"; it simply subordinated "specific experience" to the experience he associated with "assumptions." According to Mill, political economy

> reasons, and, as we contend, must necessarily reason, from assumptions, not from facts. It is built upon hypotheses, strictly analogous to those which, under the name of definitions, are the foundation of the other abstract sciences. Geometry presupposes an arbitrary definition of a line, "that which has length but not breadth." Just in the same manner does Political Economy presuppose an arbitrary definition of man, as a being who invariably does that by which he may obtain the greatest amount of necessaries, conveniences, and luxuries, with the smallest quantity of

labour and self-denial with which they can be obtained in the existing state of knowledge. . . . Political Economy . . . reasons from *assumed* premises—from premises which might be totally without foundation in fact, and which are not pretended to be universally in accordance with it. The conclusions of Political Economy, consequently, like those of geometry, are only true, as the common phrase is, *in the abstract;* that is, they are only true under certain suppositions, in which none but general causes—causes common to the *whole class* of cases under consideration—are taken into the account. ("Definition," 325–26)

This description of method limits political economy to *modeling.* By Mill's account, it is an analytic instrument designed to investigate hypothetical cases: *if* human beings were motivated *exclusively* by the desire for wealth, then they would act like this. Of course no "political economist ever imagine[d] that real men had no object of desire but wealth," Mill continues ("Definition," 327). Knowing that his assumptions are purely functional (that is, they serve the purpose of his science; they are precise but not necessarily accurate), the political economist can protect himself from blame by admitting the limitations of the kind of knowledge he produces: "All that is requisite is, that he be on his guard not to ascribe to conclusions which are grounded upon an hypothesis a different kind of certainty from that which really belongs to them. They would be true without qualification, only in a case which is purely imaginary" ("Definition," 326).

Such statements notwithstanding, Mill was committed to representing political economic knowledge as somehow related to observed particulars. In "On the Definition of Political Economy" he elaborated this relation in two ways: by allowing for the influence of what he called "disturbing causes" upon the "principles of Political Economy," and by arguing that what political economists *described* were not particular objects but *tendencies.* The first elaboration led him again to the bewildering complexity of social objects, for while he imagined that one could approach descriptive accuracy by adding and subtracting the effects of disturbing causes to the effect of principles, he also repeatedly acknowledged that the "high order of complexity" that distinguished social objects finally "def[ies] our limited powers of calculation" ("Definition," 330; *Logic,* 50, 63 [6:5]). The second elaboration led Mill to imagine a new measure of scientific adequacy, for if the aim of science was not simply to produce general laws but to describe tendencies, then some sciences might be both "exact" and "hypothetical."

This curious combination—"exact" and "hypothetical"—characterizes the science that Mill placed at the heart of the social science: "Ethology, or the Science of Character." Ethology, Mill explained, was

the ulterior science which determines the kind of character produced in conformity to those general laws [the elementary laws of mind], by any set of circumstances, physical and moral. . . .This science of Ethology may be called the Exact
Science of Human Nature; for its truths are not, like the empirical laws which depend on them, approximate generalisations, but real laws. It is, however, (as in all
cases of complex phenomena,) necessary to the exactness of the propositions that
they should be hypothetical only, and affirm tendencies, not facts. They must not
assert that something will always or certainly happen, but only that such and such
will be the effect of a given cause, so far as it operates uncounteracted. (*Logic,* 54–55
[6:5])

According to Mill, then, ethology takes one beyond the empirical laws that
Herschel described, because it does not aspire to describe observed particulars.
Mill could say that ethology is "exact" because he measured exactness only in
relation to his founding hypothesis and the tendencies that hypothesis pre-
dicted; and he could claim a "hypothetical" status for the knowledge ethology
generated without denigrating these results because, he explained, positing hypotheses enabled the philosopher to see what was otherwise invisible. What was
otherwise invisible, in turn, included all those abstractions that Mill considered
the proper objects of the new social sciences he was struggling to invent: character, society, *homo economicus.* For Mill, then, the "exact" and "hypothetical"
science of ethology corrected Bacon's unrealistic emphasis on observed
particulars; in so doing, he claimed, this prototypical social science "solved" the
problem of induction by subordinating distracting particulars to the "tendencies" and abstractions that hypothesis could uniquely illuminate (*Logic,* 56–57
[6:5]).[26]

   Although ethology's emphasis on hypothesis might have solved—or at least
defused—the problem of induction, it is difficult to see how this science could
have avoided the charge increasingly leveled against political economists: that
what the philosopher claimed were impartial hypotheses actually served the interests of those who devised them. Nor is it clear how Mill thought he could
avoid the complaint that such sciences, which inevitably focused on "mankind
in the average, or *en masse*" (*Logic,* 58 [6:5]), obliterated the kind of moral responsibility that had typically been associated with individual free will. Mill
tried to disarm this last criticism by arguing that "the doctrine of the Causation
of human actions" did not imply that *every* case was determined by the law of
"regularity *en masse,*" but his enthusiasm for statistics—the mode of representation that made this regularity visible—threatened to accelerate his argument's
drift toward what many contemporaries considered the black hole of determinism.[27] To avoid this particular form of blasphemy, which would have con-

demned his philosophy absolutely in the eyes of many readers, Mill concluded his *Logic* with a celebration of those exceptional individuals whose accomplishments seemed to (but did not) defy the laws of causation.

> Though the varieties of character among ordinary individuals neutralise one another on any large scale, exceptional individuals in important positions do not in any given age neutralise each other; there was not another Themistocles, or Luther, or Julius Caesar, of equal powers and contrary dispositions, who exactly balanced the given Themistocles, Luther, and Caesar, and prevented them from having any permanent effect. Moreover, for aught that appears, the volitions of exceptional persons, or the opinions and purposes of the individuals who at some particular time compose a government, may be indispensable links in the chain of causation by which even the general causes produce their effects. (*Logic*, 126–27 [6:11])

In such passages we see the new emphasis that the problem of induction was given by the mathematical variant of statistics that Mill endorsed: whereas British philosophers since Hume had asked how one could reason from observed particulars to final causes or from observed particulars to general laws, after statistics began to be equated with the law of large numbers, philosophers as well as ordinary readers began to ask how one could conceptualize free will, given that the regularities that emerged from "numerical calculations" seemed to leave so little room for volition, for morality, or for ethics of any kind.

POEMS AND SYSTEMS: THE EMERGENCE OF
THE POSTMODERN FACT

Despite the various oppositions we have seen writers invoke to distinguish a mode of theory- or value- or politics-free representation from its devalued twin, early nineteenth-century poets were as concerned with the problem of induction as were the champions of political economy. In this sense at least, and despite the undeniable link between the old, now discredited rhetoric and poetry, figures of arithmetic had more in common with figures of speech in the early nineteenth century than most of the practitioners of either "science" or "art" liked to acknowledge.[28] Like political economists, poets aspired both to capture observed particulars—what Wordsworth called "a simple produce of the common day"—and to produce general, possibly even universal, knowledge.[29] Unlike political economists, however—at least according to Percy Shelley—poets were not hampered by the debilitating disciplinary need to accumulate facts and calculate outcomes. To Shelley, political economy erred in limiting knowledge to observed particulars, in equating pleasure with the "transitory and particular," and in pursuing the narrow goal of "banishing the importunity of the wants of our animal nature." If one sought "durable, universal, and permanent" pleasure, as Shelley thought everyone should, then "the poetry of life" would prove more useful than all the calculations of political economy.[30] He considered poetry the "centre and circumference of knowledge" because it revealed the indwelling essence of things in its depictions of the things themselves.

> Poetry is indeed something divine. It is at once the centre and circumference of knowledge; it is that which comprehends all science, and that to which all science must be referred. It is at the same time the root and blossom of all other systems of thought: it is that from which all spring, and that which adorns all; and that which, if blighted, denies the fruit and the seed, and withholds from the barren world the nourishment and the succession of the scions of the tree of life. It is the perfect and consummate surface and bloom of things; it is as the odour and the colour of the rose to the texture of the elements which compose it, as the form and the splendour of unfaded beauty to the secrets of anatomy and corruption. What were Virtue, Love, Patriotism, Friendship &c.—what were the scenery of this beautiful Universe which we inhabit—what were our consolations on this side of the grave—and what were our aspirations beyond it—if Poetry did not ascend to bring light and fire from those eternal regions where the owl-winged faculty of calculation dare not ever soar? ("Defence," 503)

In this passage Shelley proposes that poetry works its alchemy by *replacing* the observed particulars of the phenomenal world with its own linguistic particulars ("it is the perfect and consummate bloom of things"). In the language world that is the poem, general knowledge simply coincides with particulars, because the particulars have been taken up and transformed through the medium of metaphor.

For most romantic poets, as for Shelley, Wordsworth, and (especially) Coleridge, the general and systematic nature of the knowledge that poetry produced was at least as important as was poetry's ability to convey the appearance of the phenomenal world. For John Keats, by contrast, even when the quest for systematic knowledge was detached from political economic calculation and taken over by poets, it retained its stultifying tendency. In his view, "any irritable reaching after fact & reason"—any search for system, explanation, or philosophical solution—blunted the poet's exclusive capacity, which consisted of the tolerance for "being in uncertainties, Mysteries, doubts." Keats called this capacity "negative capability" and assigned it, and not the ability to generate systematic knowledge or to observe particulars, the central role in the production of the only kind of knowledge that counted.[31]

Theorists of poetry (or literature more generally) have continued to argue that this form of writing offers a unique solution to the problem of induction, either because the literary text constitutes what W. K. Wimsatt called "a concrete universal" or because, as Steven Knapp has more recently contended, "the object of literary interest is a special kind of representational structure, each of whose elements acquires, by virtue of its connection with other elements, a network of associations inseparable from the representation itself."[32] I am less interested in assessing the grounds, much less the adequacy, of such accounts than in pointing out that, in claiming literary texts uniquely solve the problem of induction, such theorists of literature implicitly acknowledge that literary texts share science's imperative to address this issue. If this question constitutes the heart of the epistemology that has distinguished the long modern period, as I have argued throughout this book, then it is hardly surprising that poets and scientists, as well as philosophers, would worry about how to produce general knowledge from observed particulars. Only if one could renounce the desire (or need) for systematic knowledge, after all—only if one were "capable of being in uncertainties, Mysteries, doubts"—would this problem lose its ability to vex and to inspire new attempts to solve it.

Unless, of course, one were to abandon not the desire for systematic knowledge but the need to yoke knowledge systems to observed particulars. This, I believe, is the "solution" adumbrated by the romantic poets' turn away from phenomenal particulars and toward the mind that contemplates those things. When a neoformalist critic like Knapp argues that the literary text constitutes a "special kind of representational structure" whose specialness resides in its autonomy, in other words, he is carrying to its logical extreme the claim Shelley made in 1817: "A Poet participates in the eternal, the infinite, and the one; as far as relates to his conceptions, time and place and number are not" ("Defence," 483). When one imagines that the literary text creates a model that is internally consistent and whose relation to history or phenomenal particulars is incidental, one sets aside the problem of induction in favor of a mode of analysis that prefers description of the self-contained system to attempts to explain the system's possible links to the world.

Paradoxically, the neoformalism Knapp practices constitutes another face of the postmodernism that has begun to transform knowledge production in the West. Whether it takes the form of Ferdinand de Saussure's claim that signs are arbitrary, Jacques Lacan's definition of the ego as lack, Jean Baudrillard's fascination with simulation's ability to end all original reference, or Slavoj Žižek's celebration of the "meaningless traces" that thrust meaning production onto analysis itself, the postmodernist conviction that the systems of knowledge humans create constitute the only source of meaning is gradually displacing both the problem of induction and all the variants of the modern fact that I have discussed in this book. Of course, just as the Baconian variant of the modern fact—the deracinated particular—constituted a "solution" that simply set aside the problematic commonplaces of ancient philosophy, so postmodernism and the postmodern fact simply set aside the problem of induction. And just as the Baconian fact did not suddenly or completely displace its ancient predecessor, so the postmodern fact has not wholly triumphed over either the production of modern facts or the longing for them. As with any epistemological revolution, the one that late twentieth-century citizens of the global information village are now experiencing will take a long time to unfold. As with any epistemological revolution, moreover, including the one I have described in this book, the roots of this revolution reach deep into the epistemological paradigm that preceded, produced, and now coexists with it. The imbrication of this past with our present seems to me to justify the kind of history I have written here, even if there still remains much to be said, both about the past I have tried to describe and about the present that has made it possible for me to imagine undertaking this daunting task.

Introduction

1. See, for example, James Phillips Kay, *The Moral and Physical Condition of the Working Classes Employed in the Cotton Manufacture in Manchester*, 2d ed. enlarged (London: James Ridgway, 1832), and my essay on this text in Mary Poovey, *Making a Social Body: British Cultural Formation, 1830–1864* (Chicago: University of Chicago Press, 1995), 73–97.

2. [G. Robertson], "Exclusion of Opinions," *London and Westminster Review* 61 (April 1838): 37.

Chapter One

1. Lorraine Daston has argued in print and, more recently, at a conference held at Stanford University in January 1996 that "the category of the factual has a history." "'Facts' come into being and pass away because 'facts' are modes of sifting and ordering what counts as experience," Daston commented at the Stanford conference. Her thesis was that the prototypical scientific fact changed in the period from 1660 to 1730, from the singular event to a large class of events that could be reproduced at will. In contrast to Baconian rarities, she continued, the latter were deliberately bland and contributed to the universalizing of "nature" (Lorraine Daston, "Description by Omission: Nature Enlightened and Observed," presented at conference "Regimes of Description: In the Archive of the Eighteenth Century," 11–14 January 1996). Although most of this book was written before I heard Daston's provocative paper, my revisions have been influenced by her historical analysis of the category of the factual. As I note below, I find it more useful to define one long period for "the modern fact" than to distinguish as sharply as she does between what I would call variants of a single form of this epistemological category. Since so little attention has thus far been given to the category of the factual, however, disputes about periodization seem to me disagreements between allies, not fundamental differences.

2. Particularly provocative treatments of the gradual emergence of the postmodern fact out of its modern antecedent (although they do not use this terminology) include Philip Mirowski, "The When, the How and the Why of Mathematical Expression in the History of Economic Analysis," *Journal of Economic Perspectives,* 5, no. 1 (1991): 145–57; Lorraine Daston and Peter Galison, "The Image of Objectivity," *Representations* 40 (fall 1992): 81–128; Nancy Cartwright, *How the Laws of Physics Lie* (Oxford: Clarendon Press, 1983), especially "Introduction" and "Fitting Facts to Equations" (128–42); and Jean Baudrillard, *Simulations,* trans. Paul Foss, Paul Patton, and Philip Beitchman (New York: Semiotext[e], 1983), esp. 100, 136. Naomi Oreskes discusses some

of the potentials and limitations of computer modeling in "Representation vs. Refutability: A Dilemma for Models of Complex Natural Systems," *Scientific American* (forthcoming). I am grateful to the author for permission to read this unpublished manuscript. For a helpful discussion of the problems inherent in mathematical modeling, see Naomi Oreskes, Kristin Shrader-Frechette, and Kenneth Belitz, "Verification, Validation, and Confirmation of Numerical Models in the Earth Sciences," *Science* 263 (4 February 1994): 641–46.

3. See Michael Clanchy, *From Memory to Written Record: England, 1066–1307* (1979; 2d ed., Cambridge: Harvard University Press, 1993), esp. chap. 2.

4. I owe both of these examples to Janel Mueller. I am also grateful to the combined Renaissance, Eighteenth-Century, and Victorian Workshops at the University of Chicago for a helpful discussion of this aspect of my argument.

5. Scholarly studies of numbers and their uses include Alfred W. Crosby, *The Measure of Reality: Quantification and Western Society, 1250–1600* (Cambridge: Cambridge University Press, 1997); Brian Rotman, *Signifying Nothing: The Semiotics of Zero* (London: Macmillan, 1987); Frank J. Swetz, *Capitalism and Arithmetic: The New Math of the Fifteenth Century* (La Salle, Ill.: Open Court, 1987); Jens Hoyrup, "Sub-scientific Mathematics: Observations on a Pre-modern Phenomenon," *History of Science* 28, no. 22 (1989): 63–87; John Brewer, *The Sinews of Power: War, Money, and the English State, 1688–1783* (Cambridge: Harvard University Press, 1990); and Theodore M. Porter, *Trust in Numbers: The Pursuit of Objectivity in Science and Public Life* (Princeton: Princeton University Press, 1995). The literature on the history of mathematics is enormous; good introductions (for nonmathematicians) include W. W. Rouse Ball, *A Short Account of the History of Mathematics* (New York: Dover, 1960); Carl B. Boyer, *A History of Mathematics* (Princeton: Princeton University Press, 1985); John D. Barrow, *Pi in the Sky: Counting, Thinking, Being* (Boston: Little, Brown, 1992); and Lloyd Motz and Jefferson Hane Weaver, *The Story of Mathematics* (New York: Avon, 1995); I have found only a few treatments of numeracy; these include Keith Thomas, "Numeracy in Early Modern England," *Transactions of the Royal Historical Society,* 5th ser., 37 (1977): 103–32; Alexander Murray, *Reason and Society in the Middle Ages* (Oxford: Clarendon Press, 1978); Patricia Cline Cohen, *A Calculating People: The Spread of Numeracy in Early America* (Chicago: University of Chicago Press, 1982); and Cohen, "Reckoning with Commerce: Numeracy in Eighteenth-Century America," in *Consumption and the World of Goods,* ed. John Brewer and Roy Porter (New York: Routledge, 1993), 320–34. The best treatment of mathematical instruments remains E. G. R. Taylor, *The Mathematical Practitioners of Tudor and Stuart England* (Cambridge: Cambridge University Press, 1954); but see also Stephen Johnston, "Mathematical Practitioners and Instruments in Elizabethan England," *Annals of Scholarship* 48 (1991): 321–41, and A. J. Turner, "Mathematical Instruments and the Education of Gentlemen," *Annals of Science* 30 (1973): 51–88.

6. I discuss the epistemological category of "domain" in Poovey, *Making a Social Body,* 5–6. The concept of the domain, which I have adapted from Foucault, signals the way cultural formation proceeds by drawing boundaries around sets of practices and ideas, which once seemed to belong to a larger, undifferentiated continuum of practices, so as to create new and more specialized conceptual entities, characterized by a logic or rationality specific to them. Domains are both conceptual and material; indeed, as they are materialized in the form of institutions, domains divide up and organize our everyday lives. Later in this chapter I discuss in more detail the relation between *Making a Social Body* and *A History of the Modern Fact.* For now it is sufficient to say that *A History of the Modern Fact* both focuses on a smaller epistemological unit than the domain and contributes to the same history of disciplinary disaggregation of which *Making a Social Body* constitutes a chronologically later chapter.

7. Among the foundational texts in this general area are Joseph A. Schumpeter, *A History of Economic Analysis* (New York: Oxford University Press, 1954); Ronald Meek, *The Economics of Physiocracy* (Cambridge: Harvard University Press, 1963), and Meek, *"Economics and Ideology" and Other Essays* (London: Chapman and Hall, 1967); Philip Abrams, *Origins of British Sociology, 1834–1914* (Chicago: University of Chicago Press, 1968); and William Letwin, *The Origins of Scientific Economics: English Economic Thought, 1660–1776* (New York: Methuen, 1963). See also R. A. Nisbet, *The Sociological Tradition* (London: Heinemann, 1967), and Mark Blaug, "Economic Theory and Economic History in Great Britain, 1650–1776," *Past and Present* 68 (1964): 111–16. More specialized studies published in the 1970s and 1980s include Ronald Meek, *Social Science and the Ignoble Savage* (Cambridge: Cambridge University Press, 1976); Joyce Oldham Appleby, *Economic Thought and Ideology in Seventeenth Century England* (Princeton: Princeton University Press, 1978); and Peter Buck, "Seventeenth-Century Political Arithmetic: Civil Strife and Vital Statistics," *Isis* 68 (1977). 67–85. During those decades, statistics was the subject of the following studies: Michael J. Cullen, *The Statistical Movement in Early Victorian Britain: The Foundations of Empirical Research* (New York: Barnes and Noble, 1975); David Elesh, "The Manchester Statistical Society: A Case Study in Discontinuity in the History of Empirical Social Research," *Journal of the History of the Behavioral Sciences* 8 (1972): 280–301, 407–17; Victor L. Hilts, *"Aliis Exterendum, or* The Origins of the Statistical Society of London," *Isis* 69 (1978): 21–43; J. M. Eyler, *Victorian Social Medicine: The Ideas and Influence of William Farr* (Baltimore: Johns Hopkins University Press, 1979), 13–36; Martin Shaw and Ian Miles, "The Social Roots of Statistical Knowledge," in *Demystifying Social Statistics,* ed. John Irvine, Ian Miles, and Jeff Evans (London: Pluto Press, 1979): 27–38; Lawrence Goldman, "The Origins of British 'Social Science': Political Economy, Natural Science and Statistics, 1830–1835," *Historical Journal* 26, no. 3 (1983): 587–616; and Theodore M. Porter, *The Rise of Statistical Thinking, 1820–1900* (Princeton: Princeton University Press, 1986). More recently, social science has been treated as a disciplinary whole in Richard Olson, *The Emergence of the Social Sciences, 1642–1792* (New York: Twayne, 1993); Eileen Janes Yeo, *The Contest for Social Science: Relations and Representations of Gender and Class* (London: Rivers Oram Press, 1996); and Donald Levine, *Visions of the Sociological Tradition* (Chicago: University of Chicago Press, 1995). Of special interest to readers of this book are also a few treatments of numbers, numeracy, quantification, and numerical representation: Paul F. Lazarsfeld, "Notes on the History of Quantification in Sociology—Trends, Sources and Problems," *Isis* 52 (1961): 277–333; Thomas, "Numeracy in Early Modern England," 103–32; Murray, *Reason and Society in the Middle Ages,* esp. part 2; Cohen, *Calculating People;* and Porter, *Trust in Numbers.*

8. Richard Tuck, *Philosophy and Government, 1572–1651* (Cambridge: Cambridge University Press, 1993); Maurizio Viroli, *From Politics to Reason of State: The Acquisition and Transformation of the Language of Politics, 1250–1600* (Cambridge: Cambridge University Press, 1992); Quentin Skinner, *Reason and Rhetoric in the Philosophy of Hobbes* (Cambridge: Cambridge University Press, 1996); Peter N. Miller, *Defining the Common Good: Empire, Religion, and Philosophy in Eighteenth-Century Britain* (Cambridge: Cambridge University Press, 1994); Stefan Collini, Donald Winch, and John Burrow, *That Noble Science of Politics: A Study in Nineteenth-Century Intellectual History* (Cambridge: Cambridge University Press, 1983); and Donald Winch, *Riches and Poverty: An Intellectual History of Political Economy in Britain, 1750–1834* (Cambridge: Cambridge University Press, 1996).

9. Graham Burchell, Colin Gordon, and Peter Miller, eds., *The Foucault Effect: Studies in Governmentality, with Two Lectures and an Interview with Michel Foucault* (Chicago: University of Chicago Press, 1991); in this volume see especially Colin Gordon, "Governmental Rationality: An Introduction," 1–52; Michel Foucault, "Governmentality," 87–104; and Graham Burchell,

"Peculiar Interests: Civil Society and Governing 'The System of Natural Liberty,'" 119–50. See also Mitchell Dean, *The Constitution of Poverty: Toward a Genealogy of Liberal Governance* (London: Routledge, 1991).

10. See Nicholas Phillipson, "The Pursuit of Virtue in Scottish University Education: Dugald Stewart and Scottish Moral Philosophy in the Enlightenment," in *Universities, Society, and the Future,* ed. Nicholas Phillipson (Edinburgh: University of Edinburgh Press, 1983), 82–101; Richard B. Sher, *Church and University in the Scottish Enlightenment: The Moderate Literati of Edinburgh* (Princeton: Princeton University Press, 1985); Andrew Skinner, "Economics and History—the Scottish Enlightenment," *Scottish Journal of Political Economy* 12 (February 1965): 1–22; Duncan Forbes, "'Scientific Whiggism': Adam Smith and John Millar," *Cambridge Journal* 7 (1953–54): 643–70; Knud Haakonssen, "Natural Jurisprudence in the Scottish Enlightenment: Summary of an Interpretation," in *Enlightenment, Rights, and Revolution,* ed. Neil McCormick and Zenon Bankowski (Aberdeen: Aberdeen University Press, 1989): 36–49; and Istvan Hont, "Commercial Society and Political Theory in the Eighteenth Century: The Problem of Authority in David Hume and Adam Smith," in *Main Trends in Cultural History: Ten Essays,* ed. William Melching and Wyger Velema (Amsterdam: Editions Rodopi, 1994), 54–94, and Hont, "The Political Economy of the 'Unnatural and Retrograde' Order: Adam Smith and Natural Liberty," in *Franzosische Revolution und politische Okonomie,* ed. Maxine Berg et al. (Trier, Germany: Schriften aus dem Karl-Marx-Haus, 1989), 122–49.

11. Lorraine Daston, "The Moral Economy of Science," in *Constructing Knowledge in the History of Science,* edited by Arnold Thackray (Chicago: University of Chicago Press, 1995); originally published in *Osiris,* 2d ser., 10 (1995): 24. For Daston's other contributions to this genre, see Daston and Galison, "Image of Objectivity," 81–128; Daston, "Baconian Facts, Academic Civility, and the Prehistory of Objectivity," *Annals of Scholarship,* 8, nos. 3–4 (1991): 337–64; and Daston, "Objectivity and the Escape from Perspective," *Social Studies of Science* 22 (November 1992): 597–618. I would also define Peter Dear's work, which I discuss below, as historical epistemology; but because Dear has not self-consciously adopted this disciplinary assignation, I leave him to choose his own affiliations. See especially Peter Dear, *Discipline and Experience: The Mathematical Way in the Scientific Revolution* (Chicago: University of Chicago Press, 1995).

12. Both quoted in Daston, "Baconian Facts," 341. See also Daston, "Moral Economy of Science," 12–13.

13. Daston, "Baconian Facts," 343, 338.

14. Ibid., 344.

15. Dear, *Discipline and Experience,* 25.

16. Alfred Crosby also attributes considerable importance to double-entry bookkeeping in his history of quantification, although his emphasis falls on this practice's ability to make commerce visualizable rather than on double-entry bookkeeping's disciplining of observed particulars within a formal system of writing. See Crosby, *Measure of Reality,* chap. 10.

17. Daston cites academics' desire to end the bitter quarrels that swirled around rival theories as one cause of the scientific revolution; see "Baconian Facts," 350–56. Charles Webster assembles a range of causal factors in *The Great Instauration: Science, Medicine, and Reform, 1626–1660* (New York: Holmes and Meier, 1976), esp. chap. 1.

18. See John Bender and David E. Wellbery, "Rhetoricality: On the Modernist Return of Rhetoric," in *The Ends of Rhetoric: History, Theory, Practice,* ed. John Bender and David E. Wellbery (Stanford: Stanford University Press, 1990), 6–7.

19. See Steven Shapin and Simon Schaffer, *Leviathan and the Air-Pump: Hobbes, Boyle and the Experimental Life* (Princeton: Princeton University Press, 1985), esp. 342. See also Amos Funkenstein, *Theology and the Scientific Imagination from the Middle Ages to the Seventeenth Century* (Prince-

ton: Princeton University Press, 1986), 21: "In a certain sense, all science, every scientific argument or procedure, has an ideal—and, if you wish, fictional—aspect to it."

20. "Thus, for example, *epagôge* could refer to a rhetorical induction where a general claim is made plausible by the presentation of a few purportedly typical examples. It also referred to an induction by complete enumeration of instances. But most revealingly, perhaps, it could cover the recognition of an essential, necessary truth regarding some class of things when the mind grasps the universal in the particular by inspection: thus, for example, one might soon realize, after having encountered a few triangles, that not only do the internal angles of each add up to two right angles, but that this must be true of all triangles whatever, by the very nature of a triangle. Aristotle's belief in the reality of universals as entities existing above and beyond their individual instances played a critical part in establishing this very influential sense of 'induction'" (Dear, *Discipline and Experience,* 26). See also John R. Milton, "Induction before Hume," *British Journal for the Philosophy of Science* 38, no. 1 (1987): 49–74.

21. Daston, "Baconian Facts," 344.

22. I discuss modern abstraction in *Making a Social Body.* "The version of abstraction that dominated truth-claims by the end of the eighteenth century had begun to manifest three salient characteristics. First, it tended to be instantiated—by which I mean both related, in some fashion or another, to the concrete instances of the phenomenal world and institutionalized as codified practices that are confirmed and then naturalized through the social relations established by people working together. Second, and largely as a consequence of this instantiation, modern abstraction tends to be susceptible to the process of vivification that nineteenth-century writers like Marx and Freud referred to variously as reification, commodification, and fetishization. This vivification, in turn, is the basis for what has gradually become the almost complete domination of representation—of appearances—in modern mass culture. Third, modern abstraction tends to generate norms that are typically defined as such by numerical calculation. Modern abstraction, in other words, as the product of the spatialization of Euclidean geometry . . . , derives from the imposition of a conceptual grid that enables every phenomenon to be compared, differentiated, and measured by the same yardstick. Such comparisons and measurements, of course, produce some phenomena as normative—ostensibly because they are more numerous, because they represent an average, or because they constitute an ideal towards which all other phenomena move" (9).

23. See Stephanie H. Jed, *Chaste Thinking: The Rape of Lucretia and the Birth of Humanism* (Bloomington: Indiana University Press, 1989).

24. Michel Foucault, *The Order of Things: An Archaeology of the Human Sciences* (New York: Random House, 1970); Foucault, "The Order of Discourse," in *Untying the Text: A Post-structuralist Reader,* ed. Robert Young (Boston: Routledge and Kegan Paul, 1981), 48–78; and Foucault, "Governmentality," 87–104.

25. See Dean, *Constitution of Poverty,* especially the introduction, and Ian Hunter, *Culture and Government: The Emergence of Literary Education* (London: Macmillan, 1988). For one of Foucault's numerous commentaries on method, see "Questions of Method," in Burchell, Gordon, and Miller, *Foucault Effect,* 73–86.

26. Foucault, "Politics and the Study of Discourse," in Burchell, Gordon, and Miller, *Foucault Effect,* 54–55.

27. Ibid., 54.

28. Ibid., 55. For one example of the strictures imposed by such limitation, see the essays in Andrew Barry, Thomas Osborne, and Nikolas Rose, eds., *Foucault and Political Reason: Liberalism, Neo-liberalism, and Rationalities of Government* (Chicago: University of Chicago Press, 1996).

29. Shapin and Schaffer, *Leviathan and the Air-Pump,* 342.

30. Bruno Latour, *We Have Never Been Modern,* trans. Catherine Porter (Cambridge: Harvard University Press, 1993), 40, 41, and chap. 1 in general. Another science studies project that usefully bridges the gulf between the history of science per se and a sociological study of treatments of science and scientific issues is Robert N. Proctor, *Value-Free Science? Purity and Power in Modern Knowledge* (Cambridge: Harvard University Press, 1991).

31. Latour, *We Have Never Been Modern,* 26.

32. Latour's most influential studies have been almost completely limited to what happens in the laboratory or in science more generally understood, for example. See Bruno Latour and Steve Woolgar, *Laboratory Life: The Construction of Scientific Facts* (Princeton: Princeton University Press, 1979), and Bruno Latour, *The Pasteurization of France* (Cambridge: Harvard University Press, 1988). Emily Martin helped me specify Latour's place within science studies.

33. Latour, *We Have Never Been Modern,* 28.

34. Latour identifies as the best example of the new "sociology of criticism" Luc Boltanski and Laurent Thevenot's *De la justification: Les économies de la grandeur* (Paris: Gallimard, 1991). See *We Have Never Been Modern,* 44–45.

35. Latour, *We Have Never Been Modern,* 45, 44–45. Until recently, Latour writes, "critical unmasking appeared to be self-evident. It was only a matter of choosing a cause for indignation and opposing false denunciations with as much passion as possible. To unmask: that was our sacred task, the task of us moderns. To reveal the true calculations underlying the false consciousness, or the true interests underlying the false calculations. Who is not still foaming slightly at the mouth with that particular rabies?" By contrast, and in the wake of Boltanski and Thevenot's book, "instead of a resource, the critical spirit becomes a topic, one competence among others, the grammar of our indignations. Instead of practising a critical sociology the authors quietly begin a sociology of criticism" (44).

36. In thinking about the possibility—and advantages—of this kind of reading, I have been influenced by the methodological statements in Collini, Winch, and Burrow, *That Noble Science of Politics,* "Prologue," and Winch, *Riches and Poverty.* chap. 1 and p. 236. It must be said, however, that I do not believe, as Winch seems to, that a late twentieth-century reader of *The Principle of Population* can actually "see the issues through Malthus's eyes" (*Riches and Poverty,* 25).

37. This is the same process that Peter Dear associates with Hume's formulation of the problem of induction in the 1740s. See *Discipline and Experience,* 15–21.

38. My essays include Mary Poovey, "The Social Constitution of 'Class': Toward a History of Classificatory Thinking," in *Rethinking Class: Literary Studies and Social Formations,* ed. Wai Che Dimock and Michael T. Gilmore (New York: Columbia University Press, 1994), 15–56; "Aesthetics and Political Economy in the Eighteenth Century: The Place of Gender in the Social Constitution of Knowledge," in *Aesthetics and Ideology,* ed. George Levine (New Brunswick: Rutgers University Press, 1994), 79–105; and "Accommodating Merchants: Accounting, Civility, and the Natural Laws of Gender," *Differences* 8, no. 5 (1996): 1–20. John Barrell's essays are collected in *The Birth of Pandora and the Division of Knowledge* (Philadelphia: University of Pennsylvania Press, 1992). See also Michael McKeon, "Historicizing Patriarchy: The Emergence of Gender Difference in England, 1660–1760," *Eighteenth-Century Studies* 28, no. 3 (1995): 295–322; and Clifford Haynes Siskin, *The Work of Writing: Literature and Social Change in Britain, 1700–1830* (Baltimore: Johns Hopkins University Press, 1997), esp. chap. 4.

39. Susan Dwyer Amussen, *An Ordered Society: Gender and Class in Early Modern England* (New York: Basil Blackwell, 1988); Gretchen Gerzina, *Black England: Life before Emancipation* (London: J. Murray, 1995); Brian Maidment, *The Poorhouse Fugitives: Self-Taught Poets and Poetry in Victorian Britain* (Manchester: Carcanet, 1987).

40. Among the tantalizing possibilities I found as I worked on these materials, two in partic-

ular cry out for further attention. The first was raised by the structural position shared by risk and women in the system of double-entry bookkeeping: if both were excluded from this system of writing, does this mean that the exclusion of women stood in for the exclusion of risk? This is the suggestion I make in the version of "Accommodating Merchants" that appears in *Differences,* but to substantiate this claim one would need more research on attitudes toward risk and women in the sixteenth and seventeenth centuries.

The second intriguing idea I have not pursued is the similarity between the optimistic providentialism I attribute to Dugald Stewart in chapter 6 and the stance that Tania Modeleski has recommended feminists take. Modeleski argues that feminist criticism must be both performative and utopian: feminism must assume (and enact) what it also takes as its goal. In Stewart's work this stance looks like nostalgia; yet in late twentieth-century feminism it looks like a way to hold open alternatives that might otherwise close. Theorizing the terms in which such a similarity might be assessed—or explained—would be a challenging and important task. See Tania Modeleski, *Feminism without Women: Culture and Criticism in a "Postfeminist" Age* (New York: Routledge, 1991), 14, 15.

41. Among recent accounts of the relation between eighteenth-century imaginative writing and early economic writing, the following studies are particularly provocative: Edward Copeland, "Money in the Novels of Fanny Burney," *Studies in the Novel* 8 (1976): 24–37; Marc Shell, *The Economy of Literature* (Baltimore: Johns Hopkins University Press, 1978); Shell, *Money, Language, and Thought: Literary and Philosophical Economies from the Medieval to the Modern Era* (Berkeley: University of California Press, 1982); Peter de Bolla, *The Discourse of the Sublime: Readings in History, Aesthetics, and the Subject* (Oxford: Basil Blackwell, 1989); Thomas Kavanagh, *Enlightenment and the Shadows of Chance: The Novel and the Culture of Gambling in Eighteenth-Century France* (Baltimore: Johns Hopkins University Press, 1993); Colin Nicholson, *Writing and the Rise of Finance: Capital Satires of the Early Eighteenth Century* (Cambridge: Cambridge University Press, 1994); Catherine Gallagher, *Nobody's Story: The Vanishing Acts of Women Writers in the Marketplace, 1670–1820* (Berkeley: University of California Press, 1994); and James Thompson, *Models of Value: Eighteenth-Century Political Economy and the Novel* (Durham: Duke University Press, 1996). For a recent attempt to describe the specificity of literary writing in the eighteenth century, see Siskin, *Work of Writing.*

42. See Albert O. Hirschman, *The Passions and the Interests: Political Arguments for Capitalism before Its Triumph* (Princeton: Princeton University Press, 1977).

Chapter Two

1. Thomas Hobbes, *Leviathan,* ed. C. B. Macpherson (1651; Harmondsworth: Penguin Books, 1968), pt. 1, chaps. 4 and 5. "Reason, in this sense, is nothing but *Reckoning* (that is, Adding and Substracting) of the Consequences of generall names agreed upon, for the *marking* and *signifying* of our thoughts" (111).

2. Francis Hutcheson, *An Essay on the Nature and Conduct of the Passions and Affections, with Illustrations on the Moral Sense,* ed. Paul McReynolds, facsimile ed. (1729; Gainesville, Fla.: Scholars' Facsimiles and Reprints, 1969), 40.

3. In 1792, for example, Dugald Stewart compared the advantage the philosopher enjoyed over the layman to the advantage that "the expert algebraist possesses over the arithmetical accountant" (*Elements of the Philosophy of the Human Mind* [1792; Boston: James Munroe, 1847], 135). In 1874 W. Stanley Jevons invoked bookkeeping again: Bacon's "notion of scientific method was a kind of scientific bookkeeping," Jevons complained. "Facts were to be indiscriminately gathered from every source, and posted in a ledger, from which emerges in time a balance

of truth. It is difficult to imagine a less likely way of arriving at great discoveries" (Jevons, *The Principles of Science* [1874; New York: Dover, 1958], bk. 4, chap. 26, sec. 2).

4. In distinguishing between ancient and modern "facts," I am drawing on the theoretical assumptions that inform what Lorraine Daston has called "historical epistemology." Daston argues that not all modern "facts" are alike, however; the Baconian rarity was quite different from the Enlightenment regularity. Although I will discuss these historical developments in the chapters that follow, I make the crude distinction between ancient and modern facts here to signal the difference between the metaphysical realism of Aristotle and the natural realism that Bacon heralded in the seventeenth century. See Daston, "Moral Economy of Science." Readers should remember that throughout this book I place "fact" and "facts" in imaginary quotation marks to signal their socially constituted nature.

5. One of these roots, according to Paul F. Lazarsfeld, is the practice of numerical data collection that William Petty called political arithmetic; the other is the Germanic development of a system of representation designed to make descriptive information, which was not typically numerical in form, useful to the government. Lazarsfeld describes these roots as developing simultaneously and with no apparent connection to each other. He does not link them either to double-entry bookkeeping or to accounting more generally. See Lazarsfeld, "Quantification in Sociology," 283–92.

6. In her useful discussion of the relation between mercantile writing and Italian humanism, Jed describes the process by which "the concrete activity of *ragione* as figuring on paper seems to disappear from the abstract understanding of *ragione* as a new kind of narrative logic, and knowledge, in this process, is purified of its relation to the bookkeeping it serves" (Jed, *Chaste Thinking*, 104).

7. The tendency of modern accounting studies to emphasize accounting's capacity to calculate, predict, administer, and discipline is best exemplified by the work sometimes called critical accounting studies and often published in or associated with the journal *Accounting, Organizations, and Society*. Drawing variously on theories developed by Michel Foucault, Bruno Latour, and Ian Hacking, practitioners of critical accounting studies seek to understand both the genealogy and the sociology of "the vast machine of economic calculation that is accounting." This phrase comes from Peter Miller, "Accounting and Objectivity: The Invention of Calculating Selves and Calculable Spaces," *Annals of Scholarship* 9, nos. 1–2 (1992): 61. For a brief treatment of critical accounting studies, see Theodore Porter, "Quantification and the Accounting Ideal in Science," *Social Studies of Science* 22, no. 4 (1992): 634–35. See also Anthony G. Hopwood, "The Archaeology of Accounting Systems," *Accounting, Organizations, and Society* 12, no. 3 (1987): 207–34; Grahame Thompson, "Is Accounting Rhetorical? Methodology, Luca Pacioli and Printing," *Accounting, Organizations, and Society* 16, nos. 5–6 (1991): 573–80; Gareth Morgan, "Accounting as Reality Construction: Towards a New Epistemology for Accounting Practice," *Accounting, Organizations, and Society* 13, no. 5 (1988): 477–85; and the essays collected in Anthony G. Hopwood and Peter Miller, eds., *Accounting as Social and Institutional Practice* (Cambridge: Cambridge University Press, 1994), esp. Peter Miller, "Accounting as Social and Institutional Practice: An Introduction" (1–40), and Keith Hoskin and Richard Macve, "Writing, Examining, Disciplining: The Genesis of Accounting's Modern Power" (67–97). After this chapter, *A History of the Modern Fact* does not deal extensively with accounting, and the story of why accounting was not used extensively for calculation or prediction until the nineteenth century remains to be written.

The difficulty that historians of political economy and statistics have had in narrating the relation between these modern sciences and political arithmetic materialized as early as 1825, when J. R. McCulloch recognized that William Petty played a central role in the "rise and progress" of political economy but was stymied because Petty's labor and monetary theories (not to mention his politics) did not accord with nineteenth-century British assumptions about laissez-faire. In twentieth-century histories of statistics and political economy, the gulf that separates seventeenth-century political arithmetic from its nineteenth-century descendants does not typically receive much attention, since even our best historians pay relatively little attention to the substantial differences between political arithmetic and more modern theories of wealth. See J. R. McCulloch, *Discourse on the Rise, Progress, Peculiar Objects, and Importance of Political Economy: Containing an Outline of a Course of Lectures on the Principles and Doctrines of That Science* (Edinburgh: Archibald Constable, 1825), esp. 37; Cullen, *Statistical Movement in Early Victorian Britain,* 1–16; and Porter, *Rise of Statistical Thinking,* 3–6, 18–23. Paul F. Lazarsfeld comes closest to identifying the discipline in which we can see the representational method associated with political arithmetic being taken over and transformed by the more immediate disciplinary ancestor of political economy, but he cannot describe the place that Scottish moral philosophy occupied in this genealogy. See "Quantification in Sociology," 283. I address this issue in chapters 4 and 5.

8. On the emergence of structural differentiation as a marker of "modernity," see Niklaus Luhmann, *The Differentiation of Society,* trans. Steven Holmes and Charles Larmore (New York: Columbia University Press, 1982). I discuss the emergence of modern "domains" in Poovey, *Making a Social Body,* 5–15.

9. John Mellis, *A Briefe Instruction and Maner How to Keepe Bookes of Accompts after the Order of Debitor and Creditor* (London: John Windet, 1588). This loose translation of Pacioli's treatise on accounting was a version of an older English double-entry textbook, no copies of which survive. Mellis says that he is "but the renuer and reviver of an auncient old copie printed here in London the 14. of August, 1543" by Hugh Oldcastle (2–3). Oldcastle's book was titled *Profitability Treatyce.* Mellis's text has no page numbers; I have added them for convenience of citation.

10. See, for example, Bruce G. Carruthers and Wendy Nelson Espeland, "Accounting for Rationality: Double-Entry Bookkeeping and the Rhetoric of Economic Rationality," *American Journal of Sociology* 97, no. 1 (1991): 30–67; Donald N. McCloskey, "The Rhetoric of Economic Expertise," in *The Recovery of Rhetoric: Persuasive Discourse and Disciplinarity in the Human Sciences,* ed. R. H. Roberts and J. M. M. Good (Charlottesville: University Press of Virginia, 1993), 137–47; and McCloskey, *If You're So Smart: The Narrative of Economic Expertise* (Chicago: University of Chicago Press, 1990).

11. This phrase is Lewes Roberts's description of accounting. See *Merchants Mappe of Commerce* (1638), 18.

12. See, for example, Raymond de Roover, "New Perspectives on the History of Accounting," *Accounting Review* 30 (July 1955): 408; B. S. Yamey, "The Functional Development of Double-Entry Bookkeeping," *Accountant,* November 1940, 333, 334–35; and Yamey, "Accounting and the Rise of Capitalism: Further Notes on a Theme by Sombert," *Journal of Accounting Research* 2 (1964): 117–36.

13. See James A. Aho, "Rhetoric and the Invention of Double-Entry Bookkeeping," *Rhetorica* 3, no. 3 (1985): 21–43; Grahame Thompson, "Early Double-Entry Bookkeeping and the Rhetoric of Accounting Calculation," in *Accounting as a Social and Institutional Practice,* ed. Anthony G. Hopwood and Peter Miller (Cambridge: Cambridge University Press, 1994), 40–66; and Hoskin and Macve, "Writing, Examining, Disciplining," 67–97.

14. I find the argument that early accounting was associated with university-trained men persuasive because of Luca Pacioli's role in codifying double-entry bookkeeping. Nevertheless, I am puzzled by Hoskin and Macve's tendency to misstate their sources' emphasis. For example, they quote Alexander Murray as saying that the merchant's role in accounting history was "not that of a pioneer or even that of a patron of pioneers," but they neglect to point out that Murray is dis-

cussing the history of mathematics and that he does not debunk the "counting-house theory" but supplements it. More telling, they quote Michael Clanchy as stating that "the great majority of clerks and accountants . . . were trained at universities" without including the first part of this sentence: "It is possible that." Elsewhere Clanchy states that "evidence is insufficient to establish for certain that most such clerks were university men," although he notes evidence that suggests they were. See Hoskin and Macve, "Writing, Examining, Disciplining," 73; Murray, *Reason and Society in the Middle Ages,* 193–94; and Michael Clanchy, "*Moderni* in Education and Government in England," *Speculum* 50 (1975): 685.

15. In the fourth century B.C. Xenophon defined "economy" as the management of a man's estate (his "house"). See *Oeconomicus,* trans. H. G. Dakyns as "The Economist," in *The Works of Xenophon* (London: Macmillan, 1897), vol. 3. For a discussion of Xenophon's influence on the fifteenth-century Italian architect Leon Battista Alberti, see Mark Wigley, "Untitled: The Housing of Gender," in *Sexuality and Space,* ed. Beatriz Colomina (New York: Princeton Architectural Press, 1992), 334, 338–40.

16. On the heterogeneous form of early account books, see Yamey, "Functional Development of Double-Entry Bookkeeping," 333. Yamey points out that these records included "pious mottoes, rules for moral conduct, and treatments for coughs and other ailments" as well as details about credit dealings (335). See also Madeleine Foisil, "The Literature of Intimacy," in *Passions of the Renaissance,* ed. Roger Chartier, vol. 3 of *A History of Private Life,* ed. Philippe Ariès and Georges Duby (Cambridge: Harvard University Press, 1989), 331–32, and Edward Peragallo, *The Origin and Evolution of Double Entry Bookkeeping* (New York: American Institute, 1938), 18–19.

17. See Wigley, "Untitled," 342–50.

18. On the creation of the study, see Orest Ranum, "The Refuges of Intimacy," in Chartier, *Passions of the Renaissance,* 225–27; on the use of the study, see Roger Chartier, "The Practical Impact of Writing," in Chartier, *Passions of the Renaissance,* 138–44.

19. Wigley, "Untitled," 348, 347. Wigley's study draws heavily on Jed's *Chaste Thinking,* chap. 3. Jed quotes Alberti: "I wanted to keep only the books, my writings, and the writings of my ancestors shut up, then and ever after, in such a way that my wife could never even see them, much less read them. I always kept my writings not in the sleeves of my clothes, but shut off and organized in my study almost as a sacred and religious thing. I never gave my wife permission to enter this place, either with me or alone, and I often commanded her that if ever she ran across some of my writing, to hand it over to me immediately" (80).

20. The challenges to Wigley's claims have come most directly from Alan Stewart, who argues that a male householder would almost always have been accompanied in his study by a male servant or secretary. To substantiate his point, Stewart refers to descriptions of early modern studies that show two desks and to early modern discourses about the role and responsibilities of the secretary. Homosocial activities in the study were a source of considerable concern in the period, Stewart argues, and early modern writing about the secretary reveals understandable anxiety about how his access to "secret" papers would affect political decisions and the transmission of these papers to subsequent generations. See Alan Stewart, "The Early Modern Closet Discovered," *Representations* 50 (spring 1995): 76–100.

21. Wigley, "Untitled," 340, 341.

22. In addition to Stewart's article, see Richard Rambuss, *Spenser's Secret Career* (Cambridge: Cambridge University Press, 1993), esp. chap. 2, and Jonathan Goldberg, *Writing Matter: From the Hands of the English Renaissance* (Stanford: Stanford University Press, 1990), 258, 267.

23. Wigley, "Untitled," 343. It would be interesting to pursue the possible connections between the constitution of early modern "risk" as excessive to writing and the feminizing of those uncontrollable factors lumped together as "Fortuna." J. G. A. Pocock has offered some provoca-

tive comments on the gendering of excess and risk in the eighteenth century, but this is one of many areas that still calls for more research. See Pocock, "The Mobility of Property and the Rise of Eighteenth-Century Sociology," in *Virtue, Commerce, and History: Essays on Political Thought and History, Chiefly in the Eighteenth Century* (Cambridge: Cambridge University Press, 1985), 103–24.

24. Grahame Thompson emphasizes the importance of printing in the codification and dissemination of double-entry bookkeeping. See Thompson, "Early Double-Entry Bookkeeping and the Rhetoric of Accounting Calculation," 55–62.

25. Yamey, "Functional Development of Double-Entry Bookkeeping," 335.

26. Wigley offers some provocative comments on this subject. See "Untitled," 349–89.

27. See Aho, "Rhetoric," 21–43. Aho points out that Pacioli was twenty-five when he met the sixty-six-year-old Alberti, and that he went to live with Alberti in 1479. It is interesting to speculate—although it must remain speculation—whether Pacioli and Alberti discussed the implications of removing accounting from the household, as publishing a textbook most surely did. On Pacioli's relation to Alberti, see also R. Emmett Taylor, *No Royal Road: Luca Pacioli and His Times* (Chapel Hill: University of North Carolina Press, 1942), chap. 8.

28. Thompson, "Early Double-Entry Bookkeeping," 51.

29. Bender and Wellbery, "Rhetoricality," 7. Other helpful studies of early modern rhetoric include Brian Vickers, *In Defense of Rhetoric* (Oxford: Oxford University Press, 1988), esp. chaps. 1, 4, and 5; Thomas M. Conley, *Rhetoric in the European Tradition* (Chicago: University of Chicago Press, 1990), esp. chaps. 1, 4, and 5, and Terence Cave, *The Cornucopian Text: Problems of Writing in the French Renaissance* (Oxford: Clarendon Press, 1979), chap. 1.

30. Viroli, *From Politics to Reason of State,* 3.

31. On the "old" or Ciceronian rhetoric, see Tuck, *Philosophy and Government,* chap. 1. I am grateful to Jeffrey Minson for calling my attention to Tuck's important work.

32. See Tuck, *Philosophy and Government,* 1–6. The generalizations I offer about Renaissance humanism pertained first and foremost to an elite group of men, for these men were the state's citizens, the "body politic." The fact of female monarchs presented certain challenges to theorists of Renaissance humanism.

33. Erasmus published *De Copia* in 1512. See Cave, *Cornucopian Text,* 3–4.

34. See Vickers, *In Defense of Rhetoric,* 4, 80–82, 284, and Erich Auerbach, "Figura," in *Scenes from the Drama of European Literature,* ed. Erich Auerbach, Theory and History of Literature, vol. 9 (Minneapolis: University of Minnesota Press, 1984), 11–76.

35. Erasmus, *On Copia of Words and Ideas,* trans. Donald B. King and H. David Rix (Milwaukee: Marquette University Press, 1963), 13.

36. Erasmus, *On Copia,* 11, 15. On *brevitas,* see 13. Cave comments that "the importance of the balance between *copia* and *brevitas* is that it enables Erasmus to suggest that true plenitude of language is to be found not in simple extension, but in invention and imaginative richness" (*Cornucopian Text,* 21).

37. The most significant secondary materials on the probabilistic revolution include Henry G. van Leeuwen, *The Problem of Certainty in English Thought, 1630–1690* (The Hague: Martinus Nijhoff, 1963); Barbara J. Shapiro, *Probability and Certainty in Seventeenth-Century England: A Study of the Relationships between Natural Science, Religion, History, Law, and Literature* (Princeton: Princeton University Press, 1983); Lorraine Daston, *Classical Probability in the Enlightenment* (Princeton: Princeton University Press, 1988); and Peter Dear, "From Truth to Disinterestedness in the Seventeenth Century," *Social Studies of Science* 22 (1992): 619–31. I return to this subject in chapter 3.

38. In 1651 Richard Dafforne tried to introduce more books into the English system. He ar-

gued that two kinds of waste books, which were used to record bank money and running accounts, and the monthly book, which functioned as a calendar, would help the English merchant transact business in Italy, where these books were already in use. See Dafforne, *The Merchants Mirrour* (London: Nicolas Bourn, 1651), preface.

39. All the early modern accounting manuals I have consulted stress the importance of this conversion. See Mellis, *Briefe Instruction,* 37, and Roberts, *Merchants Mappe,* 15–16, 18–19. The money of account was also referred to in this period as "imaginary money." On this important subject, see Luigi Einaudi, "The Theory of Imaginary Money from Charlemagne to the French Revolution," in *Enterprise and Secular Change: Readings in Economic History,* ed. Frederic Chapin Lane (Homewood, Ill.: Richard D. Irwin, 1953), 229–61.

40. So critical are numbers to all forms of accounting that one modern analyst has called number the dominant metaphor of accounting. See Keith Robson, "Accounting Numbers as 'Inscription': Action at a Distance and the Development of Accounting," *Accounting, Organizations and Society* 17, no. 7 (1992): 685, and Morgan, "Accounting as Reality Construction," 477–85.

41. For the negative connotations of number, see Taylor, *Mathematical Practitioners,* 4, and Swetz, *Capitalism and Arithmetic,* 248–49. Even Queen Elizabeth's patronage of the esteemed mathematician John Dee did little to dispel the association between number and black magic. On Dee, see Webster, *Great Instauration,* 119.

42. The distinction between precision and accuracy is not an old one; the two terms are cited in the *Oxford English Dictionary* as synonyms. Lorraine Daston has made this distinction in "Moral Economy," 8. David S. Landes also discusses the distinction; see Landes, *Revolution in Time: Clocks and the Making of the Modern World* (Cambridge: Harvard University Press, 1983), 78–83.

43. Aho describes this system; see "Rhetoric," 32. Mellis uses a system of cross-indexing to signify the transfer of entries: when the accountant entered a statement as a credit, the appropriate page number of the ledger was to be entered above a horizontal line in the left margin of the journal entry; when he transferred the entry to the debit page of the ledger, that page number should be entered below the horizontal line in the journal (*Briefe Instruction,* 43).

44. See Mellis, *Briefe Instruction,* 81.

45. This convention disappeared only in the nineteenth century. On its disappearance, see Aho, "Rhetoric," 36–37. It is noteworthy that this disappearance dovetailed with the professionalizing of accounting in Britain—as if the responsibility once "assumed" by these metaphors was transferred to the system of credentialing and certifying by which accountants' reliability was underwritten. On the nineteenth-century professionalizing of accounting, see Edgar Jones, *Accountancy and the British Economy, 1840–1980: The Evolution of Ernst and Whinney* (London: B. T. Batsford, 1981).

46. See Peter Miller and Ted O'Leary, "Accounting and the Construction of the Governable Person," *Accounting, Organizations and Society* 12, no. 3 (1987): 256–61, and Robson, "Accounting Numbers," 700. The concept of the "responsibilization of individuals" is related to the concept of "calculating selves." On "calculating selves" see Miller, "Accounting and Objectivity," 61–86. See also Aho, "Rhetoric," 38.

47. Jed discusses the secrecy of mercantile writing and the relation between this secrecy and the constitution of privacy in *Chaste Thinking;* see 78–89, 119–20.

48. [John Browne], *The Merchants Avizo* (London: J. Norton, 1607), 4–5.

49. In his *Essays upon Several Projects* (written in 1697), Daniel Defoe made explicit the role that risk always played in long-distance commercial transactions. See my discussion of this text in chapter 4.

50. Roberts, *Merchants Mappe,* 16–17.

51. "The new science envisioned by Bacon bespeaks a shift in the relationships of power that define discourse. Scientific discourse is no longer embedded within the array of relative power positions that characterizes a stratified or hierarchical social structure; it withdraws from this interpersonal fray and takes as its opposite nature itself, over which it endeavors to establish a total command. The subject who holds this new form of power is no longer an individual leader or a hegemonic group, but rather mankind in general, a neutral or abstracted subject, a role that can be represented by whoever attains to the neutrality requisite for exercising it" (Bender and Wellbery, "Rhetoricality," 8).

52. I have deliberately modeled this phrase, "economic matters of fact," on Steven Shapin and Simon Schaffer's "natural matters of fact" to signal that the former, like the latter, was the product of a specific social practice that involved not the discovery of immutable truths but the constitution of some kinds of knowledge *as truths*. See Shapin and Schaffer, *Leviathan and the Air-Pump*.

53. Malynes's and Misselden's texts were both published in 1623. Mun's *Englands Treasure by Forraign Trade* was not published until 1664, but Supple dates it to the 1622–23 debates. See B. E. Supple, *Commercial Crisis and Change in England, 1600–1642: A Study in the Instability of a Mercantile Economy* (Cambridge: Cambridge University Press, 1959), 211–12.

54. The most compelling discussion of this difficult and weighty subject appears in Jean-Christophe Agnew, *Worlds Apart: The Market and the Theater in Anglo-American Thought, 1550–1750* (Cambridge: Cambridge University Press, 1986), esp. chap 1. Agnew argues that a variety of attempts "to envisage a social abstraction—commodity exchange—that was lived rather than thought" were formulated between 1600 and 1650. In these texts, Britons were "feeling their way round a *problematic* of exchange; that is to say, they were putting forward a coherent and repeated pattern of problems or questions about the nature of social identity, intentionality, accountability, transparency, and reciprocity in commodity transactions—the who, what, when, where, and why of exchange" (9). Agnew also claims that by the sixteenth century "market" had become an abstraction that referred to all acts of buying and selling and to the price or exchange values of services and goods (41). He traces this progressive abstraction from the market*place* to the *process* of market transactions to the *principle of market exchanges* and exchange value (52–53). In this progression, Agnew notes a double movement: "On the one hand, linguistic usage indicates that exchange had moved outward, as the expanded circulation of commodities; on the other, it suggests that exchange had moved inward, as a subjective standard of commensurability against which the world itself could be judged" (53). Having described this development in general terms, Agnew targets the commercial crisis of the 1620s as a critical moment: under the pressures of this crisis, he writes, "the constituents of a new definition of economy—as an autonomous and self-regulating process of resource allocation—[began] to coalesce in the minds of commercial and courtly policy makers" (55). My argument departs from Agnew's only in the sense that I propose that the development of a writing style that signified economic "expertise"—that established the authority to make credible claims about economic matters of fact— was a critical component of the "coalescence" of this "new definition of economy."

55. Gerald de Malynes, *The Center of the Circle of Commerce, or A Refutation of a Treatise, Intitled "The Circle of Commerce, or The Ballance of Trade," Lately Published by E. M.* (London: William Jones, 1623), 10, 128; 53, 68, 120–21. Hereafter cited in the text as *Center*.

56. Edward Misselden, *The Circle of Commerce, or The Ballance of Trade* (London: John Dawson, 1623), 93–96, 7–11. Hereafter cited in the text as *Circle*.

57. Thomas Mun, *Englands Treasure by Forraign Trade, or The Ballance of our Forraign Trade is the Rule of our Treasure* ([1662 or 1623] London: Thomas Clark, 1664), 5. Mun does represent "forraign Trade" as part of a larger system; it is a means of increasing "wealth and treasure"; and it is

fed by supply and demand—"vent and consumption of all sides" (5, 18). Future references will be cited in the text as *Treasure.*

58. Br. Suviranta points out these exclusions; see *The Theory of the Balance of Trade in England: A Study in Mercantilism* (Helsinki: Suomal, Kirjall, Seuran Kirjap, 1923), 127–34. It is worth noting in passing that the word *capital* first appears in English discussions of commerce in Mellis's *Briefe Instruction,* but in the modern sense of that term, capital was not a prominent feature of the early seventeenth-century English economy.

59. See D. C. Coleman, "Mercantilism Revisited," *Historical Journal* 23 (1980): 773–91.

60. See Viroli, *From Politics to Reason of State,* esp. introduction and chap. 6, and Tuck, *Philosophy and Government,* esp. chaps. 2, 3.

61. Tuck, *Philosophy and Government,* 82.

62. The best accounts of these debates have been supplied by Supple, *Commercial Crisis,* 198–219, and Appleby, *Economic Thought,* chap. 2. Both Supple and Appleby tend to repeat many of the methodological operations of the seventeenth-century writers, however. In particular, Supple (like Mun) acknowledges that the numerical trade figures from the period are inaccurate and must be manipulated to be useful. That is, for ease of collation and comparison, Supple converts the figures supplied by the port books into a common denominator, the so-called notional shortcloths, which, he explains, constituted a "fictional 'cloth' of 24 yards" into which customs authorities translated the "bewildering multitude of textiles." In addition to using this "fictional" standard, Supple also rounds figures off; it is no wonder, he admits, that his statistics differ from those published by other modern scholars. See Supple, *Commercial Crisis,* appendix A, esp. 257, 258.

63. Supple, *Commercial Crisis,* 56–57.

64. See J. Thomas Kelly, *Thorns on the Tudor Rose: Monks, Rogues, Vagabonds, and Sturdy Beggars* (Jackson: University Press of Mississippi, 1977), esp. chaps. 7, 8; and A. L. Beier, *Masterless Men: The Vagrancy Problem in England, 1560–1640* (New York: Methuen, 1985), chaps. 1, 2.

65. Richard Halpern, *The Poetics of Primitive Accumulation: English Renaissance Culture and the Genealogy of Capital* (Ithaca: Cornell University Press, 1991), 69–75.

66. Appleby comments on the importance of the public—that is, nonofficial—nature of these pamphlets. See *Economic Thought,* 51.

67. Supple, *Commercial Crisis,* 56, 66–67.

68. Supple quotes an unpublished port report from 1621: "Our English monies are sold for more in other countries than they are valued for here at home, [which] causes the export of our coins . . . so that both decay of trade and want of money grew both from one and the same cause, which is undervalue of our money" (*Commercial Crisis,* 175).

69. This select committee consisted of Sir Robert Cotton, Sir Ralph Maddison, John Williams, William Sanderson, and Gerald Malynes. Supple, *Commercial Crisis,* 186.

70. Ibid., 186–88.

71. Appleby also makes this point about modern economists' complaints. See *Economic Thought,* 22. For William Letwin, these writers do not meet the criterion of theorists because their work was not "scientific," that is, systematic. See Letwin, *Origins of Scientific Economics,* viii.

72. The *OED* cites an older usage of *interested,* which derived from *interesse*'s original juridical meaning, dating from about 1665: "concerned, affected; having an interest or share in something." *Disinterested* has an equivalent earlier usage, from about 1612, meaning "without being interested or concerned." The change I note here marks these words' passage from the nonevaluative status of legal descriptors to the highly evaluative realm in which personal investments were attributed to actors. *Disinterested* bears a complex relation to the modern connotations of *objective,* a subject that Lorraine Daston addresses in "Moral Economy of Science," 2–24.

73. Giovanni Botera published his *Delle cause della grandezza della città* in 1588; in 1589 he published *Ragion de stato.* On Botera, see Tuck, *Philosophy and Government,* 65 67.

74. Dear, *Discipline and Experience,* 4, 23.

75. Although Supple voices considerable respect for Mun, for example, he dismisses Misselden's ideas as unsophisticated and Malynes's as outdated. Supple, *Commercial Crisis,* 201, 211.

76. This debate is discussed by Supple, *Commercial Crisis,* 198–211, and by Appleby, *Economic Thought,* 41–48.

77. My analysis in what follows is indebted to, but elaborates on, Appleby's interpretation. See *Economic Thought,* chap. 1. In my discussion of *raison d'état,* I follow Tuck and Viroli.

78. [Sir Robert Cotton], "A Speech . . . Touching the Alteration of Coin" (1626; printed 1651), in *Old and Scarce Tracts on Money,* ed. J. R. McCulloch (1856; reprint, London: P. S. King, 1933), 127. Cotton relates the distinction between extrinsic and intrinsic value to the distinction between the role gold played as a measure or standard of value and its status as a commodity (although he confusingly reverses the terms of the parallel he intends to establish): "I must distinguish the Monies of Gold and Silver, as they are Bullion or Commodities, and as they are Measure: One the extrinsick Quality, which is at the King's pleasure, as all other Measures to name; the other the intrinsick Quantity of pure Mettal, which is in the *Merchant* to value" (128).

79. Suviranta, *Theory,* 9.

80. Jean-Christophe Agnew argues that during the commercial crisis of the 1620s, contemporaries resorted to one of two interpretations, "one visible and secular; the other, invisible and divine. Both the mercantilist and the Puritan versions of events rooted the catastrophe in a theory of debasement and corruption, and both accordingly sought to purify, clarify, and rectify what each regarded as the critical conditions of belief: commercial and religious" (*Worlds Apart,* 54). Misselden, of course, targeted the monarch's "debasement" of coin.

81. B. E. Supple, "Currency and Commerce in the Early Seventeenth Century," *Economic History Review,* 2d ser., 10 (1957); 245 46. Suviranta argues that critics of the so-called mercantilists, beginning with Adam Smith, misinterpreted their concern with a favorable balance of trade as a belief that wealth consists of precious metals. This misinterpretation, Suviranta thinks, may stem from the mercantilists' failure to define or consistently distinguish among *riches, treasure, wealth,* and *money.* See *Theory,* 415–22.

82. Suviranta, *Theory,* 93–95; Supple, "Currency," 241–48.

83. Misselden's contribution to the debate about exchange also helped redefine "intrinsic" value; in keeping with Misselden's celebration of the market, Sir Robert Cotton asserted that the intrinsic *quantity* of metal in a coin was opposed to its extrinsic *quality* (presumably a reference to the metal's fineness) and that the former "is in the *Merchant* to value." Cotton, "Speech," 128. Rice Vaughan also defines intrinsic and extrinsic value in this way. See *A Discourse of Coin and Coinage* (London: Th. Dawks, 1675; reprinted in J. R. McCulloch, *Old and Scarce Tracts on Money* [1856; reprint, London: P. S. King, 1933]), 12.

84. On Aquinas's theory of justice and its relation to accounting, see Aho, "Rhetoric," 34–35.

85. For a historical discussion of the laws of nature, see John R. Milton, "The Origin and Development of the Concept of the 'Laws of Nature,'" *Archives Européennes de Sociologie* 22 (1981): 173–95. Milton argues that the modern understanding of the laws of nature was introduced in the early fourteenth century but not universally accepted until the seventeenth.

86. Two strategies enable Mun to confer on even explicitly figurative language the quality of transparency more typically associated with numerals. First, the trope that dominates his text is the simple prepositional figure that draws on a foundational analogy but does not make it explicit. Thus "decay of trade," "yoke of Spanish slavery," "canker of war," and "body of the trade" all allude to an organic metaphor that is never made explicit as such. The second way Mun under-

mines the inherently figurative nature of all language is by explicitly rejecting or explaining other people's figures of speech. Implicit in these strategies is the point that figurative language always carries theoretical arguments; to deflate a rival's trope is thus to expose the (fallacious) argument it smuggles into public opinion. It is said that "Mony is the Life of Trade," Mun scoffs; but "we know that there was a great trading by way of commutation and barter when there was little mony stirring in the world" (*Treasure*, 17). "Although *Treasure is said to be the sinews of War*, yet this is so because it doth provide, unite & move the power of men, victuals, and munition where and when the cause doth require; but if these things are wanting in due time, what shall we then do with our mony?" (70).

87. John R. Milton points out that natural philosophers thought of nature as law-governed *before* they "discovered" the laws of nature. Only with Hooke, he states, do we find empirically determined regularities being called laws. See "Concept of the 'Laws of Nature,'" 182.

88. Quoted in Suviranta, *Theory*, 42.

89. "In my poor opinion," Child declared in *Brief Observations concerning Trade*, "the enquiry, whether we get or lose, does not so much deserve our greatest pains and care, as how we may be sure to get; the former being of no use; but in order to the latter." See Sir Josiah Child, *Brief Observations concerning Trade, and Interest of Money* (London: Elizabeth Calvert, 1668; facsimile ed. *Sir Josiah Child, Merchant Economist, with a Reprint of "Brief Observations,"* ed. William Letwin [Cambridge: Harvard University Press, 1959]), 44.

90. Halpern, *Primitive Accumulation*, 32.

91. Quoted in ibid., 19.

92. Accounts of the reduction and displacement of rhetoric can be found in Bender and Wellbery, "Rhetoricality," 8–22, and Vickers, *In Defense of Rhetoric*, chaps. 3–5. I am indebted to John Guillory for help with this subject.

93. See Cave, *Cornucopian Text*, chap. 2.

94. For an elaboration of this argument, see Tuck, *Philosophy and Government*, 21–45, and the essays of Maurice Croll in J. Max Patrick and Robert O. Evans, eds., *Style, Rhetoric, and Rhythm* (Princeton: Princeton University Press, 1966). Peter N. Miller dissents from this mainstream view, arguing that "there was no clear break between a Ciceronian and Tacitean culture separating a republican Renaissance from a princely Baroque." Instead of a clear break, Miller argues that the "*new* emphasis on Tacitus" was accompanied by philosophers' placing a different emphasis on Cicero's work. See Miller, *Defining the Common Good*, 21, 22.

95. Scipione Ammirato, an influential reviser of Botero, stressed that the prince's advisers should be experts in this sense. See Viroli, *From Politics to Reason of State*, 273–74. On the original secrecy of reason-of-state arguments, see Viroli, 252 and chap. 6 in general.

96. Ibid., 3–4.

97. Ibid., 255.

98. Ibid., 262.

99. Fernandez de Villareal, *El politico cristiano o discursos politicos*, quoted in Miller, *Defining the Common Good*, 33.

100. See Viroli, *From Politics to Reason of State*, 273–80.

101. The assumption that politics could not possess a theory or even a method that could be applied like a grammar lies behind the theorists of state's rejection of Aristotle. Ibid., 262.

102. John Wheeler, *A Treatise of Commerce. Wherein are Shewed that Commodities Arising by a Well Ordered and ruled Trade, such as that of the Societie of Merchants Aduenturers is proued to be* (London: John Harison, 1601; facsimile ed. New York: Columbia University Press, 1931), 5–6.

103. Ibid., 26.

104. Ibid., 7, 26.

105. Mun, *Englands Treasure*, 2.

Chapter Three

1. The kind of knowledge considered capable of compelling assent, of course, was mathematical demonstration. The revision in ideas about the nature of truth belongs to what scholars have called the probabilistic revolution. Although the complexities of this subject have merited extensive scholarly treatment, for my purposes it is sufficient to note that during the seventeenth century, in both England and France, a philosophical language began to be developed that stressed degrees of certainty pertinent to kinds of knowledge instead of seeking to formulate some fixed and universal relation between words and ideas or things. The most significant secondary materials on the probabilistic revolution include Daston, *Classical Probability;* Shapiro, *Probability and Certainty;* Leeuwen, *Problem of Certainty;* and Dear, "From Truth to Disinterestedness," 619–31.

2. For a discussion of the poor reputation of merchants throughout the seventeenth century, see Letwin, *Origins of Scientific Economics,* 38–41, 87–89.

3. For biographical details of Petty's life and an early assessment of his work, see Charles Henry Hull, "Petty's Life," in William Petty, *The Economic Writings of Sir William Petty,* ed. Charles Henry Hull (Cambridge: Cambridge University Press, 1899; reprint, Fairfield, N.J.: Augustus M. Kelley, 1986), 1:xviii–xxxiii. In the early nineteenth century, J. R. McCulloch identified Petty as one of the first "liberal" economists, even though Adam Smith had dismissed Petty's political arithmetic in *Wealth of Nations.* See McCulloch, introduction to *An Inquiry into the Nature and Causes of the Wealth of Nations* (Edinburgh: Adam and Charles Black and William Tait, 1828); and Smith, *An Inquiry into the Nature and Causes of the Wealth of Nations,* ed. Edwin Cannan (1776; reprint, New York: Modern Library, 1937). For modern interpretations of Petty's importance, see Letwin, who dismisses him as insufficiently mathematical (*Origins of Scientific Economics,* 114); Ian Hacking, who also relegates Petty to a position of relative unimportance (*The Emergence of Probability: A Philosophical Study of Early Ideas about Probability* [Cambridge: Cambridge University Press, 1975], 102, 205, 209); and Alessandro Roncaglia, who has provided a modern revaluation of Petty's importance (*Petty: The Origins of Political Economy* [Armonk, N.Y.: M. E. Sharpe, 1985]).

4. The circle surrounding the friar Mersenne included Descartes, Fermat, Pascal, and Gassendi. The English group, which founded the "Invisible College," then the Royal Society, included Samuel Hartlib, Boyle, John Wilkins, Jonathan Goddard, John Wallis, Ralph Bathurst, and Thomas Willis. The secondary literature on these groups is extensive. Particularly helpful is Peter Dear, *Mersenne and the Learning of the Schools* (Ithaca: Cornell University Press, 1988).

5. Daston, "Baconian Facts," 337–63.

6. Ibid., 345.

7. Quoted in ibid., 351.

8. Dear, *Discipline and Experience,* 25. See also Peter Dear, "Jesuit Mathematical Science and the Reconstitution of Experience in the Early Seventeenth Century," *Studies in the History and Philosophy of Science* 18 (1987): 133–75.

9. Dear, *Discipline and Experience,* 25.

10. [Robertson], "'Exclusion of Opinions," 57.

11. Thomas Kuhn, *The Structure of Scientific Revolutions* (Chicago: University of Chicago Press, 1962), 52; my emphasis.

12. See Francis Bacon, *Novum Organum,* trans. and ed. Peter Urbach and John Gibson

(Chicago: Open Court, 1994), 49 (part 1, preface). Future references will be cited in the text by page number.

13. Bacon, "Preparation towards a Natural and Experimental History," in *Novum Organum*, 303. All other references to this work will be cited in the text as "Preparation."

14. See Julian Martin, *Francis Bacon, the State, and the Reform of Natural Philosophy* (Cambridge: Cambridge University Press, 1992), chap. 3. Thomas Cromwell, who initiated those government reforms to which both Francis Bacon and his father were committed, had adopted Italian political arguments during his travels on the Continent between 1500 and 1512. Cromwell began to implement policies based on reason-of-state arguments as early as the 1530s. See Martin, 11–15.

15. In the most extensive discussion of "experience," "experiment," and "induction" to date, Peter Dear points out that the practices developed by Robert Hooke and Isaac Newton—just to name two of Bacon's more prominent followers—differed considerably. See *Discipline and Experience*, 22. On the range of interpretations of Bacon made by various members of the Royal Society, see also K. Theodore Hoppen, "The Nature of the Early Royal Society," part 1, *British Journal for the History of Science* 9 (1976): 3–6.

16. The citation is from Aristotle's *Metaphysics* (1027a20–27) and is quoted by Daston in "Moral Economy of Science," 13. See also Dear, *Discipline and Experience*, chap. 1; Dear, "*Totius in Verba:* Rhetoric and Authority in the Early Royal Society," *Isis* 76 (1985): 145–61, and Dear, "Jesuit Mathematical Science," 133–75.

17. Thus we get phrases like "simple experience" and "ordinary experience," which Bacon associates with "bad demonstration"; and we find passages like the following, which make it clear that the distinctions he sought to establish could not be signified by words but could be shown only in practice: "The best demonstration by far is experience, so long as it holds fast to the experiment itself. . . .But the method of learning from experience in current use is blind and silly, so that while men roam and wander along without any definite course, merely taking counsel of such things as happen to come before them, they range widely, yet move little further forward. . . .Whereas in the true course of experience, one that will bring new works, divine wisdom and order should be the pattern before us. For God on the first day of creation created light only, devoting to that task an entire day, in which He created no material substance. In the same way and from experience of every kind, we should first of all discover causes and elicit true axioms; and seek experiments that bring light, not fruit" (*Novum Organum*, 78–79 [part 1, aphorism 70]).

18. For two useful discussions of the problematic relation between "experience" and "experiment" in Bacon's method, see Martin, *Francis Bacon*, chap. 6, and Peter Urbach, *Francis Bacon's Philosophy of Science: An Account and a Reappraisal* (La Salle, Ill.: Open Court, 1987), chap. 6. Martin writes that "Bacon used the single term 'experiment' to describe very different sorts of activity by the natural historian: the passive reporting both of observed craft practices and techniques, and of particular inquiries conducted by other men; the 'artificial' investigations he carried out himself; and any subsequent, 'more subtle,' investigations" (155).

19. Bacon describes the empirical school in *Novum Organum*, 70 (part 1, aphorism 64). He associates "experiment" with reading in the *New Atlantis* (see Martin, *Francis Bacon*, 137).

20. Bacon, *Advancement of Learning*, 165. This is contained in *The Works of Francis Bacon*, ed. and trans. James Spedding, Robert L. Ellis, and Douglas D. Heath (London: Longman, 1857–58), 3:165.

21. See Miller, *Defining the Common Good*, 32–33.

22. Ibid., 43–51.

23. Clarendon noted that "the same maxim of *Salus populi suprema lex,* which had been used to the infringing the liberties of the one, [had been] made use of for the destroying the rights of the other" (quoted in ibid., 42).

24. Quentin Skinner points out that the idea that interest might be a power greater than reason was "scarcely to be found in English political literature before the 1640s" (*Reason and Rhetoric,* 427–28). See also J. A. W. Gunn, *Politics and the Public Interest in the Seventeenth Century* (London: Routledge and Kegan Paul, 1969), 1.

25. The standard work on the intellectual ferment of this period is Webster, *Great Instauration,* esp. 67–77, 88–99. See also Wolfgang van den Daele, "The Social Construction of Science: Institutionalisation and the Definition of Positive Science in the Latter Half of the Seventeenth Century," in *The Social Production of Scientific Knowledge,* ed. Everett Mendelsohn, Peter Weingart, and Richard Whitley (Boston: Reidel, 1977), 27–54, and Pamela H. Smith, *The Business of Alchemy: Science and Culture in the Holy Roman Empire* (Princeton: Princeton University Press, 1994), 3–13.

26. Thomas Hobbes, *Leviathan,* ed. C. B. Macpherson (1651; London: Penguin Books, 1968), 186. All subsequent page numbers are to this edition.

27. "[Men] appeale from custom to reason, and from reason to custome, as it serves their turn; receding from custome when their interest requires it, and settling themselves against reason, as oft as reason is against them: Which is the cause, that the doctrine of Right and Wrong, is perpetually disputed, both by the Pen and the Sword· Whereas the doctrine of Lines, and Figures, is not so; because men care not, in that subject what be truth, as a thing that crosses no mans ambition, profit, or lust. For I doubt not, but if it had been a thing contrary to any mans right of dominion, *That the three Angles of a Triangle should be equall to two Angles of a Square;* that doctrine should have been, if not disputed, yet by the burning of all books of Geometry, suppressed, as farre as he whom it concerned was able" (*Leviathan,* 166).

28. See, for example, Victoria Kahn, who struggles with, then "resolves" the apparent contradiction between rhetoric and logic in *Leviathan* by deconstructing it; in her account there is no opposition between rhetoric and logic because Hobbes's "logic" is finally "only" rhetoric. See Kahn, *Rhetoric, Prudence, and Skepticism in the Renaissance* (Ithaca: Cornell University Press, 1985). For Hobbes's diatribes against metaphors, similes, and other rhetorical devices, see *Leviathan,* 109, 110, 115, 136–37.

29. Skinner, *Reason and Rhetoric,* 3, and conclusion, 426–37. Although I agree with Skinner's analysis, I think more can be said about the extent to which Hobbes's turn to rhetoric constituted a form of "interested" analysis.

30. Hobbes associated "facts" with history, which was of two kinds: natural history and civil history. He opposed both of these data-gathering modes of knowledge production to "science." See *Leviathan,* 147–48.

31. Skinner, *Reason and Rhetoric,* 381. Skinner meticulously identifies the places in the English and Latin texts of *Leviathan* where Hobbes uses this device (382).

32. Peter Dear describes the problems that the Aristotelian method encountered in the seventeenth century as follows: "If a science cannot confirm its own principles, how are those principles, on which the certainty of the science depends, to be established?" Dear also quotes Clavius's paraphrase of Aristotle's description of method: "Every doctrine, and every discipline, is produced from pre-existing knowledge, as Aristotle says, and demonstrates its conclusions from certain assumed and conceded principles" ("Jesuit Mathematical Science," 141).

33. Peter Dear points out that the Jesuit Niccolò Cabeo also borrowed mathematicians' rhetorical devices—the *formal* structure capable of legitimating constructed experiences—in his *Philosophia Magnetica* of 1629 (Dear, "Jesuit Mathematical Science," 168). I owe the point that

Hobbes used mathematics illustratively—as a trope, or a placeholder for certainty—to John Guillory. Evidence for this point comes from Hobbes's repeated denigration of the professional groups that actually practiced arithmetic and accounting (another referent of "reckoning"). Thus Hobbes directs scorn toward accountants (112) and merchants (281–82), although he writes approvingly of the sovereign as a *metaphorical* accountant "obliged by the Law of Nature, . . . to render an account thereof to God" (376).

34. "And as in Arithmetique, unpractised men must, and Professors themselves may often erre, and cast up false; so also in any other subject of Reasoning, the ablest, most attentive, and most practised men, may deceive themselves, and inferre false Conclusions; . . . But no mans Reason, nor the Reason of any one number of men, makes the certaintie; no more than an account is therefore well cast up, because a great many men have unanimously approved it" (Hobbes, *Leviathan,* 111). Hobbes uses the fallibility of such modes of knowledge production to justify setting up an absolute monarch as arbiter or judge.

35. See Albert O. Hirschman, *Passions and the Interests,* esp. 31–56. Hirschman notes that Hume used the terms "passion of interest" and "interested affections" as synonyms for "the love of gain" (37). Although I have found Hirschman's classic study an invaluable guide to this subject, I want to supply more detail to his assertion that there was a "semantic drift of the term 'interests' toward money-making" (40–41).

36. Wolfgang van den Daele has discussed this aspect of the scientific revolution. See "Social Construction," 41.

37. Thomas Sprat, *The History of the Royal-Society of London, for the Improving of Natural Knowledge* (London: Printed by T. R., 1667), 141–42.

38. Shapin and Schaffer, *Leviathan and the Air-Pump.*

39. Steven Shapin, *A Social History of Truth: Civility and Science in Seventeenth-Century England* (Chicago: University of Chicago Press, 1994).

40. Boyle, "The Origins of Forms and Qualities according to the Corpuscular Philosophy," in *Selected Philosophical Papers of Robert Boyle,* ed. M. A. Stewart (1666; reprint, Indianapolis: Hackett, 1991), 92. Steven Shapin argues that Boyle's readiness to acknowledge the intervention of accidents and the relatively high rate of failure experienced by early experimentalists was a strategy for establishing his own credibility as a historian of these events. See *Social History of Truth.* Lorraine Daston also discusses Boyle's preference for particularized narrative, but she focuses more specifically on his contribution to the constitution of the modern fact. See "Moral Economy of Science," 12–18.

41. The best discussion of the institutions of collective and virtual witnessing is Shapin and Schaffer, *Leviathan and the Air-Pump,* chap. 7.

42. Sprat, *History of the Royal-Society,* 67. As usual, Joseph Glanvill offers a stronger statement of the Society's principled practice of discrimination. Although he declares that all men "who were arriv'd to maturity of Understanding, and a good capacity to seek after Truth, might at length be permitted to *judge for themselves,*" the Society did not grant this privilege "to *immature* Youth; or to *illiterate* or *injudicious* Men, who are not to be *trusted* to conclude for themselves in things of *difficult* Theory." Glanvill notes that the Society "advised *such,* to submit to their Instructors, and so practice the plain thing they are taught, without busie intermedling in Speculative Opinions, and things beyond their reach" ("Anti-fanatical Religion, and Free Philosophy in a Continuation of the New Atlantis," in *Essays on Several Important Subjects in Philosophy and Religion,* ed. Richard H. Popkin [1676; facs. reprint, New York: Johnson Reprint Company, 1970], 12). Robert Boyle also included "housewives" among his list of unreliable observers of natural phenomena (*The Works of the Honourable Robert Boyle,* 6 vols. [London: J. and F. Rivington, 1772], 1:110). On the exclusion of women from natural philosophical knowledge production, see Deb-

orah Taylor Bazeley, "An Early Challenge to the Precepts and Practices of Modern Science: The Fusion of Fact, Fiction, and Feminism in the Works of Margaret Cavendish, Duchess of Newcastle (1623–1673)" (Ph.D. diss., University of California, San Diego, 1990), esp. appendix A, 229–64. Margaret Cavendish was the only woman allowed to attend a meeting of the Royal Society in this period, and to contemporaries her 1667 visit seemed noteworthy more for her costume than for her participation in natural philosophical discussions.

43. See Michael Hunter, *The Royal Society and Its Fellows, 1660–1700: The Morphology of an Early Scientific Institution*, 2d ed. (Oxford: Alden Press, 1994).

44. John Guillory's classroom discussions of Sprat's *History* have been instrumental in helping me clarify my argument about the Royal Society. For another discussion of the revisionism of Sprat's *History*, see Hoppen, "Nature of the Early Royal Society," 1–10.

45. Quoted in Dear, "*Totius in Verba*," 156.

46. Dear, "From Truth to Disinterestedness," 619–31.

47. Sprat, *History of the Royal-Society*, 86. Sprat just hints at the all-important relation between the rise of natural philosophy and England's imperial project when he calls London "the head of a *mighty Empire*, the greatest that ever commanded the *Ocean*" (87).

48. "Register" is a crucial concept for Sprat, for "register" was both the term for the person who took notes in the Society meetings (*History*, 94) and the word for the literal record of the Royal Society's experimental knowledge. As the written record of the experiments, the register was the Society's legacy, the form in which its knowledge and reputation could outlive its members. Significantly, moreover, the conventions by which such records were written aspired to efface those conventions themselves, to erase the very conditions of representation by which this kind of knowledge could be distinguished from its rivals. "The *Society* had reduc'd its principal observations, into one *common-stock,* and laid them up in publique *Registers,* to be nakedly transmitted to the next Generation of Men; and so from them, to their Successors. And as their purpose was, to heap up a mixt Mass of *Experiments,* without digesting them into any perfect model: so to this end, they confin'd themselves to no order of subjects; and whatever they have recorded, they have done it, not as compleat Schemes of opinions, but as bare unfinish'd Histories" (*History*, 115). The Society's register, then, served the same function, and observed some of the same conventions, as did the ledgers of the double-entry bookkeeping system.

49. See Francis Christensen, "John Wilkins and the Royal Society Reform of Prose Style," *Modern Language Quarterly* 7, no. 3 (1946): 279–90; Richard F. Jones, "Science and Language in England of the Mid-Seventeenth Century," *Journal of English and Germanic Philology* 31 (1932): 315–31; and Murray Cohen, *Sensible Words: Linguistic Practice in England, 1640–1785* (Baltimore: Johns Hopkins University Press, 1977), 1–42.

50. See William Petty, *The History of the Survey of Ireland, Commonly Called the Down Survey,* ed. Thomas Aiskew Larcom (Dublin: Irish Archaeological Society, 1851; reprint, New York: Augustus M. Kelly, 1967), 340, and T. C. Bernard, "Sir William Petty, Irish Landowner," in *History and Imagination: Essays in Honour of H. R. Trevor-Roper,* ed. Hugh Lloyd-Jones, Valerie Pearl, and Blair Worden (London: Duckworth, 1981), 201–17.

51. Bernard, "Sir William Petty," 201.

52. For biographical details, see note 3 above; see also Lord Edmond Fitzmaurice, *Life of Sir William Petty, Chiefly from Private Documents Hitherto Unpublished* (London: John Murray, 1895), and Tony Aspromourgos, "The Life of William Petty in relation to His Economics: A Tercentenary Interpretation," *History of Political Economy* 20, no. 3 (1988): 337–40.

53. On the inadequacies of Worsley's survey, see T. W. Moody, F. X. Martin, and F. J. Byrne, eds., *Early Modern Ireland, 1534–1691,* vol. 3 of *A New History of Ireland* (Oxford: Clarendon Press, 1976), 371, and Petty, *History of the Down Survey,* chaps. 1–4.

54. On the terms of the Act of Settlement (1652), see Moody, Martin, and Byrne, *Early Modern Ireland*, 361–66.

55. For the first interpretation, see Nicholas Canny, *From Reformation to Resistance: Ireland, 1534–1660* (Dublin: Helicon, 1987), esp. 38–50, 113–31, and 159–72. For the second, see Roncaglia, *Petty*, 5.

56. Beyond a pen-and-ink drawing of the island, no general map of Ireland existed in 1652, and, although the country had witnessed four major forfeitures and redistributions of land since the mid-sixteenth century, only one of these had produced anything like an exact measurement of estates. See Victor Morgan, "The Cartographic Image of 'The Country' in Early Modern England," *Transactions of the Royal Historical Society*, 5th ser., 29 (1979): 137 n. 21, for descriptions of existing maps of Ireland, and J. H. Andrews, "Appendix: The Beginnings of the Surveying Profession in Ireland—Abstract," in *English Map-Making, 1500–1650*, ed. Sarah Tyacke (London: British Library, Reference Division Publications, 1983), 20–21, for an account of the 1586–89 survey. For an account of the making of a general map of Ireland (at the end of the eighteenth century), see W. A. Seymour, ed., *A History of the Ordnance Survey* (Folkestone: Dawson, 1980), 79–80.

57. Petty, *History of the Down Survey*, 13–14.

58. Ibid., 97.

59. For an account of these incidents, see Larcom's preface to the Down Survey (*History of the Down Survey*). Petty wrote two responses to these charges, the *History of the Down Survey* and the more vituperative "Reflections upon Some Persons and Things in Ireland," probably composed in 1660. For my analysis of the place of the Down Survey in Petty's contribution to economic rationality, see Poovey, "Social Constitution of 'Class,'" esp. 20–32.

60. Historians disagree about how central a role Petty played in the creation of modern economic theory. Scholars like William Letwin and Ian Hacking, who are primarily interested in "scientific economic thought" and statistics, have dismissed him (Letwin, *Origins of Scientific Economics*, 114, and Hacking, *Emergence of Probability*, 102, 205, 209). By contrast, the Marquise of Lansdowne, editor of *The Petty Papers* (and a descendant of Petty), declares that Petty "was the real founder and inventor of the science of statistics" because "he was the first to attempt the systematic collection of social facts and figures and to base upon them deductions both political and economical" (*The Petty Papers: Some Unpublished Writings of Sir William Petty* [London: Constable, 1927] 1:xvii]).

61. Patricia Coughlan briefly discusses Petty's use of Ireland as a laboratory. See "'Cheap and Common Animals': The English Anatomy of Ireland in the Seventeenth Century," in *Literature and the English Civil War*, ed. Thomas Healy and Jonathan Sawday (Cambridge: Cambridge University Press, 1990), esp. 220–21. T. C. Bernard argues that even though Petty hoped to use his Kerry estates as a laboratory, the impediments he encountered there—both physical and economic—kept him from doing so. See Bernard, "Sir William Petty," 201–17.

62. According to Theodore M. Porter, these descriptors help explain why "the language of quantity" has been embraced by modern governments. See Porter, *Trust in Numbers*, ix. Although Porter's outstanding study helps explain the modern affinity between governments and numerical representation, he does not pursue this relationship to its early phase when, I argue, numbers had to *be invested with* the connotations of impartiality and rigor.

63. Sir William Petty, *The Economic Writings of Sir William Petty*, ed. Charles Henry Hull (Cambridge: Cambridge University Press, 1899; reprint, Fairfield, N.J.: Augustus M. Kelley, 1986), 26, 36, 32–33, 34–36, 49, 94–95. Future references will be cited in the text by page number.

64. See G. N. Clark, *Guide to English Commercial Statistics, 1696–1782* (London: Royal Historical Society, 1938).

65. See Joan Thirsk, *Economic Policy and Projects: The Development of a Consumer Society in Early Modern Europe* (Oxford: Clarendon Press, 1978), 138.

66. See D. W. Jones, *War and Economy in the Age of William III and Marlborough* (Oxford: Basil Blackwell, 1988), 43.

67. See Kristof Glamann, "The Changing Patterns of Trade," in *The Economic Organization of Early Modern Europe,* ed. E. E. Rich and C. H. Wilson, vol. 5 of *The Cambridge Economic History of Europe* (Cambridge: Cambridge University Press, 1977), 189–93, 263–64. On English shipping, see Jones, *War and Economy,* 44–49.

68. For Petty, merchants were not the center of the circle of commerce but were simply agents who redistributed "superfluous" goods. In his scheme, keeping reliable accounts would not enhance the power of merchants by privileging mercantile instruments like double-entry bookkeeping but would enable the king to regulate the number of merchants, so as to strengthen the true source of England's wealth—the laboring poor. "By good Accompts of our growth, Manufacture, Consumption, and Importation, it might be known how many Merchants were able to mannage the Exchange of our superfluous Commodities with the same of other Countries. . . . Upon these grounds I presume a large proportion of these also might be retrenched, who properly and originally earn nothing from the Publick, being onely a kinde of Gamesters, that play with one another for the labours of the poor; yielding of themselves no fruit at all, otherwise then as veins and arteries, to distribute forth and back the blood and nutritive juyces of the Body Politick, namely, the product of Husbandry and Manufacture" (Petty, *Economic Writings,* 28).

Petty's emphasis on labor as the source of production (and therefore wealth) constitutes one of his major theoretical contributions to what would become classical economics. By emphasizing labor, Petty laid the groundwork for a redefinition of poverty, the emergence of more new concepts—including those of surplus labor and a labor theory of value—and the commodification of aggregate productivity, which is the cornerstone of the concept of "national wealth."

69. Whereas Malynes and (to a lesser extent) Misselden measured wealth by money, and whereas all the mercantilists treated gold and silver as measures of value, Petty was relatively uninterested in money. For a discussion of Petty's views on money, see Roncaglia, *Petty,* chap. 3. One sign of the decreased importance he assigned to money was his repudiation of the ban on usury. Petty defined interest as "a Reward for forbearing the use of your own Money for a Term of Time agreed upon, whatsoever need your self may have of it in the meanwhile." Of such interest, he commented that "wherefore when a man giveth out his money upon condition that he may not demand it back until a certain time to come, whatsoever his own necessities shall be in the mean time, he certainly may take a compensation for this inconvenience which he admits against himself" (*Treatise of Taxes,* in *Economic Writings,* 47).

70. "Our Silver and Gold we call by severall names, as in *England* by pounds, shillings, and pence, all which may be called and understood by either of the three. But that which I would say upon this matter is, that all things ought to be valued by two natural Denominations, which is Land and Labour; that is, we ought to say, a Ship or a garment is worth such a measure of Land, with such another measure of Labour; forasmuch as both Ships and Garments were the creatures of Lands and mens labours thereupon: This being true, we should be glad to finde out a natural Par between Land and Labour, so as we might express the value by either of them alone as well or better then by both, and reduce one into the other as easily and certainly as we reduce pence into pounds" (Petty, *Economic Writings,* 44–45).

71. Roncaglia comments that Petty's various attempts to find a natural equivalence between land and labor were unsuccessful. "What is lacking in Petty's attempts to resolve the problem of the determination of relative prices is a perception of the simple fact that the problem is integrally

related to the operation of the economic system as a whole and not to a single productive sector of the economy" (*Petty*, 84). As we will see, what appears to Roncaglia as a "simple fact" was made possible by analyses such as Petty's, even though Petty did not draw from his model the conclusions Roncaglia draws.

72. Along with John Graunt, Petty had been involved in compiling the first modern bills of mortality for London, but these were partial, limited to the city, and difficult to interpret. Partly as a consequence, the London mortality bills could be used to "support" a priori assumptions, as Petty does here. "Observations upon the Bills of Mortality" can be found in Petty, *Economic Writings*, 314–435.

73. Porter suggests this when he says that the "weakest point" of quantitative reasoning is "the contact between numbers and the world" (*Trust in Numbers*, 5).

74. The analyst whose interpretation of Petty's method most closely approaches my own is A. M. Endres, "The Functions of Numerical Data in the Writings of Graunt, Petty, and Davenant," *History of Political Economy* 17, no. 2 (1985): 245–64. Endres argues that Petty treated numerical data as "indicators," although in doing so he reinforces the larger theoretical point I have been making: that the modern fact could be interpreted either as separate from theory or as imbricated in theory. "When data are interpreted they become indicators. Use of data presupposes intentions, purposes, interests, and value judgments on the part of the interpreter. The data are used to *mean* something. . . .In the process of use, boundaries between 'raw' data and the sense of reality that they stand in for becomes blurred. The data coalesce with this unobservable sense of reality and function as indicators" (246).

75. We can see the difference between Hobbes's method and Petty's by contrasting Hobbes's statement about the "value of a man" with Petty's efforts to quantify this concept. Whereas Hobbes's formula for deriving the "*Value*, or WORTH of a man" drew primarily upon *social* factors and avoided figures of arithmetic—even as tropes—Petty's formula was arithmetical. Hobbes's formula begins as if it will involve numbers, but its terms immediately become ethical: "The *Value*, or WORTH of a man, is as of all other things, his Price; that is to say, so much as would be given for the use of his Power; and therefore it is not absolute; but a thing dependant on the need and judgment of another" (*Leviathan*, 151). Petty's calculation, by contrast, assigned numbers and deployed the rules of arithmetic; in so doing, he avoided altogether the ethical implications that might be associated with commodifying human beings. See Petty, *Economic Writings*, 106, 152, 267, 454, 512, 564.

76. Here is Petty's comment on the experiments he wanted to conduct in Ieland: "Wherefore matters being not as yet prepared for these Experiments, I can say nothing clearly of them; Only, That it seems by the best Estimates and Approaches that I have been able to make, that *London* is more healthful than *Dublin* by 3 in 32" (*Economic Writings*, 172). Petty does claim that the conclusions he has just presented ("that at *Dublin* the Wind blows 2 parts of 5 from the South-West to the West," etc.) *should* and—under the right conditions, *could*—be verified by actual experiments, which could be conducted using the instruments he has already described ("an Instrument to measure the motion of the Wind, and consequently its strength," etc.; *Economic Writings*, 172, 171). Because these experiments had not been conducted, however, the numbers that pepper his six propositions constitute informed guesses. This means that the numbers in his conclusion, like the numbers that ground his computation of the value of a people, have been derived—albeit by logical deduction—from numbers that were essentially conjectural.

77. Petty, *Economic Writings*, 244. The *Political Arithmetick*, probably written in 1671 or 1672, seems to be a direct rejoinder to Roger Cole's two-part treatise, published in 1671. The first part is titled *A Treatise Wherein is Demonstrated, That the Church and State of England, are in Equal Danger*

with the Trade of It; the second is called *Reasons of the Increase of the Dutch Trade. Wherein is Demonstrated from What Causes the Dutch Govern and Manage Trade Better Than the English; Whereby They Have So Far Improved Their Trade above the English.* See Petty, *Economic Writings,* 242 n. 4. For the dating of the *Political Arithmetick,* see *Economic Writings,* 235–36.

78. For a discussion of these events, see Bernard, "Sir William Petty," esp. 204, 207–10.

79. See Nicholas Canny, *Kingdom and Colony: Ireland in the Atlantic World, 1560–1800* (Baltimore: Johns Hopkins University Press, 1982), 115–16.

80. Peter Buck notes that "from the 1660s on, an increasing number of civil servants were drawn into the Royal Society, attracted by a sense of sharing common intellectual orientations with its scientist members. These men were the principal architects of a steady expansion and strengthening of the British state's central administration, and in the work of Graunt and Petty they found an enormously appealing approach to the problems of government. Political arithmetic provided the intellectual underpinnings for their conviction that the bases of stable and effective rule were not to be laid simply by limiting or redistributing political power but required instead increased administrative efficiency and therefore an expanded role for practical knowledge" (Buck, "Seventeenth-Century Political Arithmetic," 80–81).

81. Mario Biagioli, *Galileo, Courtier: The Practice of Science in the Culture of Absolutism* (Chicago: University of Chicago Press, 1993), chap. 1. Biagioli argues that its usefulness in warfare also enhanced the prestige of mathematics.

82. Henry VIII appointed Kratzer king's astrologer and horologer in 1519, at least partly because he had created an ingenious timepiece; John Rotz solicited Henry's patronage in 1542 by promising to bring the king a new compass; John Dee served Queen Elizabeth in many capacities, but his contributions were epitomized by the significant improvements he made in navigational instruments; and Thomas Bedwell, who was never a court mathematician, successfully parlayed his improved gunner's rule into a position as keeper of the ordnance store, through the patronage of the earl of Warwick. See Taylor, *Mathematical Practitioners,* 12–13, 15–20, and Mordechai Feingold, *The Mathematicians' Apprenticeship: Science, Universities, and Society in England, 1560–1640* (Cambridge: Cambridge University Press, 1984), 77. On Bedwell, see Johnston, "Mathematical Practioners," 321–30.

83. The most comprehensive study of the mathematical practitioners is still Taylor, *Mathematical Practitioners;* see esp. 21–27. See also Feingold, *Mathematicians' Apprenticeship,* chap. 6. On Thomas Hood, see Johnston, "Mathematical Practitioners," 332–41.

84. Taylor, *Mathematical Practioners,* 40–41; Johnston, "Mathematical Practitioners," 334.

85. If one accepts the implications of John Stow's *Survey of London* (1598), the Privy Council failed to renew the lectureship primarily because Stapler's Chapel was needed for other purposes: in 1592 the government commandeered the chapel, along with the rest of Leadenhall, to store the plunder that Walter Ralegh seized from a Portuguese ship. Johnston speculates, however, that Hood's preference for astronomy, geography, surveying, and hydrography over the more obviously military applications of mathematics contributed to the demise of the government's support. See "Mathematical Practitioners," 332–34.

86. Johnston, "Mathematical Practioners," 336.

87. Ibid., 334–35, 330.

88. See Turner, "Mathematical Instruments," 52, 53 n. 3.

89. Taylor, *Mathematical Practitioners,* 45–48; Feingold, *Mathematicians' Apprenticeship,* 195–206.

90. See Turner, "Mathematical Instruments," 61–64. On the various motives gentlemen might have had for forming such collections and for the history of instrument collections in par-

ticular, see also Gerard L'E. Turner, "The Cabinet of Experimental Philosophy," in *The Origins of Museums: The Cabinet of Curiosities in Sixteenth- and Seventeenth-Century Europe,* ed. Oliver Impey and Arthur MacGregor (Oxford: Clarendon Press, 1985), 214–22.

91. Johnston, "Mathematical Practitioners," 335. For a description of these instruments, see Turner, "Mathematical Instruments," 68–88.

92. Quoted in Turner, "Mathematical Instruments," 61. In the 1650s Robert Boyle noted that the same pattern obtained in the arena of experimental knowledge more generally: "There are many ingenious persons, especially among the nobility and gentry, who, having been first drawn to this new way of philosophy by the sight of some experiments, which for their novelty or prettiness they were much pleased with, or for their strangeness they admired, have afterwards delighted themselves to make or see variety of experiments, without having ever had the opportunity to be instructed in the rudiments or fundamental notions of that philosophy whose pleasing or amazing productions have enamoured them of it" (quoted in Feingold, *Mathematicians' Apprenticeship,* 193–94).

93. Joseph Glanvill, "Of the Modern Improvements of Useful Knowledge," in Glanvill, *Essays,* 23. The instruments Glanvill singled out for notice were the telescope, the microscope, the thermometer, and the air pump.

94. Quoted in Buck, "Seventeenth-Century Political Arithmetic," 81.

95. John Arbuthnot, "Essay on the Usefulness of Mathematical Learning," in *The Life and Works of John Arbuthnot,* ed. George A. Aitken (Oxford: Clarendon Press, 1892), 421–22.

96. Peacham could not "see how a gentleman, especially a soldier and commander, may be accomplished without geometry, though not to the height of perfection, yet at the least to be grounded and furnished with the principles and privy rules hereof." Henry Peacham, *"The Complete Gentleman," "The Truth of Our Times," and "The Art of Living in London,"* facsimile ed., ed. Virgil B. Heltzel (1622; Ithaca: Cornell University Press, 1962), 89.

97. From Charles Davenant, *The Political and Commercial Works of that Celebrated Writer, Charles Davenant . . . collected and revised by Sir Charles Whitworth,* 5 vols. (London, 1771), 1:131, 135; quoted in Brewer, *Sinews of Power,* 224.

98. Quoted in Turner, "Mathematical Instruments," 54. This comment comes from an address Jackson delivered to a mathematical society in 1719 and thus undoubtedly contains an element of self-promotion.

By 1723 Edward Stone confidently credited instruments with promoting what amounted to a revolution in the reputation of mathematics: "Mathematics are now become a popular Study, and make a part of the Education of almost every Gentleman," Stone explained. "Mathematical Instruments are the means by which those Sciences are rendered useful in the Affairs of Life. By their assistance it is that subtile and abstract Speculation is reduced into Art. They connect as it were, the Theory and the Practice, and turn what was bare Contemplation, to the most substantial Uses" (quoted in Turner, "Mathematical Instruments," 51).

99. See Brewer, *Sinews of Power,* 65–66. Brewer notes that the first figure would have been low because it did not include many lower-level administrators. The number of government tax collectors increased dramatically because, as Petty recommended, tax collection was removed from the hands of tax farmers; this occurred for customs in 1672 and for excise in 1683. See Jones, *War and Economy,* 66.

100. See Brewer, *Sinews of Power,* 226–27. The first printed source of parliamentary information, *Votes and Proceedings,* was available from the 1680s. In the 1740s the Commons voted to have their *Journals* put into print; complete back sets of these documents were available by the late 1760s.

101. See Peter Buck, "People Who Counted: Political Arithmetic in the Eighteenth Century," *Isis* 73 (1982): 28–45.

102. See Brewer, *Sinews of Power,* 225.

Chapter Four

1. [Sir Richard Steele], *The Spectator,* ed. Gregory Smith (New York: Dutton, 1970), 2:16, 17.

2. Ibid., 2:19. Sir Andrew associates "prudence" with numbers by this logic: "Numbers are so much the Measure of every thing that is valuable, that it is not possible to demonstrate the Success of any Action, or the Prudence of any Undertaking, without them. . . .When I have my Returns from Abroad, I can tell to a Shilling by the Help of Numbers the Profit or Loss by my Adventure; but I ought also to be able to shew that I had Reason for making it, either from my own Experience or that of other People, or from a reasonable Presumption that my Returns will be sufficient to answer my Expence and Hazard; and this is never to be done without the Skill of Numbers" (2:18).

3. As Peter N. Miller describes the situation, when loyalty to an opposition party could so easily be depicted by the party in power as *dis*loyalty to the nation, the opposition's only recourse "was to claim that *they,* in fact, had the best interest of the nation at heart while the governors were moved solely by desire for personal gain. The identification of personal with national interest made licit the existence of party" (Miller, *Defining the Common Good,* 89. On the political changes that led to the growth of political parties, see J. H. Plumb, *The Growth of Political Stability in England, 1675–1720* (London: Macmillan, 1967); John Cannon, *Aristocratic Century* (Cambridge: Cambridge University Press, 1984); Johnathan Clark, *Dynamics of Power: The Structure of Politics* (Cambridge: Cambridge University Press, 1985); and Linda Colley, *In Defiance of Oligarchy* (Cambridge: Cambridge University Press, 1983).

4. For a cornucopia of examples of this kind of fact, see Barbara Maria Stafford, *Artful Science: Enlightenment Entertainment and the Eclipse of Visual Education* (Cambridge: MIT Press, 1994), chaps. 1, 2.

5. On rational recreations and other popular forms of knowledge production, see, in addition to Stafford's *Artful Science,* Simon Schaffer, "Self-Evidence," in *Questions of Evidence: Proof, Practice, and Persuasion across the Disciplines,* ed. James Chandler, Arnold I. Davidson, and Harry Harootunian (Chicago: University of Chicago Press, 1994), 56–91, and Steven Shapin, "The Audience for Science in Eighteenth-Century Edinburgh," *History of Science* 12 (1974): 95–121. On the growth of political parties, see the citations in note 3 above.

6. Colin Gordon explains that what Foucault meant by "governmentality" or "rationality of government" was "a way or system of thinking about the nature of the practice of government (who can govern; what governing is; what or who is governed), capable of making some form of that activity thinkable and practicable both to its practitioners and to those upon whom it was practised. . . .Foucault was interested in the philosophical questions posed by the historical, contingent and humanly invented existence of varied and multiple forms of such a rationality" (Gordon, "Governmental Rationality," 3).

7. Although I do think that Petty's political arithmetic belongs to the "science of police," I argue that eighteenth-century efforts to apply vestiges of the political arithmetic method in Britain significantly departed from those applications of "police" that were implemented in France and the Germanic states. For a discussion of these Continental practices, and "police" more generally, see Gordon, "Governmental Rationality," 10–14, and Pasquale Pasquino, "Theatrum Politicum: The Genealogy of Capital—Police and the State of Prosperity," in *The Foucault Effect: Studies in*

Governmentality, with Two Lectures and an Interview with Michel Foucault, ed. Graham Burchell, Colin Gordon, and Peter Miller (Chicago: University of Chicago Press, 1991), 105–18.

8. For a discussion of the liberal mode of governmentality, see Burchell, "Peculiar Interests," 119–51. On fashion and consumption in the eighteenth century, see Neil McKendrick, "The Commercialization of Fashion," in *The Birth of a Consumer Society: The Commercialization of Eighteenth-Century England,* ed. Neil McKendrick, John Brewer, and J. H. Plumb (Bloomington: Indiana University Press, 1985), 34–98. The best poststructuralist theories of government by fashion can be found in Jean Baudrillard, *For a Critique of the Political Economy of the Sign,* trans. Charles Levin (St. Louis, Mo.: Telos Press, 1981); and Roland Barthes, *The Fashion System,* trans. Matthew Ward and Richard Howard (New York: Hill and Wang, 1983).

9. The classic text on the financial revolution is P. G. M. Dickson, *The Financial Revolution: A Study in the Development of Public Credit, 1688–1756* (London: Macmillan, 1967). The most influential treatments of the imaginative effects of the new system of public credit are Pocock, "Mobility of Property," 103–24, and J. G. A. Pocock, *The Machiavellian Moment: Florentine Political Thought and the Atlantic Republican Tradition* (Princeton: Princeton University Press, 1975), chap. 14. See also Nicholson, *Writing and the Rise of Finance,* introduction and chap. 2.

10. Defoe, *Review,* 7, nos. 137, 104 (quoted in Nicholson, *Writing and Finance,* 16). For a provocative discussion of Defoe's effort to discriminate between "good" and "bad" credit and the association he made between financial credit and credibility more generally, see Simon Schaffer, "Defoe's Natural Philosophy and the Worlds of Credit," in *Nature Transfigured: Science and Literature, 1700–1900,* ed. John Christie and Sally Shuttleworth (New York: Manchester University Press, 1989), 13–44.

11. See Pocock, *Machiavellian Moment,* chap. 14.

12. Nicholson points out that among the subscribers to the *Spectator* "were directors of the Bank of England . . . goldsmiths, private bankers or moneylenders. Eight were East India Company directors, and twenty were directors of the South Sea Company. Moreover, the single largest group of subscribers included the 'great body of secretaries, commissioners, clerks, and agents in the various branches of government, civil and military, required to carry on the war abroad and manage affairs at home.' Given such a constituency," Nicholson concludes, "it became a natural concern of Addison and Steele to promote a polite and civilising sense of participation in the new society being developed, acclimatising its readers to market priorities and procedures and familiarising them with codes and conventions of recognition and self-recognition appropriate to their place in a burgeoning world. More than anything else, a main function of *The Spectator* was to make its readers feel like they *were* a social constituency and one of growing significance, for whom polite culture could mediate as a validating and confidence-building network of relationships" (*Writing and Finance,* 55).

13. Ronald Paulson, *The Beautiful, Novel, and Strange: Aesthetics and Heterodoxy* (Baltimore: Johns Hopkins University Press, 1996), esp. introduction and chaps. 1, 2, 3. Of Addison, Paulson remarks, "The more accommodating Addison . . . modulates the austere virtue of civic humanism into politeness, and extends the amenities across a broader spectrum of society, noting that the 'man of a Polite Imagination' feels 'greater Satisfaction in the Prospect of Fields and Meadows, than another does in the Possession'—precisely because it is *not* his own property and he sees in it the perspective of commerce and paper money, rather than inheritance, upkeep, and tenantry" (50–51).

14. Paulson, *Beautiful, Novel, and Strange,* 2.

15. From Shaftesbury, *Inquiry concerning Virtue and Merit* (1699, 1711); quoted in Paulson, *Beautiful, Novel, and Strange,* 3.

16. As Paulson points out, Shaftesbury's claim that the gentleman-connoisseur was indifferent

to private interest did not go uncontested in the early eighteenth century. Mandeville was perhaps Shaftesbury's most persistent critic, but Hogarth soon took up the cause. "As it appeared to Mandeville, Shaftesbury had politicized disinterestedness; his hidden agenda was that Whig lords—rather than, as prior to 1688, a monarch—should rule England, politically and in matters of taste. Mandeville's strategy in the 1724 edition of *The Fable of the Bees* was to expose the desire, economic and sexual, under the supposed disinterestedness of Shaftesbury's civic humanism. Following Mandeville, Hogarth begins in *A Harlot's Progress* to reveal beneath the supposed disinterestedness and benevolence of the Shaftesburian man of taste (the connoisseur and collector) a subtext of ownership, control, and desire; and, in *A Rake's Progress,* beneath the idealized classical images, their use by Whig politicians for purposes of status and power" (Paulson, *Beautiful, Novel, and Strange,* 28).

17. See my essay "Social Constitution of 'Class,'" 41–42, and John Barrell, "The Public Prospect and the Private View: The Politics of Taste in Eighteenth-Century Britain," in Barrell, *Birth of Pandora,* 42, 45. The most consequential studies of aesthetics and disinterestedness in the domain of literature and art include, in addition to Paulson's *Beautiful, Novel, and Strange,* John Darrell, "'The Dangerous Goddess': Masculinity, Prestige, and the Aesthetic in Early Eighteenth-Century Britain," in Barrell, *Birth of Pandora,* 63–88; Jerome Stolnitz, "On the Origins of 'Aesthetic Disinterestedness,'" *Journal of Aesthetics and Art Criticism* 20, no. 2 (1961): 131–43; and Martha Woodmansee, "The Interests in Disinterestedness: Karl Philipp Moritz and the Emergence of the Theory of Aesthetic Autonomy in Eighteenth-Century Germany," *Modern Language Quarterly* 45, no. 1 (1984): 22–47.

18. See, for example, William Derham's Boyle Lectures of 1713, titled *Physico-theology, or A Demonstration of the Being and Attributes of God, from the Works of Creation* (London, 1714).

19. In 1678 Petty represented the king's failure to implement some policies that Petty had recommended as a violation of the "natural laws" of trade. "I have lately perused all the Acts relating to Trade and Manufacture which are of force in Ireland," Petty lamented to Southwell, "and could without Tears see them all repealed as Incroachments upon the Laws of Nature. For Trade will endure no other Laws, *nec volunt res mali administrari*" (Marquise of Lansdowne, ed., *Petty-Southwell Correspondence, 1676–1687* [London: Constable, 1928], 1:59).

For a brief discussion of physicotheology's transformation of political arithmetic, see Buck, "Seventeenth-Century Political Arithmetic," 83–84.

20. Luca Pacioli celebrated harmony, measure, and proportion in his *Summa,* the text that contains his treatise on double-entry bookkeeping. "If you take away from everything harmony, measure, and proportion, everything ceases to exist, and the man who has calculated falsely is not much more than stupid. Mathematical discussion does not fall clearly in our ears in any way; yet there is nothing which is not contained in its power" (quoted in Taylor, *No Royal Road,* 193).

21. Cohen, *Calculating People,* 32–34. In addition to the two government-sponsored initiatives, a few individuals, many of them self-professed admirers of Petty, also proposed projects inspired by political arithmetic. The individual efforts Cline mentions are set out in Charles Davenant's *Discourses on the Public Revenue* (1698); Gregory King's *Two Tracts: Natural and Political Observations and Conclusions upon the State and Condition of England* (1696); Edmund Halley, *Two Papers on the Degrees of Mortality in Mankind* (1693); and John Arbuthnot, "An Argument for Divine Providence, taken from the constant Regularity observ'd in the Births of both Sexes" (1711).

22. See Cohen, *Calculating People,* 34–35. Cohen notes that "had it been fully implemented, [this enumeration] would have produced a remarkable and comprehensive annual census of England a full century before decennial censuses became an accepted practice and a half-century before any other European country instituted regular censuses" (34). It seems a fitting irony that the

law passed in 1694 reversed the gender politics of Petty's suggestions for taxing nonreproductive citizens. In his proposals for the "Multiplication of Mankind," Petty wanted to tax nonreproductive women, not men. See William Petty, "Of Marriages &c.," in Marquise of Lansdowne, *Petty Papers*, 2:49–51, and my essay "Social Constitution of 'Class,'" esp. 30–32.

23. Cohen, *Calculating People*, 35–40. On various forms of resistance to numeracy and enumeration, see also Thomas, "Numeracy in Early Modern England," 103–32, and Buck, "People Who Counted," 28–45.

24. Daniel Defoe, *Essays upon Several Projects, or Effectual Ways for Advancing the Interests of the Nation* (London: Booksellers of London and Westminster, 1702), 13, 4.

25. The best recent treatment of the relation between political arithmetic and the administration of the poor in this period is Dean, *Constitution of Poverty*, esp. chaps. 1–4. Another example of a political arithmetic scheme devoted to the poor is Henry Fielding's *Proposal for Making an Effectual Provision for the Poor . . .* (1753), in *The Works of Henry Fielding*, ed. Edmund Gosse (Westminster: Archibald Constable; New York: Charles Scribner's Sons, 1899), 12:62–157.

It should be noted that although Defoe applies the method of political arithmetic only to the poor, he admits that the central problem of government obtains among the well-to-do as well. In discussing highway improvement, for example, he says that landowners should simply recognize that improving roads will benefit the nation: "Would not any man acknowledge, that putting this country into a condition for carriages and travellers to pass, would be a great work?" He also provides for coercion, however, in the form of a parliamentary act, for he realizes that reason does not always triumph over interest: "An act of parliament is omnipotent with respect to titles and tenures of land," he grimly reminds his readers, "and can empower lords and tenants to consent to what else they would not" (Defoe, *Essays*, 20).

26. Defoe, *Essays*, 26. The text Defoe alludes to is "Observations upon the Bills of Mortality" (1662), which was probably composed by John Graunt, although it is often attributed to William Petty. This text is reprinted in Petty, *Economic Writings*, 314–435.

27. This uncertainty appears again in the *Essays* in Defoe's plan for an insurance scheme for widows. Even though Defoe claims to take his plan of calculation from Sir William Petty's "ingenious calculation" about mortality rates, he immediately throws out Petty's figures because Petty had included infants and old people in his calculations, whereas his own plan would involve "none but the middling age of the people, which is the only age wherein life is anything steady; and if that be allowed, there cannot die by his computation, above one in eighty of such people every year; but because I would be sure to leave room for casualty, I'll allow one in fifty shall die out of our number subscribed" (*Essays*, 24).

28. Although Defoe did not explicitly recommend institutionalizing censorship, Richard Steele did playfully advance just such a recommendation in 1709. By having Mr. Bickerstaff confer upon himself the office of "Censor of Great Britain," Steele both mocks the excesses of fashion and calls attention to the problem of government it raised. In modeling his censor on Roman censors, moreover, Steele seems to yoke moral judgment to counting, for one of the Roman censor's duties was to make "frequent reviews of the people, in casting up their numbers, ranging them under their several tribes, disposing them into proper classes, and subdividing them into their respective centuries." See [Sir Richard Steele], *The Tatler*, ed. Donald F. Bond, 3 vols. (Oxford: Clarendon Press, 1987), 2:403 (162 [22 April 1710]). The kind of categories Steele sets up, however—"the Dappers and the Smarts, the natural and affected Rakes," etc.—imply that his censor would have institutionalized discrimination, not counting or prohibition.

29. The campaign to differentiate between communicative and persuasive language use was inaugurated by John Locke. See *Essay concerning Human Understanding*, ed. John W. Yolton, 2 vols. (London: Dent, 1961), vol. 2, bk. 3, chap. 10. I owe this observation to John Guillory.

30. Daniel Defoe, *The Complete English Tradesman* (1726; Gloucester: Alan Sutton, 1987), 4. Future references will be cited in the text by page number.

31. The stylistic complex that Defoe designates the "trading style" also includes bookkeeping. He introduces bookkeeping in the first chapter of *The Complete English Tradesman,* for he considers bookkeeping an instrument of rectitude, as critical as a timepiece or a seaman's helm. "A tradesman's books are his repeating clock," Defoe announces, "which upon all occasions are to tell him how he goes on, and how things stand with him in the world. . . .If they [the books] are not duly posted, and if every thing is not carefully entered in them, the debtor's accounts kept even, the cash constantly balanced, and the credits all stated, the tradesman is like a ship at sea, steered without a helm . . . in a word, he can give no account of himself to himself, much less to any body else" (15).

32. Simon Schaffer discusses Defoe's relation to credit in "Defoe's Natural Philosophy," 13–44. Schaffer comments that "the use of the term 'credit' was recent. In its sense of trust, authority or honour, it gained currency at the start of the seventeenth century. In its sense of commercial worth or solvency it began to be used from the end of the century. Defoe was one of the first to use 'credit' to refer to sums held on a bank account, and the first to use it to refer to the credit side of an account" (14).

33. Defoe recommends the "middle way of discoursing" on page 178, where he also notes that the morality of commerce is not tainted solely by lady customers—"for sometimes, I must say, the men customers are every jot as impertinent as the women."

34. David Hume, "Of the Standard of Taste," in *Essays Moral, Political, and Literary,* ed. Eugene F. Miller (Indianapolis: Liberty Classics, 1987), 226. Future references will be cited in the text by page number.

35. See Phillipson, "Pursuit of Virtue," 81–101, and P. B. Wood, "Science and the Pursuit of Virtue in the Aberdeen Enlightenment," in *Studies in the Philosophy of the Scottish Enlightenment,* ed. M. A. Stewart (Oxford: Oxford University Press, 1990), 127–50. See in particular Wood's critique of Phillipson, 128.

36. Richard B. Sher, "Professors of Virtue: The Social History of the Edinburgh Moral Philosophy Chair in the Eighteenth Century," in *Studies in the Philosophy of the Scottish Enlightenment,* ed. M. A. Stewart (Oxford: Oxford University Press, 1990), 90–92. Sher comments that "if, as Principal Robertson stated publicly at the close of the century, the primary purpose of the university was not only to impart knowledge and train young men for professional careers, but also to instil a 'love of religion and virtue' in every student, then moral philosophy lay at the heart of the curriculum. . . .[Moral philosophy] was . . . a means of integrating piety, politeness, propriety, and knowledge, with a view to producing learned, genteel, virtuous young men whose religious, social, and political views would prepare them for happiness and success in post-Revolution, post-Union, presbyterian Scotland" (91).

37. See ibid., 87–126.

38. Anthony, Earl of Shaftesbury, *Characteristics of Men, Manners, Opinions, Times,* ed. John M. Robertson, 2 vols. bound as one (1711; Indianapolis: Bobbs-Merrill, 1964), 1:105–6. Future references will be cited in the text by volume and page number.

39. Ronald Paulson identifies two contradictory tendencies in Shaftesburian disinterestedness. "For Shaftesbury, usually given credit for the concept of disinterestedness, the term has at least two contradictory aspects. On the one hand, reacting against Locke's system of social contracts, Shaftesbury draws attention to the impossibility of true virtue within a pragmatic system of rewards and punishments. Disinterestedness removes virtue from considerations of profit, thus in effect (the deist aspect) dissociating it from Judeo-Christian religious beliefs. On the other hand, disinterestedness makes virtue the province of an elite, and distinguishes the lower orders—

set off primarily on the basis of education—as the people to whom a religion of rewards and punishments could be useful in order to keep them happy and society safe and whole" (Paulson, *Beautiful, Novel, and Strange,* 23–24). See also Lawrence Klein, *Shaftesbury and the Culture of Politeness: Moral Discourse and Cultural Politics in Early Eighteenth-Century England* (Cambridge: Cambridge University Press, 1994), chap. 7.

40. To grasp this point, it may be helpful to remember that Shaftesbury elevated philosophy over mere observation precisely because the former privileged universal principles while the latter focused on what could occasion dispute. In privileging philosophy over the observation associated with natural science, he was therefore reversing the argument advanced by members of the Royal Society, who also wanted to end dispute, but who assumed that claims about observation, not theory, could achieve this end. Shaftesbury was not appealing to (mere) theory when he referred to mathematical principles; he was appealing to what he assumed were Platonic essences. Thus in "The Moralists," Shaftesbury has Theocles instruct his auditors: "Whilst men are at odds about the subjects, the thing itself is universally agreed. For neither is there agreement in judgments about other beauties. 'Tis controverted, 'which is the finest pile, the loveliest shape or face': but without controversy 'tis allowed 'there is a beauty of each kind.' This no one goes about to teach: nor is it learnt by any, but confessed by all. All own the standard, rule, and measure: but in applying it to things disorder arises, ignorance prevails, interest and passion breed disturbance" (Shaftesbury, *Characteristics,* 2:137–38).

41. Mandeville's *Fable of the Bees* (various editions: 1705, 1714, 1723, 1724) implied that the hidden agenda of Shaftesburian disinterestedness was to promote the cause of Whig lords and, equally important, to sanction the economic greed that Tories associated with moneyed men, the credit economy, and Whig rule. Hogarth's *A Harlot's Progress* and *A Rake's Progress* focused on the fantasies of ownership, control, and desire that lurked within Shaftesburian disinterestedness. See Paulson, *Beautiful, Novel, and Strange,* 24–45.

42. Bernard Mandeville, *The Fable of the Bees, or Private Vices, Publick Benefits,* ed. Phillip Harth (London: Penguin Books, 1970), 141 (remark L).

43. Shaftesbury, *Characteristics,* 1:336–37. In "The Moralists," Shaftesbury allows one of his characters to take this idea even further. Theocles exclaims, "Strange! that there should be in Nature the idea of an order and perfection which Nature herself wants! That beings which arise from nature should be so perfect as to discover imperfection in her constitution, and be wise enough to correct that wisdom by which they were made!" (ibid., 2:62–63).

44. "Some modern zealots appear to have no better knowledge of truth, nor better manner of judging it, than by counting noses. By this rule, if they can poll an indifferent number out of a mob; if they can produce a set of Lancashire noddles, remote provincial headpieces, or visionary assemblers, to attest a story of a witch upon a broomstick, and a flight in the air, they triumph in the solid proof of their new prodigy, and cry, *magna est veritas et praevalebit!*" (ibid., 1:98).

45. Hutcheson, *Essay on the Nature and Conduct of the Passions and Affections.* Statements like the following complicate the nature of the ontological claims Hutcheson was making: "Let it be observ'd, that by Absolute or Original Beauty, is not to be understood any Quality suppos'd to be in the Object, which should of itself be beautiful, without relation to any Mind which perceives it: For Beauty, like other Names of sensible Ideas, properly denotes the Perception of some Mind. . . . Were there no Mind with a Sense of Beauty to contemplate Objects, I see not how they could be call'd Beautiful" (Hutcheson, *An Inquiry into the Original of Our Ideas of Beauty and Virtue* [1728; reprint, Charlottesville, Va.: Lincoln-Rembrandt, 1993], 10). David Paxman associates Hutcheson's complex epistemological position with aesthetics. Paxman calls the product of this stance "knowledge without certainty," because Hutcheson repeatedly insists that knowledge acquired through the moral senses eludes, or even transcends, the kind of calculation asso-

ciated with mathematics. I assign mathematics considerably more importance that does Paxman, although I have found his comments helpful. See David Paxman, "Aesthetics as Epistemology, or Knowledge without Certainty," *Eighteenth-Century Studies* 26 (winter 1992–93): 285–306.

46. Shaftesbury's influence on Hutcheson is much noted in the critical literature. See, for example, Peter Kivy, *The Seventh Sense: A Study of Francis Hutcheson's Aesthetics and Its Influence in Eighteenth-Century Britain* (New York: Burt Franklin, 1976), chap. 1, and T. D. Campbell, "Francis Hutcheson: 'Father' of the Scottish Enlightenment," in *The Origins and Nature of the Scottish Enlightenment*, ed. R. H. Campbell and Andrew S. Skinner (Edinburgh: John Donald, 1982), 167–85.

47. In his introduction to Hutcheson's *Essay*, Paul McReynolds lists the following as contemporary attempts to provide a mathematical or quasi-mathematical account of virtue: William Wollaston, *The Religion of Nature Delineated* (1726); John Maxwell, *A Treatise of the Laws of Nature* (1727; this was a translation of Richard Cumberland's *De Legibus Naturae*, 1672); Archibald Campbell, *An Inquiry into the Original of Moral Virtue* (1728); and James Long (or possibly John Gay), *An Inquiry into the Origin of the Human Appetites and Affections* (1747). McReynolds considers Hutcheson's contribution to the effort to have been the "most sophisticated" (introduction to Hutcheson, *Essay*, xiv). He also points out that Hutcheson had been interested in mathematics while he was a university student at Glasgow (xiii). Donald N. Levine also briefly discusses Hutcheson's use of mathematical language. See *Visions of the Sociological Tradition*, 130.

48. James G. Buickerood has administered a much-needed corrective to the imprecise, and often anachronistic, references to "Newton's method" and "Newtonianism" popularized by Ernst Cassirer and Peter Gay, among others. See Buickerood, "Pursuing the Science of Man: Some Difficulties in Understanding Eighteenth-Century Maps of the Mind," *Eighteenth-Century Life* 19 (May 1995): 1–17.

49. In his notebook of 1704, Shaftesbury wrote that the pursuit of knowledge was "not ye Work of Speculation merely. 'Tis not a Newton, or Archimedes, yt excell in this. or can give help to others. There is stubborn Will to work on. Appetites & Humours, Passions & Desires are to be dealt with. and this is a Province those Philosophers are as much strangers to, as much at a Loss in, as ye Vulgar" (quoted in Klein, *Shaftesbury*, 82).

50. The distinction between contemporaneous interpretations of Newton's method appears in Buickerood, "Pursuing the Science of Man," 6. In the *Principia*, Newton had written that "hypotheses, whether metaphysical or physical, whether of occult qualities or mechanical, have no place in experimental philosophy. In this philosophy particular propositions are inferred from the phenomena, and afterwards rendered general by induction." This passage is quoted in Peter Achinstein, "Newton's Corpuscular Query and Experimental Philosophy," in *Philosophical Perspectives on Newtonian Science*, ed. Phillip Bricker and R. I. G. Hughes (Cambridge: MIT Press, 1990), 136. In this essay Achinstein attempts to sort out what Newton meant by "hypothesis" instead of focusing, as Buickerood does, on how his contemporaries interpreted his (often contradictory) statements. For another attempt to sort out the various "Newtonianisms" of the eighteenth century (and across Europe), see Robert E. Schofield, "An Evolutionary Taxonomy of Eighteenth-Century Newtonianisms," in *Studies in Eighteenth-Century Culture*, vol. 7, ed. Roseann Runte (Madison: University of Wisconsin Press, 1978), 177.

51. William Emerson, quoted in L. L. Laudan, "Thomas Reid and the Newtonian Turn of British Methodological Thought," in *The Methodological Heritage of Newton*, ed. Robert E. Butts and John W. Davis (Oxford: Basil Blackwell, 1970), 104 n. 3.

52. Logical deduction and mathematical demonstration are not the same kind of analytic method, of course. The former is epitomized by the syllogism and uses a commonly held "truth" as a basis for generating other truths, which confirm the truth value of the original assumption.

The latter, following Descartes, uses analysis to break a problem into constituent parts, then applies synthesis to assemble conclusions reached about those parts into a single whole. The two methods come together in geometry, however, whose method begins with theorems, which have become commonplace truths through analysis and synthesis, then "proves" or confirms those theorems in individual instances. Peter Dear points out that Newton used "induction . . . to generalize those features [proved through experimental analysis] to all other situations deemed similar to the experimental or observational exemplar" (*Discipline and Experience,* 241), but his emphasis on generalization points to the overlap between Newton's version of induction and both deduction and demonstration.

53. Ibid., chap. 8. It is well known that Boyle did not think mathematical demonstration was particularly useful for producing natural matters of fact. He considered mathematics too difficult and abstract for the mode of public knowledge production the Royal Society favored, and he did not see how mathematical formulas could capture what he deemed the most vital features of natural phenomena—their unique and qualitatively different properties. See Steven Shapin, "Robert Boyle and Mathematics: Reality, Representation, and Experimental Practice," *Science in Context* 2 (1988): 23–58, and Shapin, "Pump and Circumstance: Robert Boyle's Literary Technology," *Social Studies of Science* 14 (1984): 481–520.

54. Query 31; Isaac Newton, *Opticks, or A Treatise of the Reflections, Refractions, Inflections, and Colours of Light,* 2d ed. (1704; London: W. Bowyer, 1717); quoted in Dear, *Discipline and Experience,* 240.

55. Dear, *Discipline and Experience,* 242. The assumption that nature was regular and constant was considered the necessary basis of any "science," since by definition science was systematic knowledge. The question whether—or to what extent—the principles developed by "scientists" actually corresponded to, or reflected, properties of the natural world was raised repeatedly in this period.

Newton's fusion of mathematical demonstration with observational experiment also addressed the objections fostered by the notorious difficulty of repeating experiments, because when one sought what Barrow called "the Truth of Principles," a single experiment—the *experimentum crucis*—could be considered conclusive. "Sometimes," Newton explained, "from the Constancy of Nature, we may prudently infer an universal Proposition even by one Experiment alone" (quoted in Dear, *Discipline and Experience,* 225). Dear comments that "Newton's point was that a particular experimental design yields a result that makes a certain conclusion logically unavoidable" (225).

56. In emphasizing Hutcheson's belief that the knowledge produced by the moral senses is certain, I differ from David Paxman, whose thesis turns on the uncertainty of Hutcheson's new knowledge. Late in his essay, however, Paxman seems to contradict the thesis he initially supported when he admits that, for Hutcheson, "what is certain and universal is the aesthetic and moral response, even though the individual taste for beauty and goodness may differ" ("Aesthetics as Epistemology," 302).

57. This last quotation comes from Alexander Gerard. Turnbull and Gerard are quoted in Wood, "Science and the Pursuit of Virtue," 138, 139. The scholarship on Turnbull is not nearly as extensive as that dealing with any of the other figures in this chapter. In addition to Wood's essay, I have found particularly helpful David Fate Norton, "George Turnbull and the Furniture of the Mind," *Journal of the History of Ideas* 36 (1975): 701–16.

58. Both Newton and Hutcheson assumed the existence of natural laws, of course, and they described their project as discovering these laws. Here, for example, is Hutcheson: "Were there no general Laws fix'd in the Course of Nature, there could be no Prudence or Design in Men, no rational Expectation of Effects from Causes, no Schemes of Action projected, or any regular Ex-

ecution. If then, according to the Frame of our Nature, our greatest Happiness must depend upon our Actions, as it may perhaps be made [to] appear it does, 'the Universe must be govern'd, not by particular Wills, but by general Laws, upon which we can found our Expectations, and project our Schemes of Action'" (Hutcheson, *Inquiry,* 68). When I say that Turnbull made this more explicit than most natural or moral philosophers, I mean to stress how shifting the emphasis slightly—from the relationship assumed to exist between observed particulars and natural laws to the laws themselves—could virtually make the particulars disappear.

59. George Turnbull, *The Principles of Moral Philosophy: An Enquiry into the Wise and Good Government of the Moral World* (London: A. Millar, 1740), ii, i, v. Future references to this edition will be cited in the text.

60. Norton, "George Turnbull," 712. Norton continues: "Or, at least it will be as soon as our faculties for moral knowledge are used in the same manner that our faculties for natural knowledge have already been used."

61. Wood, "Science and Virtue," 133. The phrase comes from Larry Laudan, who credited Turnbull's pupil, Thomas Reid, with this honor. See Laudan, "Thomas Reid and the Newtonian Turn of British Methodological Thought," 102–31.

62. From the title page of the first edition: David Hume, *Treatise of Human Nature* (London: John Noon, 1739); reprinted in *A Treatise of Human Nature,* ed. Ernest C. Mossner (London: Penguin Books, 1984), 33. Future references will be cited in the text by page number.

63. The roots of the essay lie in the English Renaissance, in Bacon's and Montaigne's work. Bacon's essays were published in 1597, and Montaigne's were translated into English in 1603. Scott Black first alerted me to the importance of the essay as another kind of eighteenth-century "experiment" in the production of knowledge. I am very grateful to him for these suggestions.

64. See Michael McKeon, *The Origins of the English Novel* (Baltimore: Johns Hopkins University Press, 1987), esp. chap. 2, and Siskin, *Work of Writing.*

65. We see both Hume's caution and his confidence in this regard in the following introductory passage: "When I am at a loss to know the effects of one body upon another in any situation, I need only put them in that situation, and observe what results from it. But should I endeavour to clear up after the same manner any doubt in moral philosophy, by placing myself in the same case with that which I consider, 'tis evident this reflection and premeditation would so disturb the operation of my natural principles, as must render it impossible to form any just conclusion from the phaenomenon. We must therefore glean up experiments in this science from a cautious observation of human life, and take them as they appear in the common course of the world, by men's behaviour in company, in affairs, and in their pleasures. Where experiments of this kind are judiciously collected and compared, we may hope to establish on them a science, which will not be inferior in certainty, and will be much superior in utility to any other of human comprehension" (*Treatise,* 45–46).

66. In his autobiographical fragment *My Own Life,* written in 1776, Hume famously commented of the *Treatise* that "never literary attempt was more unfortunate than my Treatise of Human Nature. It fell *dead-born from the press,* without reaching such distinction, as even to excite a murmur among the zealots" (Hume, *Essays,* xxxiv).

67. In the light of his refusal of the mathematical order that anchored the philosophies of Shaftesbury and Hutcheson, it is significant that Hume's final repudiation of experimental philosophy included a retraction of his earlier argument that belief is a matter of *degree.* "The second error may be found in Vol. I. page 144. where I say, that two ideas of the same object can only be different by their different degrees of force and vivacity. I believe there are other differences among ideas, which cannot be properly comprehended under these terms. Had I said, that two ideas of the same object can only be different by their different *feeling,* I shou'd have been nearer

the truth" (*Treatise,* 678). Peter Dear explains why the problem of induction did not trouble seventeenth- or eighteenth-century Scholastic-Aristotelians. See *Discipline and Experience,* 15–21.

68. See Nicholas Phillipson, *Hume* (New York: St. Martin's Press, 1989), 26–34, 53–54. On one of the most prominent Scots societies, the Select Society, see Phillipson, "The Scottish Enlightenment," in *The Enlightenment in National Context,* ed. Roy Porter and Mikulas Teich (Cambridge: Cambridge University Press, 1981), 31–32, and Roger L. Emerson, "The Social Composition of Enlightened Scotland: The Select Society of Edinburgh, 1754–1764," *Studies on Voltaire and the Eighteenth Century* 114 (1973): 291–329. For a discussion of the Select's more narrowly scientific counterpart, the Royal Society of Edinburgh, see Steven Shapin, "Property, Patronage, and the Politics of Science: The Founding of the Royal Society of Edinburgh," *British Journal for the History of Science* 7 (1974): 1–41.

69. My discussion of the role that style and rhetoric more generally play in the epistemological work of Hume's essays is indebted to Jerome Christensen's provocative analysis in *Practicing Enlightenment: Hume and the Formation of a Literary Career* (Madison: University of Wisconsin Press, 1987). See, for example, 14: "In Hume, rhetoric is both the sign of logical contradiction or inconsistency and the device for putting inconsistency to work. It is because of the inexorable failure of rationality either to work on its own terms or to account satisfactorily for the behavior of humans in society that rhetoric becomes inevitable, not merely as the expression of the failure of rationality but as the remedy for its lapse. The person whose structural function is to apply that remedy and to exemplify its virtue is the man of letters, the rhetorician of enlightenment, whose office it was 'to maintain the ordinary correspondence of life.'" See also 23 and 32.

70. For a list of the breakdown, by volume, and the publication history of Hume's essays during his lifetime, see Eugene F. Miller, foreword to Hume, *Essays,* xii–xv, esp. nn. 5–8. Recent discussions of Hume's essays include John B. Stewart, *The Moral and Political Philosophy of David Hume* (New York: Columbia University Press, 1963); Duncan Forbes, *Hume's Philosophical Politics* (Cambridge: Cambridge University Press, 1975); David Miller, *Philosophy and Ideology in Hume's Political Thought* (Oxford: Clarendon Press, 1981); Donald W. Livingston, *Hume's Philosophy of Common Life* (Chicago: University of Chicago Press, 1984); and Christensen, *Practicing Enlightenment.*

71. Hume, "Delicacy of Taste," in *Essays,* 5. Future references to the essays will be cited in the text by essay title and page number of this edition.

72. Given the latitude of "politics" in the eighteenth century and the overlap between institutions like marriage or religion and the state, one could also plausibly include the following in one's list of the "political" essays contained in this volume: "Of the Study of History," "Of Love and Marriage," "Of Superstition and Enthusiasm," and "Of the Dignity of Human Nature."

73. See especially "Whether the British Government Inclines More to Absolute Monarchy or to a Republic" (47–53), "Of Parties in General" (54–63), and "Of the Parties of Great Britain" (64–72).

74. Hume, "Parties of Great Britain," 72. Miller reproduces the revisions to this essay and its publication history in an appendix. See Hume, *Essays,* 610–16.

75. Hume explicitly articulates this position in the opening paragraphs of an essay titled "The Sceptic," which is one of four essays he wrote "to deliver the sentiments of sects" so that readers could try out these philosophical positions instead of simply reading about them. See *Essays,* 159–80.

76. See Christensen, *Practicing Enlightenment,* 10–11. Daniel Gordon also discusses Hume's use of the dialogue. See *Citizens without Sovereignty: Equality and Sociability in French Thought, 1670–1789* (Princeton: Princeton University Press, 1994), 165–67. Hume's most famous dia-

logue, of course, is his *Dialogues concerning Natural Religion,* which was not published until after his death.

77. For a nuanced discussion of Hume's attitudes toward women, see Christensen, *Practicing Enlightenment,* chap. 4.

78. For revealing treatments of men's attitudes toward women in this period, see Felicity Nussbaum, *The Brink of All We Hate: English Satires on Women, 1660–1750* (Lexington: University Press of Kentucky, 1984), and Ellen Pollak, *The Poetics of Sexual Myth: Gender and Ideology in the Verse of Swift and Pope* (Chicago: University of Chicago Press, 1985). For valuable ideas about how women writers used the terms of men's ambivalence to adapt the market to their own desires for economic and epistemological credibility, see Gallagher, *Nobody's Story,* esp. chaps. 1–4; Margaret W. Ferguson, "Juggling the Categories of Race, Class, and Gender: Aphra Behn's *Oronooko,*" *Women's Studies* 19 (1991): 159–81; and Robert L. Chibka, "'Oh! Do Not Fear a Woman's Invention': Truth, Falsehood, and Fiction in Aphra Behn's *Oronooko,*" *Texas Studies in Literature and Language* 30 (1988): 510–37. More general treatments of eighteenth-century women writers include Jane Spencer, *The Rise of the Woman Novelist. From Aphra Behn to Jane Austen* (Oxford: Basil Blackwell, 1986); Janet Todd, *The Sign of Angellica: Women, Writing and Fiction, 1660–1800* (London: Virago, 1989); and Cheryl Turner, *Living by the Pen: Women Writers in the Eighteenth Century* (New York: Routledge, 1992).

79. For examples of the *querelles des femmes,* see Katherine Usher Henderson and Barbara F. McManus, *Half Humankind: Contexts and Texts of the Controversy about Women in England, 1540–1640* (Urbana: University of Illinois Press, 1985). J. G. A. Pocock discusses the place of the Fortuna trope in anxieties about market society in "Mobility of Property," 103–23.

80. This is the argument advanced—obliquely, of course—by Adam Phillips. See especially *Terror and Experts* (Cambridge: Harvard University Press, 1995).

Chapter Five

1. See, in particular, Brewer, *Sinews of Power,* and Lazarsfeld, "Notes on the History of Quantification," 277–333. To be fair to these fine studies, both Brewer and Lazarsfeld realize that recovering the meaning of numbers is essential to any understanding of their historical use; but they do not examine the philosophical contexts in which knowledge more generally was conceptualized.

2. The British government's unsystematic approach to collecting information, especially about issues that had both (what we would call) social and economic components, can be seen in the eighteenth-century treatment of poverty. As Dorothy Marshall points out, the government did not regularly canvass the parishes even for records of expenditures on poor relief until parliamentary committees of 1775 and 1786 required parish officials to return regular schedules. See Marshall, *The English Poor in the Eighteenth Century: A Study in Social and Administrative History* (London: George Routledge, 1926), 77. This is true even though a national poor law had been in place since 1601 that mandated the production of a number of different kinds of records and certificates. What documents were kept were generally collected in parishes. The closest approximation to the schedules required by the late eighteenth-century committees were the "overseers accounts," which recorded the money spent on individual paupers, and which date from at least the early eighteenth century. A good discussion of the kinds of poor law documents dating from this period can be found in Anne Cole, *Poor Law Documents before 1834* (Oxford: Bocardo Press, 1993), 9–34.

I take up the subject of poverty more extensively in the next chapter, where I also examine the claim made by Mitchell Dean that political economy had "more in common with its German contemporary, the science of police, than with the economics of Say and Ricardo" (*Constitution of Poverty*, 62). Dean offers extremely provocative discussions of seventeenth- and eighteenth-century conceptualizations of poverty in chaps. 1–7.

3. The German science of police was a subject of particular interest to Foucault late in his career, and it has inspired interesting work by some of his followers. For discussions of police, see the following essays in Burchell, Gordon, and Miller, *Foucault Effect:*, Michel Foucault, "Governmentality," 87–104; Pasquale Pasquino, "Theatrum Politicum: The Genealogy of Capital—Police and the State of Prosperity," 105–18; and Giovanna Procacci, "Social Economy and the Government of Poverty," 151–68. Mitchell Dean also discusses police in *Constitution of Poverty*, chap. 3. Dean develops in more detail a claim that Foucault, Pasquino, and Procacci also advance, if only in passing: that Britain was *like* parts of the Continent, especially Germany and France, in developing a science of police in the eighteenth century. As this chapter and chapter 4 should make clear, I disagree with this claim. The relatively weak constitutional basis for central government in Britain, combined with a persistent resistance to such centralization, which was advanced in the name of "liberty" and cultivated in the periodical publications that circulated information in the newly emergent public sphere, meant that developing anything like a science of police in Britain in the eighteenth century would have been very difficult.

4. For a helpful discussion of the relation between Smith's theories of subjectivity and his account of liberal governmentality, see E. J. Hundert, *The Enlightenment's Fable: Bernard Mandeville and the Discovery of Society* (Cambridge: Cambridge University Press, 1994), 219–34.

5. See Miller, *Defining the Common Good,* 399–412.

6. Stewart discusses conjectural history in his *Dissertation: Exhibiting the Progress of Metaphysical, Ethical, and Political Philosophy, since the Revival of Letters in Europe* (1792), in *The Collected Works of Dugald Stewart,* ed. Sir William Hamilton, 11 vols. (Edinburgh: Thomas Constable, 1854–60), 1:3–4. See also Stewart, *Account of the Life and Writings of Adam Smith,* in *Collected Works,* 10:33–34. This text was first delivered in 1790 to the Royal Society of Edinburgh.

7. William Robertson, *The History of the Discovery and Settlement of America* (London: Jones, 1826), 93. Robertson's *History,* first published in 1777, went through at least twenty editions by 1817.

8. Laudan comments that, thanks partly to Reid, the meaning of the term "hypothesis" has been eroded, "much to the confusion of subsequent methodological inquiry. In the early seventeenth century, hypothesis had signified any general proposition which was assumed, but not known to be true. It was used especially to refer to the unproven postulates, axioms, or first principles of any science, and this meaning of the term had been common since Aristotle and Euclid. Even Newton used 'hypothesis' in this sense in the first editions of the *Principia* (1689) and it carried no particular pejorative connotations there. But the signification of the term was gradually transformed in Newton's later writings. In a constant battle with the Cartesians, Newton would often find his opponents offering theories or conjectures which were patently false when tested empirically. Thus, six of Descartes' seven laws of motion were obviously incompatible with the most elementary impact experiments. . . .Undeterred by such anomalies, many Cartesians continued to hold Descartes' laws of motion and vortex theory, defending them in terms of their *a priori* cogency. Understandably, Newton had no patience with such an approach and he tried to discredit it with methodological arguments. The 'a priorists,' he insisted, were using hypotheses; so the natural way to eliminate their non-empirical techniques was to insist, as he finally did, that 'hypotheses have no place in experimental philosophy.' Unfortunately, Newton's blanket denunciation of hypotheses not only discredited the 'a priorists,' it also left no place for an *empirical* hypothetical-deductive method, which insisted on the experimental testing of all conjectures.

What tended to happen, therefore, was that the older meaning of hypothesis (an axiom or postulate) came to be confounded with the notion of an unempirical or untestable proposition. Where before the two had been distinguished, they were now indiscriminately confused and the legitimate arguments against the *a priori* and untestable hypotheses were mistakenly used against all hypotheses whatever. By Reid's time . . . the hypothetical method was in such ill repute that Reid, although perfectly able to distinguish the two senses of hypothesis, considered them equally objectionable and demanded that both were to be avoided scrupulously. Indeed, so vague had the term hypothesis become that Reid could include 'theory' in its connotation and could claim that Newton's arguments against hypotheses were equally arguments against theories!" (Laudan, "Thomas Reid," 118–19).

9. Thomas Reid, *An Inquiry into the Human Mind,* ed. Timothy Duggan (Chicago: University of Chicago Press, 1970), 199–200.

10. Stewart, *Life of Adam Smith,* in *Collected Works,* 10:33–34.

11. The essay most explicitly devoted to identifying such characteristics is H. M. Hopfl, "From Savage to Scotsman: Conjectural History in the Scottish Enlightenment," *Journal of British Studies* 17, no. 2 (1978): 19–40; but see also Andrew Skinner, "Natural History in the Age of Adam Smith," in *Political Studies,* vol. 15, ed. Peter Campbell (Oxford: Clarendon Press, 1967), 32–48; Skinner, "Adam Smith: Philosophy and Science," *Scottish Journal of Political Economy* 19 (1972): 307–19; and Simon Evnine, "Hume, Conjectural History, and the Uniformity of Human Nature," *Journal of the History of Philosophy,* 21, no. 4 (1993): 589–606. Recently, because of these writers' tendency to accord women a position of such importance in the cultivation of manners, women's historians have begun to produce interesting new work about the conjectural histories, although they have not generally been concerned with devising a list of generic characteristics. See Jane Rendall, "Virtue and Propriety: The Scottish Enlightenment and the Construction of Femininity" (unpublished manuscript, York University, 1955), and Mary Catherine Moran, "From Rudeness to Refinement: Philosophical History and Late Eighteenth-Century Conduct Literature on the Role of Women in 'The Natural History of Mankind'" (unpublished manuscript, Johns Hopkins University, 1994).

12. John Millar, *Observations concerning the Distinction of Ranks in Society* (Dublin: T. Ewing, 1771), iv.

13. Millar and Robertson solve the problem of sources in different ways. For Millar, the most reliable witnesses are illiterate or at least uneducated, because these witnesses are least likely to be influenced by "speculative systems." "Our information . . . with regard to the state of mankind in the more uncivilized parts of the world, is chiefly derived from the relations of travellers, whose character and situation in life neither set them above the suspicion of being easily deceived, nor of endeavouring to misrepresent the facts which they have related. . . . When illiterate men, ignorant of the writings of each other, and who, unless upon religious subjects, have no speculative systems to warp their opinions, have, in different ages and countries, described the manners of people in similar circumstances, the reader has an opportunity of comparing their several descriptions, and from their agreement or disagreement is able to ascertain the credit that is due to them" (Millar, *Observations,* xiii). Robertson also lays the burden of discrimination on the historian, but he tends to privilege "the intelligent observations of the few philosophers who have visited this part of the world [America]" (*History of America,* 93). The debate about which kind of travelers to credit echoes the late seventeenth-century discussions among members of the Royal Society about what criteria would identify authoritative witnesses. See Shapin, *Social History of Truth.* The existence of an institution that could codify and enforce these criteria helps explain why this problem was solved for natural philosophers before the end of the seventeenth century, whereas for historians it remains problematic to this day.

14. Andrew Skinner discusses the Aristotelianism of the conjectural historians in "Economics and History," 1–22.

15. These abstractions have become the staple of modern sociological and economic analysis, so their constructed nature is often difficult to grasp. Recently, however, when a group of New York mathematicians unveiled a formula designed to determine "the urban happiness quotient," they acknowledged that the entity the index was designed to measure ("the city's mental and social state" or "the well-being of New Yorkers") was strictly a function of the formula devised to describe it. Asked what the formula means, D. Carole Siegel, who heads a laboratory at the Nathan S. Kline Institute for Psychiatric Research, responded that "the mental well-being index is about mental well-being. What is it? It is what we've measured" (*New York Times,* Metro Section, 10 June 1997).

16. Henry Home, Lord Kames, *Sketches of the History of Man,* 2d ed. (1774; Edinburgh: W. Strahan, 1778), 1:23–30, 65, 72–75.

17. William Robertson, *A View of the Progress of Society in Europe from the Subversion of the Roman Empire to the Beginning of the Sixteenth Century,* vol. 4 of *The Works of William Robertson* (London: W. Sharpe, 1820), 32.

18. Adam Ferguson, *An Essay on the History of Civil Society* (1764; New Brunswick, N.J.: Transaction, 1995), 122. Hiroshi Mizuta uses letters written by Hume to date Ferguson's *Essay* to the late 1750s. See "Two Adams in the Scottish Enlightenment: Adam Smith and Adam Ferguson on Progress," in *Transactions of the Fifth International Congress on the Enlightenment,* ed. Haydn Mason (Oxford: Taylor Institute, 1980), 2:813.

19. "The establishments of men, like those of every animal, are suggested by nature, and are the result of instinct, directed by the variety of situations in which mankind are placed. Those establishments arose from successive improvements that were made, without any sense of their general effect; and they bring human affairs to a state of complication, which the greatest reach of capacity with which human nature was ever adorned, could not have projected; nor even when the whole is carried into execution, can it be comprehended in its fullest extent" (Ferguson, *Essay on the History of Civil Society,* 181–82).

20. The twin notions that the arts constituted a critical instrument for improving "manners" and that what the arts improved, women cultivated and protected were central to the narratives about the "human mind" constructed by mid-eighteenth-century historians. Thus Robertson declared that "the progress of science, and the cultivation of literature, had considerable effect in changing the manners of the European nations, and introducing that civility and refinement by which we are now distinguished" (*View of the Progress of Society,* 86). John Millar argued that once the "spirit of improvement is introduced into a country," the women "seem naturally qualified to surpass the other sex by their superior proficiency in many of these arts and manufactures which then become the objects of their attention" (*Observations concerning the Distinction of Ranks,* 73).

Mark Salber Phillips has argued that "the essential point of departure [that distinguishes "modern" history writing] is the interest of eighteenth-century writers in the structures and experiences of private life, their desire to explore the *inward* lives of individuals and the *everyday* life of societies" (Phillips, "Reconsiderations on History and Antiquarianism: Arnaldo Momigliano and the Historiography of Eighteenth-Century Britain," *Journal of the History of Ideas* 57, no. 2 [1996]: 297). I agree with his conclusion—that "this revaluation of private life challenged the classical conception of history in fundamental ways, resulting in a fruitful tension between the social and sentimental interests of the age and its inherited view of history" (2). One should remember, however, that this focus on "inward lives" substituted one interpretation of what constituted the important feature of "inward lives" for another: instead of (statesmen's) intentions, the conjectural historians stressed the motivations I have been calling subjective: sentiment, imagination, and feelings. Even though they were subjective, of course, these motivations were not psychological in the modern sense, for they were considered *collective* properties of human beings as well as individual traits.

21. The argument that commerce—or "the market system"—was a natural, self-governing domain was not articulated specifically by the conjectural historians who preceded Adam Smith. Despite efforts by some modern analysts to represent these historians as champions of commerce, in fact, the mid-eighteenth-century historians did not even consistently equate commercial society with progress. (See, for example, Skinner, "Economics and History," and also Ronald Meek, "The Scottish Contribution to Marxist Sociology," in *Democracy and the Labour Movement,* ed. John Saville [London: Lawrence and Wishart, 1954].) Instead, like David Hume, Ferguson, Millar, and Lord Kames all maintained that the riches commerce conferred on humanity tended to fuel cycles of advancement and decline instead of a steady progress toward permanent improvement, much less perfection.

Like Mandeville and to a lesser extent Shaftesbury and Hutcheson, the midcentury historians were contributing to the debate about luxury that raged for most of the eighteenth century, as well as to the more general debate about the relation between virtue and commerce. On the eighteenth-century debate about luxury per se, see Christopher J. Berry, *The Idea of Luxury: A Conceptual and Historical Investigation* (Cambridge: Cambridge University Press, 1994), chap. 6. The standard discussion of the debate about virtue and commerce is Pocock, *Machiavellian Moment,* esp. chap. 14. See also the essays in Istvan Hont and Michael Ignatieff, eds., *Wealth and Virtue: The Shaping of Political Economy in the Scottish Enlightenment* (Cambridge: Cambridge University Press, 1983).

22. In Lord Kames's account, history revealed cycles of refinement and decay instead of the kind of progress that a belief in providentialism might suggest. "Skilful husbandry, producing the necessaries of life in plenty, paves the way to arts and manufactures. . . . The appetite for property becomes headstrong, and to obtain gratification tramples down every obstacle of justice and honour. . . . [Eventually,] through fondness for social intercourse, [men] patiently undergo the severe discipline, of restraining passion and smoothing manners. . . . Men improve in urbanity by conversing with women; and however selfish at heart, they conciliate favour by assuming an air of disinterestedness. Selfishness, thus refined, becomes an effectual cause of civilization. But what follows? Turbulent and violent passions are buried, never again to revive, leaving the mind totally ingrossed by self-interest. . . . Wealth, acquired whether by conquest or commerce, is productive of luxury and sensuality. As these increase, social affections decline, and at last vanish. . . . Selfishness becomes the ruling passion: friendship is no more; and even blood-relation is little regarded. . . . And thus in the progress of manners, men end as they began: selfishness is no less eminent in the last and most polished state of society, than in the first and most savage state" (*Sketches,* 1:344–47). Lord Kames's somber conclusion—"there is no remedy, but to let the natives die out, and to repeople the country with better men" (*Sketches,* 1:415–16)—is not as pessimistic as it initially seems, however, because, like the other conjectural historians, he privileged the divine order presumably being worked out over the long run of history above what even this telescoped version of actual history revealed.

23. Among those texts classified as conjectural histories by modern historians are Hume's *Natural History of Religion* and several of his essays ("Of the Origin of Government," "Of the Rise and Progress of the Arts and Sciences," and "Of the Original Contract"). See Hopfl, "From Savage to Scotsman," 21. Simon Evnine explains why Hume's *History of England* is *not* a conjectural history. See "Hume, Conjectural History, and the Uniformity of Human Nature," 594–95.

24. See David Hume, *The Natural History of Religion* (1757), in *Principal Writings on Religion,* ed. J. C. A. Gaskin (New York: Oxford University Press, 1993), 134–38.

25. Hume, *Treatise of Human Nature*, 130–31. Christensen discusses this passage in *Practicing Enlightenment*, 136–39.

26. In what follows, I draw heavily on Christensen's analysis. See *Practicing Enlightenment*, 26–34.

27. Here is Christensen's summary of this important rupture in Hume's argument: "Society will always stand slightly beyond the family group, the very naturalness of which is its flaw: there is too intimate a connection between the family unit and the individual; society depends on remote connections having the force of close ones, on the cultivation of a willing tendency in the individual to adjust to an oblique and deferred satisfaction of his needs" (*Practicing Enlightenment*, 27).

28. Ibid., 29–30.

29. Ibid., 33.

30. The most influential statement of this principle was offered by William Paley in 1785: "There cannot be design, without a designer; contrivance, without a contriver; order, without choice; arrangement, without anything capable of arranging; subserviency and relation to a purpose, without that which could intend a purpose; means suitable to an end, and executing their office, in accomplishing that end, without the end ever having been contemplated, or the means accommodated to it" (*Natural Theology* [Edinburgh, 1849], 15).

31. To be fair, I should note that Christensen represents Hume's theory as *productively incapacitating*. "The restraints of conventions under which men are naturally induced to lay themselves will remedy incapacity but *not* by capacitating man, by empowering him; rather, these restraints will do so by imposing the general rule of induction, which requires that the particular be directed into the general, the individual into the social whole" (*Practicing Enlightenment*, 31). I agree with the conclusions of Christensen's reading, but I would emphasize more strongly the legacy that this representation of system bequeathed to economic theory in particular.

32. "The feeling that one has lost in a transaction is just that, a *feeling*. . . . The belief engendered is not so much false as insufficiently true: the error lies in a narrowness or partiality that distorts perception and thereby thwarts the smooth operations of the greater economy" (ibid., 149).

33. Hume, "Of Commerce," in *Essays*. 254. This essay was originally published in 1754.

34. That individuals do not always see beyond their own interests mandates both that philosophers continue to produce knowledge for the state and that modern governments find some way to inspire their citizenry. Thus economic theory becomes a key component of what we call liberal government. For a related argument, see Burchell, "Peculiar Interests," 119–50. Hume notes that, because individuals do not always recognize that their prosperity is tied up with that of the state—because "these principles are too disinterested and too difficult to support, it is requisite to govern men by other passions, and animate them with a spirit of avarice and industry, art and luxury" ("Of Commerce," in *Essays*, 263).

35. Here is Donald Winch's explanation for Smith's reticence: "Smith was sparing in his use of the term, partly because he saw political economy merely as part of the larger inquiries on which he was engaged, and partly perhaps because [James] Steuart had used *Inquiry into the Principles of Political Oeconomy* as the title of his rival work published nearly a decade before the *Wealth of Nations*. Smith mostly uses the term when discussing the policy implications of the mercantile and agricultural systems, thereby emphasizing the connections with the art of legislation. Treated thus, his definition of the practical objectives of political economy is fairly conventional by eighteenth-century standards: it was 'to provide a plentiful revenue or subsistence for the people, or more properly to enable them to provide such a revenue or subsistence for themselves; and secondly to supply the state or common wealth with a revenue sufficient for the publick services.' At the same time, Smith defined political economy as '*a branch of* the science of a statesman or legis-

lator,' reminding us that the *Wealth of Nations* began life as those parts of Smith's lectures on natural jurisprudence that dealt with the subordinate questions of 'police, revenue and arms'" (Winch, *Riches and Poverty,* 21).

36. Adam Smith, *An Inquiry into the Nature and Causes of the Wealth of Nations,* ed. Edwin Cannan (1776; New York: Modern Library, 1937), 4.397. Future references will be cited in the text by book, chapter, and page number.

37. See Miller, *Defining the Common Good,* 399–412. Miller's work is especially helpful in reminding us of the alternative tradition that prevailed when Smith began to write, and also in restoring to visibility some of Smith's less well known contemporaries, who also argued that the market was ruled by its own "natural liberty."

38. On the role that sympathy plays in qualifying Smith's emphasis on self-interest, see Hundert, *Enlightenment's Fable,* 219–36. Hundert's account is particularly useful in providing a context for understanding what has been called "the Adam Smith problem"—the relation between *Theory of Moral Sentiments* and *Wealth of Nations* (234–36).

The classic statement of this point in Smith's work appears in the *Theory of Moral Sentiments,* ed. D. D. Raphael and A. L. Macfie (1759; Indianapolis: Liberty Classics, 1976), 1.1.9.

39. "I have no great faith in political arithmetic, and I mean not to warrant the exactness of either of these computations" (Smith, *Wealth of Nations,* 4.5.501).

40. Smith did not publish these lectures during his lifetime; in fact, the manuscript copy was destroyed in the week before his death, according to the strict instructions Smith had given both David Hume and his literary executors, Joseph Black and James Hutton. The text we now have was compiled from student notes, which date from 1762–63. Smith began to deliver these lectures at Glasgow in 1751, but he probably based his university lectures on a private course on rhetoric that he first gave in Edinburgh in 1746. For a discussion of the lectures and the surviving text, see J. C. Bryce, introduction to Adam Smith, *Lectures on Rhetoric and Belles Lettres* (Indianapolis: Liberty Fund, 1985), 1–37.

41. Adam Smith, *The Principles Which Lead and Direct Philosophical Enquiries; Illustrated by the History of Astronomy,* in *Essays on Philosophical Subjects,* ed. W. P. D. Bryce, vol. 3 of *The Glasgow Edition of the Works and Correspondence of Adam Smith,* ed. D. D. Raphael and A. S. Skinner (Oxford: Clarendon Press, 1980), 105 (my emphasis).

42. Historians of political thought have often noted the similarities between Smith and Hume, as well as the differences that separate these philosophers (along with Millar) from most other representatives of the Scottish Enlightenment. Knud Haakonssen, for example, assigns Hume, Smith, and John Millar to a single group on the basis that all three philosophers' theories of justice can be traced, more or less directly, to Hugo Grotius. By contrast, "the mainstream of Scottish moral philosophy," whose representatives include Hutcheson, Turnbull, Lord Kames, Ferguson, Reid, and Stewart, were more indebted to Samuel von Pufendorf. See Haakonssen, "Natural Jurisprudence in the Scottish Enlightenment," 36–49. Donald Winch links Hume and Smith because they were "more responsive to European problems and audiences" (*Riches and Poverty,* 19). Duncan Forbes links Smith and Millar in "'Scientific Whiggism,'" 643–70, and he stresses Hume's Continental connections in "The European or Cosmopolitan Dimension in Hume's Science of Politics," *British Journal of Eighteenth-Century Studies* 1 (1977): 57–60. Istvan Hont describes Smith's depiction of commercial society as an alternative to Christian theology and to the complacency voiced by other conjectural historians. See "Commercial Society and Political Theory," 68, 78.

43. Although he does not stress Smith's two ideas of "nature," the best discussion to date of how Smith understood "nature" is Hont, "Political Economy of the 'Unnatural and Retrograde' Order," 122–49. "'Nature' in Smith's notion, just as in Physiocracy, was to be understood in the

spirit of natural law, as the nature of rightly ordered human societies and not as a notion relating to nature as in natural science" (ibid., 125). In much of my discussion of Smith's "nature," I am indebted to Hont.

44. Johnson's *Journey* bears obvious affinities to Martin Martin's *A Description of the Western Isles of Scotland* (1703), John Macky's *A Journey through England* (1714–23), Daniel Defoe's *A Tour through the Whole Island of Great Britain* (1724–26), and Giuseppe Baretti's *Journey from London to Genoa* (1770). On travel literature in this period, see Thomas M. Curley, *Samuel Johnson and the Age of Travel* (Athens: University of Georgia Press, 1976), esp. 183–219; Charles L. Batten Jr., *Pleasureable Instruction: Form and Convention in Eighteenth-Century Travel Literature* (Berkeley: University of California Press, 1978); Percy G. Adams, *Travel Literature and the Evolution of the Novel* (Lexington: University Press of Kentucky, 1983); Pat Rogers, *Johnson and Boswell: The Transit of Caledonia* (Oxford: Clarendon Press, 1995), chap. 3; and Pat Rogers, introduction to Daniel Defoe, *A Tour through the Whole Island of Great Britain*, ed. Pat Rogers (Harmondsworth: Penguin Books, 1971), 18–22.

45. Samuel Johnson, *A Journey to the Western Islands of Scotland*, in *"A Journey to the Western Islands of Scotland" by Samuel Johnson and "The Journal of a Tour to the Hebrides with Samuel Johnson, LL.D." by James Boswell*, ed. Allan Wendt (Boston: Houghton Mifflin, 1965), 95, 43. Future references will be cited in the text by page number.

46. Rogers, *Johnson and Boswell*, 4. Despite a few provocative comments on the relation between Johnson's text and Adam Ferguson's *Essay on the History of Civil Society*, Rogers has remarkably little to say about the implications of this relation; see 216–25.

47. It would not be correct to say that none of the experimental moral philosophers were interested in collecting data that did not confirm their a priori postulates. Thomas Reid, for one, frequently recorded "uncommon facts," although he was reluctant to draw conclusions from them; and he repeatedly remarked on the importance of making numerous observations before distinguishing between "accidental conjunctions" and "natural connections." "That habit of passing without reasoning, from the sign to the things signified, which constitutes the acquired perception, must be learned by many instances of experiments; and the number of experiments serves to disjoin those things which have been accidently conjoined, as well as to confirm our belief of natural connections" (Reid, *Inquiry into the Human Mind*, 297 [chap. 6, sec. 24]). When repeated experiments would not accommodate "uncommon facts" to his assumptions about (or belief in) "natural connections," however, Reid tended to fall back on a variant of providentialism. "The wise Author of our nature intended, that a great and necessary part of our knowledge should be derived from experience, before we are capable of reasoning, and he hath provided means perfectly adequate to this intention. For first, He governs nature by fixed laws, so that we find innumerable connections of things which continue from age to age. Without this stability of the course of nature, there could be no experience; or, it would be a false guide, and lead us into error and mischief. . . . Secondly, He hath implanted in human minds an original principle by which we believe and expect the continuance of the course of nature, and the continuence of those connections which we have observed in times past" (291 [chap. 6, sec. 24]).

Adam Ferguson was the only Scottish philosopher of this period who was born in the Highlands or spoke Gaelic. Ferguson's familiarity with the feudal societies of the Highlands may account for the greater sympathy he expressed for so-called primitive societies, and it may have contributed to his adamant support for local militias. See Hopfl, "From Savage to Scotsman," 29.

48. See Rogers, *Johnson and Boswell*, 24–25.

49. The nineteenth-century crisis about method came to a head in the debate between John Stuart Mill, who considered deduction the only method appropriate for a "science of human nature," and William Whewell, who argued for the general superiority of induction. See Mill, *A*

System of Logic Ratiocinative and Inductive (1843), bk. 6 (reprinted as *The Logic of the Moral Sciences,* ed. A. J. Ayer [La Salle, Ill.: Open Court, 1988]); and Whewell, *History of the Inductive Sciences* (1837) and *Philosophy of the Inductive Sciences* (1840) (both in *Selected Writings on the History of Science* [Chicago: University of Chicago Press, 1984]). On this debate, see Richard Yeo, *Defining Science: William Whewell, Natural Knowledge, and Public Debate in Early Victorian Britain* (Cambridge: Cambridge University Press, 1993), chap. 7.

50. Rogers usefully reminds us that Johnson's age—sixty-three when he set out—would have been especially significant to a culture accustomed to thinking of the "climacteric" as "a critical stage in human life." See *Johnson and Boswell,* chap. 1.

51. Laurence Sterne, *Tristram Shandy* (1759–67) and *A Sentimental Journey into France and Italy* (1768), and MacKenzie, *The Man of Feeling* (1771). There is a sizable secondary literature on sentimentalism and the novel of sensibility. On the former, see J. G. Barker-Benfield, *The Culture of Sensibility: Sex and Society in Eighteenth-Century Britain* (Chicago: University of Chicago Press, 1992). On the latter, see Robert Markley, "Sentimentality as Performance: Shaftesbury, Sterne, and the Theatrics of Virtue," in *The New Eighteenth Century: Theory, Politics, English Literature,* ed. Felicity Nussbaum and Laura Brown (New York: Methuen, 1987), 210–30.

52. "When obliged to have recourse to the superficial remarks of vulgar travellers, of sailors, traders, buccaneers, and missionaries, we must often pause, and, comparing detached facts, endeavour to discover what they wanted sagacity to observe" (Robertson, *History of America,* 93). Because "the prejudices and erroneous judgements" of inhabitants influence accounts, "it is . . . by a comparison only of the ideas and the practice of different nations, that we can arrive at the knowledge of those rules of conduct, which, independent of all positive institutions, are consistent with propriety, and agreeable to the sense of justice" (Millar, *Observations,* v). Ferguson discusses the methodological problem in pt. 2, sec. 1 of his *Essay on the History of Civil Society* (74–81).

53. "Books are faithful repositories, which may be a while neglected or forgotten; but when they are opened again, will again impart their instruction: memory, once interrupted, is not to be recalled. Written language is a fixed luminary, which, after the cloud that had hidden it has past away, is again bright in its proper station. Tradition is but a meteor, which, if once it falls, cannot be rekindled" (Johnson, *Journey,* 83). After many attempts to find out about the bards, who were supposed to have kept memories of the Highland traditions alive, Johnson gives up. "Thus hopeless are all attempts to find traces of Highland learning. Nor are their primitive customs and ancient manner of life otherwise than very faintly and uncertainly remembered by the present race" (84).

54. Johnson voices his skepticism about Ossian in *Journey,* 86–89. His opinions were extremely controversial, for respected Scots scholars, including Hugh Blair and Lord Kames, staunchly defended the validity of these poems. Not until 1853 was a German philologist able to prove conclusively that the poems were a forgery. For treatments of Ossian that link these issues to the Scottish Enlightenment, see Maurice Colgan, "Ossian: Success or Failure for the Scottish Enlightenment?" and Leah Leneman, "The Effects of Ossian in Lowland Scotland," both in *Aberdeen and the Enlightenment,* ed. Jennifer J. Carter and Joan H. Pittock (Aberdeen: Aberdeen University Press, 1987), 344–49, 357–62.

55. Johnson was attentive to the condition of Scottish roads from the beginning of his trip. Initially he was exhilarated by the absence of tollgates, and at Loch Ness he found the roads "a source of entertainment" (*Journey,* 23). As he and Boswell traveled farther from Edinburgh and the paths of the Roman armies (who had laid most of the Scottish roads), however, Johnson became increasingly aware of how the condition (or absence) of roads affected commerce, justice, and travel itself. Near the conclusion of the journey, when Johnson and Boswell had returned to the mainland and Roman roads, Johnson remarked that roadworks, along with the standardized

measure they permit, constituted one site at which the traditional Scots and modern English cultures continued to clash. "After two days at *Inverary* we proceeded *Southward* over *Glencroe,* a black and dreary region, now made easily passable by a military road, which rises from either end of the *glen* by an acclivity not dangerously steep, but sufficiently labourious. In the middle, at the top of the hill, is a seat with this inscription, *Rest, and be thankful.* Stones were placed to mark the distances, which the inhabitants have taken away, resolved they said, *to have no new miles*" (118). For a discussion of the condition of Scottish roads during this period, see Anne Gordon, *To Move with the Times: The Story of Transport and Travel in Scotland* (Aberdeen: Aberdeen University Press, 1988), chaps. 2–5.

56. Lord Kames, *Sketches,* vol. 1, sketch I. Here Kames concludes that "there are different races of men fitted by nature for different climates" (1:23). Lord Monboddo (James Burnett) published the first volume of his six-volume *Of the Origin and Progress of Language* in 1773 (the final volume appeared in 1792) and the equally ambitious *Ancient Metaphysics* between 1779 and 1799. One of the most dramatic social events of Johnson's tour of Scotland was his meeting with Monboddo, although Johnson barely mentioned it; in his *Journal of a Tour to the Hebrides* (1785), Boswell more than made up for Johnson's silence. On the meeting with Monboddo and Monboddo's anthropological speculations more generally, see Rogers, *Johnson and Boswell,* appendix (226–31).

57. "Populousness" was the term commonly used in eighteenth-century Britain for the relative density of inhabitants in any given place. The currency of this term suggests that eighteenth-century politicians and philosophers were more interested in the density of habitation than either the absolute size or the increase (or decrease) of what we would call the "population" (although a vigorous debate about whether the British were increasing or decreasing in number raged during the second half of the century). I will return to the issue of population in the next chapter. For one contemporary contribution to the subject of "populousness," see Hume, "Of the Populousness of Ancient Nations," in *Essays,* 377–464.

58. Smith associated curiosity with the debased genre of the novel, and he considered modern novelists' tendency to resort to suspense inferior to ancient historians' use of important facts. "As newness is the only merit in a Novel and curiosity the only motive which induces us to read them, the writers are necessitated to make use of this method [suspense] to keep it [interest] up. Even the Antient Poets who had not reality on their side never have recourse to this method, the importance of the narration they trust will keep us interested" (*Lectures on Rhetoric,* 97; see 96–97).

Chapter Six

1. See Collini, Winch, and Burrow, *That Noble Science of Politics,* esp. 4–5. Donald Winch revisits and refines some of these methodological issues in *Riches and Poverty,* chap. 1, and 236.

2. Discussions of Stewart's influence on Henry Brougham, Francis Horner, James Mackintosh, and Francis Jeffrey can be found in Collini, Winch, and Burrow, *That Noble Science of Politics,* chap. 1, and Biancamaria Fontana, *Rethinking the Politics of Commercial Society: The "Edinburgh Review," 1802–1832* (Cambridge: Cambridge University Press, 1985), chap. 1, and 96–104.

3. On the importance of Stewart's pedagogy, see Phillipson, "Pursuit of Virtue," 82–101.

4. Fontana argues that McCulloch's *Discourse on the Rise, Progress, Peculiar Objects and Importance of Political Economy* is "the sole document we possess by a member of the *Edinburgh Review* grouping which offers an academic presentation of the subject matter of political economy roughly comparable with that outlined in Stewart's lectures" (*Rethinking Commercial Society,* 105). Although I agree with Fontana that McCulloch's text is "roughly comparable" with Stewart's writings, I emphasize the disciplinary differences between the two in the last section of this chap-

ter. McCulloch attended Edinburgh University from 1807 to 1811 (although he did not graduate), so it is possible that he heard Stewart lecture in the latter's last years at Edinburgh.

5. In addition to the works I have already cited, see Knud Haakonssen, "From Moral Philosophy to Political Economy: The Contribution of Dugald Stewart," in *Philosophers of the Scottish Enlightenment,* ed. V. Hope (Edinburgh: University of Edinburgh Press, 1984), 211–32. Haakonssen argues that Stewart's revisions of Smith constitute a "dissolution" of Smith's intentions (212).

6. Stewart, *Philosophy of the Human Mind,* 99. Future references will be cited in the text by page number.

7. "It is chiefly in compliance with common language and common prejudices, that I am sometimes led in the following observations, to contrast theory with experience. In the proper sense of the word theory, it is so far from standing in opposition to experience, that it implies a knowledge of principles, of which the most extensive experience alone could put us in possession" (*Philosophy,* 147).

8. Haakonssen has emphasized Stewart's attention to system and to the future. See "From Moral Philosophy to Political Economy," esp. 228–30.

9. Haakonssen attributes Smith's subjectivism to two factors: his commitment to a developmental theory of moral phenomena and his insistence that the only basis for moral judgments is the ability to identify with others through sympathy. "Moral judgements are therefore to be understood as formed under the influence of a vast complexity of previous moral judgements, either directly or as they are internalised by the individual, and normally both. And at the same time the present judgement may of course well contribute to, and maybe change, this social store of moral knowledge. In this way the theory of the development of moral phenomena becomes a necessary element in the theory of morality, and this theory of development is turned into history once we add the particular circumstances of the society in question—its relationship to other societies, its physical circumstances, its exposure to accidental factors of all sorts" ("From Moral Philosophy to Political Economy," 217).

10. Stewart, *Account of the Life and Writings of Adam Smith, LL.D.,* in *Collected Works of Dugald Stewart,* 18:58.

11. Smith, *Wealth of Nations,* 5.3.2; 734–40.

12. On Stewart's indebtedness to Condorcet and his awkward attempts to distance his work from that of French theorists, see Collini, Winch, and Burrow, *That Noble Science of Politics,* 32 44. Stewart's admiration for the French *économistes* appears in *Philosophy of the Human Mind,* 155 61.

13. The note chastises anxious Britons for confounding "the speculative doctrines of Political Economy, with those discussions concerning the first principles of Government which happened unfortunately at that time to agitate the public mind." Stewart defiantly reprinted his original text, but he implied that, however he felt about the right of philosophers to debate state policy, he did have new criticisms of Smith's system, which he promised to publish in another treatise. See *Account of the Life,* in *Collected Works of Dugald Stewart,* n. G, 87.

14. Ibid., n. I, 88–95.

15. Stewart, *Dissertation: Exhibiting the Progress of Metaphysical, Ethical, and Political Philosophy, since the Revival of Letters in Europe,* in *Collected Works,* 1:97, 477–78. Note that Stewart praises political economists for their philosophical method, not for their particular doctrines. "Whatever praise, therefore, may be due to the fathers of the modern science of political economy, belongs, at least in part . . . to those abstract studies by which they were prepared for an analytical investigation of its first and foremost principles" (478).

16. The most notable exception to this generalization is Winch, *Riches and Poverty,* esp. part 3.

17. Charles Dickens, *Hard Times* (1854; Harmondsworth: Penguin Books, 1969), 48 (chap. 2).

18. On Bentham's desire for numerical information, see John R. Poynter, *Society and*

Pauperism: English Ideas on Poor Relief, 1795–1834 (London: Routledge and Kegan Paul, 1969), 129–30. Poynter notes that Bentham made a proper system of bookkeeping a centerpiece of his planned Houses of Industry, that he was frustrated by the British government's failure to undertake a national census, and that he even tried to gather the numerical information the government had yet to supply when he asked parishes to return lists of indoor and outdoor paupers in the 1790s. To his despair, most of the parishes failed to respond. While Bentham's efforts to encourage and revalue numerical information obviously form a part of the story I want to tell, his influence was registered most powerfully through his disciples, especially Edwin Chadwick and James Mill, in the 1830s.

19. Bills of mortality had probably been kept for London since at least 1592, although no bills from before 1658 survived the Great Fire of London (1666). It is also unclear when these bills were first published, although by 1662 they were regularly printed and subscribers could purchase them for four shillings a year. Whatever their origin and however frequently they were printed, it is clear that after 1603 these records were regularly kept by parish clerks. They continued to be kept until 1849, by which time they had been superseded by the bills issued, since 1840, by the new office of the registrar-general. See Charles Henry Hull, "On the Bills of Mortality," in *The Economic Writings of Sir William Petty,* ed. Charles Henry Hull (Cambridge: Cambridge University Press, 1899; reprint, Fairfield, N.J.: Augustus M. Kelley, 1986), lxxx–lxxxiv. On the disputed authorship of the "Observations upon the Bills of Mortality," see Hull, "The Authorship of the 'Observations upon the Bills of Mortality,'" also in *Economic Writings,* xxxix–liv.

Life tables were intended to convert the data provided by the bills of mortality into information that could be used for determining the value of annuities or the price of insurance, or even for estimating the proportion of men available for military conscription. In his important account of eighteenth-century life tables, Peter Buck argues that before about 1750 these tables (however inaccurate) were primarily intended to support the needs of the state; they were, in other words, instruments of political arithmetic. After midcentury, partly because the British government was no longer willing to rely on annuities for revenue, political arithmetic concerns no longer framed the study of the bills of mortality. Instead, the importance of life tables began to turn on their use for private insurance. See Buck, "People Who Counted," 28–45. Although his work constitutes an invaluable contribution to our understanding of who collected this kind of numerical information in eighteenth-century Britain and the purposes for which it was collected, Buck does not engage the sticky question of why insurance companies did not base their price structures either on the most reliable life tables or on mathematical probability until 1785. This perplexing issue has been addressed by Lorraine Daston, "The Domestication of Risk: Mathematical Probability and Insurance, 1650–1830," in *The Probabilistic Revolution,* ed. Lorenz Kruger, Lorraine J. Daston, and Michael Heidelberger (Cambridge: MIT Press, 1987), 1:237–60, and Geoffrey Wilson Clark, "Betting on Lives: Life Insurance in English Society and Culture, 1695–1775" (Ph.D. diss., Princeton University, 1993).

20. On the scarcity and unreliability of numerical accounts of poor relief, see Poynter, *Society and Pauperism,* 19, 141, 225, 276, 281. "Even later [after 1785], when returns were more continuous, it is difficult to allow for changes in food prices, in the value of money generally, in population and pauper numbers, and for the extent to which relief was paid in lieu of wages. Mere totals do not tell us much. Late seventeenth-century estimates by King (£622,000), and Dunning (£819,000) were only guesses, as was Fielding's estimate of £1 million in 1754. The first Parliamentary returns showed an average of £698,971 for the years 1748–50, but were so imperfect that they were put aside behind the Speaker's Chair and not published until found by the assiduous Rickman and the Select Committee of 1817–18" (19).

21. On the parliamentary committees, which were appointed in 1775 and 1786, see Marshall, *English Poor,* 77. M. Dorothy George discusses the Society for Bettering the Condition of the Poor in *London Life in the Eighteenth Century* (1925; reprint, Chicago: Academy Chicago, 1984), 25. As the precedents to calls for this kind of information (although they did not all recommend *numerical* information) George cites Henry Fielding's *Proposal for Making an Effectual Provision for the Poor* (1753) and Jonas Hanway's various pleas on behalf of the infant poor in workhouses: *A Candid Historical Account of the Hospital for . . . Exposed and Deserted Young Children* (1759), and *An Ernest Appeal for Mercy to the Children of the Poor* (1766). The society founded to improve the conditions of the poor was the Society for Bettering the Condition and Increasing the Comforts of the Poor. See Poynter, *Society and Pauperism,* 91–98.

22. On the scarcity of bankruptcy information, see Sheila Marriner, "English Bankruptcy Records and Statistics before 1850," *Economic History Review,* 2d ser., 33, no. 3 (1980): 351–67, and Julian Hoppit, *Risk and Failure in English Businesses, 1700–1800* (Cambridge: Cambridge University Press, 1987), chaps. 1–2. On Josiah Wedgwood's rejection of cost accounting, see Neil McKendrick, "Josiah Wedgwood and Cost Accounting in the Industrial Revolution," *Economic History Review,* 2d ser., 23, no. 1 (1970): 45–67.

23. See Sir John Sinclair, *The Statistical Account of Scotland, 1791–1799,* ed. Donald J. Withrington and Ian R. Grant, 21 vols. (Edinburgh: EP Publishing, 1983), and Donald J. Withrington, "What Was Distinctive about the Scottish Enlightenment?" in *Aberdeen and the Enlightenment,* ed. Jennifer J. Carter and Joan H. Pittock (Aberdeen: Aberdeen University Press, 1987), 9–19.

24. See, for example, Thomas, "Numeracy in Early Modern England," 103–32; Buck, "People Who Counted," 32–35; and Cohen, *Calculating People,* chap. 1.

25. From Richard Cumberland, *De Legibus Naturae Disquisitio Philosophica* (1672; translation published in 1727 as *A Treatise of the Law of Nature*); quoted in Miller, *Defining the Common Good,* 272.

26. The classic studies of utilitarianism more generally are Élie Halévy, *La formation du radicalisme philosophique,* 3 vols. (Paris, 1901–4); Ernest Albee, *A History of English Utilitarianism* (London: G. Allen and Unwin, 1901); and John Plamenatz, *Mill's Utilitarianism, with a Study of the English Utilitarians* (Oxford: Basil Blackwell, 1949). These writers tended to focus on the nineteenth-century secular variant of utilitarianism, but as David Lieberman has recently argued, Bentham's secular utilitarianism did not displace its theological predecessor until the mid 1830s, when J. S. Mill and William Whewell began to represent Bentham as a moralist (Lieberman, "Happiness 101: William Paley, Utilitarianism and Moral Instruction in Eighteenth-Century England," unpublished manuscript, University of California, Berkeley, 1996). I am indebted to him for letting me read this essay in manuscript. Another reassessment of theological utilitarianism is Thomas A. Horne, "'The Poor Have a Claim Founded in the Law of Nature': William Paley and the Rights of the Poor," *Journal of the History of Philosophy* 23, no. 1 (1985): 51–70.

27. William Paley, *Principles of Moral and Political Philosophy,* 9th American ed. (Boston: West and Richardson, 1818), 50.

28. See Winch, *Riches and Poverty,* 243–44; Eric Heavener, "Food, Sex, and God: The Christian Social Theory of T. R. Malthus" (Ph.D. diss., Johns Hopkins University, 1992), 46–48; and S. H. Hollander, "Malthus and Utilitarianism with Special Reference to the *Essay on Population,*" *Utilitas* 1 (1989): 170–210.

29. The best collection of essays on this important subject is Tore Frangsmyr, J. L. Heilbron, and Robin E. Rider, eds., *The Quantifying Spirit in the Eighteenth Century* (Berkeley: University of California Press, 1990).

30. Thomas Robert Malthus, *An Essay on the Principle of Population,* ed. Antony Flew (1798; reprinted, London: Penguin Books, 1982), 67. Future references to this, the first edition, will be cited in the text by date ("1798") and page number.

31. See R. B. Simons, "T. R. Malthus on British Society," *Journal of the History of Ideas* 16 (1955): 73.

32. The problem of populousness encouraged several eighteenth-century British writers to offer estimates of the size of the population, even though nothing resembling a modern census had been conducted in Britain. In general these attempts were either local, like Thomas Percival's *Further Observations on the State of the Population in Manchester and Other Adjacent Places* (1773), or speculative, like Richard Price's *Essay on the Population of England from the Revolution to the Present Time* (1779).

33. Beginning in his own day, most analysts of Malthus have discounted the theodicy and have tended to downplay the importance of theological issues in his version of political economy. In the past twenty-five years, however, this has begun to change. In addition to Winch, *Riches and Poverty,* and Heavener, "Food, Sex, and God," see also J. M. Pullen, "Malthus' Theological Ideas and Their Influence on His Principle of Population," *History of Political Economy* 13, no. 1 (1981): 39–54, and M. B. Harvey-Phillips, "Malthus' Theodicy: The Intellectual Background of His Contribution to Political Economy," *History of Political Economy* 16, no. 4 (1984): 591–608.

34. For a discussion of the history of this concept, see Milton, "Origin and Development of the Concept of the 'Laws of Nature,'"173–95.

35. This criticism is implicit in Poynter's argument that Malthus's method was not empirical enough, for example. See *Society and Pauperism,* 147–48.

36. Ibid., 165–77.

37. "Preface to the Second Edition," reprinted in *Essay on the Principle of Population,* sixth edition (1826), with variant readings from the second edition (1803), in *The Works of Thomas Robert Malthus,* ed. E. A. Wrigley and David Soudan, 8 vols. (London: William Pickering, 1986), 2:iii–iv. Most of the additions that appeared in the second edition were reprinted in all subsequent editions. Because the editors of Malthus's *Works* have republished only the sixth edition (with variant readings from the second), I cite that edition in what follows.

38. See Antony Flew, introduction to Malthus, *Essay* (1798), 13; and for the bibliography, *Works,* 3:701–11.

39. On Eden, see Poynter, *Society and Pauperism,* 111–17. Poynter reports that Eden confessed his numbers were unreliable, both because the evidence was incomplete and because laborers (he claimed) tended to understate their wages (115). Note also that John Rickman, who finally succeeded in convincing Parliament to institute a regular census, was lobbying for numerical information throughout the first decades of the nineteenth century. See ibid., 245, 251–53.

40. This is Antony Flew's complaint: "From the subsistence of such a general average limit you cannot validly deduce that the same limit will be effective all the time in every particular case. On the contrary, if there is to be any point in talking of an average, there must be room for cases falling both above and below the average" (introduction to Malthus, *Essay* [1798], 36).

41. Quoted in ibid., 39. Flew also discusses the insight offered by Archbishop Richard Whately and Nassau Senior that Malthus confused two senses of "tendency": "that in which a tendency to produce something is a cause which, operating unimpeded, would produce it; and that in which to speak of a tendency to produce something is to say that that result may reasonably be expected in fact to occur" (ibid., 38).

42. When John Weyland, a rural justice of the peace who issued his first attack on Malthus in 1807, voiced two different opinions in the same pamphlet, he revealed that the same person might hold a variety of convictions about numbers and knowledge. On the one hand, Weyland

considered numerical information sufficiently important to warrant a prefatory call for readers to supply "observations, accompanied with facts, that may tend either to confirm or impugn any part of his opinions." On the other hand, however, he objected to Malthus's tendency to use numbers indiscriminately; in particular, he criticized his use of numbers to equate individuals who are essentially different in kind. It was absurd for Malthus to argue that an adult male who emigrates or dies in war encourages propagation at home, Weyland asserts: "The bare fact, as to numerical population, may approach to something near the truth; but as the animal lost by war, or emigration, is an active, perfect, proper man; and the thing got in return is not even an infant, but only the prospect (certain to be sure) that he will be replaced by one; the conclusion, as to numerical population, is certainly incorrect; and as to national power, altogether unfounded." [John Weyland], *A Short Inquiry into the Policy, Humanity, and Past Effects of the Poor Laws . . . in Which Are Included a Few Considerations on the Questions of Political Oeconomy, Most Intimately Connected with the Subject; Particularly on the Supply of Food in England* (London: J. Hatchard, 1807), xv, 52. Weyland later published a more sustained attack on Malthus, *The Principles of Population and Production as They Are Affected by the Progress of Society with a View to Moral and Political Consequences* (1820).

43. Donald Winch discusses the romantic critique of Malthus in *Riches and Poverty,* chap. 11. See also Poynter, *Society and Pauperism,* 171–85.

44. Robert Southey, "On the State of the Poor, the Principle of M. Malthus's Essay on Population, and the Manufacturing System, 1812," in *Essays, Moral and Political,* vol. 1 (London: John Murray, 1832), 87–88. This essay was originally published in the *Quarterly Review* in 1812.

45. Samuel Taylor Coleridge, cited in Thomas Allsop, ed., *Letters, Conversations and Recollections of S. T. Coleridge* (London: E. Moxon, 1836), 136–37; Coleridge uses "political empiric" in *Lay Sermons,* in *The Collected Works of Samuel Taylor Coleridge,* ed. R. J. White (London: Routledge and Kegan Paul; Princeton: Princeton University Press, 1972), 6:143, 150–55. See also Winch, *Riches and Poverty,* 289–90, 295–96.

46. Southey, "On the State of the Poor," 111, 112.

47. See Winch, *Poverty and Riches,* chap. 11.

48. Of Malthus's tendency to use the theological language of "God's gift" to pursue moral principles, Ricardo wrote: "I do not agree that in a treatise on Political Economy it should be so considered. The gift is great or little according as it is more or less, not according as it may be more or less morally useful" (quoted in Winch, *Riches and Poverty,* 350, from *Notes on Malthus,* in David Ricardo, *The Works and Correspondence of David Ricardo,* ed. Piero Sraffa and H. M. Dodd [Cambridge: Cambridge University Press for the Royal Economic Society, 1951–73], 2:210).

49. On the Malthus-Ricardo debate, see Collini, Burrow, and Winch, *That Noble Science of Politics,* chap. 2, and Winch, *Riches and Poverty,* chap. 13. For an important discussion of the methodological component of this debate, see Goldman, "Origins of British 'Social Science,'" 587–616.

50. Quoted in Collini, Burrow, and Winch, *That Noble Science of Politics,* 79, from *Quarterly Review* (January 1824).

51. [J. R. McCulloch], "Ricardo's *Political Economy,*" *Edinburgh Review* 30 (June 1818): 59–87; quotation on 64. This was the first of McCulloch's thirty-one contributions to the *Edinburgh Review;* he continued to publish economic articles there until 1837. On McCulloch's relationship with the *Edinburgh Review,* see D. P. O'Brien, *J. R. McCulloch: A Study in Classical Economics* (New York: Barnes and Noble, 1970), 34–42, and Fontana, *Rethinking the Politics,* chap. 2.

52. J. R. McCulloch, *The Principles of Political Economy: With a Sketch of the Rise and Progress of the Science* (Edinburgh: William and Charles Tait, 1825), 3. As I point out later, McCulloch's insistence that commerce plays a central role in distributing "the blessings of civilization" suggests this

more expansive definition of "values" (*Principles*, 122). Future references will be cited in the text by page number.

53. For a discussion of the tension between Ricardo's style and McCulloch's popularizing efforts, see Claudia Klaver, "Moralizing the Economy: The Constitution and Contestation of Nineteenth-Century Economic Authority" (Ph.D. diss., Johns Hopkins University, 1995). For another discussion of McCulloch's success as a popularizer of Ricardo, see S. G. Checkland, "The Propagation of Ricardian Economics in England," *Economica*, n.s., 16 (February 1949): 40–52. McCulloch was not the only or even the first would-be popularizer of political economy. In 1816 Jane Marcet published *Conversations on Political Economy* for use in the schoolroom (six additional editions had been published by 1839). In the early 1830s Harriet Martineau launched her enormously successful *Illustrations of Political Economy*, which were intended for the working classes.

54. The *OED* records the word "economist" as coming into use before the word "economy" in its modern sense (that is, as the institutionalized production, consumption, and distribution of a nation's wealth). In fact, although the *OED* records the former as appearing (in its modern sense, as a student of economics) in 1804, it has no entry for the latter; this suggests that "economy" was not used in its modern sense in the nineteenth century.

55. So reviled was political economy in 1824 that the sponsors of the Ricardo Lectures were afraid even to advertise for the subscriptions necessary to raise the speaker's fee. "You can have little notion of the dread of publicity which hangs over many of us; & of the aversion to Political Economy which yet here is almost universal," James Mill wrote to McCulloch in 1824 (O'Brien, *McCulloch*, 48).

56. It should be noted that—to a degree remarkable even in early nineteenth-century Britain—McCulloch tended to reprint huge sections of his own writing, sometimes amended, sometimes not. Although McCulloch defended this practice, arguing that publishing ideas in a newspaper was a good way to find out if they needed revision, his detractors charged him with both self-plagiarism and bilking the public: "Were all Mr M'Culloch's various publications, or reprints, collected together, they might be compressed by a skillful and honest redacteur, into the bulk of one moderate sized volume, sold for 10s 6d," wrote John Wilson of McCulloch's "stale harangue" (quoted in O'Brien, *McCulloch*, 73–74). For the modern reader this means that one of McCulloch's major theoretical texts is almost as good as any other; indeed, the text I will primarily use here, *Discourse on the Rise, Progress, Peculiar Objects, and Importance of Political Economy*, was reprinted, almost verbatim, in his *Principles of Political Economy*. Among the other works titled *Principles of Political Economy* published in the first half of the nineteenth century were texts by Malthus (1836) and John Stuart Mill (1840).

57. The first chair of political economy was established at the East India College at Haileybury in 1805 and was first held by Malthus. Collini, Burrow, and Winch point out, however, that this position was initially called the professor of general history, politics, commerce, and finance; in 1818 it was divided into the chair of general polity and the laws of England and the chair of modern history and political economy (*That Noble Science of Politics*, 67). See also Patricia James, *Population Malthus: His Life and Times* (London: Routledge and Kegan Paul, 1979), 172, 178.

58. Before 1828, only one critical edition of Smith's work had been published in Britain: in 1814 David Buchanan had attempted to "rectify what is amiss in Dr Smith; to supply omissions; to give his reasonings an application to modern times; and to exhibit, as far as the author is qualified, a complete system of political economy" (quoted in Fontana, *Rethinking the Politics*, 71; see also 69–75). Although Buchanan's edition did replace the truncated, arbitrarily abridged versions of *Wealth of Nations* printed in Britain in the eighteenth century, it was not until McCulloch's 1828 edition that Smith's work was given a context in an international history of economic writ-

ings. McCulloch's collections include *A Select Collection of Scarce and Valuable Tracts and Other Publications on Paper Currency and Banking* (1857); *A Select Collection of Scarce and Valuable Tracts and Other Publications on the National Debt and the Sinking Fund* (1857); *A Select Collection of Scarce and Valuable Tracts on Commerce* (1859); and *A Select Collection of Scarce and Valuable Economical Tracts* (1859).

59. J. R. McCulloch, *A Discourse on the Rise, Progress, Peculiar Objects, and Importance, of Political Economy: Containing an Outline of a Course of Lectures on the Principles and Doctrines of That Science* (Edinburgh: Archibald Constable, 1825), 8. Future references will be cited in the text by page number.

60. In *Wealth of Nations*, in a sentence that separated the two collectives McCulloch tended to use interchangeably ("the public," "the body of the people"), Smith recommended mandatory public education: "The public can impose upon almost the whole body of the people the necessity of acquiring those most essential parts of education" (5.3.2; 738; see also 737–40). Malthus began to recommend teaching the poor to read in the second (1803) edition of his *Essay*. See Poynter, *Society and Pauperism*, 160.

61. According to Charles Lyell, British geology reached this crisis at the beginning of the nineteenth century, and geologists' response to it was successful because it was collective and institutional. "Although the reluctance to theorize was carried somewhat to excess, no measure could be more salutary at such a moment than a suspension of all attempts to form what were termed 'theories of the earth.' A great body of new data were required, and the Geological Society of London, founded in 1807, conduced greatly to the attainment of this desirable end. To multiply and record observations, and patiently to await the result at some future period, was the object proposed by them, and it was their favourite maxim that the time was not yet come for a general system of geology, but that all must be content for many years to be exclusively engaged in furnishing materials for future generalizations. By acting up to these principles with consistency, they in a few years disarmed all prejudice, and rescued the science from the imputation of being a dangerous, or at best but a visionary pursuit" (Lyell, *Principles of Geology*, ed. Martin S. Rudwick, 3 vols. [1830–33; Chicago: University of Chicago Press, 1990], 1:71–72).

62. Poynter, *Society and Pauperism*, 247, 276–82. On the commission of 1832, see chap. 9, esp. 316 21.

63. Poynter notes Rickman's importance but says little about him. See *Society and Pauperism*, 245, 251–53, 276. See also O. Williams, *Life and Letters of John Rickman* (1911); Cullen, *Statistical Movement*, 12–23, 15, 65, 84, 156; and Theodore M. Porter, *Rise of Statistical Thinking*, 30–31.

Chapter Seven

1. On Quételet and the laws of regularity, see Porter, *Rise of Statistical Thinking*, chap. 2, and Goldman, "Origins of British 'Social Science,'" 600–604. In *Hard Times*, Dickens makes the pathetic Tom Gradgrind rationalize his bank robbery by referring to this law.

2. Sir John Sinclair, "Address to the Reader," in *Statistical Account of Scotland*, 1:15. On Sinclair, see Withrington's "General Introduction" (1:ix–xlii), and Withrington, "What Was Distinctive about the Scottish Enlightenment?" both in *Aberdeen and the Enlightenment*, ed. Jennifer J. Carter and Joan H. Pittock (Aberdeen: Aberdeen University Press, 1987), 9–19. Porter discusses the German origins of "statistics" in *Rise of Statistical Thinking*, 23–24.

3. Bisset Hawkins, *Elements of Medical Statistics* (quoted in Porter, *Rise of Statistical Thinking*, 24). By 1835, William Cooke Taylor assumed that the tabular form of statistical information was essential; see "Objects and Advantages of Statistical Science," *Foreign Quarterly Review* 16 (October 1835): 103. Future references will be cited in the text by page number.

4. Although these campaigns to define statistics as numbers based but not mathematical were influential, it must be noted that statistics was primarily considered a *descriptive* practice, whether or not it used numerical representation. This means that statistics inherited many of the ambiguities implicit in theories of description like Adam Smith's, and it helps explain why contemporaries continued to find it difficult to delineate the range of objects that statisticians should consider. In 1835, for example, when J. R. McCulloch called attention to "The State and Defects of British Statistics," he could only plead the impossibility of defining this new science, instead of clearly distinguishing between it and adjacent sciences: "By a statistical account of a country, we mean a work describing its situation and extent—its natural and acquired capacities of production—the quantity and value of the various articles of utility and convenience existing and annually produced in it—the number and classes of its inhabitants, with their respective incomes—its institutions for the government, improvement, and defence of the population; with a variety of subordinate and subsidiary statements and details. A science embracing so great a variety of objects is not easily defined or limited; nor is it in all cases possible to state absolutely what ought to be taken in, and what left out. It has many features in common with geography and politics; and embraces that sort of mongrel science that has sometimes been called political arithmetic. . . .In statistical works, a short notice of the principal divisions of the country, with reference especially to its climate, soil, native products, agriculture, manufactures, and population, is generally sufficient. This much, however, cannot be dispensed with. It is the substratum on which all the rest of the building is to stand; and the completeness of the other and more elevated parts will generally depend more on the compactness and solidity of this than on any thing else" (McCulloch, "State and Defects of British Statistics," *Edinburgh Review* 61 [April 1835]: 156–57). Future references will be cited in the text.

5. Goldman, "Origins of British 'Social Science,'" 591–92. Susan Cannon omitted Whewell and included Adam Sedgwick instead; but as Morrell and Thackray make clear, Sedgwick, who was professor of geology at Cambridge and president of the 1833 meeting of the BAAS, opposed the creation of the statistical section—at least until the advocates of statistics agreed to limit the science to the collection of numerical data. See S. F. Cannon, *Science in Culture: The Early Victorian Period* (New York: Dawson and Science History Publications, 1978), 240–44.

6. Whewell papers, Trinity College Library, Cambridge, Add. MSS c. 51 (n.d., 1830 or 1831?) (quoted in Goldman, "Origins," 595). Malthus's objections to Ricardian deduction can be found in the introduction to *Principles of Political Economy* (1820), in Malthus, *Works*, 5:1–18.

7. Goldman quotes Richard Jones, whom he identifies as the intellectual leader of this group, as criticizing those who argued that "changes in social organisation and the subjects they lead us in sight of, are not the proper objects of economical science, which is wealth and wealth alone. . . .Economical science can never . . . be successfully pursued if such subjects be wholly eschewed by its promoters. There is a close connection between the economical and social organisation of nations and their powers of production" ("Text Book of Lectures on the Political Economy of Nations, Delivered at the East India College, Haileybury, lecture 4," quoted in Goldman, "Origins," 600).

8. Cannon, *Science in Culture,* 105.

9. Here is Jones: "If we wish to make ourselves acquainted with the economy and arrangements by which different nations of the earth produce and distribute their revenues, I really know of but one way to attain our object, and that is, to look and see. We must get comprehensive views of facts, that we may arrive at principles that are truly comprehensive" ("An Introductory Lecture on Political Economy, Delivered at King's College, London, 27 February 1833," quoted in Goldman, "Origins," 596).

10. Goldman, "Origins," 599.

11. "Introduction," *Journal of the Statistical Society of London* 1 (May 1838): 1. Future references will be cited in the text by page number. See also Hilts, "*Aliis Exterendum,*" 21–43.

12. Goldman, "Origins," 612–15. On the Manchester Statistical Society, see Elesh, "Manchester Statistical Society," 280–301, 407–17, and T. S. Ashton, *Economic and Social Investigations in Manchester, 1833–1933: A Centenary History of the Manchester Statistical Society* (London: P. S. King, 1934), chap. 2.

13. See Jack Morrell and Arnold Thackray, *Gentlemen of Science: Early Years of the British Association for the Advancement of Science* (Oxford: Oxford University Press, 1981), 291–96.

14. *Reports of the British Association for the Advancement of Science* (1833), 3:xc–xci (quoted in Hilts, "*Aliis Exterendum,*" 34).

15. The "disgraceful vice" that raised this troubling question for Taylor is masturbation, although he was so worried about description's ability to incite imitation that he would not even name it. "A clergyman, the master of a very large and popular school, the locality of which, for reasons that will presently appear, we must not specify, recently informed one of his friends, that he had discovered a new pupil in the act of practising a disgraceful vice. 'Send him home to his parents and say nothing about it,' was the friend's judicious recommendation. The schoolmaster however, placed great confidence in his own eloquence and the corrective powers of birch; he assembled his boys, made an excellent harangue on the guilt of the delinquent, and gave him a sound flogging. The example of crime proved more influential than the example of punishment, and the vice spread so rapidly that the whole school was broken up in consequence. These and countless similar facts lead us to question the propriety of describing vice at all, in the moral tales designed for young persons, even though the consequent punishment be ever so strongly depicted" ("Objects," 108). I discuss this example in Poovey, "'Figures of Arithmetic, Figures of Speech': The Discourse of Statistics in the 1830s," *Critical Inquiry* 19 (Winter 1993); rpt. in *Questions of Evidence: Proof, Practice, and Persuasion across the Disciplines,* ed. James Chandler, Arnold I. Davidson, and Harry Harootunian (Chicago: University of Chicago Press, 1994), 418–20.

16. [Herman Merivale], "Moral and Intellectual Statistics of France," *Edinburgh Review* 69 (April 1839): 49–74; quotation on 51.

17. Robert Chambers, *Vestiges of the Natural History of Creation,* ed. James A. Secord (1844; Chicago: University of Chicago Press, 1994), 349–54. Secord discusses the reception of this book in his introduction, xxvi–xxxviii; see also Richard Yeo, "Science and Authority in Mid-Nineteenth Century Britain: Robert Chambers and *Vestiges of the Natural History of Creation,*" *Victorian Studies* 28 (1984): 5–31.

18. See Charles Dickens, "Full Report of the First Meeting of the Mudfog Association for the Advancement of Everything," *Bentley's Miscellany* 2 (1837): 397–413, and Thomas Carlyle, "Chartism" (1839), in *The Works of Thomas Carlyle,* 30 vols. (New York: Scribner's Sons, 1903–4), 29:124–29.

19. [G. Robertson], "Exclusion of Opinions," *London and Westminster Review* 61 (April 1838): 25, 26, 37. Future references will be cited by page number in the text.

20. In addition to the London Statistical Society (1834) and the statistical section of the BAAS (1833), the 1830s saw statistical societies founded in Manchester (1833) and Glasgow (1836). In 1832, the Board of Trade established a statistical office; this, along with the census, remained the primary site for the production of official statistical data until the creation of the Registrar General's Office in 1837.

21. McCulloch explicitly argued in 1835 that collecting numerical information on a large scale would require the "intervention of Government," and he recommended placing purpose-trained agents throughout Britain to aid this collection ("State," 177–81). As secretary to the Poor Law Commission, Edwin Chadwick brought McCulloch's dream as near to fruition as it was

possible to do in the first half of the nineteenth century. See my "Thomas Chalmers, Edwin Chadwick, and the Sublime Revolution in Nineteenth-Century Government" and "Domesticity and Class Formation: Chadwick's 1842 *Sanitary Report*," in Poovey, *Making a Social Body*, 98–114, 115–31.

22. John Herschel, *A Preliminary Discourse on the Study of Natural Philosophy*, facsimile ed., ed. Arthur Fine (1830; Chicago: University of Chicago Press, 1987), 76. Future references will be cited in the text by page number. The frontispiece of Herschel's text was a portrait of Bacon.

23. "The successful process of scientific enquiry," Herschel cautioned, "demands continually the alternate use of both the *inductive* and *deductive* method. . . .The inductive and deductive methods may be said to go hand in hand, the one verifying the conclusions deduced by the other; and the combination of experiment and theory, which may thus be brought to bear in such cases, forms an engine of discovery infinitely more powerful than either taken separately" (*Preliminary Discourse*, 174, 181).

24. See Daston and Galison, "Image of Objectivity," 81–128.

25. John Stuart Mill, *The Logic of the Social Sciences* (La Salle, Ill.: Open Court, 1987), 61, 63 (6.6). This volume reprints book 6 of Mill's *A System of Logic Ratiocinative and Inductive, Being a Connected View of the Principles of Evidence and the Method of Scientific Investigation*, first published in 1843. Future references will be cited in the text with the abbreviation *Logic;* for convenience of reference, I will include the book and chapter numbers of the longer work.

Part of this chapter reprints the definition of political economy that Mill had developed in 1831, revised in 1833, and published in 1836 as "On the Definition of Political Economy; and on the Method of Philosophical Investigation in That Science," *London and Westminster Review* 25–26 (October 1836–January 1837): 1–29. This article was reprinted as essay 5 in Mill's *Essays on Some Unsettled Questions of Political Economy* (London: J. W. Parker, 1844), 120–64. The edition I cite as "Definition" appears in *Essays on Economics and Society*, intro. Lord Robbins and ed. J. M. Robson (Toronto: University of Toronto Press; London: Routledge and Kegan Paul, 1967), 309–39.

26. "In every science, therefore, which has reached the stage at which it becomes a science of causes, it will be usual, as well as desirable, first to obtain the highest generalisations, and then deduce the more special ones from them. Nor can I discover any foundation for the Baconian maxim, so much extolled by subsequent writers, except this: That before we attempt to explain deductively from more general laws any new class of phenomena, it is desirable to have gone so far as is practicable in ascertaining the empirical laws of those phenomena, so as to compare the results of deduction not with one individual instance after another, but with general propositions expressive of the points of agreement which have been found among many instances. For if Newton had been obliged to verify the theory of gravitation, not by deducing from Kepler's laws, but by deducing all the observed planetary positions which had served Kepler to establish those laws, the Newtonian theory would probably never have emerged from the state of an hypothesis" (*Logic,* 57–58 [6.5]). Mill takes the implications of this claim one step further in his note, where he cites Whewell: "'To which,' says Dr Whewell, 'we may add, that it is certain from the history of the subject, that in that case the hypothesis would never have been framed at all'" (*Logic,* 58, note [6.5]).

27. "The facts of statistics . . . have yielded conclusions, some of which have been very startling to persons not accustomed to regard moral actions as subject to uniform laws. The very events which in their own nature appear most capricious and uncertain, and which in any individual case no attainable degree of knowledge would enable us to foresee, occur, when considerable numbers are taken into the account, with a degree of regularity approaching to mathematical. What act is that which all would consider as more completely dependent on indi-

vidual character, and on the exercise of individual free will, than that of slaying a fellow-creature? Yet in any large country, the number of murders, in proportion to the population, varies (it has been found) very little from one year to another, and in its variations never deviates widely from a certain average. What is still more remarkable, there is a similar approach to constancy in the proportion of these murders annually committed with every kind of instrument. . . . This singular degree of regularity *en masse,* combined with the extreme irregularity in the cases composing the mass, is a felicitous verification *a posteriori* of the law of causation in its application to human conduct" (*Logic* 121–22 [6.11]). Mill tried to sidestep the charge of determinism by distinguishing his use of statistics from that of the historian Henry Thomas Buckle, whose *History of Civilization in England* (London, 1857–61) Mill greatly admired. See *Logic,* 123–25 (6.11).

28. One sign among many of the disciplinary gulf that had begun to open between those practices we associate with science and those we call art was the end of the long tradition of a single individual's writing both studies of rhetoric and scientific treatises. Bacon had composed both kinds, as had Adam Smith and Dugald Stewart. As far as I can tell, Richard Whately (1787–1863) was one of the last Britons to attempt what increasingly seemed like a disciplinary crossover: Whately's *Elements of Rhetoric* appeared in 1828, and his influential *Introductory Lectures on Political Economy* was published in 1831.

29. William Wordsworth, "Preface to the Edition of 1814," *The Excursion,* in *The Prose Works of William Wordsworth,* ed. W. J. B. Owen and Jane Worthington Smyser, 4 vols. (Oxford: Clarendon Press, 1974), 3:7 (line 113).

30. Percy Shelley, "A Defence of Poetry, or Remarks Suggested by an Essay Entitled 'The Four Ages of Poetry,'" in *The Poetry and Prose of Percy Shelley,* ed. Donald H. Reiman and Sharon B. Powers (New York: W. W. Norton, 1977), 500, 502.

31. From *The Letters of John Keats,* ed. Hyde Edward Rollins, 2 vols. (Cambridge: Harvard University Press, 1958), 1:193.

32. See W. K. Wimsatt Jr., "The Concrete Universal," in *The Verbal Icon: Studies in the Meaning of Poetry* (Lexington: University Press of Kentucky, 1954), 69–84, and Steven Knapp, *Literary Interest: The Limits of Anti-formalism* (Cambridge: Harvard University Press, 1993), 104.

Abrams, Philip. *The Origins of British Sociology, 1834–1914.* Chicago: University of Chicago Press, 1968.

Achinstein, Peter. "Newton's Corpuscular Query and Experimental Philosophy." In *Philosophical Perspectives on Newtonian Science,* edited by Phillip Brickcr and R. I. G. Hughes, 135–74. Cambridge: MIT Press, 1990.

Adams, Percy G. *Travel Literature and the Evolution of the Novel.* Lexington: University Press of Kentucky, 1983.

Agnew, Jean Christophe. *Worlds Apart: The Market and the Theater in Anglo-American Thought, 1550–1750.* Cambridge: Cambridge University Press, 1986.

Aho, James A. "Rhetoric and the Invention of Double-Entry Bookkeeping." *Rhetorica* 3, no. 3 (1985): 21–43.

Albee, Ernest. *A History of English Utilitarianism.* London: G. Allen and Unwin, 1901.

Allsop, Thomas, ed. *Letters, Conversations and Recollections of S. T. Coleridge.* London: E. Moxon, 1836.

Amussen, Susan Dwyer. *An Ordered Society: Gender and Class in Early Modern England.* New York: Basil Blackwell, 1988.

Andrews, J. H. "Appendix: The Beginnings of the Surveying Profession in Ireland—Abstract." In *English Map-Making, 1500–1650,* edited by Sarah Tyacke. London: British Library, Reference Division Publications, 1983.

Appleby, Joyce Oldham. *Economic Thought and Ideology in Seventeenth Century England.* Princeton: Princeton University Press, 1978.

Arbuthnot, John. "An Argument for Divine Providence, taken from the constant Regularity observ'd in the Births of both Sexes." 1711.

———. "Essay on the Usefulness of Mathematical Learning." In *The Life and Works of John Arbuthnot,* edited by George A. Aitken, 409–35. Oxford: Clarendon Press, 1892.

Ashton, T. S. *Economic and Social Investigations in Manchester, 1833–1933: A Centenary History of the Manchester Statistical Society.* London: P. S. King, 1934.

Aspromourgos, Tony. "The Life of William Petty in relation to His Economics: A Tercentenary Interpretation." *History of Political Economy* 20, no. 3 (1988): 337–40.

Auerbach, Erich. "Figura." In *Scenes from the Drama of European Literature,* edited by Erich Auerbach, 11–76. Theory and History of Literature, vol. 9. Minneapolis: University of Minnesota Press, 1984.

Bacon, Francis. *Novum Organum*. Translated and edited by Peter Urbach and John Gibson. Chicago: Open Court, 1994.

———. *The Works of Francis Bacon*. Edited and translated by James Spedding, Robert L. Ellis, and Douglas D. Heath. London: Longman, 1857–58.

Ball, W. W. Rouse. *A Short Account of the History of Mathematics*. New York: Dover, 1960.

Barker-Benfield, J. G. *The Culture of Sensibility: Sex and Society in Eighteenth-Century Britain*. Chicago: University of Chicago Press, 1992.

Barrell, John. *The Birth of Pandora and the Division of Knowledge*. Philadelphia: University of Pennsylvania Press, 1992.

Barrow, John D. *Pi in the Sky: Counting, Thinking, Being*. Boston: Little, Brown, 1992.

Barry, Andrew, Thomas Osborne, and Nikolas Rose, eds. *Foucault and Political Reason: Liberalism, Neo-liberalism, and Rationalities of Government*. Chicago: University of Chicago Press, 1996.

Barthes, Roland. *The Fashion System*. Translated by Matthew Ward and Richard Howard. New York: Hill and Wang, 1983.

Batten, Charles L., Jr. *Pleasurable Instruction: Form and Convention in Eighteenth-Century Travel Literature*. Berkeley: University of California Press, 1978.

Baudrillard, Jean. *For a Critique of the Political Economy of the Sign*. Translated by Charles Levin. St. Louis, Mo.: Telos Press, 1981.

———. *Simulations*. Translated by Paul Foss, Paul Patton, and Philip Beitchman. New York: Semiotext(e), 1983.

Bazeley, Deborah Taylor. "An Early Challenge to the Precepts and Practices of Modern Science: The Fusion of Fact, Fiction, and Feminism in the Works of Margaret Cavendish, Duchess of Newcastle (1623–1673)." Ph.D. diss., University of California, San Diego, 1990.

Beier, A. L. *Masterless Men: The Vagrancy Problem in England, 1560–1640*. New York: Methuen, 1985.

Bender, John, and David E. Wellbery. "Rhetoricality: On the Modernist Return of Rhetoric." In *The Ends of Rhetoric: History, Theory, Practice,* edited by John Bender and David E. Wellbery, 3–42. Stanford: Stanford University Press, 1990.

Bernard, T. C. "Sir William Petty, Irish Landowner." In *History and Imagination: Essays in Honour of H. R. Trevor-Roper,* edited by Hugh Lloyd-Jones, Valerie Pearl, and Blair Worden, 201–17. London: Duckworth, 1981.

Berry, Christopher J. *The Idea of Luxury: A Conceptual and Historical Investigation*. Cambridge: Cambridge University Press, 1994.

Biagioli, Mario. *Galileo, Courtier: The Practice of Science in the Culture of Absolutism*. Chicago: University of Chicago Press, 1993.

Blaug, Mark. "Economic Theory and Economic History in Great Britain, 1650–1776." *Past and Present* 68 (1964): 111–16.

Bolla, Peter de. *The Discourse of the Sublime: Readings in History, Aesthetics, and the Subject*. Oxford: Basil Blackwell, 1989.

Boltanski, Luc, and Laurent Thevenot. *De la justification: Les économies de la grandeur*. Paris: Gallimard, 1991.

Botera, Giovanni. *Delle cause della grandezza della città.* 1588.

———. *Ragion de stato.* 1589.

Boyer, Carl B. *A History of Mathematics.* Princeton: Princeton University Press, 1985.

Boyle, Robert. "The Origins of Forms and Qualities according to the Corpuscular Philosophy." In *Selected Philosophical Papers of Robert Boyle,* edited by M. A. Stewart, 1–96. Indianapolis: Hackett, 1991.

———. *The Works of the Honourable Robert Boyle.* 6 vols. London: J. and F. Rivington, 1772.

Brewer, John. *The Sinews of Power: War, Money, and the English State, 1688–1783.* Cambridge: Harvard University Press, 1990.

[Brown, John]. *The Merchants Avizo.* London: J. Norton, 1607.

Buck, Peter. "People Who Counted: Political Arithmetic in the Eighteenth Century." *Isis* 73 (1982): 28–45.

———. "Seventeenth-Century Political Arithmetic: Civil Strife and Vital Statistics." *Isis* 68 (1977): 67–85.

Buckle, Henry Thomas. *History of Civilization in England.* London, 1857–61.

Buickerood, James G. "Pursuing the Science of Man: Some Difficulties in Understanding Eighteenth-Century Maps of the Mind." *Eighteenth-Century Life* 19 (May 1995): 1–17.

Burchell, Graham. "Peculiar Interests: Civil Society and Governing 'The System of Natural Liberty.'" In *The Foucault Effect: Studies in Governmentality, with Two Lectures and an Interview with Michel Foucault,* edited by Graham Burchell, Colin Gordon, and Peter Miller, 119–50. Chicago: University of Chicago Press, 1991.

Burchell, Graham, Colin Gordon, and Peter Miller, eds. *The Foucault Effect: Studies in Governmentality, with Two Lectures and an Interview with Michel Foucault.* Chicago: University of Chicago Press, 1991.

Campbell, Archibald. *An Inquiry into the Original of Moral Virtue.* 1728.

Campbell, T. D. "Francis Hutcheson: 'Father' of the Scottish Enlightenment." In *The Origins and Nature of the Scottish Enlightenment,* edited by R. H. Campbell and Andrew S. Skinner, 167–85. Edinburgh: John Donald, 1982.

Cannon, John. *Aristocratic Century.* Cambridge: Cambridge University Press, 1984.

Cannon, S. F. *Science in Culture: The Early Victorian Period.* New York: Dawson and Science History Publications, 1978.

Canny, Nicholas. *From Reformation to Resistance: Ireland, 1534–1660.* Dublin: Helicon, 1987.

———. *Kingdom and Colony: Ireland in the Atlantic World, 1560–1800.* Baltimore: Johns Hopkins University Press, 1982.

Carlyle, Thomas. *The Works of Thomas Carlyle.* 30 vols. New York: Charles Scribner's Sons, 1903–4.

Carruthers, Bruce G., and Wendy Nelson Espeland. "Accounting for Rationality: Double-Entry Bookkeeping and the Rhetoric of Economic Rationality." *American Journal of Sociology* 97, no. 1 (1991): 31–69.

Cartwright, Nancy. *How the Laws of Physics Lie.* Oxford: Clarendon Press, 1983.

Cave, Terence. *The Cornucopian Text: Problems of Writing in the French Renaissance.* Oxford: Clarendon Press, 1979.

Chambers, Robert. *Vestiges of the Natural History of Creation.* 1844. Edited by James A. Secord. Chicago: University of Chicago Press, 1994.

Chartier, Roger. "The Practical Impact of Writing." In *Passions of the Renaissance,* edited by Roger Chartier, 111–59, vol. 3 of *A History of Private Life,* edited by Philippe Ariès and Georges Duby. Cambridge: Harvard University Press, 1989.

Checkland, S. G. "The Propagation of Ricardian Economics in England." *Economica,* n.s., 16 (February 1949): 40–52.

Chibka, Robert L. "'Oh! Do Not Fear a Woman's Invention': Truth, Falsehood, and Fiction in Aphra Behn's *Oronooko.*" *Texas Studies in Literature and Language* 30 (1988): 510–37.

Child, Sir Josiah. *Brief Observations concerning Trade, and Interest of Money.* London: Elizabeth Calvert, 1668. Facsimile edition, *Sir Josiah Child, Merchant Economist, with a Reprint of "Brief Observations,"* edited by William Letwin. Cambridge: Harvard University Press, 1959.

Christensen, Francis. "John Wilkins and the Royal Society Reform of Prose Style." *Modern Language Quarterly* 7, no. 3 (1946): 279–90.

Christensen, Jerome. *Practicing Enlightenment: Hume and the Formation of a Literary Career.* Madison: University of Wisconsin Press, 1987.

Clanchy, Michael. *From Memory to Written Record: England, 1066–1307.* 1979. 2d ed. Cambridge: Harvard University Press, 1993.

———. "*Moderni* in Education and Government in England." *Speculum* 50 (1975): 671–88.

Clark, G. N. *Guide to English Commercial Statistics, 1696–1782.* London: Royal Historical Society, 1938.

Clark, Geoffrey Wilson. "Betting on Lives: Life Insurance in English Society and Culture, 1695–1775." Ph.D. diss., Princeton University, 1993.

Clark, Jonathan. *Dynamics of Power: The Structure of Politics.* Cambridge: Cambridge University Press, 1985.

Cohen, Murray. *Sensible Words: Linguistic Practice in England, 1640–1785.* Baltimore: Johns Hopkins University Press, 1977.

Cohen, Patricia Cline. *A Calculating People: The Spread of Numeracy in Early America.* Chicago: University of Chicago Press, 1982.

———. "Reckoning with Commerce: Numeracy in Eighteenth-Century America." In *Consumption and the World of Goods,* edited by John Brewer and Roy Porter, 320–34. New York: Routledge, 1993.

Cole, Anne. *Poor Law Documents before 1834.* Oxford: Bocardo Press, 1993.

Cole, Roger. *Reasons of the Increase of the Dutch Trade. Wherein Is Demonstrated from What Causes the Dutch Govern and Manage Trade Better Than the English; Whereby They Have So Far Improved Their Trade above the English.* 1671.

———. *A Treatise Wherein Is Demonstrated, That the Church and State of England, Are in Equal Danger with the Trade of It.* 1671.

Coleman, D. C. "Mercantilism Revisited." *Historical Journal* 23 (1980): 773–91.

Coleridge, Samuel Taylor. *The Collected Works of Samuel Taylor Coleridge.* Edited by R. J. White. 14 vols. London: Routledge and Kegan Paul; Princeton: Princeton University Press, 1972.

Colgan, Maurice. "Ossian: Success or Failure for the Scottish Enlightenment?" In *Aberdeen and the Enlightenment,* edited by Jennifer J. Carter and Joan H. Pittock, 344–49. Aberdeen: Aberdeen University Press, 1987.

Colley, Linda. *In Defiance of Oligarchy.* Cambridge: Cambridge University Press, 1983.

Collini, Stefan, Donald Winch, and John Burrow. *That Noble Science of Politics: A Study in Nineteenth-Century Intellectual History.* Cambridge: Cambridge University Press, 1983.

Conley, Thomas M. *Rhetoric in the European Tradition.* Chicago: University of Chicago Press, 1990.

Copeland, Edward. "Money in the Novels of Fanny Burney." *Studies in the Novel* 8 (1976): 24–37.

[Cotton, Sir Robert]. "A Speech . . . Touching the Alteration of Coin." 1626; printed 1651. Reprinted in *Old and Scarce Tracts on Money,* edited by J. R. McCulloch, 121–42. 1856. Reprint, London: P. S. King, 1933.

Coughlan, Patricia. "'Cheap and Common Animals': The English Anatomy of Ireland in the Seventeenth Century." In *Literature and the English Civil War,* edited by Thomas Healy and Jonathan Sawday. Cambridge: Cambridge University Press, 1990.

Crosby, Alfred W. *The Measure of Reality: Quantification and Western Society, 1250–1600.* Cambridge: Cambridge University Press, 1997.

Cullen, Michael J. *The Statistical Movement in Early Victorian Britain: The Foundations of Empirical Social Research.* New York: Barnes and Noble, 1975.

Curley, Thomas M. *Samuel Johnson and the Age of Travel.* Athens: University of Georgia Press, 1976.

Dafforne, Richard. *The Merchants Mirrour.* London: Nicolas Bourn, 1651.

Daston, Lorraine. "Baconian Facts, Academic Civility, and the Prehistory of Objectivity." *Annals of Scholarship* 8, nos. 3–4 (1991): 337–64.

———. *Classical Probability in the Enlightenment.* Princeton: Princeton University Press, 1988.

———. "Description by Omission: Nature Enlightened and Observed." Presented at conference "Regimes of Description: In the Archive of the Eighteenth Century." Stanford University, 11–14 January 1996.

———. "The Domestication of Risk: Mathematical Probability and Insurance, 1650–1830." In *The Probabilistic Revolution,* edited by Lorenz Kruger, Lorraine J. Daston, and Michael Heidelberger, 1:237–60. Cambridge: MIT Press, 1987.

———. "The Moral Economy of Science." In *Constructing Knowledge in the History of Science,* edited by Arnold Thackray. Chicago: University of Chicago Press, 1995. Originally published in *Osiris,* 2d ser., 10 (1995): 2–24.

———. "Objectivity and the Escape from Perspective." *Social Studies of Science* 22 (November 1992): 597–618.

Daston, Lorraine, and Peter Galison. "The Image of Objectivity." *Representations* 40 (fall 1992): 81–128.

Davenant, Charles. *Discourses on the Public Revenue.* 1698.

———. *The Political and Commercial Works of that Celebrated Writer, Charles Davenant . . . collected and revised by Sir Charles Whitworth.* 5 vols. London, 1771.

Dean, Mitchell. *The Constitution of Poverty: Toward a Genealogy of Liberal Governance.* London: Routledge, 1991.

Dear, Peter. *Discipline and Experience: The Mathematical Way in the Scientific Revolution.* Chicago: University of Chicago Press, 1995.

———. "From Truth to Disinterestedness in the Seventeenth Century." *Social Studies of Science* 22 (1992): 619–31.

———. "Jesuit Mathematical Science and the Reconstitution of Experience in the Early Seventeenth Century." *Studies in the History and Philosophy of Science* 18 (1987): 133–75.

———. *Mersenne and the Learning of the Schools.* Ithaca: Cornell University Press, 1988.

———. "*Totius in Verba:* Rhetoric and Authority in the Early Royal Society." *Isis* 76 (1985): 145–61.

Defoe, Daniel. *Complete English Tradesman.* 1726. Gloucester: Alan Sutton, 1987.

———. *Essays upon Several Projects, or Effectual Ways for Advancing the Interests of the Nation.* London: Booksellers of London and Westminster, 1702.

———. *Review.* 9 vols. 19 February 1704–11 June 1713.

———. *A Tour through the Whole Island of Great Britain.* Edited by Pat Rogers. Harmondsworth: Penguin Books, 1971.

Derham, William. *Physico-theology, or A Demonstration of the Being and Attributes of God, from the Works of Creation.* London, 1714.

Dickens, Charles. "Full Report of the First Meeting of the Mudfog Association for the Advancement of Everything." *Bentley's Miscellany* 2 (1837): 397–413.

———. *Hard Times.* 1854. Harmondsworth: Penguin Books, 1969.

Dickson, P. G. M. *The Financial Revolution: A Study in the Development of Public Credit, 1688–1756.* London: Macmillan, 1967.

Einaudi, Luigi. "The Theory of Imaginary Money from Charlemagne to the French Revolution." In *Enterprise and Secular Change: Readings in Economic History,* edited by Frederic Chapin Lane, 229–61. Homewood, Ill.: Richard D. Irwin, 1953.

Elesh, David. "The Manchester Statistical Society: A Case Study in Discontinuity in the History of Empirical Social Research." *Journal of the History of the Behavioral Sciences* 8 (1972): 280–301, 407–17.

Emerson, Roger L. "The Social Composition of Enlightened Scotland: The Select Society of Edinburgh, 1754–1764." *Studies on Voltaire and the Eighteenth Century* 114 (1973): 291–329.

Endres, A. M. "The Functions of Numerical Data in the Writings of Graunt, Petty, and Davenant." *History of Political Economy* 17, no. 2 (1985): 245–64.

Erasmus. *On Copia of Words and Ideas.* Translated by Donald B. King and H. David Rix. Milwaukee: Marquette University Press, 1963.

Evnine, Simon. "Hume, Conjectural History, and the Uniformity of Human Nature." *Journal of the History of Philosophy* 21, no. 4 (1993): 589–606.

Eyler, J. M. *Victorian Social Medicine: The Ideas and Influence of William Farr.* Baltimore: Johns Hopkins University Press, 1979.

Feingold, Mordechai. *The Mathematicians' Apprenticeship: Science, Universities, and Society in England, 1560–1640.* Cambridge: Cambridge University Press, 1984.

Ferguson, Adam. *An Essay on the History of Civil Society.* 1764. New Brunswick, N.J.: Transaction, 1995.

Ferguson, Margaret W. "Juggling the Categories of Race, Class, and Gender: Aphra Behn's *Oronooko.*" *Women's Studies* 19 (1991): 159–81.

Fielding, Henry. *Proposal for Making an Effectual Provision for the Poor. . . .* 1753. In *The Works of Henry Fielding,* edited by Edmund Gosse, 12:62–157. Westminster: Archibald Constable; New York: Charles Scribner's Sons, 1899.

Fitzmaurice, Lord Edmond. *Life of Sir William Petty, Chiefly from Private Documents Hitherto Unpublished.* London: John Murray, 1895.

Foisil, Madeleine. "The Literature of Intimacy." In *Passions of the Renaissance,* edited by Roger Chartier, 327–62, vol. 3 of *A History of Private Life,* edited by Philippe Ariès and Georges Duby. Cambridge: Harvard University Press, 1989.

Fontana, Biancamaria. *Rethinking the Politics of Commercial Society: The "Edinburgh Review," 1802–1832.* Cambridge: Cambridge University Press, 1985.

Forbes, Duncan. "The European or Cosmopolitan Dimension in Hume's Science of Politics." *British Journal of Eighteenth-Century Studies* 1 (1977): 57–60.

———. *Hume's Philosophical Politics.* Cambridge: Cambridge University Press, 1975.

———. "'Scientific Whiggism': Adam Smith and John Millar." *Cambridge Journal* 7 (1953–54): 643–70.

Foucault, Michel. "Governmentality." In *The Foucault Effect: Studies in Governmentality, with Two Lectures and an Interview with Michel Foucault,* edited by Graham Burchell, Colin Gordon, and Peter Miller, 87–104. Chicago: University of Chicago Press, 1991.

———. "The Order of Discourse." In *Untying the Text: A Post-structuralist Reader,* edited by Robert Young, 48–78. Boston: Routledge and Kegan Paul, 1981.

———. *The Order of Things: An Archaeology of the Human Sciences.* New York: Random House, 1970.

———. "Politics and the Study of Discourse." In *The Foucault Effect: Studies in Governmentality, with Two Lectures and an Interview with Michel Foucault,* edited by Graham Burchell, Colin Gordon, and Peter Miller, 53–72. Chicago: University of Chicago Press, 1991.

———. "Questions of Method." In *The Foucault Effect: Studies in Governmentality, with Two Lectures and an Interview with Michel Foucault,* edited by Graham Burchell, Colin Gordon, and Peter Miller, 73–86. Chicago: University of Chicago Press, 1991.

Frangsmyr, Tore, J. L. Heilbron, and Robin E. Rider, eds. *The Quantifying Spirit in the Eighteenth Century.* Berkeley: University of California Press, 1990.

Funkenstein, Amos. *Theology and the Scientific Imagination from the Middle Ages to the Seventeenth Century.* Princeton: Princeton University Press, 1986.

Gallagher, Catherine. *Nobody's Story: The Vanishing Acts of Women Writers in the Marketplace, 1670–1820.* Berkeley: University of California Press, 1994.

George, M. Dorothy. *London Life in the Eighteenth Century.* 1925. Reprint, Chicago: Academy Chicago, 1984.

Gerzina, Gretchen. *Black England: Life before Emancipation.* London: J. Murray, 1995.

Glamann, Kristof. "The Changing Patterns of Trade." In *The Economic Organization of Early Modern Europe,* vol. 5 of *The Cambridge Economic History of Europe,* edited by E. E. Rich and C. H. Wilson, 185–289. Cambridge: Cambridge University Press, 1977.

Glanvill, Joseph. *Essays on Several Important Subjects in Philosophy and Religion.* Edited by Richard H. Popkin. 1676. Facsimile edition. New York: Johnson Reprint Company, 1970.

Goldberg, Jonathan. *Writing Matter: From the Hands of the English Renaissance.* Stanford: Stanford University Press, 1990.

Goldman, Lawrence. "The Origins of British 'Social Science': Political Economy, Natural Science and Statistics, 1830–1835." *Historical Journal* 26, no. 3 (1983): 587–616.

Gordon, Anne. *To Move with the Times: The Story of Transport and Travel in Scotland.* Aberdeen: Aberdeen University Press, 1988.

Gordon, Colin. "Governmental Rationality: An Introduction." In *The Foucault Effect: Studies in Governmentality, with Two Lectures and an Interview with Michel Foucault,* edited by Graham Burchell, Colin Gordon, and Peter Miller, 1–52. Chicago: University of Chicago Press, 1991.

Gordon, Daniel. *Citizens without Sovereignty: Equality and Sociability in French Thought, 1670–1789.* Princeton: Princeton University Press, 1994.

[Graunt, John?]. *Observations upon the Bills of Mortality.* 1662. In *The Economic Writings of Sir William Petty,* 314–435. 1899. Reprint, Fairfield, N.J.: Augustus M. Kelley, 1986.

Guillory, John. *Cultural Capital: The Problem of Literary Canon Formation.* Chicago: University of Chicago Press, 1993.

Gunn, J. A. W. *Politics and the Public Interest in the Seventeenth Century.* London: Routledge and Kegan Paul, 1969.

Haakonssen, Knud. "From Moral Philosophy to Political Economy: The Contribution of Dugald Stewart." In *Philosophers of the Scottish Enlightenment,* edited by V. Hope, 211–32. Edinburgh: University of Edinburgh Press, 1984.

———. "Natural Jurisprudence in the Scottish Enlightenment: Summary of an Interpretation." In *Enlightenment, Rights, and Revolution,* edited by Neil MacCormick and Zenon Bankowski, 36–49. Aberdeen: Aberdeen University Press, 1989.

Hacking, Ian. *The Emergence of Probability: A Philosophical Study of Early Ideas about Probability.* Cambridge: Cambridge University Press, 1975.

Halévy, Elie. *La formation du radicalisme philosophique.* 3 vols. Paris, 1901–4.

Halley, Edmund. *Two Papers on the Degrees of Mortality in Mankind.* 1693.

Halpern, Richard. *The Poetics of Primitive Accumulation: English Renaissance Culture and the Genealogy of Capital.* Ithaca: Cornell University Press, 1991.

Hanway, Jonas. *A Candid Historical Account of the Hospital . . . Exposed and Deserted Young Children.* 1759.

———. *An Earnest Appeal for Mercy to the Children of the Poor.* 1766.

Harvey-Phillips, M. B. "Malthus' Theodicy: The Intellectual Background of His Contribution to Political Economy." *History of Political Economy* 16, no. 4 (1984): 591–608.

Heavener, Eric. "Food, Sex, and God: The Christian Social Theory of T. R. Malthus." Ph.D. diss., Johns Hopkins University, 1992.

Henderson, Katherine Usher, and Barbara F. McManus. *Half Humankind: Contexts and Texts of the Controversy about Women in England, 1540–1640.* Urbana: University of Illinois Press, 1985.

Herschel, John. *A Preliminary Discourse on the Study of Natural Philosophy.* 1830. Facsimile ed. Edited by Arthur Fine. Chicago: University of Chicago Press, 1987.

Hilts, Victor L. "*Aliis Exterendum,* or The Origins of the Statistical Society of London." *Isis* 69 (1978): 21–43.

Hirschman, Albert O. *The Passions and the Interests: Political Arguments for Capitalism before Its Triumph.* Princeton: Princeton University Press, 1977.

Hobbes, Thomas. *Leviathan.* Edited by C. B. Macpherson. 1651. Harmondsworth: Penguin Books, 1968.

Hollander, S. H. "Malthus and Utilitarianism with Special Reference to the *Essay on Population.*" *Utilitas* 1 (1989): 170–210.

Hont, Istvan. "Commercial Society and Political Theory in the Eighteenth Century: The Problem of Authority in David Hume and Adam Smith." In *Main Trends in Cultural History: Ten Essays,* edited by William Melching and Wyger Velema, 54–94. Amsterdam: Editions Rodopi, 1994.

―――. "The Political Economy of the 'Unnatural and Retrograde' Order: Adam Smith and Natural Liberty." In *Französische Revolution und politische Ökonomie,* edited by Maxine Berg et al., 122–49. Trier, Germany: Schriften aus dem Karl-Marx-Haus, 1989.

Hont, Istvan, and Michael Ignatieff, eds. *Wealth and Virtue: The Shaping of Political Economy in the Scottish Enlightenment.* Cambridge: Cambridge University Press, 1983.

Hopfl, H. M. "From Savage to Scotsman: Conjectural History in the Scottish Enlightenment." *Journal of British Studies* 17, no. 2 (1978): 19–40.

Höppen, K. Theodore. "The Nature of the Early Royal Society," part 1. *British Journal for the History of Science* 9 (1976): 1–24.

Hoppit, Julian. *Risk and Failure in English Businesses, 1700–1800.* Cambridge: Cambridge University Press, 1987.

Hopwood, Anthony G. "The Archaeology of Accounting Systems." *Accounting, Organizations, and Society* 12, no. 3 (1987): 207–34.

Hopwood, Anthony G., and Peter Miller, eds. *Accounting as Social and Institutional Practice.* Cambridge: Cambridge University Press, 1994.

Horne, Thomas A. "'The Poor Have a Claim Founded in the Law of Nature': William Paley and the Rights of the Poor." *Journal of the History of Philosophy* 23, no. 1 (1985): 51–70.

Hoskin, Keith, and Richard Macve. "Writing, Examining, Disciplining: The Genesis of Accounting's Modern Power." In *Accounting as a Social and Institutional Practice,* edited by Anthony G. Hopwood and Peter Miller, 67–97. Cambridge: Cambridge University Press, 1994.

Hoyrup, Jens. "Sub-scientific Mathematics: Observations on a Pre-modern Phenomenon." *History of Science* 28, no. 22 (1989): 63–87.

Hull, Charles Henry. "The Authorship of the 'Observations upon the Bills of Mortality.'" In *The Economic Writings of Sir William Petty,* ed. Charles Henry Hull. Cambridge: Cambridge University Press, 1899; reprint, Fairfield, N.J.: Augustus M. Kelley, 1986, xxxix–liv.

―――. "Petty's Life." In *The Economic Writings of Sir William Petty,* ed. Charles Henry Hull, xiii–xxxiii. Cambridge: Cambridge University Press, 1899. Reprint, Fairfield, N.J.: Augustus M. Kelley, 1986.

Hume, David. *Essays Moral, Political, and Literary.* Edited by Eugene F. Miller. Indianapolis: Liberty Classics, 1987.

————. *The Natural History of Religion.* 1757. In *Principal Writings on Religion,* edited by J. C. A. Gaskin. New York: Oxford University Press, 1993.

————. *A Treatise of Human Nature.* Edited by Ernest C. Mossner. London: Penguin Books, 1984.

Hundert, E. J. *The Enlightenment's Fable: Bernard Mandeville and the Discovery of Society.* Cambridge: Cambridge University Press, 1994.

Hunter, Ian. *Culture and Government: The Emergence of Literary Education.* London: Macmillan, 1988.

Hunter, Michael. *The Royal Society and Its Fellows, 1660–1700: The Morphology of an Early Scientific Institution.* 2d ed. Oxford: Alden Press, 1994.

Hutcheson, Francis. *An Essay on the Nature and Conduct of the Passions and Affections with Illustrations on the Moral Sense.* Facsimile ed. Edited by Paul McReynolds. 1729. Gainesville, Fla.: Scholars' Facsimiles and Reprints, 1969.

————. *An Inquiry into the Original of Our Ideas of Beauty and Virtue.* 1728. Reprint, Charlottesville, Va.: Lincoln-Rembrandt, 1993.

James, Patricia. *Population Malthus: His Life and Times.* London: Routledge and Kegan Paul, 1979.

Jed, Stephanie H. *Chaste Thinking: The Rape of Lucretia and the Birth of Humanism.* Bloomington: Indiana University Press, 1989.

Jevons, W. Stanley. *The Principles of Science.* 1874. New York: Dover, 1958.

Johnson, Samuel. *A Journey to the Western Islands of Scotland.* In *"A Journey to the Western Islands of Scotland" by Samuel Johnson and "The Journal of a Tour to the Hebrides with Samuel Johnson, LL.D." by James Boswell,* edited by Allan Wendt. Boston: Houghton Mifflin, 1965.

Johnston, Stephen. "Mathematical Practitioners and Instruments in Elizabethan England." *Annals of Science* 48 (1991): 321–41.

Jones, D. W. *War and Economy in the Age of William III and Marlborough.* Oxford: Basil Blackwell, 1988.

Jones, Edgar. *Accountancy and the British Economy, 1840–1980: The Evolution of Ernst and Whinney.* London: B. T. Batsford, 1981.

Jones, Richard F. "Science and Language in England of the Mid-Seventeenth Century." *Journal of English and Germanic Philology* 31 (1932): 315–31.

Journal of the Statistical Society of London 1 (May 1838).

Kahn, Victoria. *Rhetoric, Prudence, and Skepticism in the Renaissance.* Ithaca: Cornell University Press, 1985.

Kames, Henry Home, Lord. *Sketches of the History of Man.* 1774. 2d ed. Edinburgh: W. Strahan, 1778.

Kavanagh, Thomas. *Enlightenment and the Shadows of Chance: The Novel and the Culture of Gambling in Eighteenth-Century France.* Baltimore: Johns Hopkins University Press, 1993.

Kay, James Phillips. *The Moral and Physical Condition of the Working Classes Employed in the Cotton Manufacture in Manchester.* 2d ed. enlarged. London: James Ridgway, 1832.

Keats, John. *The Letters of John Keats.* Edited by Hyde Edward Rollins. 2 vols. Cambridge: Harvard University Press, 1958.

Kelly, J. Thomas. *Thorns on the Tudor Rose: Monks, Rogues, Vagabonds, and Sturdy Beggars.* Jackson: University Press of Mississippi, 1977.

King, Gregory. *Two Tracts: Natural and Political Observations and Conclusions upon the State and Condition of England*. 1696.

Kivy, Peter. *The Seventh Sense: A Study of Francis Hutcheson's Aesthetics and Its Influence in Eighteenth-Century Britain*. New York: Burt Franklin, 1976.

Klaver, Claudia. "Moralizing the Economy: The Constitution and Contestation of Nineteenth-Century Economic Authority." Ph.D. diss., Johns Hopkins University, 1995.

Klein, Lawrence. *Shaftesbury and the Culture of Politeness: Moral Discourse and Cultural Politics in Early Eighteenth-Century England*. Cambridge: Cambridge University Press, 1994.

Knapp, Steven. *Literary Interest: The Limits of Anti-formalism*. Cambridge: Harvard University Press, 1993.

Kuhn, Thomas. *The Structure of Scientific Revolutions*. Chicago: University of Chicago Press, 1962.

Landes, David S. *Revolution in Time: Clocks and the Making of the Modern World*. Cambridge: Harvard University Press, 1983.

Lansdowne, Marquise of, ed. *The Petty Papers: Some Unpublished Writings of Sir William Petty*. London: Constable, 1927.

————. *The Petty-Southwell Correspondence, 1676–1687*. London: Constable, 1928.

Latour, Bruno. *The Pasteurization of France*. Cambridge: Harvard University Press, 1988.

————. *We Have Never Been Modern*. Translated by Catherine Porter. Cambridge: Harvard University Press, 1993.

Latour, Bruno, and Steve Woolgar. *Laboratory Life: The Construction of Scientific Facts*. Princeton: Princeton University Press, 1979.

Laudan, L. L. "Thomas Reid and the Newtonian Turn of British Methodological Thought." In *The Methodological Heritage of Newton*, edited by Robert E. Butts and John W. Davis, 102–31. Oxford: Basil Blackwell, 1970.

Lazarsfeld, Paul F. "Notes on the History of Quantification in Sociology—Trends, Sources and Problems." *Isis* 52 (1961): 277–333.

Leeuwen, Henry G. van. *The Problem of Certainty in English Thought, 1630–1690*. The Hague: Martinus Nijhoff, 1963.

Leneman, Leah. "The Effects of Ossian in Lowland Scotland." In *Aberdeen and the Enlightenment*, edited by Jennifer J. Carter and Joan H. Pittock, 357–62. Aberdeen: Aberdeen University Press, 1987.

Letwin, William. *The Origins of Scientific Economics: English Economic Thought, 1660–1776*. New York: Methuen, 1963. Reprint, Westport, Conn.: Greenwood Press, 1975.

Levine, Donald. *Visions of the Sociological Tradition*. Chicago: University of Chicago Press, 1995.

Lieberman, David. "Happiness 101: William Paley, Utilitarianism and Moral Instruction in Eighteenth-Century England." Unpublished manuscript, University of California, Berkeley, 1996.

Livingston, Donald W. *Hume's Philosophy of Common Life*. Chicago: University of Chicago Press, 1984.

Locke, John. *Essay concerning Human Understanding*. Edited by John W. Yolton. 2 vols. London: Dent, 1961.

Long, James [or John Gay?]. *An Inquiry into the Origin of the Human Appetites and Affections*. 1747.

Luhmann, Niklas. *The Differentiation of Society.* Translated by Steven Holmes and Charles Larmore. New York: Columbia University Press, 1982.

Lyell, Charles. *Principles of Geology.* Edited by Martin S. Rudwick. 3 vols. 1830–33. Chicago: University of Chicago Press, 1990.

McCloskey, Donald N. *If You're So Smart: The Narrative of Economic Expertise.* Chicago: University of Chicago Press, 1990.

———. "The Rhetoric of Economic Expertise." In *The Recovery of Rhetoric: Persuasive Discourse and Disciplinarity in the Human Sciences,* edited by R. H. Roberts and J. M. M. Good, 137–47. Charlottesville: University Press of Virginia, 1993.

McCulloch, J. R. *Discourse on the Rise, Progress, Peculiar Objects, and Importance of Political Economy.* Edinburgh: Archibald Constable, 1825.

———, ed. *An Inquiry into the Nature and Causes of the Wealth of Nations.* Edinburgh: Adam and Charles Black and William Tait, 1828.

———, ed. *Old and Scarce Tracts on Money.* 1856. Reprint, London: P. S. King, 1933.

———. *The Principles of Political Economy: With a Sketch of the Rise and Progress of the Science.* Edinburgh: William and Charles Tait, 1825.

———. "Ricardo's *Political Economy.*" *Edinburgh Review* 30 (June 1818): 59–87.

———. *A Select Collection of Scarce and Valuable Economical Tracts.* 1859.

———. *A Select Collection of Scarce and Valuable Tracts and Other Publications on Paper Currency and Banking.* 1857.

———. *A Select Collection of Scarce and Valuable Tracts and Other Publications on the National Debt and the Sinking Fund.* 1857.

———. *A Select Collection of Scarce and Valuable Tracts on Commerce.* 1859.

———. "State and Defects of British Statistics." *Edinburgh Review* 61 (April 1835): 154–81.

McKendrick, Neil. "The Commercialization of Fashion." In *The Birth of a Consumer Society: The Commercialization of Eighteenth-Century England,* edited by Neil McKendrick, John Brewer, and J. H. Plumb, 34–98. Bloomington: Indiana University Press, 1985.

———. "Josiah Wedgwood and Cost Accounting in the Industrial Revolution." *Economic History Review,* 2d ser., 23, no. 1 (1970): 45–67.

MacKenzie, Henry. *The Man of Feeling.* 1771.

McKeon, Michael. "Historicizing Patriarchy: The Emergence of Gender Difference in England, 1660–1760." *Eighteenth-Century Studies* 28, no. 3 (1995): 295–322.

———. *The Origins of the English Novel.* Baltimore: Johns Hopkins University Press, 1987.

Maidment, Brian, ed. *The Poorhouse Fugitives: Self-Taught Poets and Poetry in Victorian Britain.* Manchester: Carcanet, 1987.

Malthus, Thomas Robert. *An Essay on the Principle of Population.* 1798. Edited by Antony Flew. Reprinted London: Penguin Books, 1982.

———. *Principles of Political Economy.* 1836.

———. *The Works of Thomas Robert Malthus.* Edited by E. A. Wrigley and David Soudan. 8 vols. London: William Pickering, 1986.

Malynes, Gerard de. *The Center of the Circle of Commerce, or A Refutation of a Treatise, Intitled "The*

Circle of Commerce, or The Ballance of Trade," Lately Published by E. M. London: William Jones, 1623.

———. *Lex Mercatoria.* 1622.

———. *The Maintenance of Free Trade.* 1622.

———. *A Treatise of the Canker of England's Commonwealth.* 1602.

Mandeville, Bernard. *The Fable of the Bees, or Private Vices, Publick Benefits.* Edited by Phillip Harth. London: Penguin Books, 1970.

Marcet, Jane. *Conversations on Political Economy.* 1816.

Markley, Robert. "Sentimentality as Performance: Shaftesbury, Sterne, and the Theatrics of Virtue." In *The New Eighteenth Century: Theory, Politics, English Literature,* edited by Felicity Nussbaum and Laura Brown, 210–30. New York: Methuen, 1987.

Marriner, Sheila. "English Bankruptcy Records and Statistics before 1850." *Economic History Review,* 2d ser., 33, no. 3 (1980): 351–67.

Marshall, Dorothy. *The English Poor in the Eighteenth Century: A Study in Social and Administrative History.* London: George Routledge, 1926.

Martin, Julian. *Francis Bacon, the State, and the Reform of Natural Philosophy.* Cambridge: Cambridge University Press, 1992.

Maxwell, John. *A Treatise of the Laws of Nature.* 1727.

Meek, Ronald. *"Economics and Ideology" and Other Essays.* London: Chapman and Hall, 1967.

———. *The Economics of Physiocracy.* Cambridge: Harvard University Press, 1963.

———. "The Scottish Contribution to Marxist Sociology." In *Democracy and the Labour Movement,* edited by John Saville. London: Lawrence and Wishart, 1954.

———. *Social Science and the Ignoble Savage.* Cambridge: Cambridge University Press, 1976.

Mellis, John. *A Briefe Instruction and Maner How to Keepe Bookes of Accompts after the Order of Debitor and Creditor.* London: John Windet, 1588.

[Merivale, Herman]. "Moral and Intellectual Statistics of France." *Edinburgh Review* 69 (April 1839): 49–74.

Mill, John Stuart. *Essays on Economics and Society.* Introduced by Lord Robbins and edited by J. M. Robson. Toronto: University of Toronto Press; London: Routledge and Kegan Paul, 1967.

———. *Essays on Some Unsettled Questions of Political Economy.* London: J. W. Parker, 1844.

———. *The Logic of the Moral Sciences.* Edited by A. J. Ayer. La Salle, Ill.: Open Court, 1988.

———. *The Logic of the Social Sciences.* La Salle, Ill.: Open Court, 1987.

———. "On the Definition of Political Economy; and on the Method of Philosophical Investigation in That Science." *London and Westminster Review* 25–26 (October 1836–January 1837): 1–29.

———. *Principles of Political Economy.* 1840.

Millar, John. *Observations concerning the Distinction of Ranks in Society.* Dublin: T. Ewing, 1771.

Miller, David. *Philosophy and Ideology in Hume's Political Thought.* Oxford: Clarendon Press, 1981.

Miller, Peter. "Accounting and Objectivity: The Invention of Calculating Selves and Calculable Spaces." *Annals of Scholarship* 9, nos. 1–2 (1992): 61–86.

————. "Accounting as Social and Institutional Practice: An Introduction." In *Accounting as Social and Institutional Practice,* edited by Anthony G. Hopwood and Peter Miller, 1–40. Cambridge: Cambridge University Press, 1994.

Miller, Peter, and Ted O'Leary. "Accounting and the Construction of the Governable Person." *Accounting, Organizations and Society* 12, no. 3 (1987): 256–61.

Miller, Peter N. *Defining the Common Good: Empire, Religion, and Philosophy in Eighteenth-Century Britain.* Cambridge: Cambridge University Press, 1994.

Milton, John R. "Induction before Hume." *British Journal for the Philosophy of Science* 38, no. 1 (1987): 49–74.

————. "The Origin and Development of the Concept of the 'Laws of Nature.'" *Archives Européennes de Sociologie* 22 (1981): 173–95.

Mirowski, Philip. "The When, the How and the Why of Mathematical Expression in the History of Economic Analysis." *Journal of Economic Perspectives* 5, no. 1 (1991): 145–57.

Misselden, Edward. *The Circle of Commerce, or The Ballance of Trade.* London: John Dawson, 1623.

Mizuta, Hiroshi. "Two Adams in the Scottish Enlightenment: Adam Smith and Adam Ferguson on Progress." In *Transactions of the Fifth International Congress on the Enlightenment,* edited by Haydn Mason, 2:812–19. Oxford: Taylor Institute, 1980.

Modeleski, Tania. *Feminism without Women: Culture and Criticism in a "Postfeminist" Age.* New York: Routledge, 1991.

Monboddo, James Burnett, Lord. *Ancient Metaphysics.* 1779–99.

————. *Of the Origin and Progress of Language.* 6 vols. 1773–92.

Moody, T. W., F. X. Martin, and F. J. Byrne, eds. *Early Modern Ireland, 1534–1691.* Vol. 3 of *A New History of Ireland.* Oxford: Clarendon Press, 1976.

Moran, Mary Catherine. "From Rudeness to Refinement: Philosophical History and Late Eighteenth-Century Conduct Literature on the Role of Women in 'The Natural History of Mankind.'" Unpublished manuscript, Johns Hopkins University, 1994.

Morgan, Gareth. "Accounting as Reality Construction: Towards a New Epistemology for Accounting Practice." *Accounting, Organizations, and Society.* 13, no. 5 (1988): 477–85.

Morgan, Victor. "The Cartographic Image of 'The Country' in Early Modern England." *Transactions of the Royal Historical Society,* 5th ser., 29 (1979): 129–54.

Morrell, Jack, and Arnold Thackray. *Gentlemen of Science: Early Years of the British Association for the Advancement of Science.* Oxford: Oxford University Press, 1981.

Motz, Lloyd, and Jefferson Hane Weaver. *The Story of Mathematics.* New York: Avon, 1995.

Mun, Thomas. *England's Treasure by Forraign Trade, or The Ballance of our Forraign Trade is the Rule of our Treasure* [1662 or 1623?]. London: Thomas Clark, 1664.

Murray, Alexander. *Reason and Society in the Middle Ages.* Oxford: Clarendon Press, 1978.

Newton, Isaac. *Opticks, or A Treatise of the Reflections, Refractions, Inflections, and Colours of Light.* 1704. 2d ed. London: W. Bowyer, 1717.

————. *Principia Mathematica.* 1687.

Nicholson, Colin. *Writing and the Rise of Finance: Capital Satires of the Early Eighteenth Century.* Cambridge: Cambridge University Press, 1994.

Nisbet, R. A. *The Sociological Tradition.* London: Heinemann, 1967.

Norton, David Fate. "George Turnbull and the Furniture of the Mind." *Journal of the History of Ideas* 36 (1975): 701–16.

Nussbaum, Felicity. *The Brink of All We Hate: English Satires on Women, 1660–1750*. Lexington: University Press of Kentucky, 1984.

O'Brien, D. P. J. R. *McCulloch: A Study in Classical Economics*. New York: Barnes and Noble, 1970.

Oldcastle, Hugh. *Profitability Treatyce*. London, [1543?].

Olson, Richard. *The Emergence of the Social Sciences, 1642–1792*. New York: Twayne, 1993.

Oreskes, Naomi. "Representation vs. Refutability: A Dilemma for Models of Complex Natural Systems." *Scientific American*, forthcoming.

Oreskes, Naomi, Kristin Shrader-Frechette, and Kenneth Belitz. "Verification, Validation, and Confirmation of Numerical Models in the Earth Sciences." *Science* 263 (4 February 1994): 641–46.

Pacioli, Luca. *De computis et scripturis*. 1494.

Paley, William. *Natural Theology*. Edinburgh, 1849.

———. *Principles of Moral and Political Philosophy*. 9th American ed. Boston: West and Richardson, 1818.

Pasquino, Pasquale. "Theatrum Politicum: The Genealogy of Capital—Police and the State of Prosperity." In *The Foucault Effect: Studies in Governmentality, with Two Lectures and an Interview with Michel Foucault*, edited by Graham Burchell, Colin Gordon, and Peter Miller, 105–18. Chicago: University of Chicago Press, 1991.

Patrick, J. Max, and Robert O. Evans. *Style, Rhetoric, and Rhythm*. Princeton: Princeton University Press, 1966.

Paulson, Ronald. *The Beautiful, Novel, and Strange: Aesthetics and Heterodoxy*. Baltimore: Johns Hopkins University Press, 1996.

Paxman, David. "Aesthetics as Epistemology, or Knowledge without Certainty." *Eighteenth-Century Studies* 26 (winter 1992–93): 285–306.

Peacham, Henry. *"The Complete Gentleman," "The Truth of Our Times," and "The Art of Living in London."* 1622. Facsimile ed. Edited by Virgil B. Heltzel. Ithaca: Cornell University Press, 1962.

Peragallo, Edward. *The Origin and Evolution of Double Entry Bookkeeping*. New York: American Institute, 1938.

Percival, Thomas. *Further Observations on the State of the Population in Manchester and Other Adjacent Places*. 1773.

Petty, William. *The Economic Writings of Sir William Petty*. Edited by Charles Henry Hull. 1899. Reprint, Fairfield, N.J.: Augustus M. Kelley, 1986.

———. *The History of the Survey of Ireland, Commonly Called the Down Survey*, edited by Thomas Aiskew Larcom. Dublin: Irish Archaeological Society, 1851. Reprint, New York: Augustus M. Kelly, 1967.

Phillips, Adam, *Terror and Experts*. Cambridge: Harvard University Press, 1995.

Phillips, Mark Salber. "Reconsiderations on History and Antiquarianism: Arnaldo Momigliano and the Historiography of Eighteenth-Century Britain." *Journal of the History of Ideas* 57, no. 2 (1996): 297–316.

Phillipson, Nicholas. *Hume.* New York: St. Martin's Press, 1989.

————. "The Pursuit of Virtue in Scottish University Education: Dugald Stewart and Scottish Moral Philosophy in the Enlightenment." In *Universities, Society, and the Future,* edited by Nicholas Phillipson, 82–101. Edinburgh: University of Edinburgh Press, 1983.

————. "The Scottish Enlightenment." In *The Enlightenment in National Context,* edited by Roy Porter and Mikulas Teich, 19–40. Cambridge: Cambridge University Press, 1981.

Plamenatz, John. *Mill's Utilitarianism, with a Study of the English Utilitarians.* Oxford: Basil Blackwell, 1949.

Plumb, J. H. *The Growth of Political Stability in England, 1675–1720.* London: Macmillan, 1967.

Pocock, J. G. A. *The Machiavellian Moment: Florentine Political Thought and the Atlantic Republican Tradition.* Princeton: Princeton University Press, 1975.

————. "The Mobility of Property and the Rise of Eighteenth-Century Sociology." In *Virtue, Commerce, and History: Essays on Political Thought and History, Chiefly in the Eighteenth Century,* 103–24. Cambridge: Cambridge University Press, 1985.

Pollak, Ellen. *The Poetics of Sexual Myth: Gender and Ideology in the Verse of Swift and Pope.* Chicago: University of Chicago Press, 1985.

Poovey, Mary. "Accommodating Merchants: Accounting, Civility, and the Natural Laws of Gender." *Differences: A Journal of Feminist Cultural Studies* 8 (fall 1996): 1–20.

————. "Aesthetics and Political Economy in the Eighteenth Century: The Place of Gender in the Social Constitution of Knowledge." In *Aesthetics and Ideology,* edited by George Levine, 79–105. New Brunswick: Rutgers University Press, 1994.

————. "'Figures of Arithmetic, Figures of Speech': The Discourse of Statistics in the 1830s." *Critical Inquiry* 19, no. 2 (1993): 256–76. Reprinted in *Questions of Evidence: Proof, Practice, and Persuasion across the Disciplines,* edited by James Chandler, Arnold I. Davidson, and Harry Harootunian, 401–21. Chicago: University of Chicago Press, 1994.

————. *Making a Social Body: British Cultural Formation, 1830–1864.* Chicago: University of Chicago Press, 1995.

————. "The Social Constitution of 'Class': Toward a History of Classificatory Thinking." In *Rethinking Class: Literary Studies and Social Formations,* edited by Wai Che Dimock and Michael T. Gilmore, 15–56. New York: Columbia University Press, 1994.

Porter, Theodore M. "Quantification and the Accounting Ideal in Science." *Social Studies of Science* 22, no. 4 (1992): 633–52.

————. *The Rise of Statistical Thinking, 1820–1900.* Princeton: Princeton University Press, 1986.

————. *Trust in Numbers: The Pursuit of Objectivity in Science and Public Life.* Princeton: Princeton University Press, 1995.

Poynter, John R. *Society and Pauperism: English Ideas on Poor Relief, 1795–1834.* London: Routledge and Kegan Paul, 1969.

Price, Richard. *Essay on the Population of England from the Revolution to the Present Time.* 1779.

Proctor, Robert N. *Value-Free Science? Purity and Power in Modern Knowledge.* Cambridge: Harvard University Press, 1991.

Pullen, J. M. "Malthus' Theological Ideas and Their Influence on His Principle of Population." *History of Political Economy* 13, no. 1 (1981): 39–54.

Rambuss, Richard. *Spenser's Secret Career.* Cambridge: Cambridge University Press, 1993.

Ranum, Orest. "The Refuges of Intimacy." In *Passions of the Renaissance,* edited by Roger Chartier, 207–63, vol. 3 of *A History of Private Life,* edited by Philippe Ariès and Georges Duby. Cambridge: Harvard University Press, 1989.

Reid, Thomas. *An Inquiry into the Human Mind.* Edited by Timothy Duggan. Chicago: University of Chicago Press, 1970.

Rendall, Jane. "Virtue and Propriety: The Scottish Enlightenment and the Construction of Femininity." Unpublished manuscript, York University, 1995.

Reports of the British Association for the Advancement of Science. 1833.

Ricardo, David. *The Works and Correspondence of David Ricardo.* Edited by Piero Sraffa and H. M. Dobb. Cambridge: Cambridge University Press for the Royal Economic Society, 1951–73.

Roberts, Lewes. *Merchants Mappe of Commerce.* 1638.

[Robertson, G.]. "Exclusion of Opinions." *London and Westminster Review* 61 (April 1838): 23–38.

Robertson, William. *The History of the Discovery and Settlement of America.* London: Jones, 1826.

———. *A View of the Progress of Society in Europe from the Subversion of the Roman Empire to the Beginning of the Sixteenth Century.* Vol. 4, *The Works of William Robertson.* London: W. Sharpe, 1820.

Robson, Keith. "Accounting Numbers as 'Inscription': Action at a Distance and the Development of Accounting." *Accounting, Organizations and Society* 17, no. 7 (1992): 685–708.

Rogers, Pat. *Johnson and Boswell: The Transit of Caledonia.* Oxford: Clarendon Press, 1995.

Roncaglia, Alessandro. *Petty: The Origins of Political Economy.* Armonk, N.Y.: M. E. Sharpe, 1985.

Roover, Raymond de. "New Perspectives on the History of Accounting." *Accounting Review* 30 (July 1955): 405–20.

Rotman, Brian. *Signifying Nothing: The Semiotics of Zero.* London: Macmillan, 1987.

Schaffer, Simon. "Defoe's Natural Philosophy and the Worlds of Credit." In *Nature Transfigured: Science and Literature, 1700–1900,* edited by John Christie and Sally Shuttleworth, 13–44. New York: Manchester University Press, 1989.

———. "Self-Evidence." In *Questions of Evidence: Proof, Practice, and Persuasion across the Disciplines,* edited by James Chandler, Arnold I. Davidson, and Harry Harootunian, 56–91. Chicago: University of Chicago Press, 1994.

Schofield, Robert E. "An Evolutionary Taxonomy of Eighteenth-Century Newtonianisms." In *Studies in Eighteenth-Century Culture,* vol. 7, edited by Roseann Runte, 175–92. Madison: University of Wisconsin Press, 1978.

Schumpeter, Joseph A. *A History of Economic Analysis.* New York: Oxford University Press, 1954.

Seymour, W. A., ed. *A History of the Ordnance Survey.* Folkestone: Dawson, 1980.

Shaftesbury, Anthony, Earl of. *Characteristics of Men, Manners, Opinions, Times.* 1711. Edited by John M. Robertson. Indianapolis: Bobbs-Merrill, 1964.

———. *Inquiry concerning Virtue and Merit.* 1699, 1711.

Shapin, Steven. "The Audience for Science in Eighteenth-Century Edinburgh." *History of Science* 12 (1974): 95–121.

————. "Property, Patronage, and the Politics of Science: The Founding of the Royal Society of Edinburgh." *British Journal for the History of Science* 7 (1974): 1–41.

————. "Pump and Circumstance: Robert Boyle's Literary Technology." *Social Studies of Science* 14 (1984): 481–520.

————. "Robert Boyle and Mathematics: Reality, Representation, and Experimental Practice." *Science in Context* 2 (1988): 23–58.

————. *A Social History of Truth: Civility and Science in Seventeenth-Century England*. Chicago: University of Chicago Press, 1994.

Shapin, Steven, and Simon Schaffer. *Leviathan and the Air-Pump: Hobbes, Boyle and the Experimental Life*. Princeton: Princeton University Press, 1985.

Shapiro, Barbara J. *Probability and Certainty in Seventeenth-Century England: A Study of the Relationships between Natural Science, Religion, History, Law, and Literature*. Princeton: Princeton University Press, 1983.

Shaw, Martin, and Ian Miles. "The Social Roots of Statistical Knowledge." In *Demystifying Social Statistics*, edited by John Irvine, Ian Miles, and Jeff Evans, 27–38. London: Pluto Press, 1979.

Shell, Marc. *The Economy of Literature*. Baltimore: Johns Hopkins University Press, 1978.

————. *Money, Language, and Thought: Literary and Philosophical Economies from the Medieval to the Modern Era*. Berkeley: University of California Press, 1982.

Shelley, Percy. *The Poetry and Prose of Percy Shelley*. Edited by Donald H. Reiman and Sharon B. Powers. New York: W. W. Norton, 1977.

Sher, Richard B. *Church and University in the Scottish Enlightenment: The Moderate Literati of Edinburgh*. Princeton: Princeton University Press, 1985.

————. "Professors of Virtue: The Social History of the Edinburgh Moral Philosophy Chair in the Eighteenth Century." In *Studies in the Philosophy of the Scottish Enlightenment,* edited by M. A. Stewart, 87–126. Oxford: Oxford University Press, 1990.

Simons, R. B. "T. R. Malthus on British Society." *Journal of the History of Ideas* 16 (1955): 60–75.

Sinclair, Sir John. *The Statistical Account of Scotland, 1791–1799*. Edited by Donald J. Withrington and Ian R. Grant. 21 vols. Edinburgh: EP Publishing, 1983.

Siskin, Clifford Haynes. *The Work of Writing: Literature and Social Change in Britain, 1700–1830*. Baltimore: Johns Hopkins University Press, 1997.

Skinner, Andrew. "Adam Smith: Philosophy and Science." *Scottish Journal of Political Economy* 19 (1972): 307–19.

————. "Economics and History—the Scottish Enlightenment." *Scottish Journal of Political Economy* 12 (February 1965): 1–22.

————. "Natural History in the Age of Adam Smith." In *Political Studies,* vol. 15, edited by Peter Campbell, 32–48. Oxford: Clarendon Press, 1967.

Skinner, Quentin. *Reason and Rhetoric in the Philosophy of Hobbes*. Cambridge: Cambridge University Press, 1996.

Smith, Adam. *The Glasgow Edition of the Works and Correspondence of Adam Smith*. Edited by D. D. Raphael and A. S. Skinner. Oxford: Clarendon Press, 1980.

————. *An Inquiry into the Nature and Causes of the Wealth of Nations*. 1776. Edited by Edwin Cannan. New York: Modern Library, 1937.

————. *Lectures on Rhetoric and Belles Lettres.* Edited by J. C. Bryce. Indianapolis: Liberty Fund, 1985.

————. *The Theory of Moral Sentiments.* 1759. Edited by D. D. Raphael and A. L. Macfie. Indianapolis: Liberty Classics, 1976.

Smith, Pamela H. *The Business of Alchemy: Science and Culture in the Holy Roman Empire.* Princeton: Princeton University Press, 1994.

Southey, Robert. "On the State of the Poor, the Principle of M. Malthus's Essay on Population, and the Manufacturing System, 1812." In *Essays, Moral and Political,* vol. 1. London: John Murray, 1832.

Spencer, Jane. *The Rise of the Woman Novelist: From Aphra Behn to Jane Austen.* Oxford: Basil Blackwell, 1986.

Sprat, Thomas. *The History of the Royal-Society of London, for the Improving of Natural Knowledge.* London: Printed by T. R., 1667.

Stafford, Barbara Maria. *Artful Science: Enlightenment Entertainment and the Eclipse of Visual Education.* Cambridge: MIT Press, 1994.

[Steele, Sir Richard]. *The Spectator.* Edited by Gregory Smith. New York: Dutton, 1970.

————. *The Tatler.* Edited by Donald F. Bond. 3 vols. Oxford: Clarendon Press, 1987.

Sterne, Laurence. *A Sentimental Journey into France and Italy.* 1768.

————. *Tristram Shandy.* 1759–67.

Stewart, Alan. "The Early Modern Closet Discovered." *Representations* 50 (spring 1995): 76–100.

Stewart, Dugald. *The Collected Works of Dugald Stewart.* Edited by Sir William Hamilton. 11 vols. Edinburgh: Thomas Constable, 1854–60.

————. *Elements of the Philosophy of the Human Mind.* 1792. Boston: James Munroe, 1847.

Stewart, John B. *The Moral and Political Philosophy of David Hume.* New York: Columbia University Press, 1963.

Stolnitz, Jerome. "On the Origins of 'Aesthetic Disinterestedness.'" *Journal of Aesthetics and Art Criticism* 20, no. 2 (1961): 131–43.

Stow, John. *Survey of London.* 1598.

Supple, B. E. *Commercial Crisis and Change in England, 1600–1642: A Study in the Instability of a Mercantile Economy.* Cambridge: Cambridge University Press, 1959.

————. "Currency and Commerce in the Early Seventeenth Century." *Economic History Review,* 2d ser., 10 (1957): 239–55.

Suviranta, Br. *The Theory of the Balance of Trade in England: A Study in Mercantilism.* Helsinki: Suomal, Kirjall, Seuran Kirjap, 1923.

Swetz, Frank J. *Capitalism and Arithmetic: The New Math of the Fifteenth Century.* La Salle, Ill.: Open Court, 1987.

Taylor, E. G. R. *The Mathematical Practitioners of Tudor and Stuart England.* Cambridge: Cambridge University Press, 1954.

Taylor, R. Emmett. *No Royal Road: Luca Pacioli and His Times.* Chapel Hill: University of North Carolina Press, 1942.

Taylor, William Cook. "Objects and Advantages of Statistical Science." *Foreign Quarterly Review* 16 (October 1835): 103–16.

Thirsk, Joan. *Economic Policy and Projects: The Development of a Consumer Society in Early Modern Europe.* Oxford: Clarendon Press, 1978.

Thomas, Keith. "Numeracy in Early Modern England." *Transactions of the Royal Historical Society,* 5th ser., 37 (1977): 103–32.

Thompson, Grahame. "Early Double-Entry Bookkeeping and the Rhetoric of Accounting Calculation." In *Accounting as a Social and Institutional Practice,* edited by Anthony G. Hopwood and Peter Miller, 40–66. Cambridge: Cambridge University Press, 1994.

————. "Is Accounting Rhetorical? Methodology, Luca Pacioli and Printing." *Accounting, Organizations, and Society* 16, nos. 5–6 (1991): 573–80.

Thompson, James. *Models of Value: Eighteenth-Century Political Economy and the Novel.* Durham: Duke University Press, 1996.

Todd, Janet. *The Sign of Angellica: Women, Writing and Fiction, 1660–1800.* London: Virago, 1989.

Tuck, Richard. *Philosophy and Government, 1572–1651.* Cambridge: Cambridge University Press, 1993.

Turnbull, George. *The Principles of Moral Philosophy: An Enquiry into the Wise and Good Government of the Moral World.* London: A. Millar, 1740.

Turner, A. J. "Mathematical Instruments and the Education of Gentlemen." *Annals of Science* 30 (1973): 51–88.

Turner, Cheryl. *Living by the Pen: Women Writers in the Eighteenth Century.* New York: Routledge, 1992.

Turner, Gerard L'E. "The Cabinet of Experimental Philosophy." In *The Origins of Museums: The Cabinet of Curiosities in Sixteenth- and Seventeenth-Century Europe,* edited by Oliver Impey and Arthur MacGregor, 214–22. Oxford: Clarendon Press, 1985.

Urbach, Peter. *Francis Bacon's Philosophy of Science: An Account and a Reappraisal.* La Salle, Ill.: Open Court, 1987.

van den Daele, Wolfgang. "The Social Construction of Science: Institutionalisation and the Definition of Positive Science in the Latter Half of the Seventeenth Century." In *The Social Production of Scientific Knowledge,* edited by Everett Mendelsohn, Peter Weingart, and Richard Whitley, 27–54. Boston: Reidel, 1977.

Vaughan, Rice. *A Discourse of Coin and Coinage.* London: Th. Dawks, 1675. Reprinted in J. R. McCulloch, *Old and Scarce Tracts on Money,* 1–120. 1856. Reprint, London: P. S. King, 1933.

Vickers, Brian. *In Defense of Rhetoric.* Oxford: Oxford University Press, 1988.

Viroli, Maurizio. *From Politics to Reason of State: The Acquisition and Transformation of the Language of Politics, 1250–1600.* Cambridge: Cambridge University Press, 1992.

Webster, Charles. *The Great Instauration: Science, Medicine, and Reform, 1626–1660.* New York: Holmes and Meier, 1976.

[Weyland, John]. *The Principles of Population and Production as They Are Affected by the Progress of Society with a View to Moral and Political Consequences.* 1820.

————. *A Short Inquiry into the Policy, Humanity, and Past Effects of the Poor Laws . . . in Which Are Included a Few Considerations on the Questions of Political Oeconomy, Most Intimately Connected with the Subject; Particularly on the Supply of Food in England.* London: J. Hatchard, 1807.

Whately, Richard. *Elements of Rhetoric.* 1828.

————. *Introductory Lectures on Political Economy.* 1831.

Wheeler, John. *A Treatise of Commerce. Wherein are Shewed that Commodities Arising by a Well Ordered and ruled Trade, such as that of the Societies of Merchants Aduenturers is proued to be.* London: John Harison, 1601. Facsimile ed. New York: Columbia University Press, 1931.

Whewell, William. *Selected Writings on the History of Science.* Chicago: University of Chicago Press, 1984.

Wigley, Mark. "Untitled: The Housing of Gender." In *Sexuality and Space,* edited by Beatriz Colomina, 327–89. New York: Princeton Architectural Press, 1992.

Williams, O. *Life and Letters of John Rickman.* 1911.

Wimsatt, W. K., Jr. *The Verbal Icon: Studies in the Meaning of Poetry.* Lexington: University Press of Kentucky, 1954.

Winch, Donald. *Riches and Poverty: An Intellectual History of Political Economy in Britain, 1750–1834.* Cambridge: Cambridge University Press, 1996.

Withrington, Donald J. "What Was Distinctive about the Scottish Enlightenment?" In *Aberdeen and the Enlightenment,* edited by Jennifer J. Carter and Joan H. Pittock, 9–19. Aberdeen: Aberdeen University Press, 1987.

Wollaston, William. *The Religion of Nature Delineated.* 1726.

Wood, P. B. "Science and the Pursuit of Virtue in the Aberdeen Enlightenment." In *Studies in the Philosophy of the Scottish Enlightenment,* edited by M. A. Stewart, 127–50. Oxford: Oxford University Press, 1990.

Woodmansee, Martha. "The Interests in Disinterestedness: Karl Philipp Moritz and the Emergence of the Theory of Aesthetic Autonomy in Eighteenth-Century Germany." *Modern Language Quarterly* 45, no. 1 (1984): 22–47.

Wordsworth, William. *The Prose Works of William Wordsworth.* Edited by W. J. B. Owen and Jane Worthington Smyser. 4 vols. Oxford: Clarendon Press, 1974.

Xenophon. *Oeconomicus.* Translated by H. G. Dakyns as "The Economist." In *The Works of Xenophon,* vol. 3. London: Macmillan, 1897.

Yamey, B. S. "Accounting and the Rise of Capitalism: Further Notes on a Theme by Sombert." *Journal of Accounting Research* 2 (1964): 117–36.

————. "The Functional Development of Double-Entry Bookkeeping." *Accountant,* November 1940, 333–42.

Yeo, Eileen Janes. *The Contest for Social Science: Relations and Representations of Gender and Class.* London: Rivers Oram Press, 1996.

Yeo, Richard. *Defining Science: William Whewell, Natural Knowledge, and Public Debate in Early Victorian Britain.* Cambridge: Cambridge University Press, 1993.

————. "Science and Authority in Mid-Nineteenth Century Britain: Robert Chambers and *Vestiges of the Natural History of Creation.*" *Victorian Studies* 28 (1984): 5–31.

Abrams, Philip, 6
abstraction, xvi, 15, 25, 28, 123, 130,
    137–38, 149, 215, 217, 223–25, 227,
    238–39, 243, 247–48, 250, 256, 270–71,
    273–74, 276, 286, 297, 300, 312, 317,
    333n. 22; Platonic, 153, 179, 195–96,
    202–3
accounting, xvi, 29–36, 281; household ori-
    gins of, 33–36. *See also* double-entry
    bookkeeping; study
Achenwell, Gottfried, 308
Act of Settlement, 121
Addison, Joseph, 152, 170, 198, 203; *Tatler,*
    152, 203
Agnew, Jean-Christophe, 341n. 54, 343n. 80
aesthetics, 152–53, 169, 206. *See also* Shaftes-
    bury, Earl
Aho, James, 37–38
Alberti, Leon Batista, 34–36, 61, 338n. 19,
    339n. 27
Amussen, Susan Dwyer, 24
Appleby, Joyce Oldham, 342n. 62
Arbuthnot, John, 142–43
Aristotle, 8, 14, 38, 40, 98–99, 106–7, 110,
    112; and analytic, 224; *Art of Rhetoric,*
    106; and commonplaces, 8–10, 98, 106;
    Metaphysics, 8; *Posterior Analytics,* 8
arithmetic, 55–56, 90, 104, 107–10, 130;
    moral, 181
Aubrey, John, 141
authorial intention, 24, 268–69

Babbage, Charles, 309–12
Bacon, Francis, xvii–xviii, xxi, 8–11, 15, 17,
    19–20, 22, 65, 68, 81, 85, 94, 96–103,

110, 112, 132, 146, 149, 150, 152, 171,
    185, 188–90, 202, 245, 247, 250, 256,
    267, 273, 276–77, 285, 299, 304, 310,
    317, 328, 341n. 51, 346nn. 14, 15; and
    experience, 97–101, 346n. 17; and exper-
    iment, 97–98, 100–102, 110; and induc-
    tion, xvii, 15, 17, 98–99, 188–89; and
    method, 10–11, 97–101, 132, 150, 272,
    277, 346n. 18; *Novum Organum,* 97, 100;
    and plain style, 97, 101; and singular or
    deviating instances, 94, 98–99, 145–46,
    149, 185
balance, xvii, 13, 38, 43, 54–55, 58, 68–69,
    76, 78, 167. *See also* double-entry book-
    keeping
balance of trade, 68–69, 73, 77–79, 83–84,
    90, 93, 210
Bank of England, 150–51, 158, 295
Barbon, Nicholas, 83–84
Barrell, John, 24
Barrow, John, 188
Battle of Culloden, xxii, 250
Baudrillard, Jean, 327
Bedwell, Thomas, 139, 353n. 82
Beier, A. L., 69
belief, xx–xxi, 27, 174, 183, 193, 199–200,
    203, 229, 287–88; a priori, 247–48
Bender, John, 12, 38, 65, 341n. 51
Bentham, Jeremy, 279, 375n. 18, 377n. 26.
    See also theological utilitarianism
Bentley, Richard, 155
bills of mortality, 280–81, 352n. 72, 376n. 19
Board of Trade, 69, 157
Boswell, James, 249, 251, 256, 260, 263,
    373n. 55, 374n. 56; *Journals,* 263

Botero, Giovanni, 85, 88, 93
Boyle, Robert, xvii–xviii, 11–12, 17, 19–20,
    22–23, 93–94, 101, 111, 115–16, 121,
    125, 132, 133, 188–90, 194, 202, 348n.
    40, 42
Boyle Lectures, 155, 183–84, 187, 194
Bristed, John, 291
British Association for the Advancement of
    Science (BAAS), 309, 311–13, 315, 322,
    382n. 5, 383n. 20. See also statistical
    movement
Brougham, Henry, 269
Buchanan, David, 380n. 58
Buck, Peter, 353n. 80, 376n. 19
Buickerood, James G., 187–88, 361n. 48
Burrow, John, 6, 266

Cannon, Susan, 309
Carlyle, Thomas, xxiv, 279, 302, 315
Cavendish, Thomas, 140, 339n. 36
certainty, 107, 109, 131, 133, 137, 181, 262,
    320; and knowledge, 104, 110, 133
Chadwick, Edwin, 281, 383n. 21
Chambers, Robert, 315
Charles I, 103
Charles II (Charles Stuart), 108, 124–25,
    132, 138
Child, Sir Josiah, 84, 298, 344n. 89
Christensen, Jerome, 209, 229–33, 364n. 69,
    370n. 31
Christian Platonism, 265, 283–84, 287, 313
Church of England, 155
Cicero, 14, 39–41, 79, 84, 86, 105. See also
    rhetoric
civility, 84, 89, 111, 258; mercantile, 87,
    89–90
Clanchy, Michael, 4
Clarendon, Earl of (Edward Hyde), 10,
    347n. 233
Clarke, Samuel, 155
Cocker, Edward, 313
Cohen, Patricia Cline, 157
Coleman, D. C., 67
Coleridge, Samuel Taylor, xxiii, 293–94,
    296, 326
Collini, Stefan, 6, 266
commerce, xvii, 11, 27, 39, 66–67, 71, 75,
    86–90, 127, 210, 212, 222, 224, 235, 239,

257–58, 275, 290, 297–98, 302, 379n. 52;
    domain of, xvii, 66, 82, 87; laws of,
    87–88, 302; morality of, 359n. 33; system
    of, 67, 78–79, 90, 92
commercial system, 66, 92, 229, 259
common sense, 106, 200
Company of Merchant Adventurers, 87–88
computation, 129, 131, 135, 141–42
Condorcet, Jean-Antoine-Nicolas Caritat,
    276, 285
conjectural history, xix, xxi–xxii, 1, 15, 24,
    219–32 passim, 239–40, 245, 248–51,
    256–57, 273, 300, 369n. 21; and cultural
    relativism, 260, 261; universals, 261–62
conjecture, xx–xxi, 76–77, 215, 218–22,
    228–29, 233, 236, 245, 250, 270, 276,
    285, 322
conversation, 150, 198, 203, 209–10, 262
Cotton, Sir Robert, 73, 343n. 78, 83
counting, 54, 172, 214, 261–62, 279–81,
    283–87, 292, 295, 313
Cox, Sir Richard, 135
credentialing, 265, 305
credit, xvi, 27, 151, 167, 359n. 32; personal,
    41
creditworthiness, xvi–xvii, 27, 59, 64, 88,
    166, 168
Crosby, Alfred, 332n. 16
Cromwell, Oliver, 120, 121, 125, 257
Cromwell, Richard, 122
cultural relativism, 218, 225, 249, 261,
    263–64, 266, 269, 285
Cumberland, Richard, 282
currency, 73–74, 298
custom, 146, 162–66 passim, 170, 199–200,
    230–31, 251–52

Dafforne, Richard, 339n. 38
Darwin, Charles, 315
Daston, Lorraine, 7, 8, 10, 14, 19, 94–95,
    319, 329n. 1, 336n. 4
Davenant, Charles, 142
Dean, Mitchell, 7, 17
Dear, Peter, 8–10, 14, 19, 70–71, 95–96, 116,
    188–89, 333n. 20, 346n. 15, 347n. 32, 33,
    362n. 55
deduction, xviii, 13, 94, 123, 132–33, 149,
    187, 223, 232, 251, 271, 283, 285, 290,

296, 307, 311–14, 317–18, 322; Hobbesian, xviii, 123, 133; logical, 361n. 52; mathematical, 309–10

Dee, John, 139, 340n. 41

Defoe, Daniel, xix, 13, 146, 150–53, 158–70, 173, 179, 340n. 49; *Complete English Tradesman,* 166, 168, 359n. 31; *Essays upon Several Projects,* xix, 146, 158, 162, 165, 358nn. 25, 27; and national pension office, 159–60; and polite learning, 159, 162, 163, 165

demonstration, mathematical, 108, 189–91, 293, 345n. 1, 361n. 52; rhetorical, 110, 115; Scholastic, 110–11

De Quincey, Thomas, 293

Derham, William, 155

Descartes, René, 94, 121, 187, 219, 220, 246, 270

description, xii, 219, 226, 237, 243–45, 248, 314, 382n. 4; and interpretation, xii, xxv; qualitative, 122

Dickens, Charles, xxiv, 279, 303, 315; *Hard Times,* 303

disciplinary division, 18–20, 25–26, 266–68, 330n. 6, 385n. 28

discourse, 18–19

discrimination, 170, 183–84, 191, 206

disinterestedness, xvi, xviii–xx, 27, 70–71, 75, 86, 94, 116, 119, 151–54 passim, 179–80, 182, 271, 280, 342n. 72, 359n. 39. *See also* interestedness

domains, 17, 330n. 6; of commerce, xvii, 66, 82, 87; differentiation of, 5, 18–20, 25, 91; economic, 75, 79; of philosophy, 93; of politics, 87–93

double-entry bookkeeping, xvi–xviii, 3, 11–12, 17, 24, 29–33, 35–66, 68, 76–78, 82, 86, 89–91, 92, 98, 107, 109, 111, 116, 119, 133, 138, 140, 167, 168; and balance, xvii, 38, 43, 54–55, 58; challenge to status hierarchy, 38, 42, 63–65, 90; codification of, 33–34, 36, 38; creation of writing positions, 42, 59–61, 65; display of mercantile virtue, xvi–xvii, 11, 30, 35, 37, 58, 64, 69, 86; effect of accuracy, 30, 42, 56, 63–64, 90, 108, 133; and excess, 61–62; exclusion of risk, 62–63, 334n. 40, 338n. 23; fiction and personification in, 11–12,

54, 57–58; formal precision of, 30, 33, 55 56, 64, 77–78, 90, 133, 145; inventory in, 42, 44–45 (fig. 1), 54, 57; journal, 43, 46–47 (fig. 2), 57–58, 61; ledger, 35, 38, 41, 43, 48–53 (figs. 3, 4, 5), 54–58; memorial, 42–43, 59; narrative, 43–43, 61–63; and numbers, 43, 54–55, 91, 138; as prototype of modern tact, 3, 10–11, 29; and rhetoric, xvi, 11–12, 30–33, 37–42, 63–65, 68, 89; as rule-governed system, xxi, 11, 30, 37, 54, 63, 66; system, xvi–xvii, 11, 42–3, 54–59, 77–78, 108. *See also* mercantile accommodation

Down Survey, 121, 123, 134, 138. *See also* Petty, William

economics, 4, 6, 123, 278

economy, 66, 224, 338n. 15, 380n. 54; national, 65, 126 27

Eden, Sir Morton Frederick, 291, 378n. 39

Edinburgh Review, 269–70, 297, 314

education, 204, 277–78, 301–2, 308, 354n. 98; schemes for, xxiii, 265

Emerson, William, 187

emigration, 259–60

empiricism, 101, 276, 285, 288–89, 310

emulation, 146–47, 150, 162–68 passim, 170

Encyclopaedia Britannica, 276

English Civil War, xviii, 102–3, 105, 111, 114, 121, 126

epistemology, xv, 17, 19, 153, 266, 270, 327–28. *See also* historical epistemology

Erasmus, 39–40, 60, 79, 84–86, 339n. 36. *See also* rhetoric

essay, xx, 198, 203–4, 206, 209–10, 212–13, 363n. 63

evidence, xxiv, 2, 95–96, 239–40, 248, 251, 255, 292, 316

exchange, 90, 341n. 54; bills of, xvi, 28, 88–89; 69–70; committees on, 69–70; rate of, 73–74, 81–82, 84

experience, 8–9, 67, 70–71, 95–96, 97–101, 105–6, 128, 130, 153, 171–73, 184, 197–205, 208–9, 212, 221, 234–35, 239, 246, 252, 256, 261–62, 272–73, 275, 285, 322; common, 8–9, 32, 70–77, 99–100, 106–7, 109, 115; extensive, 272–73; mer-

experience (*continued*)
    cantile, 32, 71, 80, 82; personal, 83,
    123–24, 215
experiment, 8, 97–98, 100–102, 104, 111,
    117, 123, 132–33, 148, 155, 174,
    186–200 passim, 208, 223, 273, 285; and
    knowledge, 97–98, 102–3, 150
experimentalism, 22, 93, 111–12, 116,
    132–33, 188–90, 193, 198–99, 215
experimental philosophy, xix–xx, xxii, 123,
    228–29, 233, 250. *See also* moral philoso-
    phy
expert, xv, xxiv, 15–16, 71, 123, 131; mer-
    chant, 78. *See also* professional
expertise, 16, 84, 91, 123, 137, 341n. 54; eco-
    nomic, 66, 94, 128; mercantile, xvii, 32,
    78, 81, 83, 89, 167; professional, 27, 268

fact: ancient, 29, 101, 336n. 4; Aristotelian,
    xvi, 2, 8, 98; and conjecture, 77, 123, 229,
    256, 270; economic matter of, xviii, 68,
    116, 128, 137, 180, 341n. 52; as evidence,
    xxiv, 2, 96; modern, xi–xxv, 1–16, 18–20,
    26, 28, 30, 92–98, 101, 105, 111, 123–24,
    134, 138, 145–46, 153, 198, 215–16, 219,
    229, 249, 270, 273, 280–81, 299–300,
    306, 307, 315, 317–18, 327–28, 329n. 1,
    336n. 4; naming of modern, xii–xiii; nat-
    ural matter of, 116, 199, 341n. 52; natural
    philosophical matter of, 111, 115; nu-
    merical, 143, 144, 156; peculiarity of
    modern, xii, xv, 1–2, 4–8, 10, 19, 20, 299;
    and political economy, 217, 237, 247–48,
    290, 299; postmodern, 3–4, 327–28; and
    systematic knowledge, xxi, xxv, 1–4,
    8–16, 31, 217, 272, 277–78; as theory-
    and value-free, xviii–xix, xxiv–xxv,
    94–97, 99, 111, 116, 128, 141, 192, 307;
    theory dependent, 130, 134, 310
Ferguson, Adam, xxi, 176, 222, 225–26, 254,
    269; *Essay on the History of Civil Society,*
    222, 368n. 17
fiction, xxi, 230, 232–33, 236, 245, 278, 313,
    374n. 58; and numbers, 54, 57–59
Fielding, Henry, 291
Fontana, Biancamaria, 374n. 4
Forbes, Duncan, 7
Foreign Quarterly Review, 313

Foucault, Michel, 6–7, 17–18, 31. *See also*
    governmentality
free will, 175, 195, 308, 314, 324
French Revolution, 266, 276, 284
Freud, Sigmund, 148
Friday, Joe, 1

Galileo, 104
Galison, Peter, 319
Gassendi, Pierre, 121
Gay, John, 282
generalization, xvi, 28, 93, 130, 170, 173,
    202, 208, 215, 229–30, 253, 260–61,
    267–68, 286, 300, 310, 318
Gentleman's Magazine, 257
Geological Society of London, 305, 381n. 61
Gerzina, Gretchen, 24
gestural mathematics, 172–74, 185–86,
    189–91, 199–200, 214, 248, 280,
    282–83, 293; definition of, 172
Glanvill, Joseph, 141, 348n. 42
Glorious Revolution, xix, 146, 157
Godwin, William, 285
Goldman, Lawrence, 309–10
government, xv–xxii, 86–91, 93–94, 104,
    125, 197, 304; and knowledge produc-
    tion, xv, xviii–xx, 2–4, 93, 102–3, 124,
    138; liberal, xvi, xix–xxi, 2, 162; projects
    of, 93–94, 102, 146–47, 155, 158, 301,
    305–6, 357n. 21; self-, xix–xx, 89,
    147–48, 152, 156, 168, 177–78, 180, 182,
    190–91, 196, 198, 207, 214, 265, 305;
    sovereign, xx, 27, 114, 162; by taste,
    157–58. *See also* reason of state
governmentality, xvii, 6, 31, 158, 177, 203,
    234, 236–37, 239; Foucault's theory of,
    6–7, 17, 31, 355n. 6; liberal, xxi, 27, 147,
    148, 150, 158, 164, 166, 170, 175, 179,
    193, 196–98, 213–14, 217, 227, 301–2,
    305
Grotius, Hugh, 282

Haakonssen, Knud, 7, 375n. 9
Halley, Edmond, 280
Halpern, Richard, 69, 84
Hanway, Jonas, 291
Hariot, Thomas, 140
Harris, John, 155

Hartlib, Samuel, 121
Hawkins, Bisset, 308–9
Hazlitt, William, 293
Henry VIII, 73, 353n. 82
Herschel, John, xxv, 3, 307, 317–22, 324,
  384n. 23; *Preliminary Discourse on the
  Study of Natural Philosophy,* xxv, 317–21
Hirschman, Albert O., 27, 348n. 35
historical epistemology, 7, 22
A History of the Modern Fact, xii–xxv passim,
  4–7, 16–28; compared with *Making a So-
  cial Body,* 25–26, 330n. 6, 333n. 20;
  methodology of, 16–26
Hobbes, Thomas, xviii, 13, 19–20, 29,
  93–94, 104–10, 114, 121, 131–33, 146,
  187, 225, 347n. 27, 29, 30, 348n. 34,
  352n. 75; and deduction, xviii, 93;
  Leviathan, 29, 104–9 passim, 114, 131,
  347n. 28; and reason of state, 93, 109; and
  reckoning, 104–5, 107, 132, 335n. 1
Hogarth, William, 180, 357n. 18
Home, Henry. *See* Kames, Lord
Homer, 252
Hont, Istvan, 7, 371n. 43
Hood, Thomas, 140
Hornor, Francis, 269
Hoskin, Keith, 337n. 14
Hues, Robert, 140
human mind, 28, 215, 223–24, 227–28, 238,
  250, 271, 273, 285, 300
human nature, xx–xxi, 16, 28, 106, 148, 156,
  172, 174–75, 205, 207–8, 215, 217,
  223–25, 229–32, 243–44, 247, 272,
  285–86, 324
Humboldt, Alexander von, 290–91, 310
Hume, David, xx–xxi, xxii, 7, 13–5, 22, 83,
  150, 170–74, 183, 191, 197–213, 218–19,
  221–22, 227–39, 245, 254, 264, 266–67,
  270, 272, 276, 283, 290, 325, 348n. 35;
  and the essay, xx, 150, 198, 203–4, 206,
  209–10, 212–13, 218; *Essays, Moral and
  Political,* 207–9; and fiction, xxi, 13–14,
  230, 232–33, 236, 245; "Of Essay-
  Writing," 204, 209–12; "Of the Delicacy
  of Taste and Passion," 204–6, 209; "Of
  the Rise and Progress of the Arts and Sciences," 209, 211–12; "Of the Standard of
  Taste," 170, 174; "On Civil Liberty," 204,

208–9; and problem of induction, xx,
  14–15, 173–74, 183, 198, 201–3, 207,
  219, 222, 229, 264; and skepticism, 174,
  197–99, 201, 222, 233, 264, 267, 276;
  "That Politics May Be Reduced to a Science," 204, 207, 209; *Treatise of Human
  Nature,* xxi, 197, 199–204, 207–8, 222,
  229–31, 236, 363n. 65, 67
Hunter, Ian, 17
Hutcheson, Francis, xx–xxii, 29, 109,
  148–49, 155, 176, 182–85, 188–92, 194,
  197, 202–3, 221, 246, 264–65, 273,
  282–83, 293, 362n. 56; *An Essay on the
  Nature and Conduct of the Passions,* 183,
  191; *An Inquiry into the Originals of Our
  Ideas of Beauty and Virtue,* 185, 191, 360n.
  45, 363n. 58
hypothesis, 1, 187–88, 194, 198, 219, 236,
  312, 314, 322–24, 361n. 50, 366n. 8

imagination, 246, 253, 271
impartiality, 123–24, 128, 131–32, 135,
  137–38, 152, 239, 271, 280, 295
induction, xvii–xviii, 13, 94, 98–99, 155,
  188–89, 222–23, 230, 232, 251, 267, 273,
  277–78, 283, 296, 307, 310, 312–13,
  317–28 passim; Baconian, xvii–xviii, 98,
  133, 150, 170–71, 189, 221–22, 228, 322;
  extensive, 372–73, 304; mathematical,
  310; problem of, xx–xxv passim, 15–16,
  30–31, 70, 149, 176–77, 198, 201–3, 207,
  215, 217, 219, 222, 229, 264, 266, 270,
  277–78, 286, 282, 300, 302, 304–8 passim, 303, 315, 317, 320–21, 325–28 passim. *See also* Hume, David
interest, xvi, xviii–xxi, 27, 70, 86, 88, 90,
  93, 95, 102–3, 107–10, 119–20, 123,
  134–36, 145, 151, 159–60, 209, 219;
  common, 86, 89, 103, 109, 342n. 72,
  347n. 24, 348n. 35; economic, 70, 120,
  151; national, 109, 136, 145, 151–52;
  party, xx, 145, 311; political, 132, 135; of
  ruler, 86, 88, 134, 138; self-, xx, xxiv, 71,
  86, 88–89, 103, 107, 109, 120, 124–25,
  130–38 passim, 151, 160, 180, 209, 220,
  227, 235–36, 301; state, 103
interestedness, 70, 154. *See also* disinterestedness

interpretation, xii, 126, 129–30, 155, 237, 240; and calculation, 129; and description, xii
Interregnum, 142
introspection, 156, 181–82, 199, 224, 229, 238–39, 314

Jackson, John, 142
James I, 69–70
James II, 125, 138, 147, 150, 157, 162
James, Richard, 309–11
Jed, Stephanie H., 17, 30, 336n. 6
Jeffrey, Francis, 269
Jevons, W. Stanley, 335n. 3
Johnson, Samuel, xxii, 26, 198, 218, 249–64, 266, 310, 373n. 50, 53–55, 374n. 56; and cultural relativism, 218, 249, 256, 260–64, 266; *Idler,* 257; *Journey to the Western Islands of Scotland,* xxii, 26, 218, 249–64; *Rambler,* 257; and systematic knowledge, 249–50; and universalist assumptions, xxii, 261
Johnson, Stephen, 140
Jones, Richard, 382n. 7, 8
judgment, 170, 172–73, 182, 205–6, 208, 234

Kahn, Victoria, 347n. 28
Kames, Lord (Henry Home), xxi, 203, 219, 221–28 passim, 233, 246, 261, 274, 369n. 22, 374n. 56; *Sketches of the History of Man,* 222, 227
Keats, John, 307, 326
Kelly, J. Thomas, 69
Knapp, Steven, 327
knowledge, 144, 204, 208, 254, 267–68; abstract, 154, 212; Aristotelian, 8, 92, 98; crises in sixteenth- and seventeenth-century, 97, 104, 110; general, xviii, 97, 207–9, 213, 228–29, 233, 239, 249, 255, 266–67, 268, 292, 304–7, 310, 312, 325–27; and interest, 4, 11, 71, 93, 111; public and private, 34–37, 39; and women, 209–12, 227, 334n. 40, 338n. 23. *See also* systematic knowledge
Kratzer, Nicholas, 139
Kuhn, Thomas, 96–7

labor, 127–29, 351n. 68, 71; division of, 210, 238, 275, 283
Lacan, Jacques, 327
Latour, Bruno, 19–21, 23, 334n. 35
Laudan, L. L., 220, 366n. 8
laws of nature, 75–76, 107, 196–97, 207, 214, 277, 279, 289, 296, 299, 302, 317, 344n. 87, 362n. 58
Lazarsfeld, Paul F., 336n. 5, 7
Letwin, William, 6
Levine, Donald, 6
Licensing Act, 146, 152, 162
Lieberman, David, 377n. 76
Locke, John, 153, 171, 199, 243, 358n. 29; ethics, 153
London and Westminster Review, xxiv
London Philosophical Society, 121
Longinus, 313
Long Parliament, 122
Lyell, Charles, 381n. 61

Macaulay, Thomas, 270
McCulloch, J. R., xxiii–xxiv, 3, 16, 22, 234, 264–65, 267, 269, 278, 295–306, 307, 313–15, 320–22, 336n. 7, 374n. 4, 379n. 52, 380n. 53, 56–58, 381n. 60, 382n. 4, 383n. 21; *Discourse on the Rise, Progress, Peculiar Objects, and Importance of Political Economy,* 297–99, 301–5; and education, xxiii, 301–2; *Principles of Political Economy,* 298; and professionalism, xxiv, 3, 305–6; problem of induction, xxiv, 278, 302, 304–5; and statistics, xxiv, 305–6, 307
McKeon, Michael, 24
MacKenzie, Henry, 253
Mackintosh, James, 269
Macpherson, James, 256, 373n. 54
Macve, Richard, 337n. 14
Malthus, Thomas, xxiii–xxiv, 218, 234, 248, 264–65, 269, 278–96 passim, 300, 302–6, 309–10, 378n. 41, 381n. 60; *Essay on the Principle of Population,* xxiii, 265, 280, 283, 285–6, 288–92, 295–96; Malthus-Ricardo debate, 295–97, 379n. 48; and moral questions, xxiii, 290, 294; and numerical representation, 248, 279–80, 287, 289–95, 309, 378n. 42; and political economy, xxiii, 279, 290–91, 294, 310;

and providentialism, xxiii, 265, 283–84, 287–90, 294–95; theodicy, 288–90, 293, 378n. 33

Malynes, Gerald de, 66–9, 71–74, 76–82, 84, 90, 239; *Center of the Circle of Commerce,* 66, 71–72, 79; debate with Misselden, 68, 72, 74, 76–78, 84; and Ciceronian rhetoric, 68, 79–80, 84; *Lex Mercatoria,* 72; *Maintenance of Free Trade,* 72; support for sovereign power, 68, 72–73; *A Treatise of the Canker of England's Commonwealth,* 72

Malyneux, Emery, 141

Manchester Statistical Society, 311

Mandeville, Bernard, 152, 180, 182, 357n. 16, 360n. 41

manners, 152, 209, 222, 227, 257

manufacturing system, 294–96, 309

market, xx–xxi, 297, 302; economy, 147, 169, 197; society, 147, 158–59, 161, 165, 169, 193, 207, 214, 260; system, xx, 66–67, 75, 79, 116, 216–17, 224, 228–29, 238–39, 242, 247–48, 250, 271, 300, 369n. 21

Martin, Julian, 98

mathematical formulas, 308, 320

mathematical instruments, xix, 138, 141, 319–20

mathematical models, 319–20, 322

mathematical language, 156, 188, 282, 312–13, 325

mathematicals, xix, 138–39

mathematics, xviii–xix, xxv, 4–5, 104–5, 107–8, 123–24, 130–31, 133, 138–39, 141, 154, 157, 179, 181–82, 185–88, 191, 194, 198, 223, 269, 271, 283, 295–96, 311, 313, 354n. 98; as trope, 107, 131, 183, 312–13, 325

Meek, Ronald, 6

Mellis, John, 32, 41–59, 61, 167, 337n. 9, 340n. 43, 342n. 58; *A Briefe Instruction and Maner How to Keepe Bookes of Accompts,* 32, 41, 61

memory, 144–45, 255–56

mercantile accommodation, xvi, 68, 86–91; as model of good government, 88–89, 91

mercantile expertise, xvii, 32, 78, 81, 83, 89, 167

mercantile virtue, xvi–xvii, 11, 20, 35, 37–38, 64, 88, 91, 144

mercantile writing, xviii, 12, 17, 66, 166–69

mercantilists, 67, 83; influence on government, 67–68

merchant apologists, xvii, 15–16, 81, 87–88, 91, 167, 298

merchants, xvi–xviii, 11, 2, 35, 39, 71, 87–91, 115–20, 128; credibility of, xvii, 41, 59, 63–64, 89, 116, 159; creditworthiness, xvii, 59, 64, 68; as fact-gathering instruments, xviii, 93, 116; social status of, 11, 32, 39, 42, 89–91, 93–4

The Merchants Avizo, 59–61

Mercenne Circle, 94, 121, 345n. 4

Merivale, Herman, 314

method, 309, 317, 320, 322–23; scientific, 307, 311, 317, 320. *See also* Bacon, Francis, and method

Mill, James, 297

Mill, John Stuart, xxv, 3, 137, 292, 307, 317, 322–25; ethology, 323–24, 384n. 26, 27; *Logic,* xxv, 317, 323–25

Millar, John, 222–23, 254; *Observations concerning the Distinction of Ranks in Society,* 223, 367n. 13, 368n. 20

Miller, Peter N., 6, 344n. 94, 355n. 3

Milne, Joshua, 291

Milton, John R., 344n. 87

Misselden, Edward, xvii, 16, 66–72, 74–82, 90, 93–94, 126, 126, 239, 343n. 83; debate with Malynes, 68, 72, 74, 76–78, 84; and Ciceronian rhetoric, 68, 79–80, 84; *Circle of Commerce,* 66, 71–72, 79; *Free Trade,* 72

Modeleski, Tania, 334n. 40

Monboddo, Lord, 225, 261, 374n. 56

moral philosophy, xiv–xv, xxi–xxii, 16, 263, 265, 278, 281, 300, 308, 310; experimental, 15, 19, 28, 133, 146, 148, 154, 174–77, 184, 191–94, 198, 202–6, 221, 223, 225, 248, 251, 256–57, 261–62, 269, 300, 302; Scottish, xix–xx

Morrell, Jack, 309, 311

motivation, 147–48, 156, 163–64, 168, 174–75, 180, 186, 194, 206

Mun, Thomas, xvii, 13, 16, 32–33, 66–72, 79–84, 87–88, 90–91, 92–94, 126, 128,

Mun, Thomas (*continued*)
239, 341n. 57, 343n. 86; *England's Treasure by Forraign Trade,* 66, 72, 80–81; and plain style, 68, 79–82

Napoleonic Wars, 295
natural history, 99, 100, 225, 234
natural philosophy, xiv, xviii, 4, 8–9, 11–13, 15–16, 19–20, 31, 65, 94–95, 97, 101–3, 111–12, 114–17, 119, 125, 128, 141, 175, 183, 193–94, 198, 256, 269, 288, 300, 307, 317, 322; and merchants, 12
natural science, 310
necromancy, 239
Newton, Sir Isaac, 150, 154, 177, 186–94 passim, 197, 200, 219, 220, 246, 267, 317, 366n. 8; *History of Astronomy,* 246; and method, 155, 174, 183, 187–89, 361n. 48, 362n. 55; *Opticks,* 187–88; *Principia,* 187, 219, 361n. 50
Nicholson, Colin, 356n. 12
nominalism, 270–71
North, Dudley, 298
Norton, David Fate, 197
numbers xi–xv, 4–6, 13, 16, 18, 29, 43, 54–55, 90–91, 108, 124, 128–30, 131–35, 138, 144–45, 157, 172, 214, 240–43, 246–47, 249, 262, 265, 278, 290, 292–93, 304, 309, 313–14, 317, 320, 340n. 40, 41, 382n. 4; and analysis, xii–xv; and facts, xii, 4–5, 29, 214, 216; law of large, 314, 325; as transparent, 5, 13, 80, 119, 239; as value- and theory-free, xxii–xxiii, xxv, 4–5, 6, 13, 128, 131, 135, 239, 283, 287, 292–94, 306
numerical information, 5, 214–17, 242–43, 279–83, 289–91, 306, 308–9, 318–21, 375n. 18, 376n. 19; collection of, xix, 4, 137, 142–43, 158, 242, 283, 289–91, 305, 308–9, 316–17, 321, 365n. 2, 378n. 39
numerical representation, xi, xiii, xix, xxii, xxv, 13, 26, 29, 32, 77, 84, 89, 119, 123–24, 130, 138, 141–43, 213, 239, 248, 265, 279–80, 284, 287–95, 304, 308, 312–13; and figurative language, 6, 26. *See also* gestural mathematics
numerical tables, xi, 241–42, 280, 283, 290, 308

observation, 106, 155–56, 170, 171, 176, 178, 181, 184–85, 187–90, 192, 194–95, 197, 202, 210, 227, 229, 234, 241, 247, 249, 256, 262, 271, 273, 285, 287, 289, 299, 310, 316, 318–19; personal, 82, 90, 92, 171, 173, 206
opinion, 172–73, 209, 219, 311, 315–16; public, 159–62, 301
Oughtred, William, 139

Pacioli, Luca, xvi, 24, 32–33, 37–39, 155, 337n. 9, 14, 339n. 27, 357n. 20; *De Computis et Scripturis,* 32, 37
Paley, William, 279, 282–83, 370n. 30; *Principles of Moral and Political Philosophy,* 282
particulars, xxii, xxvii–xviii, xxiv, 1, 91, 111, 202, 208, 234, 245, 270, 277, 286, 299–300, 326; Bacon and, xvii–xviii, 9–10, 15, 81, 97–100, 245; and conjecture, 222; deracinated, xvii, 9–10, 95, 130, 189, 272–73, 299, 328; discrete, xii, 14, 246, 268, 310; as evidence, 9–10, 92, 208, 277; observed, xv–xviii, xxi–xxii, 3–4, 9–10, 15, 41, 81, 92–93, 97, 99, 109, 149, 155–56, 194, 202, 214–16, 221–22, 225, 229, 238, 248–49, 261–62, 267, 272, 278, 285–86, 292, 300, 304, 307, 310, 318, 323–27; and systematic knowledge, xii, xv, xvii–xviii, 1–3, 9–10, 207
party, 144–45, 150, 207–9, 277, 355n. 3
Paulson, Ronald, 152–53, 359n. 39
Peace of Amiens, 290
Pell, John, 121
Petty, William, xviii–xix, xxi, xxiii, 13, 16, 93–94, 110, 120–38, 141–43, 145–65 passim, 172, 214, 239–40, 280, 298, 336n. 7, 350n. 59, 60, 351n. 68–71, 74–76, 357n. 19; and calculation, 129–31, 135, 142; *History of the Down Survey,* 122; interests in Ireland, xii–xix, 120, 122, 125, 134; and numerical representation, 123–24, 128–32, 134, 141–42, 164, 239, 241, 352n. 72; *Political Anatomy of Ireland,* 134–36; and political arithmetic, xviii, 13, 94, 134, 137, 142, 158, 160–61, 214, 355n. 7, 358n. 27; *Political Arithmetick,* 132–34; and taxes, 124–29, 160, 354n. 98; *A Treatise of Ireland,*

134–36; *A Treatise of Taxes and Contributions,* 124–25, 129, 132, 134
Phillipson, Nicholas, 7, 176, 203
physicotheology, 154–55
plain style, 68, 79–82, 97, 101, 167, 169
pneumatology, 191–92, 220
Pocock, J. G. A., 151, 338n. 23
poetry, 308, 313, 325–27
political arithmetic, xv, xix, xxii, 13, 19, 27, 31, 94, 134, 137, 142–43, 146–47, 157–61 passim, 165, 198, 261–62, 361, 336n. 7, 358n. 25
political correctness, 21, 268
political economy, xix, xxi–xiv, xiii, 15–6, 19, 27, 31, 146, 171, 192, 216–19, 229, 233, 236–40, 244, 247–48, 265–314 passim, 320–26, 336n. 7; as amoral, xxiii–xxiv, 278–79, 283
Political Economy Club, 6
political science, 267, 304
Polizeiwissenschaft (science of police), 147, 214, 355n. 7, 366n. 3
poor law, 143, 295, 306, 316–17, 365n. 2; New Poor Law, 25
Pope, Alexander, 212
population, 286–87, 290–92, 300, 322
Population Abstract, 291
Porter, Theodore, 308, 350n. 62
Poynter, John R., 306, 375n. 18, 376n. 20
Price, William, 280, 290
privacy and publicity, 34–35, 37, 58–59,
professionalism, xv, xxiv, 3, 16, 278, 305
proportion, 153, 155, 179–82, 185, 191–92, 195, 286
protopsychology, 170, 174, 216–17, 244
providential design, 182, 184, 190, 192, 197, 202, 223, 227–29, 233, 235–36, 246, 249, 275–78, 283–84, 286, 288, 290, 296, 303
providentialism, xxi–xxiii, 15, 149, 177, 197, 228, 264–66, 273, 278, 287–89, 294, 296, 302–3
psychology (and psychoanalysis), 28, 147–48, 213, 244–45
Pufendorf, Samuel von, 109, 282

quantification, xii, xxi, 108, 110, 122, 138
Quetelet, Lambert Adolphe, 308–9, 311–12, 314, 317

ragion di stato, 66, 68, 85. *See also* reason of state
Ralegh, Sir Walter, 140, 353n. 85
reason of state, xvii–xviii, xxi, 32, 66, 71, 84–89, 91, 93, 109–11, 125, 136, 180, 217, 226, 237. *See also* government
Reid, Thomas, 176, 219–20, 222, 269, 271, 366n. 8, 372n. 47
Renaudot, Theophraste, 95
Restoration, 110, 114, 123, 142
rhetoric, xiv, xvi, xviii, 4, 11–12, 26, 30–33, 37–42, 63–65, 67, 79–85, 88–89, 105–7, 109, 115, 118–19, 132, 165, 167, 174, 177, 188, 243–44, 248, 267, 308, 312–13; Ciceronian, 40, 68, 79, 85; and *copia,* 40, 85, 90, 339n. 36; cultural prestige of, 33, 39, 91; history of, 38–40; reproduction of status hierarchy, 11–12, 38. *See also* Cicero; numerical representation; plain style
Ricardo, David, 16, 295–97, 308–11, 313; Malthus-Ricardo debate, 295–97, 379n. 48; *Principles of Political Economy,* 295, 298
Ricardo Lectures, xxiii, 297, 304
Richardson, Samuel, 212
Rickman, John, 306, 378n. 39
Roberts, Lewis, 62
Robertson, G., xxiv–xxv, 96, 315–16
Robertson, William, xxi, 215, 219–26, 238, 271; *History of the Discovery and Settlement of America,* 219, 223–25, 367n. 13, 368n. 20; and human mind, 238; *View of the Progress of Society,* 226
Rogers, Pat, 249–50
Roncaglia, Alessandro, 351n. 71
Rotz, John, 139, 353n. 82
Royal Society, xviii, 84, 93–94, 97, 102, 104, 110–20, 123–24, 141, 152, 165, 189, 291, 345n. 4, 348n. 42, 353n. 80; *Philosophical Transactions,* 142, 291
Royal Society of Edinburgh, 275

Sanky, Sir Jerome, 122
Saussure, Ferdinand de, 327
Schaffer, Simon, 13, 19–20, 111, 341n. 52
Schumpeter, Joseph A., 6
science of wealth, 265–66, 269, 275, 279–80, 297, 300, 307; and society, xxii, 4, 20, 29, 65, 94, 215–17

Scholastic philosophy, 79, 82, 85, 90, 95–98,
    100–101, 104, 107, 110, 116, 177,
    188–89, 219, 233, 236
Scottish highlands, xxii, 250–62. *See also*
    Johnson, Samuel
Sedgwick, Adam, 311–12
Senior, Naussau, 309
sentiment, 243–44, 254. *See also* Smith,
    Adam
Seybert, Adam, 291
Shaftesbury, Earl (Anthony Ashley Cooper),
    152–54, 177–83, 185–87, 190–92, 195,
    202, 206, 243, 265, 282–83, 357n. 16;
    Characteristics, 177; "The Moralists," 178,
    360n. 40, 43
Shapin, Steven, 13, 19–20, 111, 341n. 52,
    348n. 40
Shelley, Percy, 307, 325–27; "Defense of Po-
    etry," 325–27
Sher, Richard B., 7, 176, 183, 359n. 36
Sinclair, John, 281, 291; *Statistical Account of
    Scotland,* 308
Siskin, Clifford Haynes, 24
skepticism, 15, 101, 105, 174, 181, 191,
    197–99, 201, 222, 233, 236, 264–65, 267,
    276–77, 297
Skinner, Andrew, 7
Skinner, Quentin, 6, 105, 347n. 24, 29
Smith, Adam, xix, xxi–xxiii, 7, 19, 67,
    83–84, 109, 215–24 passim, 236–48, 254,
    263–67, 270–71, 273–76, 283–84, 290,
    294, 298, 300, 302, 304–5, 310; and de-
    scription, 26, 243, 248, 273; *Lectures on
    Rhetoric and Belles Lettres,* 243–47, 371n.
    40; and numerical representation,
    xxi–xxii, 217, 239–40; and political
    economy, xix, xxi–xxii, 216–17, 237,
    240, 247–48, 310, 370n. 35, 374n. 58,
    385n. 9, 380n. 58, 381n. 60; and subjec-
    tivism, 274; *Theory of Moral Sentiments,*
    217, 238; *Wealth of Nations,* xxiii, 216–17,
    237–42, 246–48, 298, 302
sociality, 155, 204, 206, 227, 264
social science, xv, xxv, 4, 6, 307, 312, 317
society, 230–34, 236, 322
Southey, Robert, xxiii, 293–96
Spanish Armada, 139–40
Sprat, Thomas, 11–13, 93, 101, 111–20, 132,

146, 152, 349n. 48; *History of the Royal So-
    ciety,* 11, 101, 112, 116–17
Stafford, Barbara Maria, 145
standardization, 36, 169–70, 283, 319
statistical movement, 25, 309–16
Statistical Society of London, xxiv, 309–16,
    383n. 20
statistics, xii–xiv, xxiv–xxv, 13, 15, 19, 30,
    261, 266–67, 304–17, 321, 324–25,
    336n. 7, 382n. 4, 383n. 20, 384n. 27; and
    method, 309, 317, 320, 322–23; and
    probability, 314; and progress, 315
Steele, Richard, 144–45, 152, 157, 170, 203;
    Spectator, 144–45, 152, 203, 356n. 12,
    358n. 28
Sterne, Lawrence, 253
Stewart, Alan, 338n. 28
Stewart, Dugald, xxii–xxiii, 219–22, 228,
    245, 264–65, 269–78, 284–85, 288, 300,
    303–5, 318, 334n. 40, 335n. 3; *Account of
    the Life and Writings of Adam Smith,* 221,
    275–77; *Dissertation,* 276–77; *Elements of
    Philosophy of the Human Mind,* 270–75,
    277; and moral philosophy, xxii, 277; and
    objectivism, 274; and political economy,
    xxii, 269, 275–77, 374n. 4, 375n. 13, 15;
    providentialism of, xxii–xxiii, 273–75,
    278, 284, 288, 303
study, the, 34–36
style, 97, 101, 118, 132; problems of writing,
    168, 201, 204, 209, 212–13, 243–44
subject, economic, 136; universal, 16, 65,
    107, 133–35, 146, 150, 286
subjectivity, xix–xx, 16, 20, 27–28, 114, 120,
    146–48, 153, 155–56, 166, 168, 179,
    195–97, 199, 203–4, 207, 213–14, 217,
    223, 227, 238, 244, 254, 256, 299, 301, 318
Supple, B. S., 69, 342n. 62
Suviranta, Br., 343n. 81
swearing, 163–64
Swift, Jonathan, 150, 212
sympathy, 152, 238, 375n. 9
system, xvi, 111, 234–38, 242, 244, 247, 274
systematic knowledge, xii, xv, xviii–xx, xxv,
    1–4, 8–16, 27, 31, 77–78, 148, 170, 177,
    195, 203–4, 213, 216, 221, 233, 235, 237,
    244, 249–50, 262, 268, 271, 297,
    299–300, 326–27; ancient, 8

Tacitus, 68, 85, 204
taste, 150, 152, 156–57, 169–73, 191, 203–6, 210, 232. *See also* government
taxonomy, 175, 193, 256–57, 266–68, 304–5, 321
Taylor, William Cooke, 313–14, 383n. 15
Thackray, Arnold, 309, 311
theological utilitarianism, 279, 282–84, 377n. 26
theory, xx–xxi, 219, 227, 232–34, 241, 243, 247–49, 262, 273, 276–77, 305, 312, 315–16, 322
Thirty Years' War, 147
Thomas Aquinas, Saint, 75
Thompson, Grahame, 38
Tooke, William, 291
trade, xvii, 27–28, 66–67, 69–70, 72–85, 90, 116, 126–27, 143, 239, 276; committees on, 69–70; foreign, 81–82, 341n. 57
Tuck, Richard, 6, 67–68
Turnbull, George, xx–xxii, 148–49, 155, 176, 191–97, 202, 221, 228, 264–65, 273; *Principles of Moral Philosophy,* 192

Union of 1707, 257
universality, 155, 175, 257
universals, xvi, xxii, 8, 14, 28, 107, 113, 115, 146, 149, 156, 180, 214, 247, 250, 253, 256–57, 261–62

value, 2, 40, 72–74, 128–31, 134–35, 137, 343n. 78, 83, 351n. 68, 69, 352n. 75,

379n. 52; moral, 110, 279–80, 282, 297; national, 71–72
Viroli, Maurizio, 6, 39, 67–68
virtue, 139, 166, 168, 173–78, 180–82, 185, 193, 195, 197, 278, 288, 295; mercantile, xvi–xvii, 11, 20, 35, 37–38, 55–56, 43, 64, 88, 91, 144; public, 151–52, 157
Voltaire, 233

Webster, Charles, 10
Wedgwood, John, 281
Wellbery, David E., 12, 38, 65, 341n. 51
Weyland, John, 378n. 42
Wheeler, John, 87–88
Whewell, William, 309–10, 322
Wigley, Mark, 34–36, 338n. 20
Wilkins, John, 114, 118, 188
William of Orange, 150, 52, 162
Wimsatt, W. K., 327
Winch, Donald, 6, 237, 266, 370n. 35
Wollstonecraft, Mary, 23
Wood, P. B., 176, 197
Wordsworth, William, 293, 307, 325–26
Worsley, Benjamin, 121
Wright, Edward, 140

Xenophon, 35, 338n. 15

Yeo, Eileen Janes, 6

Žižek, Slavoj, 327

David Ropeik risk commin
Ropeik Harvard
book on risk   ToTN 9/3/04